Structure and Chemistry of Crystalline Solids

Bodie E. Douglas
Shih-Ming Ho

Structure and Chemistry of Crystalline Solids

Springer

Bodie E. Douglas
Shih-Ming Ho
University of Pittsburgh
Pittsburgh, PA
USA

Library of Congress Control Number: 2005927929

ISBN-10: 0-387-26147-8
ISBN-13: 978-0387-26147-8

Printed on acid-free paper.

Printed in the United States of America. (SPI/EB)

9 8 7 6 5 4 3 2 1

springeronline.com

Preface

Books on crystal structures have long recognized the importance of close packing in many crystal structures and the roles of tetrahedral and octahedral sites in such structures are well known. However, none has recognized that there is a general scheme involving packing (**P**), octahedral (**O**), and tetrahedral (**T**) sites occurring in layers, in the sequence **PTOT**. The spacing is also quite regular, determined by the geometry of octahedra and tetrahedra. The **PTOT** system and the notation presented are important in teaching crystal structures in any area concerned with such structures and with solid state applications. Earlier books have recognized only part of nature's system for efficient packing in crystals. Those working with crystal structures in chemistry, mineralogy, geology, metallurgy, and material science need a simple, widely applicable, system. This book is intended for these individuals. The simple notation gives important information about the structure.

Approximately 300 structures are described using the **PTOT** notation. These structures are encountered for many thousands of compounds. The system and the notation reveal similarities and differences among structures. Figures have been selected and prepared carefully. For many of these, labels for layers or atoms have been added to identify important features and aid in visualizing the structure.

This book presents a unifying scheme using a very simple system of notation. The scheme and notation are explained and applied. This approach provides insight into similarities and differences among structures. The notation system emphasizes the environment of individual ions or molecules. The chemistry and properties are dependent on the symmetry of sites and the coordination number of atoms or ions. Small deformations in a crystal structure can change the space group and the crystallographic notation so that the relationship between a slightly deformed structure and the original structure is not apparent. The deformation commonly makes little change in the **PTOT** notation and the relationships can be seen.

Background is presented in early chapters for those interested in solids who are not experts in crystallography. Chapters 1 to 3 cover crystal systems and classes, symmetry, and terminology used in describing crystal structures. Chapter 4 treats metals, other elements and compounds with structures based on close-packing of **P** layers. Chapter 5 deals with structures (**PO**) based on close-packed ions (or atoms) in **P** layers and other ions (or atoms) in **O** layers. Chapter 6 deals with structures based on close-packed ions (or atoms) in **P** layers with other ions (or atoms) in one or both tetrahedral layers (**PT** or **PTT**). Compounds in Chapter 7 have ions (or atoms) in various combinations of **P**, **T** and **O** layers, including cases where all **PTOT** layers are occupied. Chapter 8 deals with compounds involving unusual combinations of layers and multiple layers, cases where one layer is occupied by more than one of the *A*, *B* or *C* packing positions. Intermetallic compounds are covered in Chapter 9, silica and silicates are covered in Chapter 10, and selected organic compounds are covered in Chapter 11. Chapter 12 provides a summary and covers the interpretation of structures and assignment of notation. Two appendices cover the literature, general considerations of solids, and predictions of structures.

Earlier books have missed the beauty of the general scheme for crystal structures. Considering structures of metals, including those at high temperature and pressure, the body-centered cubic (*bcc*) structure is as common as the cubic close-packed (*ccp*) and hexagonal close-packed (*hcp*) structures. The common *bcc* structure of metals has been considered as an anomaly, an exception to close packing. This view, based on packing hard spheres, is too limited because metal atoms are not hard spheres. In some cases the *bcc* structure has about the same or even greater density than *ccp* or *hcp* forms of the same metal. In the **PTOT** system *bcc* is shown to be an expected example of the general system based on close packing. It is an elegant example of the case where all layers are fully occupied.

Structures involving full packing layers and one tetrahedral layer (**PT**) are about as common for *ccp* (*ABC* sequence) as for *hcp* (*AB* sequence) structures. The sodium chloride structure (**PO**), with chloride ions *ccp* and sodium ions in octahedral layers, is the most common MX ionic structure. The *hcp* counterpart is the NiAs structure, an unusual and uncommon structure because the octahedral sites are poorly screened. Fluorite, CaF2, and Li2O are the common structures for MX2 and M2X structures. These are based on *ccp* structures involving packing layers and both tetrahedral layers, all filled (**PTT**). The counterpart in the *hcp* system is rarely encountered and is found only for unusual compounds, because the adjacent **T** sites are too close for full occupancy without bonding. The severe limitations in **PO** and **PTT** *hcp* structures have not been recognized. Many examples of compounds involving partially filled layers of each type are included here and those, based on the *hcp* system, are found with sites in positions occupied staggered to avoid interferences.

Layered silicates reveal two types of filled close-packed layers of oxide ions. Those bonded to Al^{3+} or Mg^{2+} ions in octahedral sites are the usual close-packed layers, oxide ions form a network of hexagons with an oxide ion at the centers. The oxide layers forming the bases of tetrahedra have oxide ions which form smaller hexagons without oxide ions at the centers. This is also the pattern found for layers 2/3 filled, but those oxide layers in silicates are filled. For many of the layered silicates the repeating units, based on packing positions, requires stacking as many as three unit cells.

The manuscript was almost finished when we learned of the *CrystalMaker*® computer program. Many structures in the *CrystalMaker* library were added and many figures were redone. The availability of many silicate structures resulted in Chapter 10, including structures of silica and some silicates moved from earlier chapters. Later thousands of other structures were made available in *CrystalMaker* by Professor Yoshitaka Matsushita of the University of Tokyo. Many of these were added or structures from other sources were replaced.

Many structures are included from Wyckoff's volumes, *Crystal Structures,* and Pearson's *The Crystal Chemistry and Physics of Metals and Alloys.* Refinements are available for some of these structures. The descriptions and figures of these and other some older sources are good for visualizing and understanding the structures. That is important and *CrystalMaker* has been a great aid in this regard.

Figures in the book are limited in size and to grayscales. The addition of a CD, based on *CrystalMaker,* makes it possible to use color and larger figures. It provides an opportunity to show steps which can be taken to interpret and understand structures. The detailed examination of layered silicate structures shows general features not noted earlier.

The CD with the book is intended as an educational tool and a resource of structures using *CrystalMaker*. It should help visualization of structures and seeing relationships among similar structures. The first three parts of the CD are in slide-show format. Part I includes various types of structures. Part II illustrates the determination of relative positions of atoms in layers in a structure. Part III involves the interpretation of the structures of two complex metal silicates. Part IV includes representative structures in *CrystalMaker*. The user can manipulate structures in many ways using *CrystalMaker* demos included for Mactinosh and Windows. The figures were selected, prepared, and organized by Bodie Douglas, but the CD was created by Stephen B. Douglas, Jr. Carol Fortney, a doctoral student, helped solve some problems with CrystalMaker files for PCs.

We appreciate the help of Dr. Darel Straub, retired from the University of Pittsburgh, for critical review of the manuscript. Senior Editor David Packer of Springer obtained a reviewer who had the background in chemistry and crystal structures to see the value of the proposed book. They helped with the title and Appendix B. The book was produced by Chernow Editorial Services under the supervision of Barbara Chernow. Our objective is to maintain the high standard maintained by Springer and help those interested in crystal structures and solid state applications.

Bodie Douglas
Shih-Ming Ho
Pittsburgh PA

Contents

Preface		v
Chapter 1	Introduction	1
Chapter 2	Classification of Crystals, Point Groups, and Space Groups	6
Chapter 3	Close Packing and the PTOT System	21
Chapter 4	Crystal Structures of the Elements and Some Molecular Crystals	34
Chapter 5	Structures Involving P and O Layers	63
Chapter 6	Crystal Structures Involving P and T Layers	117
Chapter 7	Crystal Structures Involving P, T, and O Layers	147
Chapter 8	Structures with Multiple Layers	172
Chapter 9	Crystal Structures of Some Intermetallic Compounds	195
Chapter 10	Crystal Structures of Silica and Metal Silicates	233
Chapter 11	Structures of Organic Compounds	279
Chapter 12	Predicting Structures and Assigning Notations	292
Appendix A	Further Reading	306
Appendix B	Polyhedra in Close-Packed Structures	309

Subject Index 317

Minerals and Gems Index 325

Formula Index 327

Chapter 1

Introduction

Investigators from many disciplines are interested in crystal structures, but with very different perspectives. X-ray crystallographers provide the detailed structures that are the bases for those used in other disciplines. Mineralogists are interested in identification, morphology, and chemical composition of minerals. Geologists are interested in identification, chemical composition, and chemical environments of ions in minerals. They are concerned with the formation and transformation of minerals during the changes in the earth over the ages. Chemists are interested in the classification of crystals, the coordination number and local symmetry of ions in crystals (not only minerals), and their relationship to those of similar structure or similar chemical formula. Material scientists are interested in the chemical and physical properties of substances. Properties greatly depend on structures. No notation proposed satisfies the needs of all investigators. The notation used for X-ray crystallography serves well for describing structures in complete detail, but is not well suited for the interests of many investigators other than crystallographers.

When we consider crystal structures we usually think of the pattern and symmetry of the packing of the atoms, ions, or molecules in building the lattice based on X-ray crystallography. However, detailed descriptions of crystals and their classification are much older. The seven systems of crystals and the 32 classes of crystal symmetry were recognized by 1830. The 14 **Bravais Lattices** were presented by A. Bravais in 1848.

The early work of great scientists is inspiring and impressive considering the information available to them. The German astronomer J. Kepler was impressed by the regular and beautifully shaped snow flakes (Figure 1.1). He expressed his emotional response in a paper in 1611 entitled "A New Year's present; on hexagonal snow." This inspired him to the idea that this regularity might be due to the regular geometrical arrangement of minute equal units. Later, he considered close packing of spheres and drew a lot of pictures we would now consider as space lattices. The inspiration from the packing of tiny units to build beautiful crystals to the vast dimensions of his field of

Figure 1.1. Snow crystal.
(*Source*: L. V. Azaroff, *Introduction to Solids*, 1960, McGraw-Hill, New York, p. 2.)

astronomy led him to believe that the material world (or universe) was the creation of a "spirit," delighting in harmony and mathematical order. The wonderful developments in almost 4 centuries since, just confirm his insight.

The French mineralogist René J. Haüy published a paper in 1782, in which he concluded, from studies of cleavage of calcite ($CaCO_3$) and other crystals, that crystals are produced by regular packing of units he called "parallelepipeds." Figure 1.2 is a reproduction of one of his figures illustrating the construction of the "dog-teeth" habit of calcite from rhombohedral units. Like Kepler, he concluded that crystals are geometrical structures. His work has been admired, and he has been cited as the father of crystallography.

The English physicist William Barlow began as a London business man; later he became interested in crystal structures and devoted his life to that study. In 1894, he published his findings of the 230 space groups. It is amazing that from consideration of symmetry three scientists in different countries arrived at the 230 space groups of crystals at about this time. Barlow then worked with ideas of close packing. He pictured the atoms in a crystal as spheres, which, under the influence

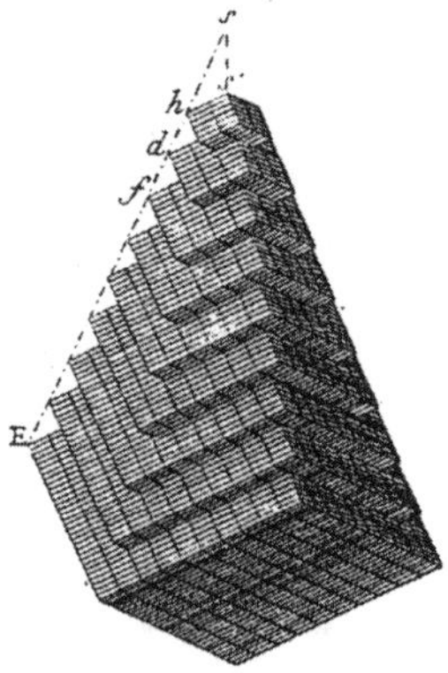

Figure 1.2. Geometrical building of a calcite crystal from rhombohedral units.
(*Source*: F.C. Phillips, *An Introduction to Crystallography*, Longmans, Green, London, 1955, p. 35.)

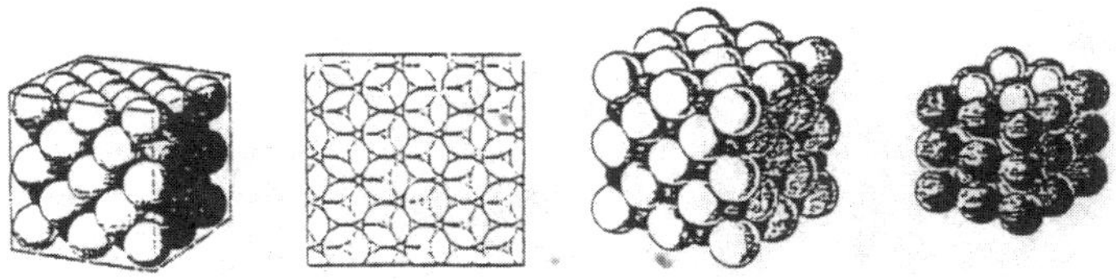

Figure 1.3. Arrangements of spheres for *(a) ccp*, *(b) hcp*, *(c)* NaCl, and *(d)* CsCl as described by W. Barlow.

(*Source*: L. Bragg, The Crystalline State, G. Bell and Sons, London, 1949 p. 270.)

of attractions and repulsions, packed together into the smallest volume possible. He distinguished cubic and hexagonal close-packed structures. Four of his figures are shown in Figure 1.3. The two close-packed arrangements of like spheres are (*a*) cubic and (*b*) hexagonal close-packed structures, common structures of metals. Figures 1.3*c* and *d* show two ways of packing spheres of two different sizes. These are the (*c*) NaCl and (*d*) CsCl structures. This work, published in 1897, set the foundation of structures based on close packing, and provided a great deal of insight for X-ray crystallography to follow.

Obviously, much of the development of crystallography predates the discovery of diffraction of X-rays by crystals. Early studies of crystal structures were concerned with external features of crystals and the angles between faces. Descriptions and notations used were based on these external features of crystals. Crystallographers using X-ray diffraction are concerned with the unit cells and use the notation based on the symmetry of the 230 space groups established earlier.

The notation for simplest descriptions of crystal habits and the much more complicated notation for the 230 space groups fail to provide information about the primary concerns of the chemist. Chemists are interested in the immediate environment of atoms or ions, such as coordination number and symmetry. Structures are often referred to by the chemical name or the common names of minerals: sodium chloride, cesium chloride, zinc blende, wurtzite, rutile, perovskite, etc. To describe another crystal structure as a variation of rutile or perovskite requires the visualization of rutile or perovskite to understand the relationship of the other structure to the reference structure. The notation for space groups is not adequate for simple visualization of such variations of a structure. A simple notation that can be visualized easily is needed. Such a simple notation based on close packing is presented here. Packing positions are designated **P** positions or layers, and interstitial positions **T** (tetrahedral) and **O** (octahedral) also occur in layers between **P** layers. Between any two close-packed layers there are two **T** layers and precisely halfway between the **P** layers there is one **O** layer, giving the sequence **PTOT**. Thus, this is known as the **PTOT** notation or system. The sequence and relative spacing of **PTOT** layers is exactly the same for cubic close-packed (*ccp*) or hexagonal close-packed (*hcp*) structures or even uncommon, more complex, close-packed structures.

Simple examples illustrate the system before we present more detailed descriptions. The very common structure of sodium chloride can be regarded as having layers of Cl^- (or Na^+) in all **P** (packing) positions in a *ccp* arrangement with Na^+ (or Cl^-) in all **O** (octahedral) positions. Thus, the repeating sequence is $|\mathbf{P}_A\mathbf{O}\mathbf{P}_B\mathbf{O}\mathbf{P}_C\mathbf{O}|\ldots$ The three packing positions for *ccp* are indicated by *A*, *B*, and *C*. Because there are six layers in the repeating sequence the notation is **3·2PO**. One of the structures of ZnS is zinc blende, described here as a *ccp* arrangement of S^{2-} in **P** layers with Zn^{2+} in all **T** sites of one **T** layer between the **P** layers, giving the sequence $|\mathbf{P}_A\mathbf{T}\mathbf{P}_B\mathbf{T}\mathbf{P}_C\mathbf{T}|\ldots$ or **3·2PT**. The other ZnS structure is wurtzite, with an *hcp* arrangement of S^{2-} in **P** layers with Zn^{2+} in all **T** sites of one layer between **P** layers giving $\mathbf{P}_A\mathbf{T}\mathbf{P}_B\mathbf{T}\ldots$ or **2·2PT**. There are only two packing positions, *A* and *B*, for *hcp*. The relationship between the two structures **3·2PT** and **2·2PT** is easily visualized if we remember the *ccp* and *hcp* structures. The first number of the index indicates the close-packed structure, **3** for *ccp* and **2** for *hcp*. The product is the total number of layers in the repeating sequence.

The most common structures for metals are *ccp* and *hcp*, but more than a dozen metals have the body-centered cubic (*bcc*) structure as the primary structure (under normal conditions). Even more than this number of metals have the *bcc* structure at high temperature or pressure. The *bcc* structure is not truly close packed, and the packing is not as efficient in space filling for *hard spheres*. The common occurrence of *bcc* for metals was considered anomalous. It can be considered as a simple example of filling all **P, O,** and both **T** layers by the same atoms in a *ccp* arrangement of the **P** layers giving the notation **3·2PTOT**. For *bcc*, the structure must expand for hard spheres because the atoms in the **O** and **T** positions are larger than the **O** and **T** voids. As we will see (Section 4.3.1), the *bcc* and *ccp* structures are very closely related, and are interchangeable by compression or expansion.

The common name of the structure of CaF_2 is fluorite, the name of the mineral. The **PTOT** notation for fluorite is **3·3PTT**. Because there is the same number of sites in each **P, O,** and **T** layer, the stoichiometry of CaF_2 indicates that in **3·3PTT** Ca^{2+} ions are in **P** layers and F^- ions fill all **T** sites in both **T** layers. The index product indicates that there are nine layers in the repeating unit or $|\mathbf{P}_A\mathbf{T}\mathbf{T}\mathbf{P}_B\mathbf{T}\mathbf{T}\mathbf{P}_C\mathbf{T}\mathbf{T}|$ giving a *ccp* arrangement of Ca^{2+} ions ($\mathbf{P}_A\mathbf{P}_B\mathbf{P}_C$) with F^- ions filling both **T** layers. The structure of Li_2O is called the antifluorite structure. In the **PTOT** system, it is **3·3PTT**, the same as that of CaF_2. The stoichiometry requires that the O^{2-} ions are in **P** layers and Li^+ ions in all sites of both **T** layers.

We have seen examples of the 1:1 stoichiometry corresponding to NaCl (**3·2PO**) and ZnS (**3·2PT** or **2·2PT**). Rutile (TiO_2) has **P** layers filled by O^{2-} with Ti^{4+} in **O** sites. The stoichiometry is 1:2 and this is clear from the notation $\mathbf{2{\cdot}2PO_{1/2}}$. Only 1/2 of the sites in the **O** layer are filled and the **P** layers are *hcp*. Later we will examine a more systematic treatment of applications of the **PTOT** notation. It applies to thousands of inorganic crystal structures including more complex structures. It is not limited to metals and MX, MX_2 and M_2X compounds. These few examples are used here to illustrate the simple **PTOT** system.

All **P**, **O**, and **T** layers have the same hexagonal close-packed arrangement within each layer. The two **T** layers are equivalent for *ccp* and *hcp*, and for *ccp*, only **P** and **O** layers are interchangeable, and together they are equivalent to the two **T** layers (considered together). Because of these similarities, *ccp*, *hcp*, the simple cubic structure, and even *bcc* structures can be handled in the **PTOT** system. It also applies to much more complex structures. The **PTOT** system provides a framework for considering the *mechanism* of formation and transformation of crystal structures. The transformations of structures of metals, *ccp*, *hcp*, and *bcc*, are of particular interest. These are considered in detail in Chapter 4.

In this book we are particularly interested in simple descriptions of structures that are easily visualized and providing information of the chemical environment of the ions and atoms involved. For metals, there is an obvious pattern of structures in the periodic table. The number of valence electrons and orbitals are important. These factors determine electron densities and compressibilities, and are essential for theoretical band calculations, etc. The first part of this book covers classical descriptions and notation for crystals, close packing, the **PTOT** system, and the structures of the elements. The latter and larger part of the book treats the structures of many crystals organized by the patterns of occupancies of close-packed layers in the **PTOT** system.

Chapter 2

Classification of Crystals, Point Groups, and Space Groups

2.1. Seven Crystal Systems and the 14 Bravais Lattices

From the study of angles between faces and cleavage planes of crystals T. Bergman and R. Haüy concluded, independently, around 1780, that all crystals consist of a masonry-type arrangement of equal parallelepipedal building bricks. Later it was recognized that there are seven crystallographic unit cells of crystals. The **Seven Systems of Crystals** are now usually grouped in terms of their axes of symmetry:

1. **Cubic System** with four threefold axes (along cubic diagonals)
2. **Tetragonal System** with one fourfold axis (through the centers of the two square faces).
3. **Orthorhombic System** with three twofold axes (through the centers of opposite faces).
4. **Trigonal (or Rhombohedral) System** with one threefold axis.
5. **Hexagonal System** with one sixfold axis.
6. **Monoclinic System** with one twofold axis.
7. **Triclinic System** with no axes of symmetry.

The trigonal system can be considered as a subdivision of the hexagonal unit. On this basis there would be only six different crystal systems, but conventionally, the trigonal system (also called the rhombohedral system) is retained separately. Figure 2.1 shows two rhombohedral cells within a hexagonal cell.

A crystal lattice is an array of points arranged according to the symmetry of the crystal system. Connecting the points produces the lattice that can be divided into identical parallelepipeds. This parallelepiped is the **unit cell**. The **space lattice** can be reproduced by repeating the unit cells in three dimensions. The seven basic primitive space lattices (P) correspond to the seven systems. There are variations of the primitive cells produced by lattice points in the center of cells (**body-centered cells,** I) or in the center of faces (**face-centered cells,** F). Base-centered orthorhombic and monoclinic lattices are designated by C. Primitive cells contain one lattice point ($8 \times 1/8$). Body-centered cells

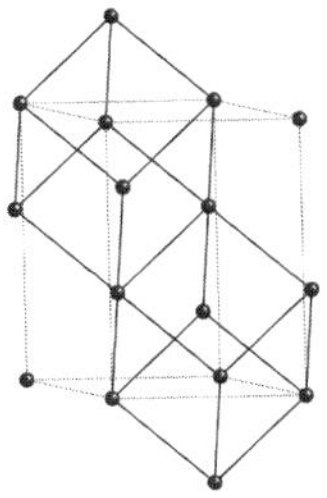

Figure 2.1. Two rhombohedral cell are shown within a hexagonal cell (lighter lines).

(*Source*: *Crystal Maker*, by David Palmer, *Crystal maker* Software Ltd., Beg Groke Science Park, Bldg. 5, Sandy Lane, Yarnton, Oxfordshire, OX51PK, UK.)

contain two lattice points $(1 + 8 \times 1/8)$. Face-centered cells contain four lattice points $(8 \times 1/8 + 6 \times 1/2)$. Base-centered cells contain two lattice points $(8 \times 1/8 + 2 \times 1/2)$. It might seem that many space lattices are possible, but A. Bravais demonstrated that there are only 14 space lattices, now known as the **14 Bravais Lattices**. These **Seven Systems of Crystals** and **14 Bravais Lattices** are shown in Table 2.1 and Figure 2.2.

2.2. Point Groups

Symmetry is the fundamental basis for descriptions and classification of crystal structures. The use of symmetry made it possible for early investigators to derive the classification of crystals in the seven systems, 14 Bravais lattices, 32 crystal classes, and the 230 space groups *before* the discovery of X-ray crystallography. Here we examine symmetry elements needed for the point groups used for discrete molecules or objects. Then we examine additional operations needed for space groups used for crystal structures.

The presence of an n-fold axis (C_n) of symmetry requires that an object be invariant to rotation through $360°/n$. The square planar $PtCl_4^{2-}$ (Figure 2.3) has a C_4 axis through the center of and perpendicular to the plane, two equivalent C_2 axes through the centers of the edges and Pt, and two equivalent C_2 axes through diagonally arranged Cl atoms and Pt. There are two equivalent symmetry (mirror) planes (σ) through the centers of the edges and Pt and two equivalent symmetry planes through diagonal Cl atoms and Pt. Such symmetry planes through the major (highest order) C_n axis are called vertical planes (σ_v). There is another symmetry plane through Pt and all Cl atoms. Such a plane perpendicular to the major axis is called a horizontal plane (σ_h). The Pt atom is located at a center of symmetry (i) because inversion of any Cl atom (or any point) through the inversion center (the origin) gives an equivalent Cl atom (or an equivalent point).

There is another symmetry element or operation needed for discrete molecules, the improper rotation, S_n. A tetrahedron (the structure of

Table 2.1. Crystal systems and the 14 Bravais Lattices.

Seven systems	Axes and angles	14 Bravais Lattices	Lattice symbols
Cubic	$a = b = c$	Primitive	P
	$\alpha = \beta = \gamma = 90°$	Body centered	I
		Face centered	F
Tetragonal	$a = b \neq c$	Primitive	P
	$\alpha = \beta = \gamma = 90°$	Body centered	I
Orthorhombic	$a \neq b \neq c$	Primitive	P
	$\alpha = \beta = \gamma = 90°$	Body centered	I
		Base face centered	C
		Face centered	F
Trigonal or rhombohedral	$a = b = c$		
	$\alpha = \beta = \gamma \neq 90°$	Primitive	R
Hexagonal	$a = b \neq c$		
	$\alpha = \beta = 90°$	Primitive	P
	$\gamma = 120°$		
Monoclinic	$a \neq b \neq c$	Primitive	P
	$\alpha = \beta = 90° \neq \gamma$	Base face centered	B or C
	or ($\alpha = \beta = 90° \neq \gamma$)		
Triclinic	$a \neq b \neq c$	Primitive	P
	$\alpha \neq \beta \neq \gamma \neq 90°$		

CH_4, Figure 2.3) has C_3 axes through each apex (along one C—H bond) and C_2 axes through the centers of the tetrahedral edges. There are also symmetry planes through each of these C_2 axes. In addition, if we rotate about one of the C_2 axes of CH_4 by 90° and then reflect through a mirror plane perpendicular to the axis we have an equivalent arrangement of CH_4. This is an improper rotation, S_n, here S_4. The S_n operation involves rotation by $360°/n$ followed by reflection through the plane perpendicular to the axis of rotation. All of these symmetry operations are used to assign *point groups* to objects or discrete molecules.

Point groups are designated by Schoenflies symbols. Groups involving proper rotations only are $\mathbf{C}_n$ groups, **T** (tetrahedral), **O** (octahedral), or **I** (icosahedral) groups. Although the **T**, **O**, and **I** groups have no other symmetry operations, simple molecules or ions such as CH_4 ($\mathbf{T}_d$), SF_6 ($\mathbf{O}_h$), and $B_{12}H_{12}^{2-}$ ($\mathbf{I}_h$) have many other symmetry elements. The $\mathbf{S}_n$ groups involve only proper and improper rotations, although S_2 is equivalent to *i*. Dihedral groups ($\mathbf{D}_n$, $\mathbf{D}_{nh}$, $\mathbf{D}_{nd}$) have n C_2 axes perpendicular to the major C_n axis. The trigonal BF_3 molecule (Figure 2.3) has C_3 as the major axis and three C_2 axes perpendicular to C_3. These are the symmetry operations for the $\mathbf{D}_3$ point group, but there is a symmetry plane perpendicular to C_3 (σ_h, through all atoms) so the point group is $\mathbf{D}_{3h}$. The point group for $PtCl_4^{2-}$ is $\mathbf{D}_{4h}$. Ferrocene, $Fe(C_5H_5)_2$ (Figure 2.3),

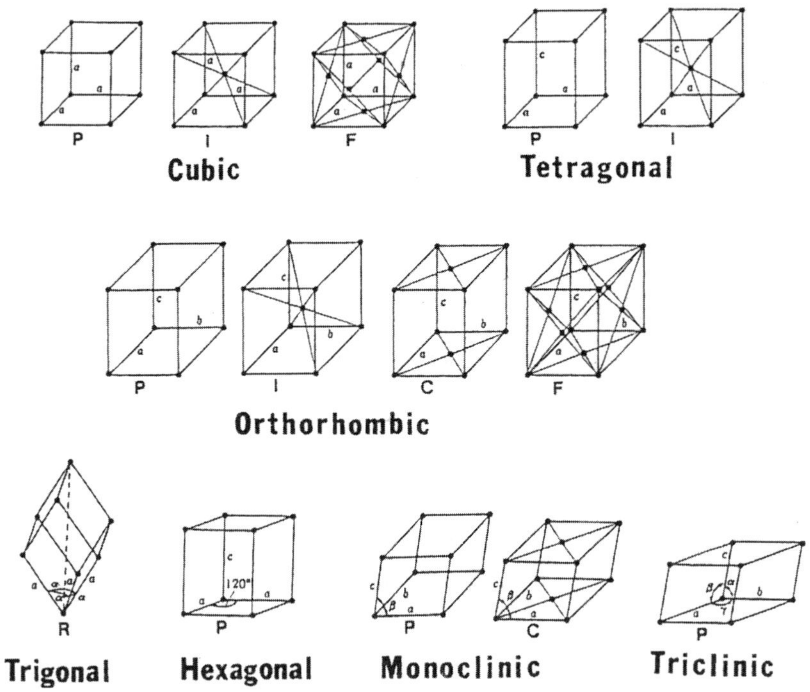

Figure 2.2. The seven Crystal Systems and the 14 Bravais Lattices.

has two dicyclopentadienyl rings: one planar ring below, and one above the Fe atom. The C_5H_5 rings are in parallel planes and are staggered when viewed along the C_5 axis. There are five C_2 axes perpendicular to the C_5 axis. There are five vertical mirror planes, each bisecting the angle between C_2 axes. These are called dihedral planes, a special case of vertical planes, and the point group for $Fe(C_5H_5)_2$ is $\mathbf{D}_{5d}$.

Chemists use the Schoenflies Notation and rotation-reflection for improper rotation in assigning point groups. Crystallographers

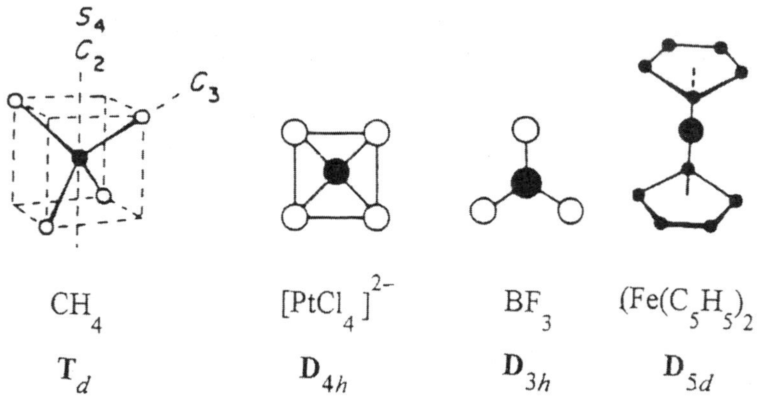

Figure 2.3. Molecular structures of CH_4, $[PtCl_4]^{2-}$, BF_3, and ferrocene $[Fe(C_5H_5)_2]$.

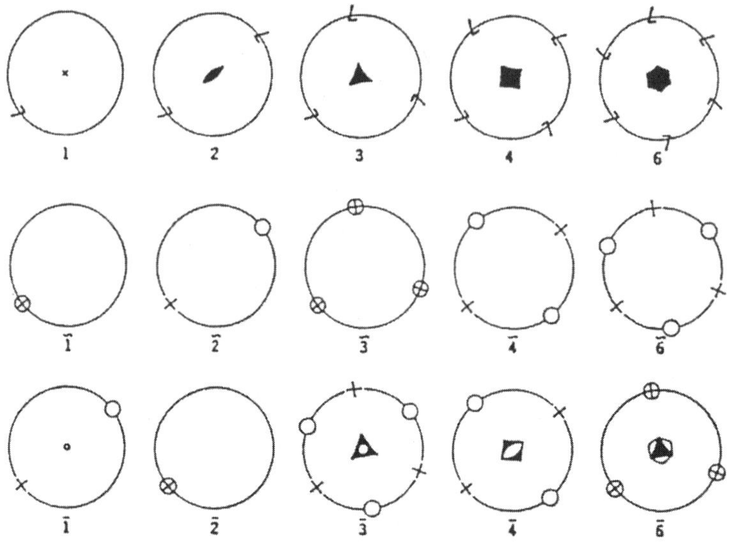

Figure 2.4. Results of proper and improper rotations.

use the International (Hermann-Maguin) Notation and rotation-inversion as improper rotation in assigning space groups. Proper rotations C_1, C_2, C_3, C_4, and C_6 are designated by the numbers 1, 2, 3, 4, and 6 in the International System. Improper rotations (rotation-inversion) are indicated by a bar over the number, $\bar{1}$, $\bar{2}$, $\bar{3}$, $\bar{4}$, and $\bar{6}$ Rotation-reflections are indicated by $\tilde{1}$, $\tilde{2}$, $\tilde{3}$, $\tilde{4}$, and $\tilde{6}$. The result of these operations on a point in a plane is shown in Figure 2.4. The results of proper rotations are shown by an unsymmetrical symbol 7. The results of both types of improper rotations are shown acting on a marker appearing as X on top and O on the bottom so that reflection in *m* or inversion turns it over. The rotation-inversion and rotation-reflection axes are equivalent in pairs, as shown in Table 2.2. Conventionally, these improper axes are designated as $\bar{1}$, *m*, $\bar{3}$, $\bar{4}$, and $\bar{6}$ where a mirror plane is indicated by *m*.

If a mirror plane contains a proper rotation axis it is designated 2*m*, 3*m*... If the mirror plane is perpendicular to the axis, it is 2/*m*, 3/*m*... Only the minimum number of symmetry elements is used, omitting other equivalent elements. The $\mathbf{D}_2$ group is 222, but $\mathbf{D}_{2h}$ is *mmm* because 2/*m* 2/*m* 2/*m* is redundant, as the perpendicular planes require C_2 axes. The group $\mathbf{C}_{4v}$ is 4*mm*, rather than 4*m*, because there are two distinguishable mirror planes. The point groups **I** and $\mathbf{I}_h$ are not encountered in crystallographic groups.

The symmetry elements, proper rotation, improper rotation, inversion, and reflection are required for assigning a crystal to one of the 32 crystal systems or crystallographic point groups. Two more symmetry elements involving translation are needed for crystal structures—the **screw axis**, and the **glide plane**. The **screw axis** involves a combination of a proper rotation and a confined translation along the axis of rotation. The **glide plane** involves a combination of a proper reflection and a confined translation within the mirror plane. For a unit cell

TABLE 2.2. **Conventional designation of improper axes.**

Rotation-inversion axes	Equivalent rotation-reflection	Conventional designation
$\bar{1}$	$\tilde{2}$	$\bar{1}$
$\bar{2}$	$\tilde{1}$	m
$\bar{3}$	$\tilde{6}$	$\bar{3}$
$\bar{4}$	$\tilde{4}$	$\bar{4}$
$\bar{6}$	$\tilde{3}$	$\bar{6}$

requiring a screw axis or a glide plane for its symmetry description, the corresponding crystallographic point group is obtained by setting translation equal to zero.

2.3. Miller Indices

Early investigators were interested in designating the faces of crystals. We choose an arbitrary lattice point as the origin and give the intercepts of the plane with the three crystallographic axes in units of *a, b,* and *c*. These three intercepts are known as the **Weiss index** of the plane. The numbers in the Weiss index can be positive or negative, 1 for a plane cutting at $1a$, $\frac{1}{2}$ for a plane cutting at $\frac{1}{2}a$ (or another fraction, etc.), or infinity for a plane parallel to an axis. The more convenient system is to use the **Miller index**, the reciprocal of the Weiss index. This eliminates fractions and infinity because the Miller index for a plane parallel to an axis is zero. The general Miller index is *hkl* for a face or a set of planes, where *h* is the number of equal divisions of *a* in the unit cell, *k* the divisions of *b*, and *l* the divisions of *c*. In Figure 2.5, two planes corresponding to two faces of an octahedron are shown, one cutting *a, b,* and *c* at 1, giving a 111 plane. The lower face or plane sharing the *a–b* line has intercepts at $1a$, $1b$, and $-1c$, giving the Miller index $1\,1\,\bar{1}$ (negative values are given as the number with a bar over it). Other planes are shown in Figure 2.6.

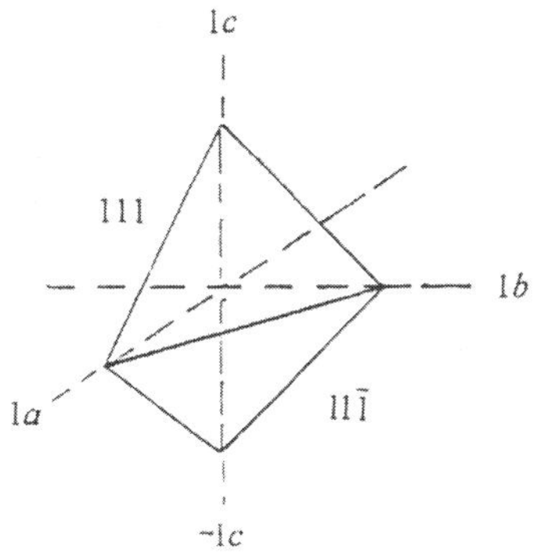

Figure 2.5. The 111 and 11$\bar{1}$ faces of an octahedron.

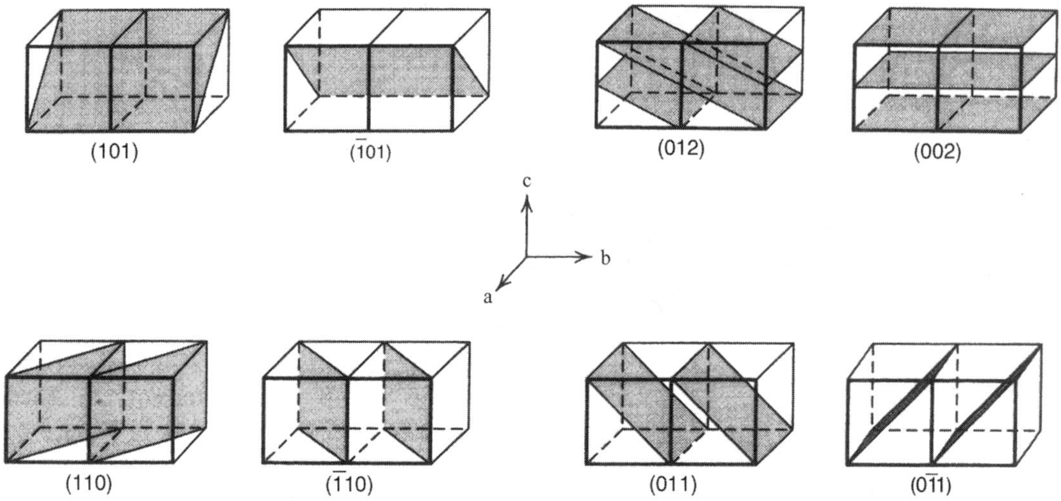

Figure 2.6. Miller indices for planes in cubic cells.

2.4. Stereographic Projections

Most crystals are imperfect, and the full symmetry is not apparent. Projection techniques are used to find the true symmetry of a crystal and its classification among the 32 crystallographic point groups. We choose the center of the crystal as the origin and construct a sphere centered at the origin. A radius is drawn from the origin normal to each face and marked as a dot on the sphere. The symmetry of all of these points belongs to one of the crystallographic point groups. The spherical projection is reduced to a more convenient planar **stereographic projection.** The sphere is divided into hemispheres above and below the horizontal plane. The points of the "northern" hemisphere are connected by rays to the "south pole." The intersection of each of these rays with the horizontal plane is marked by a solid dot. The intersection with the horizontal plane with each ray from a point in the "southern" hemisphere to the "north pole" is marked by an open circle. The highest symmetry axis is chosen as vertical. The great circle is a thick line if there is a symmetry plane in this plane; otherwise, it is represented by a thinner line. Vertical symmetry planes are represented as lines (diameters) through the center. In the cubic groups there are symmetry planes tilted relative to σ_v and σ_h planes. These are shown as portions of ellipses, projections on the great circles representing the tilted planes. Twofold axes in the equatorial plane are shown as a line connecting the symbols for C_2.

Special positions are those falling on the symmetry elements of the crystal or cell. Stereograms indicate these positions by the following symbols:

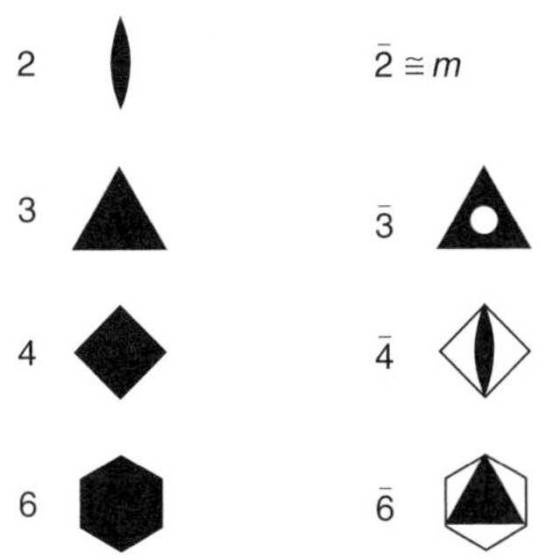

The symbol for each axis appears in the great circle where its points of intersection with the sphere would project on the horizontal plane. These positions are at the center or on the great circle except for threefold and twofold axes of the cubic groups. The stereographic projections for a crystallographic point group are shown as two circles: one showing the symmetry elements, and one showing the array of points on the sphere resulting from the symmetry operations of the group acting on an initial general point. The projections of these points are solid dots for points on the upper hemisphere and are open circles for points on the lower hemisphere. A solid dot in an open circle indicates two points on the same vertical line. The two circles for stereographic projections can be combined unless the result is too cluttered for clarity.

2.5. The 32 Crystal Classes

A crystal is a well-tailored network or lattice of atoms. The construction of the lattice is directed stringently by the symmetry elements of the crystal. We can choose a central point and consider the periodical stacking around this point. Combinations of symmetry elements limit the arrangements to only 32 patterns—the 32 point groups or the 32 **classes of crystals**.

The five proper rotations 1, 2, 3, 4, and 6 give the corresponding point groups ($\mathbf{C}_1$, $\mathbf{C}_2$, $\mathbf{C}_3$, $\mathbf{C}_4$, and $\mathbf{C}_6$) represented by the stereograms for the rotational $\mathbf{C}_n$ groups in Figure 2.7. The rotation-inversion axes shown in Table 2.2 produce the five point groups $\bar{1}(\mathbf{C}_i)$, $\bar{2}(m = \mathbf{C}_s)$, $\bar{3}(\mathbf{S}_6)$, $\bar{4}(\mathbf{S}_4)$, and $\bar{6}(\mathbf{C}_{3h} = \mathbf{S}_3)$, where $\mathbf{C}_s = \mathbf{S}_1$ and $\mathbf{C}_i = \mathbf{S}_2$ shown in Figure 2.7. Combinations of proper rotation axes produce six more rotational point groups ($\mathbf{D}_2$, $\mathbf{T}$, $\mathbf{D}_4$, $\mathbf{D}_3$, $\mathbf{D}_6$, and $\mathbf{O}$), as diagrammed in Figure 2.8. These are dihedral groups ($\mathbf{D}_n$) plus the rotational groups $\mathbf{T}$ and $\mathbf{O}$. Combinations of proper axes perpendicular to a mirror plane produce five point groups, $1/m = m = \bar{2}$, $2/m$, $3/m = \bar{6}$, $4/m$, and $6/m$. Two of these are identical with $\bar{2}$ and $\bar{6}$ shown in Figure 2.7, leaving three additional point groups, $2/m\,(\mathbf{C}_{2h})$, $4/m\,(\mathbf{C}_{4h})$, and $6/m\,(\mathbf{C}_{6h})$ represented in Figure 2.8.

Combination of a proper rotation with a vertical mirror plane (m or σ_v) produces the $\mathbf{C}_{nv}$ groups, ($1m = m$), $2mm$ ($= mm$), $3m$, $4mm$, and $6mm$. The point group m ($\mathbf{C}_s$) is included in Figure 2.7, leaving four

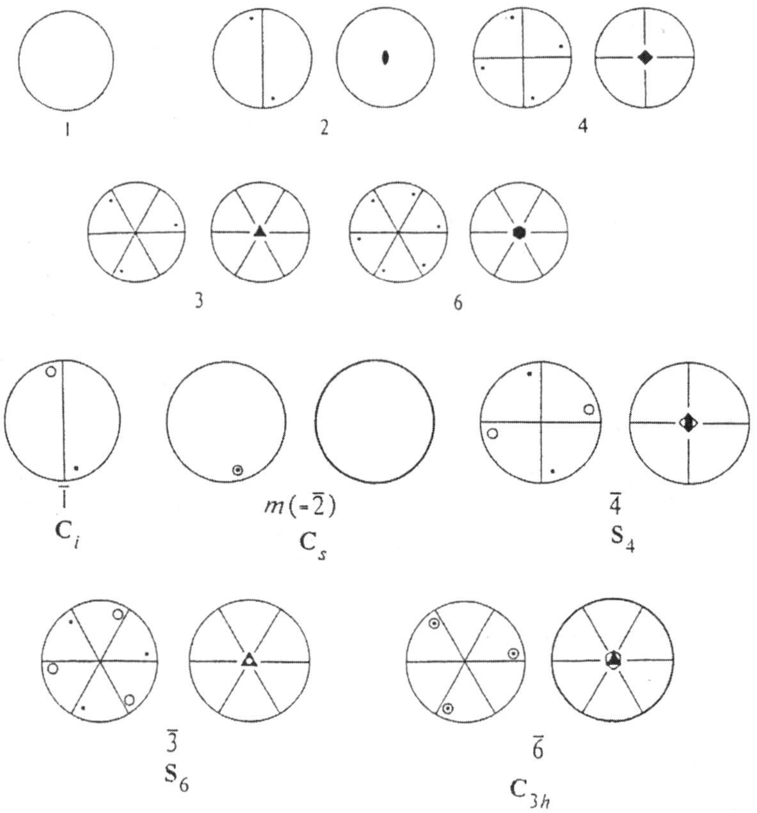

Figure 2.7. Stereograms for the six $\mathbf{C}_n$ point groups produced by proper rotations and those for the point groups produced by improper rotations.

$\mathbf{C}_{nv}$ groups in Figure 2.8. Combination of a proper rotation with a horizontal mirror plane (m or σ_h) and a vertical plane produces six more point groups: $2/m\ 2/m\ 2/m$ (= mmm) ($\mathbf{D}_{2h}$), $3/m\ m2$ (= $\bar{6}m2$) ($\mathbf{D}_{3h}$), $4/m\ 2/m\ 2/m$ (= $4/m\ mm$) ($\mathbf{D}_{4h}$), $6/m\ 2/m\ 2/m$ (= $6/m\ mm$) ($\mathbf{D}_{6h}$), $2/m\,\bar{3}$ (= $m\,\bar{3}$) ($\mathbf{T}_h$), and $4/m\bar{3}2/m$ (= $m\bar{3}m$) ($\mathbf{O}_h$). These are shown in Figure 2.9.

These proper rotations and improper rotations and combinations of proper rotations and mirror planes produce 29 point groups. The remaining three point groups are obtained by combinations of improper axes $\bar{3}$ and $\bar{4}$ with proper rotations. The products are point groups $\bar{4}2m$ ($\mathbf{D}_{2d}$), $\bar{3}m$ ($\mathbf{D}_{3d}$), and $\bar{4}3m$ ($\mathbf{T}_d$) as shown in Figure 2.9. These complete the 32 crystallographic point groups or crystal classes. These are grouped by **Crystal Systems** in Table 2.3.

2.6. The 230 Space Groups

The 32 crystallographic point groups result from combinations of symmetry based on a fixed point. These symmetry elements can be combined with the two translational symmetry elements: the screw

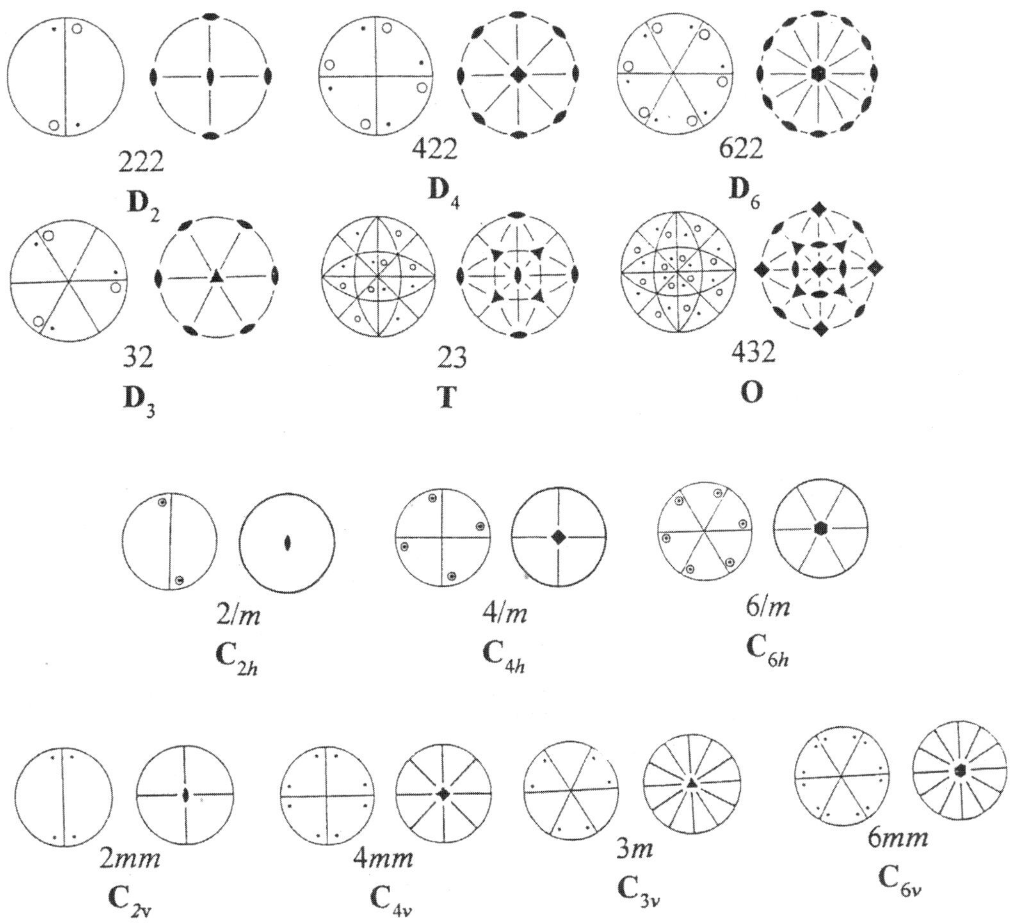

Figure 2.8. Stereograms for the six $\mathbf{D}_n$, **T**, and **O** point groups from combinations of proper rotations. Those for the three $\mathbf{C}_{nh}$ point groups from combinations of proper rotations and a horizontal mirror plane [in addition to $\bar{2}$ (m or $\mathbf{C}_{1h}$) and $\bar{6}$ ($\mathbf{C}_{3h}$) shown in Figure 2.7]. Those for the $\mathbf{C}_{nv}$ point groups from combinations of proper rotations and a vertical mirror plane.

axis, and the glide plane. The results of these combinations give the rest of the 230 crystallographic space groups. Thirty-two of the space groups are the same as the point groups. In the crystals belonging to the 32 point groups each unit is reproduced without alteration. In the other space groups each unit repeats with translation and rotation or reflection.

Three investigators from different countries independently arrived at the 230 space groups around 1890:

E. S. Federov (1853–1919, Russian)
A. M. Schoenflies (1853–1928, German)
W. Barlow (1845–1934, English)

This is an impressive example of a major advancement occurring in three places at the right time. These men had to have thorough knowledge of the crystallography of the time, mathematics, and great insight.

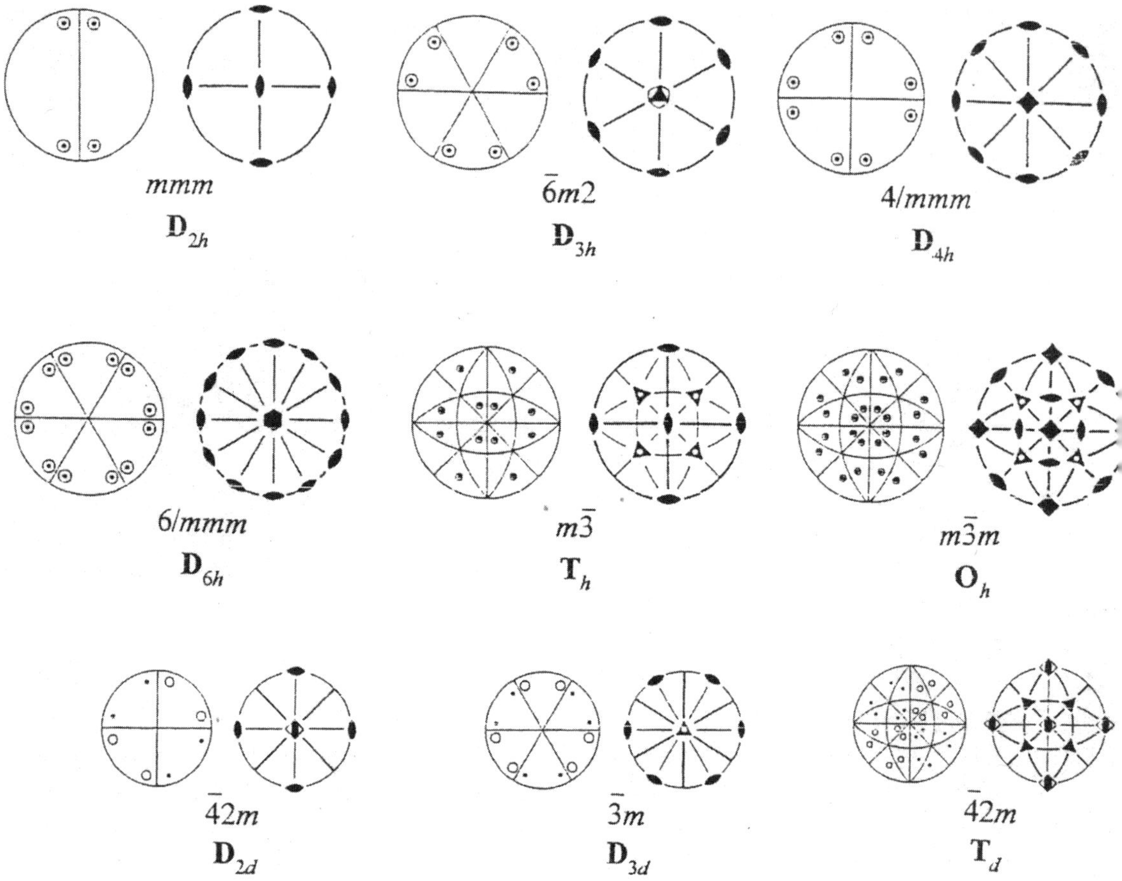

Figure 2.9. Stereograms for point groups from proper rotations and a plane. Stereograms for the point groups from combinations of $\bar{4}$ of improper rotations $\bar{3}$ and $\bar{4}$ with proper rotations.

Table 2.4 shows the crystal systems, point groups, and the corresponding space groups. The numbers for space groups are those as derived and numbered by Schoenflies. The space groups isomorphous to each point group are indicated by a superscript (*e.g.*, Number 194, $\mathbf{D}_{6h}^{4}$).

2.7. Hermann-Mauguin Notation and Classification

Designations for individual space groups use the notation proposed by Hermann and Mauguin in 1935. This notation has been adopted by the International Union of Crystallography and is in general use. The symbols provide more information than just the symmetry elements for a crystal. The first symbol is a capital letter: P for a simple or primitive lattice, I for a body-centered lattices, and F for face-centered lattices with all faces centered. Lattices with one face centered are designated A (100 planes), B (010 planes), or C (001 planes). R is the symbol for a face-centered trigonal lattice for which a primitive rhombohedral unit cell is chosen. After the general description of the unit cell a number identifies the major axis of rotation. For example,

TABLE 2.3. **The seven crystal systems and the 32 crystal classes.**

Crystal system	Schoenflies symbol	International symbol	Crystal system	Schoenflies symbol	International symbol
Triclinic	C_1	1	Rhombohedral	C_3	3
	C_i	$\bar{1}$		S_6	$\bar{3}$
				C_{3v}	$3m$
Monoclinic	C_s	m		D_3	32
	C_2	2		D_{3d}	$\bar{3}m$
	C_{2h}	$2/m$	Hexagonal	C_{3h}	$\bar{6}$
				C_6	6
Orthorhombic	C_{2v}	$2mm$		C_{6h}	$6/m$
	D_2	222		D_{3h}	$\bar{6}m2$
	D_{2h}	mmm		C_{6v}	$6mm$
				D_6	622
Tetragonal	C_4	4		D_{6h}	$6/m\ mm$
	S_4	$\bar{4}$			
	C_{4h}	$4/m$	Cubic	T	23
	C_{4v}	$4mm$		T_h	$m\bar{3}$
	D_{2d}	$\bar{4}2m$		T_d	$\bar{4}3m$
	D_4	422		O	432
	D_{4h}	$4/m\ mm$		O_h	$m\bar{3}m$

P$2/m$ is a primitive cell with a twofold axis and a mirror plane perpendicular to the axis.

Major axes are indicated by positive numbers for C_n and barred numbers, $\bar{2}$, $\bar{3}$, etc., for improper axes of rotation. Screw axes are indicated by subscripts such as 2_1, 3_2, etc. A 4_1 screw axis involves translation of 1/4 upward for an anticlockwise rotation, 4_2 involves translation by 1/2 (2/4), an 4_3 involves translation by 3/4. Mirror planes (m) and glide planes are indicated by letters, using the letters corresponding to translation by the fractions along a particular direction as follows

Translation	*Symbol*
$\frac{a}{2}$	a
$\frac{b}{2}$	b
$\frac{c}{2}$	c
$\frac{a+b}{2}, \frac{b+c}{2}$ or $\frac{c+a}{2}$	n
$\frac{a+b}{4}, \frac{b+c}{4}$ or $\frac{c+a}{4}$	d

Examples are given in Table 2.4.

Table 2.4. Space groups, crystal systems, point groups, and the Hermann-Mauguin Symbols.

Space group number	Crystal system	Point group: International symbol	Point group: Schoenflies symbol	Hermann-Mauguin symbol
1	Triclinic	1	$\mathbf{C}_1{}^1$	P1
2		$\bar{1}$	$\mathbf{C}_i{}^1$	P$\bar{1}$
3–5		2	$\mathbf{C}_2{}^1$ (No. 3)	P2
6–9	Monoclinic	m	$\mathbf{C}_s{}^1$ (No. 6)	Pm
10–15		$2/m$	$\mathbf{C}_{2h}{}^1$ (No. 10)	P$2/m$
16–24		222	$\mathbf{D}_2{}^1$ (No. 16)	P222
25–46	Orthorhombic	$mm2$	$\mathbf{C}_{2v}{}^1$	P$mm2$
47–74		mmm	$\mathbf{D}_{2h}{}^{28}$ (No. 74)	Imma
75–80		4	$\mathbf{C}_4{}^1$ (No. 75)	P4
81–82		$\bar{4}$	$\mathbf{S}_4{}^1$	P$\bar{4}$
83–88		$4/m$	$\mathbf{C}_{4h}{}^1$	P$4/m$
89–98	Tetragonal	422	$\mathbf{D}_4{}^1$	P422
99–110		$4mm$	$\mathbf{C}_{4v}{}^1$	P$4mm$
111–122		$\bar{4}2m$	$\mathbf{D}_{2d}{}^1$	P$\bar{4}2m$
123–142		$4/m\ mm$	$\mathbf{D}_{4h}{}^1$ (No. 123)	P$4/m\ mm$
143–146		3	$\mathbf{C}_3{}^1$ (No. 143)	P3
147–148		$\bar{3}$	$\mathbf{S}_6{}^1$	P$\bar{3}$
149–155	Trigonal	312	$\mathbf{D}_3{}^1$	P312
156–161		$3m$	$\mathbf{C}_{3v}{}^1$	P$3m1$
162–167		$\bar{3}m$	$\mathbf{D}_{3d}{}^6$ (No. 167)	R$\bar{3}$c
168–173		6	$\mathbf{C}_6{}^1$ (No. 168)	P6
174		$\bar{6}$	$\mathbf{C}_{3h}{}^1$	P$\bar{6}$
175–176		$6/m$	$\mathbf{C}_{6h}{}^1$	P$6/m$
177–182	Hexagonal	622	$\mathbf{D}_6{}^1$	P622
183–186		$6mm$	$\mathbf{C}_{6v}{}^1$	P$6mm$
187–190		$\bar{6}m2$	$\mathbf{D}_{3h}{}^1$	P$6m2$
191–194		$6/m\ mm$	$\mathbf{D}_{6h}{}^4$ (No. 194)	P$6_3/m\ mc$
195–199		23	$\mathbf{T}^1$ (No. 195)	P23
200–206		$m\bar{3}$*	$\mathbf{T}_h{}^1$ (No. 200)	P$m\bar{3}$
207–214	Cubic	432	$\mathbf{O}^1$ (No. 207)	P432
215–220		$\bar{4}3m$	$\mathbf{T}_d{}^1$ (No. 215)	P$\bar{4}3m$
221–230		$m\bar{3}m$*	$\mathbf{O}_h{}^1$ (No. 221)	P$m\bar{3}m$

*These are commonly used abbreviations, the full H-M symbol for $\mathbf{T}_h$ is $2/m\bar{3}$ (also sometimes abbreviated as $m3$), and the full H-M symbol for $\mathbf{O}_h$ is $4/m\ \bar{3}2/m$ (also sometimes abbreviated as $m3m$. $\mathbf{T}_h$ and $\mathbf{O}_h$ have centers of symmetry thus $3 + I$ is equivalent to $\bar{3}$.

2.8. Representative Designation of Crystals

Commonly, many compounds with the same formula type have the same type of structure. Usually the most common compound is chosen as representative, but it could be the first structure of the type studied. For example, hundreds of binary compounds of the MX type have the same crystal structure as NaCl. Other compounds can be described as having the NaCl (or halite, the name of the mineral) structure. If the structure being considered has slight differences, these differences can be described in terms of the reference structure. One often sees statements such as a compound has a disordered spinel ($MgAl_2O_4$) type structure or an inverse spinel structure. This requires knowledge of the spinel structure because "inverse" or "disordered" terms describe variations of occupancies of octahedral and tetrahedral sites.

2.9. Pearson's Simplified Notation

Many systems of notation and classification have been proposed. The well-known books by R. W. G. Wyckoff, A. F. Wells, F. C. Phillips, L. Bragg, M. J. Buerger, L. V. Azakoff, D. M Adams, and W. B. Pearson (Appendix A, Further Reading) have discussed these proposals. These proposals include close packing of atoms, nets, or prism connections, stacking of coordination polyhedra and even a crystal-algebra method. Application of most of these proposals requires familiarity with the features of many structures. Only specialists can be expected to have

Table 2.5. Pearson's Structure Symbols.

Crystal system	Lattice symbol	Structure symbol*
Triclinic, a	P	aPx
Monoclinic, m	P	mPx
	C	mCx
Orthorhombic, o	P	oPx
	C	oCx
	F	oFx
	I	oIx
Tetragonal, t	P	tPx
	I	tIx
Hexagonal, h and	P	hPx
Trigonal, h		
Rhombohedral, h	R	hRx
Cubic, c	P	cPx
	F	cFx
	I	cIx

*x = Number of atoms in the unit cell.

all of the necessary information at their fingertips. None of these has general approval or wide use.

Professor Pearson (see Appendix A) adopted simple notation providing useful information. His **Structure Symbol** gives the crystal system by a lower case letter (*e.g.*, m for monoclinic), the lattice type by a capital letter (*e.g.*, P for primitive), and the number of atoms in the unit cell. The structure symbols are given in Table 2.5. For example, the fluorite (CaF_2) structure is described as cF12 for a face-centered cubic unit cell containing 12 ''atoms'' ($8F^-$ and $(1/8 \times 8) + (6 \times 1/2) = 4Ca^{2+}$). The symbol for Al_7Cu_2Fe alloy is tP40, primitive tetragonal unit cell containing 40 atoms or $4Al_7Cu_2Fe$ units.

Chapter 3

Close Packing and the PTOT System

3.1. Close Packing

Laves' principles, *(a) the space principle, (b) symmetry principle*, and *connection principle*, are very important in packing spheres: the spheres are expected to pack together in a crystal structure of *highest symmetry, highest coordination*, and with *densest packing*. If we let hard spheres of uniform diameter, such as marbles or ball bearings, arrange themselves in one layer on the bottom of a box, they form a *two-dimensional close-packed layer*. Each sphere is surrounded by a hexagon of six spheres at equal distance, and each of these spheres is at the center of such a hexagon (except at edges, of course). Figure 3.1***a*** shows this close-packed layer expanded so we can see the positions for added layers. The atoms touch in a close-packed layer. We can begin a second close-packed layer by dropping an identical sphere on the first layer. This sphere will roll into an indentation—any one, they are all equivalent. To keep track of positions in subsequent layers, let us label the positions of the spheres in the first layer *A* positions. Following the alphabetical order, we call the position of the sphere beginning the second layer position *B*. These letters label the positions as viewed from above the layers as in Figure 3.1. Adding more spheres to complete the second layer, without moving the first one at position *B*, gives the *B* layer. It is identical to the first layer except shifted. The choice of the position of the first sphere of the second layer is arbitrary. Figure 3.1***b*** shows the positions of the two layers.

When we begin a third layer there are two choices:

(1) The spheres of the third layer can be directly above those in the first layer to give another *A* layer or the sequence *ABA*. Continuation of this sequence *ABABAB*... gives a structure known as *hexagonal close packing* (*hcp*).

(2) The spheres of the third layer can be in the indentations of the second layer that are *not* directly above the first layer. These are *C* positions. Continuation of this sequence *ABCABCABC*... gives a structure known as *cubic close packing* (*ccp*).

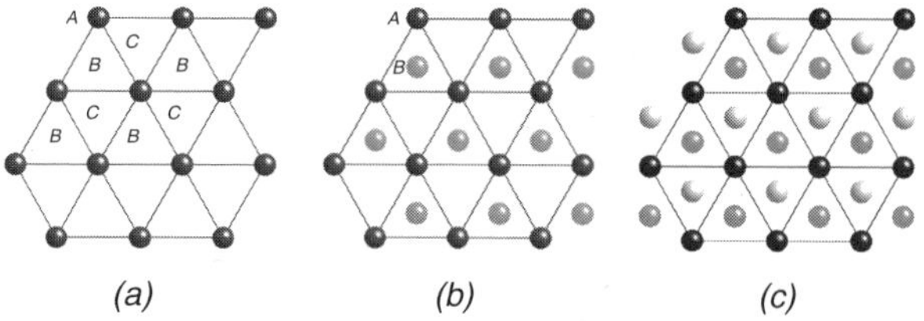

Figure 3.1. ***(a)*** Positions of atoms for one close-packed layer of spheres. ***(b)*** Positions for two layers in projection ***(c)*** three layers (*A*, *B*, and C) in projection.

(*Source*: *CrystalMaker*, by David Palmer, *CrystalMaker* Software Ltd., Begbroke Science Park, Bldd. 5, Sandy Lane, Yarnton, Oxferdshire, OX 5/PF, UK.)

The result of adding a second layer (*AB*) and then adding a third layer to give *ccp* is shown in Figure 3.1***c***. In a three-dimensional close-packed structure there are only three positions; *A*, *B*, and *C*. We call them packing positions (**P**, $\mathbf{P}_A$, etc.). The only restriction is that two adjacent close-packed layers must have different positions, *AB*, *BC*, *AC*, etc., because the spheres of the second layer must be in one set of the indentations of the first layer. Complex sequences such as *ABACBCA*... or random sequences are possible, but it is very rare to find any sequence other than *ccp* and *hcp*. Nature prefers the highest symmetry. More than 80 elements have *ccp* or *hcp* structures. We will encounter the *ABACABAC*... sequence, known as *double hexagonal*, in a few structures.

3.2. T and O Layers (Layers of Interstitial Sites)

To begin a second close-packed layer, a sphere rolls into an indentation formed by three spheres of the first layer, forming a regular tetrahedron (Figure 3.2***a***). When the second layer is completed and a third layer is added, each sphere of the second layer is at the apex of a tetrahedron with three spheres of the first layer and another tetrahedron with three spheres of the third layer. For *hcp*, the bases of these two tetrahedra are eclipsed (the spheres for both bases of the two tetrahedra are at *A* positions; see Figure 3.2***b***). For *ccp*, the bases of these two tetrahedra are staggered (*A* and *C* positions for the bases, see Figure 3.2***c***). At the center of each tetrahedron there is a small opening that could be occupied by a small atom. These interstitial sites are called *tetrahedral sites* (**T**, or tetrahedral holes). The layers of these **T** sites are in arrangements identical to the **P** layers. For any close-packed structure there are two types of tetrahedral sites and two layers of **T** sites (or we say there are two **T** layers). In one case, between adjacent **P** layers, the sphere of the second **P** layer is at the apex of a tetrahedron formed with three spheres of the first layer. The apex is upward, so we designate the tetrahedral site as $\mathbf{T}^+$. The same sphere of the second **P** layer is part of a tetrahedron pointed downward ($\mathbf{T}^-$) with three spheres of the third layer. In most cases, we do not need to make the distinction between $\mathbf{T}^+$ and $\mathbf{T}^-$. In Figure 3.2***b*** and ***c*** only the

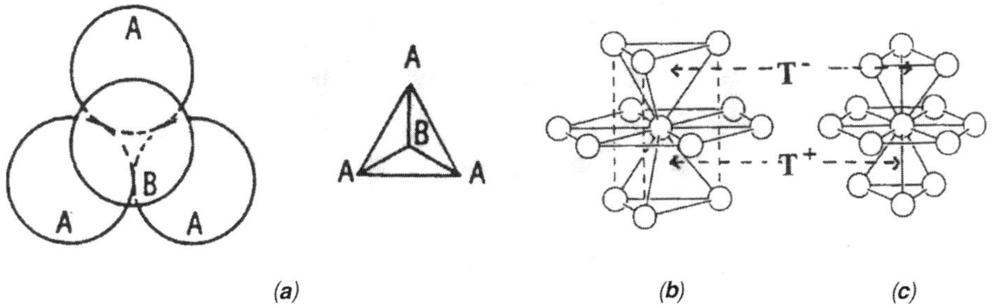

Figure 3.2 *(a)* A tetrahedral site in a close-packed structure. Orientation of tetrahedra in adjacent layers for *(b) hcp* and *(c) ccp* structures.

(*Source*: *CrystalMaker*, by David Palmer, *CrystalMaker* Software Ltd., Begbroke Science Park, Bldg. 5, Sandy Lane, Yarnton, Oxfordshire, OX51PK, UK.)

tetrahedra with apices of the second layer are shown. There are also tetrahedra with the bases in layer $\mathbf{P}_B$ with apices in **P** layers (1 and 3).

It is important to note that each sphere is part of two tetrahedra, so there are twice as many **T** sites as **P** sites or between *any two* close-packed **P** layers there are two **T** layers, $\mathbf{T}^+$ and $\mathbf{T}^-$. The positions (*A*, *B*, or *C*) of the **T** sites correspond to the apex **P** positions. For *hcp*, these are *A* and *B*. In the sequence from the $\mathbf{P}_A$ layer, for the first $\mathbf{T}^+$ layer each tetrahedron has its apex in the second **P** layer ($\mathbf{P}_B$), and in the second $\mathbf{T}^-$ layer, each tetrahedron has its apex in the first layer ($\mathbf{P}_A$) giving $|\mathbf{P}_A\mathbf{T}_B\mathbf{T}_A\mathbf{P}_B\mathbf{T}_A\mathbf{T}_B|\mathbf{P}_A\ldots$. Figure 3.2*c* shows three **P** layers and the tetrahedra between these with apices in the $\mathbf{P}_B$ layer. The bases are staggered ($\mathbf{P}_A$ and $\mathbf{P}_C$).

If we examine three touching spheres of one layer, they touch three spheres below (and also above). These two staggered triangles of two touching layers are opposite faces of an octahedron (Figure 3.3). We can tip the figure so that four spheres are in a plane with one above and one below—the usual representation for visualization of an octahedron. There is one octahedral site (**O**) for each **P** position, and hence, there is one layer of **O** sites between any two close-packed layers. The **O** positions (*A*, *B*, or *C*) differ from those of the two adjacent **P** positions so that for *hcp* we get $\mathbf{P}_A\mathbf{O}_C\mathbf{P}_B\mathbf{O}_C\mathbf{P}_A\ldots$ Note that for *hcp* the **O** sites are *only* in *C* positions. For *ccp* we get $|\mathbf{P}_A\mathbf{O}_C\mathbf{P}_B\mathbf{O}_A\mathbf{P}_C\mathbf{O}_B|\mathbf{P}_A\mathbf{O}_C\mathbf{P}_B\ldots$; the **O** positions are *A*, *B*, and *C*

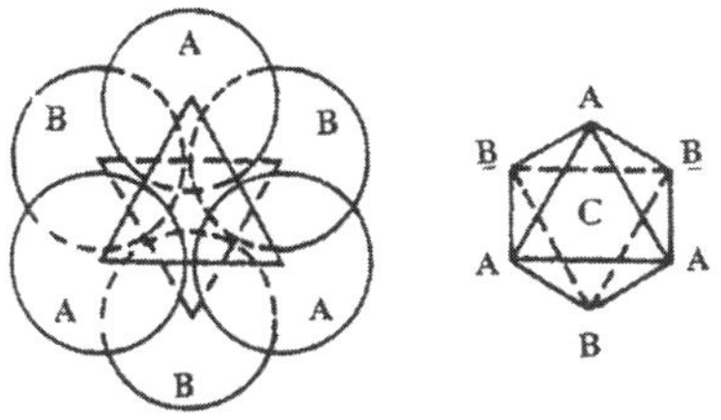

Figure 3.3. An octahedral site (*C* position) between *A* and *B* layers in a close-packed structure.

positions, as for the **P** positions. The full sequence is $|\mathbf{P}_A\mathbf{TO}_C\mathbf{TP}_B\mathbf{TO}_C\mathbf{T}|\ldots$ for *hcp* and $|\mathbf{P}_A\mathbf{TO}_C\mathbf{TP}_B\mathbf{TO}_A\mathbf{TP}_C\mathbf{TO}_B\mathbf{T}|\ldots$ for *ccp*.

3.3. Formation of the PTOT Framework

There are two common close-packed structures, *ccp* and *hcp*, but the sequence and spacings of **T** and **O** sites are the same between any two close-packed layers. We can see the arrangement in Figure 3.4. This is a cross-section perpendicular to the packing layers (looking at the edge of the sandwich). To see the relationships, we choose positions for the **P** layers; here we use $\mathbf{P}_A$ and $\mathbf{P}_B$. The **P** positions are occupied by atoms touching so that the **P—P** distance is $2r_\mathbf{P}$. This is the distance shown by lines, not the distance between **P** layers. The lower three **P** positions are closer to the viewer than the higher two **P** positions. Two lower **P** positions and one higher **P** connected by lines form one face of a tetrahedron. The center of the tetrahedron is directly below the higher $\mathbf{P_B}$, so it is a $\mathbf{T}_B$ site. The tetrahedron points upward so it can be labeled $\mathbf{T}^+$. The center lower **P** position, the two higher ones and another **P** in the *B* layer (not shown, but closer to the viewer) form a tetrahedron. This **T** site is directly above the center $\mathbf{P}_A$ so it is $\mathbf{T}_A$. The tetrahedron points downward so the site can be labeled $\mathbf{T}^-$.

The trigonal group of the two upper **P** positions and a third one in the same layer form a face of an octahedron. The opposite staggered trigonal octahedral face is formed by the central **P** position in the $\mathbf{P}_A$ layer and two more positions in the same layer. The center of the octahedron (**O**) is behind the other sites in the figure. It is exactly halfway between the **P** layers. The *A*, *B*, and *C* positions refer to positions viewed from above. The **O** site is in a *C* position, but for the view shown it is in line with **T** and **P** sites. The sequence is $\mathbf{P}_A\mathbf{T}^+\mathbf{O}_C\mathbf{T}^-\mathbf{P}_B$ or $\mathbf{P}_A\mathbf{T}_B\mathbf{O}_C\mathbf{T}_A\mathbf{P}_B$. For *hcp* it is $|\mathbf{P}_A\mathbf{T}_B\mathbf{O}_C\mathbf{T}_A\ \mathbf{P}_B\mathbf{T}_A\mathbf{O}_C\mathbf{T}_B|$, and the unit cell has a total of eight packing layers. For *ccp* it is $|\mathbf{P}_A\mathbf{T}_B\mathbf{O}_C\mathbf{T}_A\ \mathbf{P}_B\mathbf{T}_C\mathbf{O}_A\mathbf{T}_B\ \mathbf{P}_C\mathbf{T}_A\mathbf{O}_B\mathbf{T}_C|$, and the unit cell has a total of 12

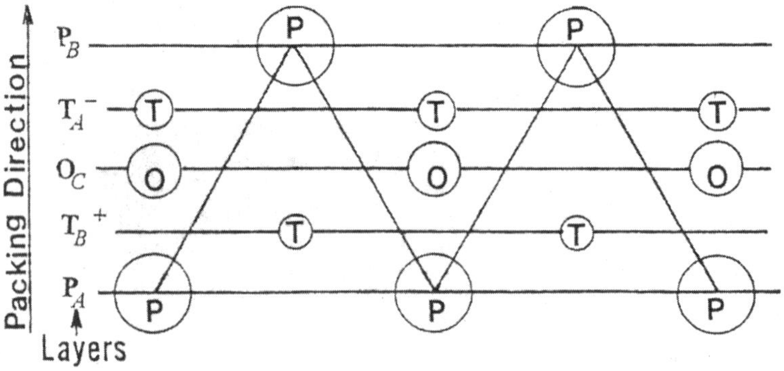

Figure 3.4. Construction of the basic close-packed unit (two packing layers for any close-packed structure), **P**, a packing atom; **O**, an octahedral site; **T**, a tetrahedral site; *A*, *B*, and *C* are the three relative packing positions in projection.

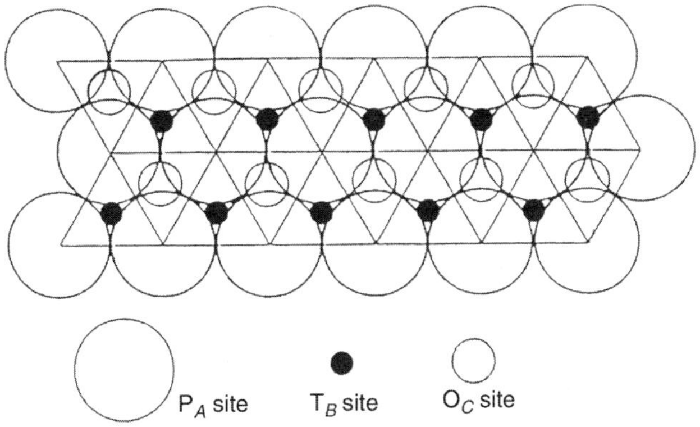

Figure 3.5. A close-packed *A* layer with **T** sites in *B* positions and **O** sites in C positions.

packing layers. The notation for most structures is more simple than this, because there are usually some empty layers that are omitted in the notation.

The spacing of the **T** layers is determined by the distance of the center of a tetrahedron from its base (p. 315). We have noted that the **O** layer is halfway between **P** layers. The spacing for all sites for any two close-packed layers is $\mathbf{PT}_{0.25}\mathbf{O}_{0.5}\ \mathbf{T}_{0.75}$ relative to the closest **P—P** distance taken as 1.000 (see Figures Appendices B.10 and B.11). Each **T** layer is halfway between a **P** layer and an **O** layer. The ratio of the radius of an atom in an O site to that of the packing atom is 0.414, and for a **T** site it is 0.225. Thus, an octahedral site is nearly twice the size of a tetrahedral site for close-packed **P** positions (touching). Figure 3.5 shows a close-packed **P** layer and **T** and **O** layers. The circles touching represent **P** positions, the smaller open circles represent **O** positions and the smallest filled circles represent **T** positions. Here the layers shown are $\mathbf{P}_A$, $\mathbf{T}_B$, and $\mathbf{O}_C$. The next layer would be $\mathbf{P}_B$, and just below it there is a $\mathbf{T}_A$ layer. It is important to remember that the hexagonal arrangements within a layer are *exactly* the same for each **P**, **O**, and **T** layer.

The sequence and spacings given for close packing are not artificial descriptions or approximations, as these are determined by geometry. The **PTOT** system is the most detailed and definitive treatment presented for close-packed structures, and many other structures can be described in this system.

3.4. Notation in the PTOT System

The objective is to use simple, clear, and informative notation for designating the close-packing type and layers occupied, including partial occupancy. From this information and knowledge of close-packed structures we can determine coordination numbers and local symmetries. For these purposes there are two parts of the notation; the index, and the layers occupied (**L**):

$a{\cdot}b$	PTOT
Index	L

The index a is usually the number 2 or 3, indicating *hcp* or *ccp* packing, respectively, and b is a number such that the product $a{\cdot}b$ gives the total number of close-packed layers in the unit cell. The total number of close-packed layers includes **P**, **T**, and **O** layers. The index b can be a fraction, but the product $a{\cdot}b$ must be a whole number.

The index usually indicates the sequence: cubic (**3**), hexagonal (**2**), or double hexagonal (**4**) for the *ABAC* sequence. The Pearson symbols (Table 2.5) can clarify cases such as hexagonal structures with an *ABC* sequence and $a = \mathbf{3}$ for the index. The symbols t for tetragonal, o for orthorhombic, m for monoclinic, and h for hexagonal, rhombohedral, or trigonal indicate the type of distortion of an idealized structure. Without distortion, $a = \mathbf{3}$ is for a cubic structure and $a = \mathbf{2}$ is for a hexagonal (or rhombohedral) structure.

The layer portion (**L**) of the notation indicates which close-packed **P**, **T**, and/or **O** layers are occupied and the extent of occupancy, $\mathbf{T}_{1/2}$, etc. Thousands of crystal structures can be described in this way. Here we consider the combinations of layers commonly encountered. The major portion of this book deals with individual elements, compounds, and their crystal structures.

Throughout the book we talk about close-packed layers. We saw in Figure 3.5 that the arrangement is the same for spheres touching in **P** layers or for smaller spheres in **O** or **T** layers even though they do not touch. In most crystals of inorganic compounds such as NaCl and Li_2O we usually assign the larger anions to **P** layers and the smaller cations to **O** layers (Na^+) or **T** layers (Li^+). Commonly the ions in **O** or **T** sites are larger than those interstitial sites in a close-packed structure, and the structure must expand to accommodate the larger ions. Even if the anions in **P** layers do not touch, because counterions are larger than **T** and **O** sites, the arrangement is unchanged and relative spacing is retained. It is realistic to consider such a structure as in the close-packed **PTOT** system, because the relative spacing and local environments are the same.

3.5. Major Types of Structures in the PTOT System

3.5.1. Structures with Only P Layers Filled

In solid metals we expect nondirectional bonding. Thus, packing of identical spheres should produce a close-packed structure; *ccp* or *hcp*. These are the common structures for metals. In *ccp* and *hcp* structures the *CN* for each atom is 12, six nearest neighbors in the same layer, three above, and three below (see Figure 3.2). The efficiency of packing (space occupancy) is 74% for *ccp* and *hcp*. Many metals adopt more than one structure, depending on temperature and pressure. Many other elements, including the noble gases and those forming discrete molecules, commonly have *ccp* and *hcp* structures.

The other structure encountered for more than 40 metals (under appropriate conditions) is the body-centered cubic (*bcc*) structure. This is not truly a close-packed structure for hard spheres, with only 68% of the volume occupied. In the *bcc* structure each atom has eight nearest neighbors (at the corners of the cube) and six next-nearest neighbors, through the cubic faces, only slightly (~15%) farther away. The similarity of the stability of these structures is shown by iron, for which the *bcc* structure is stable at low temperature, *ccp* at intermediate temperature, and again, *bcc* is more stable at high temperature. The *bcc* structure can be handled in the **PTOT** system, with the same atoms in all **P**, **O**, and **T** layers, giving **3•2PTOT**. Placing hard spheres of equal size in **O** and **T** sites as well as **P** sites cause the *bcc* structure to be less dense. However, atoms are not hard spheres, and as we shall see (Section 4.3.3), the *bcc* structure has about the same density or it is more dense than *hcp* or *ccp* structures in many cases. A mechanism for the transformation between the *bcc* and *ccp* structures in terms of the **PTOT** system will be discussed later.

3.5.2. Structures with Two Types of Layers Occupied

The **PO** structures involve **P** and **O** layers with both **T** layers empty. The most common structure of **PO** type is the NaCl structure described by **3•2PO**. The larger Cl^- ions are in the *ccp* **P** layers and Na^+ ions fill all **O** layers (Figure 3.6). In a *ccp* **PO** structure the **P** and **O** layers are identical. We could reverse the labels in the figure, letting the light balls represent Cl^- and the dark ones Na^+. NiAs is an example of an *hcp* **PO** structure or **2•2PO**, but this structure is not common for *hcp* because of interactions of atoms in **O** layers, all in *C* positions (see Section 5.2.1).

The **PT** structure has filled **P** layers and one **T** layer filled between **P** layers. Zinc sulfide has two modifications: zinc blende or sphalerite

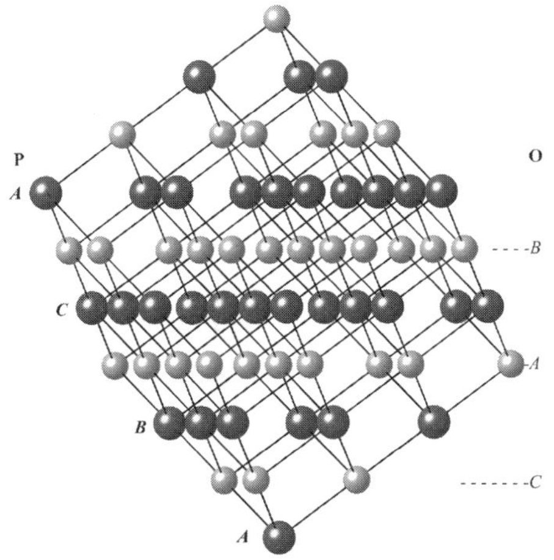

Figure 3.6. The structure of NaCl (**3•2PO**) showing the positions of **P** and **O** layers.

(*Source: CrystalMaker*, by David Palmer, *CrystalMaker* Software Ltd., Begbroke Science Park, Bldg. 5, Sandy Lane, Yarnton, Oxfordshire, OX51PF, UK.)

(**3•2PT**) and wurtzite (**2•2PT**). Zinc blende has *ccp* S^{2-} ions with Zn^{2+} ions in one of two **T** layers. The sequence is $|\mathbf{P}_A\mathbf{T}_B\mathbf{P}_B\mathbf{T}_C\mathbf{P}_C\mathbf{T}_A|$. Both **P** and **T** sites are in *A*, *B*, and *C* positions. Wurtzite has *hcp* S^{2-} ions with Zn^{2+} ions in one of two **T** layers. The sequence is $|\mathbf{P}_A\mathbf{T}_B\mathbf{P}_B\mathbf{T}_A|$. Here, **P** and **T** sites are only in *A* and *B* positions. The **O** layers are vacant so the *C* positions are empty.

3.5.3. Structures with Three and Four Layers Occupied

The **PTT** structure has **P** and both **T** layers filled with the **O** layer empty. Li_2O and fluorite (CaF_2) have the **3•3PTT** structure (later, p. 169, we will consider an alternative description). For Li_2O, oxide ions are in **P** layers with Li^+ ions in **T** layers, and for CaF_2, Ca^{2+}ions occupy **P** layers with F^- in **T** layers. The corresponding *hcp* structure **2•3PTT** is rare, pp. 139–144. The sequence for *hcp* would be $|\mathbf{P}_A\mathbf{T}_B\mathbf{T}_A\mathbf{P}_B\mathbf{T}_A\mathbf{T}_B|$. Strong interaction would result between very close $\mathbf{T}_A$ sites just above and below $\mathbf{P}_B$ sites ($\mathbf{T}_A\mathbf{P}_B\mathbf{T}_A$ or $\mathbf{T}_B$ sites just above and below $\mathbf{P}_A$).

The **2•4PTOT** structure is an *hcp* structure with eight (**2•4**) close-packed layers in the unit cell. The full sequence is $|\mathbf{P}_A\mathbf{T}_B\mathbf{O}_C\mathbf{T}_A\mathbf{P}_B\mathbf{T}_A\mathbf{O}_C\mathbf{T}_B|$. This full structure, shown in Figure 3.7, with all layers labeled is not known. Note that only the **O** sites (white balls) are in *C* sites. This accounts for the few known examples of a structure with full occupancy of **O** sites for an *hcp* structure. Structures with only half of the **O** sites filled, or with only alternate layers occupied, avoid the interaction. When the **O** layers are omitted *any hcp* structure has open channels along the *C* positions.

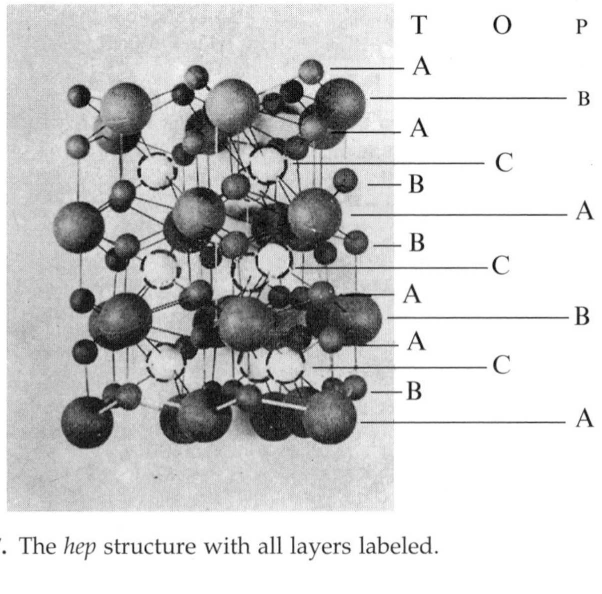

Figure 3.7. The *hep* structure with all layers labeled.

In Figure 3.7, we can see that **P** and **T** layers occupy only *A* and *B* positions. Let us focus on a $\mathbf{P}_B$ layer. Just below and above the $\mathbf{P}_B$ layer there are $\mathbf{T}_A$ layers. These **T** sites are very close, with *no* shielding. For *hcp* structures no examples are encountered for **PTT** or **PTOT** with both **T** layers filled without unusual features (pp. 139–144). Partial filling of both **T** layers avoids repulsion involving adjacent **T** sites. There are many variations of the **3•4PTOT** structure based on a *ccp* structure or $\mathbf{P}_A\mathbf{P}_B\mathbf{P}_C$. The full sequence is $|\mathbf{P}_A\mathbf{T}_B^+\mathbf{O}_C\mathbf{T}_A^-\mathbf{P}_B\mathbf{T}_C^+\mathbf{O}_A\mathbf{T}_B^-\mathbf{P}_C\mathbf{T}_A^+\mathbf{O}_B\mathbf{T}_C^-|$. The **O** layers are in a *ccp* sequence (*CABC*...) and the **T** layers have a *ccp* sequence (for $\mathbf{T}^+$ *BCA*... and for $\mathbf{T}^-$ *ABC*...). Because each **P** layer has two close **T** layers (e.g., $\mathbf{T}_A\mathbf{P}_B\mathbf{T}_C$) and each **O** layer has two close **T** layers (e.g., $\mathbf{T}_A\mathbf{O}_B\mathbf{T}_C$) with exactly the same relationships, we can reverse designations of **P** and **O** layers. This is true for *ccp*, but not for *hcp*. The full **3•4PTOT** *ccp* structure is shown in Figure 3.8. Note the equivalence of **P** and **O** layers, exactly as for NaCl. We could also reverse the roles of pairs of **T** layers ($\mathbf{T}^+$ and $\mathbf{T}^-$) with the roles of **P** and **O** together. The first **T** layer above a **P** layer is $\mathbf{T}^+$, and this becomes **O** and $\mathbf{T}^-$ becomes **P**. **P** and **O** layers become **T** layers:

$$|\mathbf{P}_A\mathbf{T}_B\mathbf{O}_C\mathbf{T}_A\mathbf{P}_B\mathbf{T}_C\mathbf{O}_A\mathbf{T}_B\mathbf{P}_C\mathbf{T}_A\mathbf{O}_B\mathbf{T}_C|$$

$$\vdots \quad \vdots \quad \vdots \quad \vdots \quad \vdots$$

$$|\mathbf{P}_A\mathbf{T}_B\mathbf{O}_C\mathbf{T}_A\mathbf{P}_B\mathbf{T}_C\mathbf{O}_A\mathbf{T}_B\mathbf{P}_C\mathbf{T}_A\mathbf{O}_B\mathbf{T}_C|$$

With only **P** and **O** layers filled, the structure is **3•2PO** for NaCl. If all **P** and **O** layers are filled with the same atoms it is a simple cubic structure. Look at Figure 3.6 and imagine all balls to be identical—it is simple cubic. The structure $\mathbf{3{\cdot}2T^+T^-}$ with the same atoms in all **T** sites (both layers) is the same simple cubic structure. Thus, a simple cubic structure is easily represented in the **PTOT** system. Because the system is based on a close-packed arrangement, we prefer to begin with **P** positions and use **3•2PO** in preference to $\mathbf{3{\cdot}2T^+T^-}$. The *ccp* arrangement without filling **O** and **T** layers is $\mathbf{P}_A\mathbf{P}_B\mathbf{P}_C$ or **3P**.

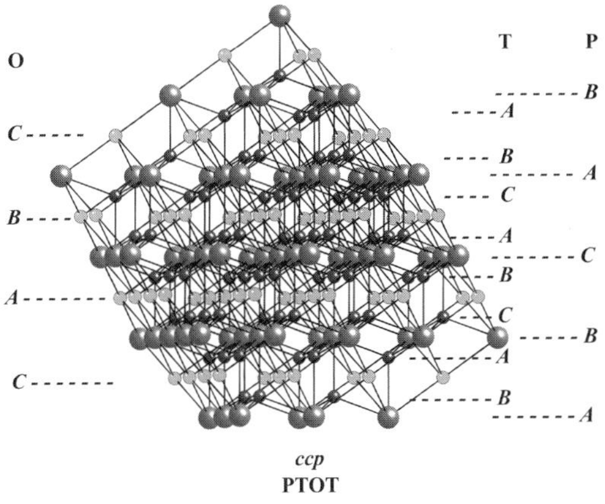

Figure 3.8. The *ccp* structure of **P** layers with all layers labeled.

(*Source: CrystalMaker*, by David Palmer, *CrystalMaker* Software Ltd., Begbroke Science Park, Bldg. 5, Sandy Lane, Yarnton, Oxfordshire, OX51PF, UK.)

If we fill all **P, O,** and **T** layers with identical atoms we can see from Figure 3.8 that the resulting structure is body-centered cubic (*bcc*). We saw that a simple cubic structure is **3·2PO** or $\mathbf{3{\cdot}2T^{+}T^{-}}$, so interpenetrating these two simple cubic structures gives **3·2PTOT**, the *bcc* structure. The notation is **3·2PTOT** rather than **3·4PTOT** because **P** and **O** sites are identical for *ccp*. With identical atoms in all sites we can see the sequence by replacing **O** by **P**, $|\mathbf{P}_A\mathbf{TO}_C\mathbf{TP}_B\mathbf{T}_C\mathbf{O}_A\mathbf{TP}_C\mathbf{TO}_B\mathbf{T}| \rightarrow |\mathbf{P}_A\mathbf{TP}_C\mathbf{TP}_B\mathbf{T}|\mathbf{P}_A\mathbf{TP}_C\mathbf{TP}_B\mathbf{T}|$. There are six layers in the repeating unit giving **3·2PTOT**. The fact that the sequence is *ACB* is not significant, any sequence (*CBA, CAB*, etc.) of the three different positions corresponds to *ccp*. If there are different atoms in **P** and **O** layers, the full sequence is **3·4PTOT**, with 12 layers in the repeating unit.

3.5.4. Partially Filled Layers

In the **PTOT** scheme some **P, O,** and **T** layers can be empty or partially filled. This is determined by the stoichiometry of the compound, relative sizes of ions or atoms, and preferences for coordination number (*CN*). Each **P** site has 12 nearest **P** sites ($CN = 12$). An **O** site has six near **P** sites, and a **T** site has four nearest **P** sites.

The partial filling of a **P, O,** or **T** layer can be shown as $\mathbf{L}_{1/2}$, $\mathbf{L}_{2/3}$, etc. This can result from empty sites or in cases where more than one kind of atom form the same layer. The common cases are described below.

3.5.4.1. *$L_{1/3}$ and $L_{2/3}$ Close-Packed Layers*

$\mathbf{L}_{1/3}$ and $\mathbf{L}_{2/3}$ layers, with hexagonal patterns, are the most common cases of partially filled layers encountered in close-packed structures. Figure 3.9***a*** shows a complete close-packed layer, and Figure 3.9***b***

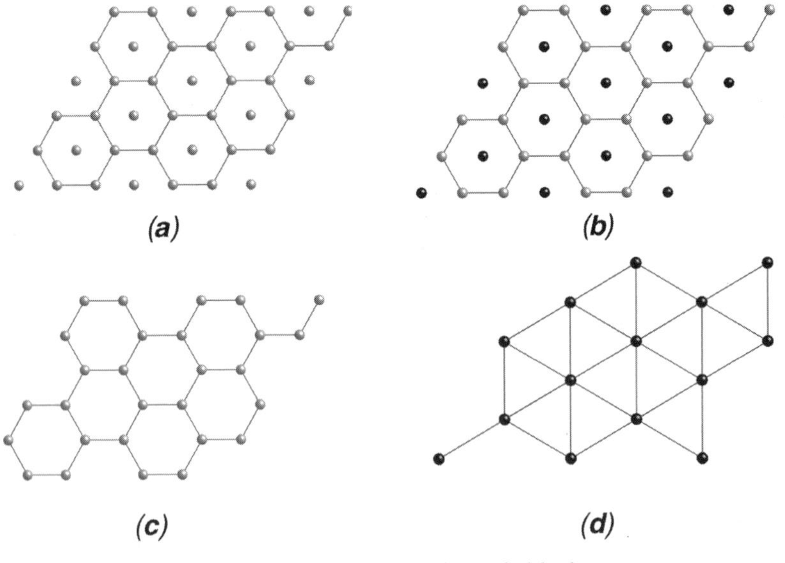

Figure 3.9. A close-packed layer *(a)* and *(b)* with black atoms at centers of hexagons; and hexagonal networks corresponding to $\mathbf{L}_{2/3}$ *(c)* and $\mathbf{L}_{1/3}$ *(d)* for partially occupied layers.

(*Source: CrystalMaker*, by David Palmer, *CrystalMaker* Software Ltd., Begbroke Science Park, Bldg. 5, Sandy Lane, Yarnton, Oxfordshire, OX51PF, UK.)

shows the hexagonal network with the centers of the hexagons marked by •. Omission of the • sites gives the planar hexagonal network in Figure 3.9*c*. This corresponds to $\mathbf{L}_{2/3}$. Omitting the light circles of Figure 3.9*b* gives the planar hexagonal network shown in Figure 3.9*d*. It corresponds to $\mathbf{L}_{1/3}$. These two patterns have higher symmetry than others involving partial filling of layers.

The beautiful $\mathbf{L}_{2/3}$ layer has been adopted for fine art patterns, architectural designs, and even by bees for honey comb. Later we will consider examples of $\mathbf{L}_{2/3}$ layers involving graphite structures (see Section 4.4.3).

3.5.4.2. *$L_{1/4}$ and $L_{3/4}$ Close-Packed Layers*

There are many possibilities for partial filling of a close-packed layer, but those of high symmetry are more common. In Figure 3.10*a* 1/4 of the close-packed sites are marked by •. If these are vacant, we have $\mathbf{L}_{3/4}$, a 3/4 filled close-packed layer retaining the hexagonal pattern. If the light-colored sites are vacant and • sites are filled, we have $\mathbf{L}_{1/4}$ with larger hexagons (twice as large) with an occupied site at the center of each hexagon. There is another pattern shown in Figure 3.10*b*. Here the third horizontal row is shifted by one unit to the left. There are no regular hexagons, but a tetragonal unit outlined. Again, $\mathbf{L}_{3/4}$ corresponds to • sites empty and $\mathbf{L}_{1/4}$ corresponds to only • sites occupied. These patterns have lower symmetry than those shown in Figure 3.10*a*, and are less common.

3.5.4.3. *$L_{1/2}$ Close-Packed Layers*

Again, there are many possibilities for a half-filled layer, but more symmetric arrangements are common. In Figure 3.11*a* alternate horizontal rows are vacant. Filling alternate rows parallel to the other edges gives the same arrangement. This $\mathbf{L}_{1/2}$ layer has lower symmetry than those for $\mathbf{L}_{1/4}$, $\mathbf{L}_{1/3}$, $\mathbf{L}_{2/3}$, and $\mathbf{L}_{3/4}$. In Figure 3.11*b* pairs of sites are omitted in each row. The resulting layer has still lower symmetry than that in Figure 3.11*a*.

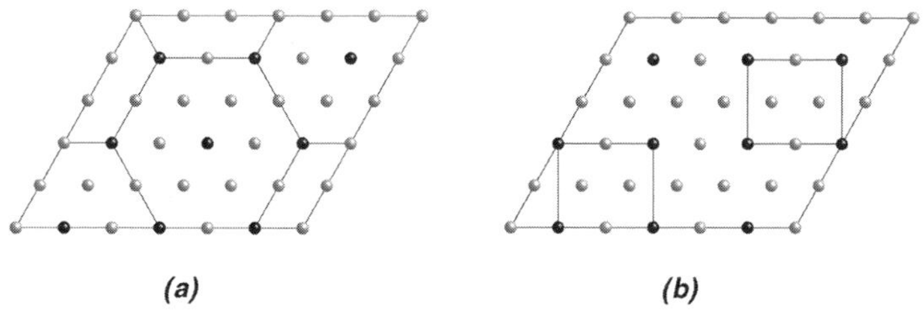

Figure 3.10. A partially filled close-packed layer corresponding to *(a)* $\mathbf{L}_{3/4}$ with • sites vacant or $\mathbf{L}_{1/4}$ with only • sites occupied. *(b)* Another arrangement for $\mathbf{L}_{3/4}$ and $\mathbf{L}_{1/4}$ occupancies.

(*Source: CrystalMaker*, by David Palmer, *CrystalMaker* Software Ltd., Begbroke Science Park, Bldg. 5, Sandy Lane, Yarnton, Oxfordshire, OX51PF, UK.)

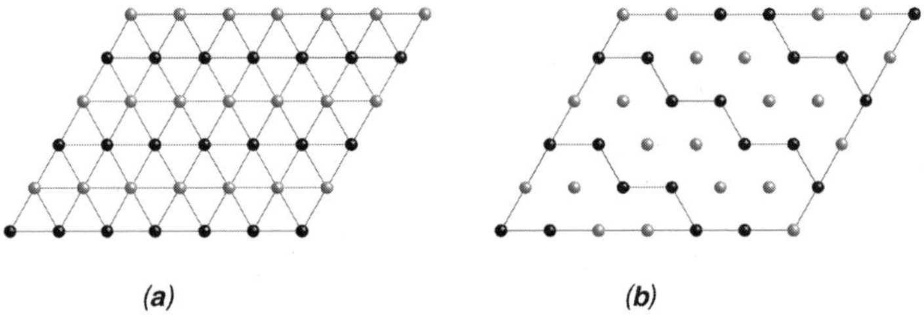

Figure 3.11. Two arrangements of $L_{1/2}$ occupancies.

(*Source: CrystalMaker*, by David Palmer, *CrystalMaker* Software Ltd., Begbroke Science Park, Bldg. 5, Sandy Lane, Yarnton, Oxfordshire, OX51PF, UK.)

3.5.5. Summary

Here we have briefly considered the structures for various combinations of **P**, **T**, and **O** layers. All patterns for *ccp* structures are shown in Figure 3.12. The face-centered cube is the same as the cubic close-packed structure. The two names stress particular features. The unit cell of a *ccp* structure is *fcc*. The packing direction, the direction of stacking the close-packed layers, is along the body diagonals of the cube. The packing direction is vertical in Figures 3.6, 3.7, and 3.8.

The zinc blende structure of ZnS (**3•2PT**) has S^{2-} filling *ccp* **P** layers and Zn^{2+} filling one **T** layer. For Li_2O (**3•3PTT,** the antifluorite structure), the O^{2-} ions fill *ccp* **P** layers with Li^+ ions filling *both* **T** layers. The NaCl structure has the **3•2PO** structure, with Cl^- ions filling *ccp* **P** layers and Na^+ ions filling **O** layers. Cadmium chloride ($CdCl_2$) has a

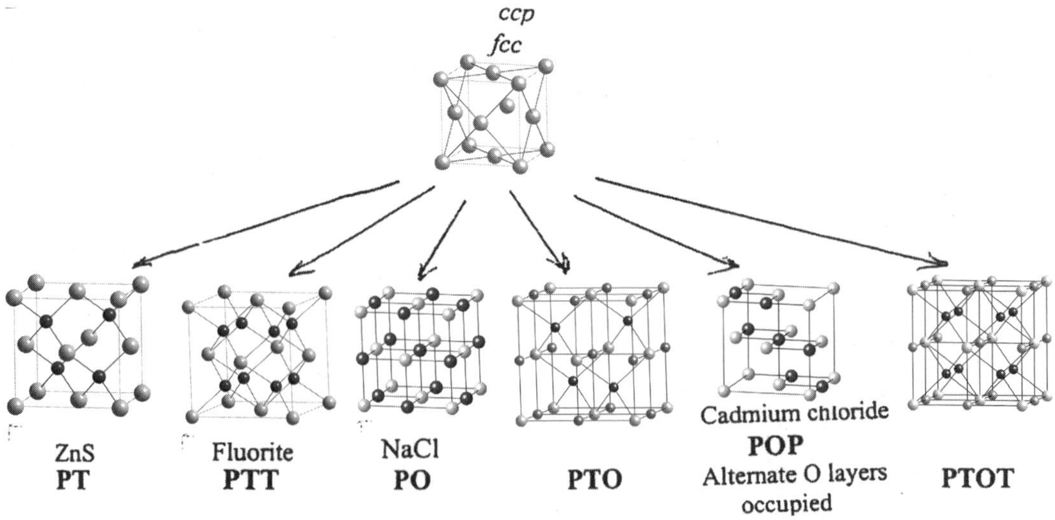

Figure 3.12. Inorganic structures derived from a cubic close-packed arrangement by filling tetrahedral and/or octahedral sites.

(*Source: CrystalMaker*, by David Palmer, *CrystalMaker* Software Ltd., Begbroke Science Park, Bldg. 5, Sandy Lane, Yarnton, Oxfordshire, OX51PF, UK.)

layer structure. The Cl^- ions fill *ccp* **P** layers with Cd^{2+} ions filling *alternate* **O** layers (**POP**). The *ccp* structure with *all* layers filled is illustrated by BiF_3, with Bi filling the *ccp* **P** layers and with F filling the **O** layer and both **T** layers.

Not all of these patterns are encountered for *hcp* structures because there are strong interactions between **O** sites because they are all in *C* positions without shielding. There is also strong interaction between close adjacent **T** layers preventing full occupancy of both **T** layers for *hcp*. These problems are avoided for *ccp* structures because **P**, **O**, and **T** sites are staggered at *A*, *B*, and *C* positions.

Next, we will consider in more detail crystal structures of elements, and crystal structures for the various combinations of layers, including multiple-layer structures.

Chapter 4

Crystal Structures of the Elements and Some Molecular Crystals

In the introductory chapters the examples using the **PTOT** notation have been simple inorganic compounds. Most of these structures have a close-packed arrangement of anions with cations occupying **T** and/ or **O** sites. Most of the elements are monatomic, and close packing is expected. The structures of most elements with diatomic molecules and even those with larger molecules can be described also in the **PTOT** system.

4.1. Noble Gases

The noble gases are monatomic. The atoms have closed electron shells so there is only nondirectional van der Waals interaction between atoms. All six noble gases have the **3P** (*ccp*) crystal structure with four atoms per unit cell. As noted earlier, the unit cell for *ccp* is the same as the face-centered cubic cell (see Figure 4.1). The noble gases are ideal for close packing, and *ccp* is expected to be favored over *hcp* because of the higher symmetry of *ccp*. Helium has two additional allotropic structures: *hcp* (**2P**) and *bcc*. The *hcp* structure is close packed (Figure 4.2), and the *bcc* is described by the notation **3·2PTOT** with all **P**, **T**, and **O** sites occupied by He atoms (see Figure 4.3).

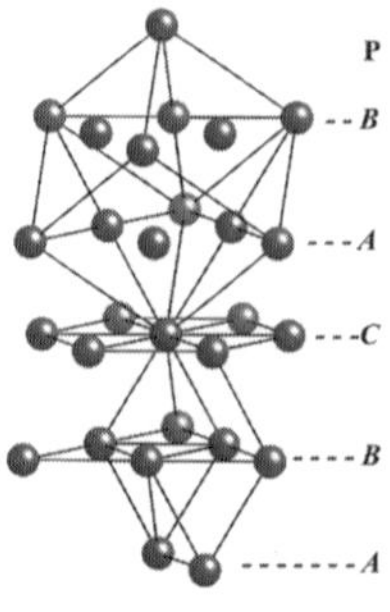

Figure 4.1. The **3P** structure (*ccp* or face-centered cubic). The cubic unit cell is outlined by double lines.

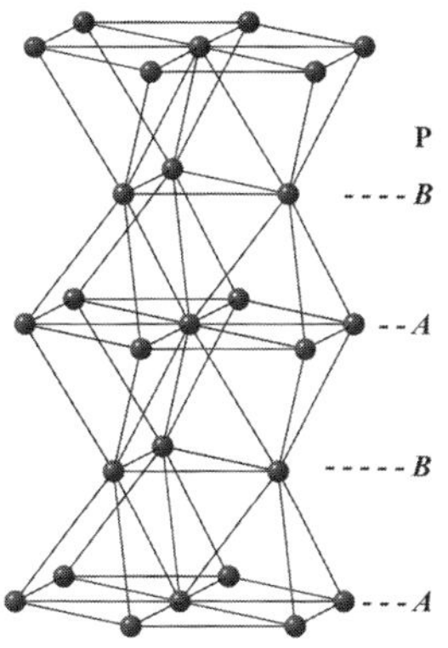

Figure 4.2. The **2P** structure (hexagonal close-packed, *hcp*). The hexagonal unit cell is outlined by double lines.

4.2. Metals

4.2.1. Structures of Metals

Of the 109 elements 23 of these are classified as nonmetals—in the upper right corner of the long form of the periodic table. Some of the elements along the diagonal between the metals and nonmetals have been classified as metalloids, having some properties of metals and some of nonmetals. There are more than 80 metals, and most of them have close-packed structures, *ccp* (**3P**) or *hcp* (**2P**). Many of the metals have one stable structure under ordinary conditions and a second structure at higher temperature and/or higher pressure. A few metals have three different structures under the conditions studied. At least three metals include four different structures at various temperatures and pressures. The fourth structure encountered for La, Ce, Pr, and Am is double hexagonal, designated **4P**. It is a close-packed structure

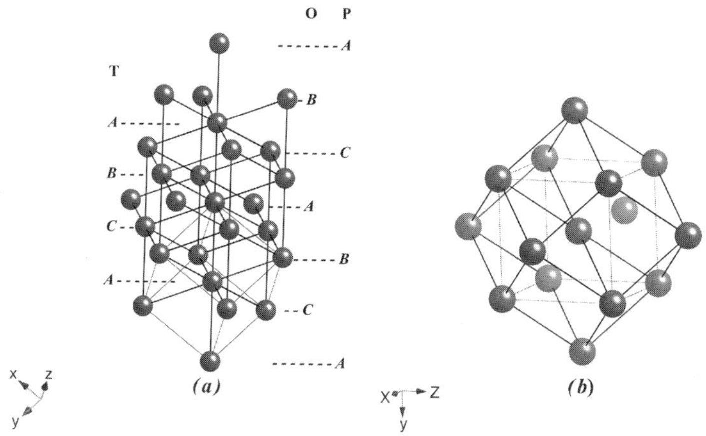

Figure 4.3 ***(a)*** The **3·2PTOT** structure (*bcc*). The cubic unit cell is outlined by double lines. ***(b)*** The *bcc* cube and the **3·2PTOT** structure showing nearest neighbors.

(*Source: CrystalMaker*, by David Palmer, *CrystalMaker* Software Ltd., Begbroke Science Park, Bldg. 5, Sandy Lane, Yarnton, Oxfordshire, OX51PF, UK.)

with an unusual sequence of layers. For **2P**, the sequence is $|\mathbf{P}_A\mathbf{P}_B|$, for **3P** it is $|\mathbf{P}_A\mathbf{P}_B\mathbf{P}_C|$, and for **4P** the sequence is $|\mathbf{P}_A\mathbf{P}_B\mathbf{P}_A\mathbf{P}_C|$. Six allotropic forms have been reported for plutonium (Donohue, Appendix A). The alkali metals and 13 more metals have the **3·2PTOT** (*bcc*) structure as the stable structure under ordinary conditions. Structures of the elements are shown in Table 4.1. We appreciate the suggestions of Professor Leo Brewer of additional cases of *bcc* structures for many metals.

The *ccp* and *hcp* structures are truly close packed with the same efficiency of space filling and, consequently, the same density for the same spheres. Each has the same coordination number (***CN***), 12. In Figure 4.2, the packing direction for *hcp* is vertical, perpendicular to the close-packed layers. This is the unique direction for *hcp*. In Figure 4.1, the packing direction for *ccp* is vertical and along one of the three body diagonals of the face-centered cubic unit cell. Metals with the **3P** (*ccp*) structure are more malleable and ductile than those with the **2P** (*hcp*) structure. For *hcp* the close-packed layers can slide over another in planes parallel to these layers. Along any other direction there are no smooth layers. For *ccp*, the structure is identical along each of the three body diagonals of the face-centered cube. The layers perpendicular to each of these directions are close-packed layers, and one can slide over another easily. The *ccp* structure has higher symmetry.

Iron is a good example, showing the similarity of the stabilities of these structures, in this case **3P** (*ccp*) and **3·2PTOT** (*bcc*). Below 910° C, the structure of α-Fe is **3·2PTOT**, between 910° and 1390°C the stable structure of γ-Fe is **3P** (*ccp*), and from 1390° to the melting point **3·2PTOT** is stable (δ-Fe). The *bcc* structure is more open for *hard spheres of uniform size* than *ccp* because the atoms in the **O** and **T** sites are larger than these sites in *ccp* or *hcp*. This opening causes "wasted" space. Hydrogen or carbon atoms can occupy the interstitial sites. Carbon is more soluble in *ccp* γ-Fe, occupying **O** sites, and the presence of carbon favors the *ccp* structure, lowering the temperature for the transition from α-Fe to γ-Fe. The slightly more open *bcc* structure allows lattice vibrations of greater amplitude and energy, favoring this structure at high temperature for several metals.

The ***CN*** for *ccp* and *hcp* is 12. For *bcc*, there are eight closest neighbors and six more neighbors at slightly greater distance. Both sets of neighbors must interact directly with the central atom. A metal atom has too few orbitals and too few electrons for many directed bonds such as those in covalent molecules. Bonding in metals has been considered in terms of the valence-bond theory and the band theory. The band theory is a molecular orbital approach treating the metal lattice as a unit. Without directional interaction we expect the structures for metals to be *ccp* or *hcp*. Strong directional sp^3 bonding in tin results in the diamond structure (**3·2PT**) in gray Sn, like C, Si, and Ge. The next element in the family is Pb with the **3P** (*ccp*) structure.

The **3P** (*ccp*) structure is centrosymmetric, and there are three equivalent packing directions along the cubic diagonals. The six neighbors of an atom in any close-packed layer for **2P** (*hcp*) or **3P** (*ccp*) are at the corners of a centrosymmetric hexagon. For *hcp*, the six neighbors in

TABLE 4.1. Crystal structures of the elements.

1H_2 **2P**																	2 He **3P** **2P** **PTOT**
3 Li **PTOT** **2P** **3P**	4 Be **2P** **PTOT**											5 B_{12} **3P′**	6 C **3·2PT** **2·2PT** **2IIP** **3IIP**	7 N_2 **3P** **2P**	8 O_2 **3P**	9 F_2 **3P′**	10 Ne **3P**
11 Na **PTOT** **2P**	12 Mg **2P**											13 Al **3P**	14 Si **3·2PT**	15 P **2IIP**	16 S_6 **3P′**	17 Cl_2 **3P′**	18 Ar **3P**
19 K **PTOT**	20 Ca **3P** **PTOT** **2P**	21 Sc **2P** **PTOT** **3P**	22 Ti **2P** **PTOT**	23 V **PTOT**	24 Cr **PTOT** **3P** **2P**	25 Mn **PTOT** **3P′**	26 Fe **PTOT** **3P** **2P**	27 Co **2P** **3P**	28 Ni **3P** **2P** **PTOT**	29 Cu **3P**	30 Zn **2P**	31 Ga **2IIP**	32 Ge **3·2PT**	33 As **3·2(PO)′**	34 Se **3P′**	35 Br_2 **3P′**	36 Kr **3P**
37 Rb **PTOT**	38 Sr **3P** **2P** **PTOT**	39 Y **2P** **PTOT**	40 Zr **2P** **PTOT**	41 Nb **PTOT**	42 Mo **PTOT** **3P**	43 Tc **2P**	44 Ru **2P**	45 Rh **3P**	46 Pd **3P**	47 Ag **3P**	48 Cd **2P**	49 In **3P′**	50 Sn **3·2PT** **3·2PT′**	51 Sb **3·2(PO)′**	52 Te **3P′**	53 I_2 **3P′**	54 Xe **3P**
55 Cs **PTOT**	56 Ba **PTOT**	71 Lu **2P** **PTOT**	72 Hf **2P** **PTOT** **3P**	73 Ta **PTOT** **3P**	74 W **PTOT** **3P**	75 Re **2P**	76 Os **2P**	77 Ir **3P**	78 Pt **3P**	79 Au **3P**	80 Hg **3(PO)′** **PTOT**	81 Tl **2P** **PTOT**	82 Pb **3P**	83 Bi **3·2(PO)′**	84 Po **3PO**	85 At	86 Rn **3P**
	88 Ra **PTOT**																

57 La **2P** **3P** **4P** **PTOT**	58 Ce **2P** **3P** **4P** **PTOT**	59 Pr **2P** **3P** **4P** **PTOT**	60 Nd **4P** **3P** **PTOT**	61Pm **4P**	62 Sm **9P** **PTOT**	63 Eu **PTOT**	64 Gd **2P** **PTOT**	65 Tb **2P** **PTOT**	66 Dy **2P** **PTOT**	67 Ho **2P** **PTOT**	68 Er **2P** **PTOT**	69 Tm **2P** **PTOT**	70 Yb **3P** **2P** **PTOT**
89 Ac **3P**	90 Th **3P** **PTOT**	91 Pa **PTOT** **3P**	92 U **PTOT**	93 Np **PTOT**	94 Pu **3P** **PTOT**	95 Am **4P** **3P**	96 Cm **4P** **3P**	97 Bk **3P** **4P**					

2P = *hcp*
3P = *ccp*
4P = *ABAC*
PTOT = *bcc*
3PO = simple cubic
3·2(PO)′ = distorted simple cubic

IIP = double P layer
′ : means distorted

adjacent layers form a hexagonal prism that is not centrosymmetric (see Figure 3.2). For a hexagonal structure the ideal *c/a* ratio is 1.633. Mg and Co have *c/a* ratios very close to this value (also for the second structure of Li, Na, and Sr). For most metals with the *hcp* structure the ratio *c/a* is less than 1.63. Only Zn ($c/a = 1.856$) and Cd (1.886) have a ratio greater than 1.63. These deviations for *c/a* indicate significant differences in bonding for the two sets of neighbors. For Zn and Cd, the *d* orbitals are filled and should not be involved in bonding. Here, unlike the metals to the left of Zn and Cd, *p* orbitals are most important in bonding. For Mg, the configuration is $3s^2$ so hybridization involving s and *d* or *p* orbitals is not expected. For the transition metals, *d* orbitals are involved in bonding, and for the lanthanides, bonding involves *d* and/or *f* orbitals.

The *bcc* (**3·2PTOT**) structure as the stable structure at ordinary temperature is encountered for the alkali metals (ns^1), V family [$(n-1)d^3 ns^2$or$(n-1)d^4ns^1$], Mn($3d^54s^2$), Fe($3d^64s^2$), and Eu ($4f^74s^2$). These are metals with half filled or nearly half-filled orbitals. The greater the number of singly occupied orbitals, the greater the number of directed bonds. The *bcc* structure has eight atoms at the corners of a cube and an octahedral arrangement of six atoms through the cubic faces, forming an octahedron. Both arrangements have $\mathbf{O}_h$ symmetry. The *bcc* structure is favored by transition metals with half-filled subshells.

The group IIA metals have the outer electron configuration ns^2 with no singly occupied orbital for bonding. The structures are *hcp* for Be and Mg and *ccp* for Ca and Sr. The structures are *bcc* for Ba and Ra, for which promotion to $(n-1)d^1ns^1$ is more favorable than for the other metals of the family. Metals with filled or nearly filled orbitals (Co, Ni, Pt metals, Cu family, Zn, and Cd) have *ccp* (**3P**) or *hcp* (**2P**) structures.

Among the lanthanides and actinides there are several metals with the unusual **4P** (*ABAC*) structure, and Sm has the strange **9P** (*ABAB CBCAC*) structure. There are uncertainties of the structures of some of the metals beyond U. For most of these metals only small samples are available, purity is a problem, and in some cases samples are deposited on a filament. Impurities and deposition on another metal can change the structure.

For Pu, for which large amounts of information have been available, six allotropic forms have been reported. These are described as follows:

$$\alpha \xrightarrow{122^\circ C} \beta \xrightarrow{206^\circ C} \gamma \xrightarrow{319^\circ C} \delta \xrightarrow{451^\circ C} \delta' \xrightarrow{476^\circ C} \varepsilon$$

α	β	γ	δ	δ′	ε
Monoclinic	Monoclinic	Orthorhombic	*ccp*	body-centered tetragonal	*bcc*
$P2_1/m$	$I2/m$	$Fddd$	$Fm3m$	$I4/mmm$	$Im3m$
$\mathbf{C}_{2h}{}^2$	$\mathbf{C}_{2h}{}^3$	$\mathbf{D}_{2h}{}^{24}$	$\mathbf{O}_h{}^5$	$\mathbf{D}_{4h}{}^{17}$	$\mathbf{O}_h{}^9$
Z = 16	Z = 34	Z = 8	Z = 4	Z = 2	Z = 2
d = 19.84	d = 17.70	d = 17.14	d = 16.26	d = 16.00	d = 16.87

Z is the number of atoms per unit cell. The density (d) is given in g/cm^3. Only powder data are available for β- and γ-Pu. γ-Pu is orthorhombic, but the cell can be described as a face-centered cell with four **T** sites occupied. This is a distorted **3·2PT** cell, the distorted diamond structure. The crystal structure of γ-Pu is unique among metals with *CN* 10. Six neighbors are in a planar hexagon with two atoms above and below, corresponding to opposite edges of an elongated tetrahedron. Densities in Table 4.2 are given for δ-Pu (*ccp*, **3P**) and ε-Pu (*bcc*, **3·2PTOT**).

Mercury (α-Hg) has a rhombohedral structure. The structure can be described as a simple cubic arrangement distorted so that the interaxial angle is decreased from 90° to 70.5°, shown as (**3PO**)′ in Table 4.1. The cubes share Hg atoms so that each Hg has six nearest neighbors, three of each cube. These six close neighbors form a trigonal antiprism. Each Hg atom has six more distant hexagonally arranged neighbors in the close-packed layer. Below 79K β-Hg has a tetragonal structure.

Gallium is an unusual metal, melting at 29.8°C (it is also the element with the greatest liquidus range, *bp ca.* 2,250°C). In the solid there is one very short distance (2.44 Å) with six other neighbors in pairs at 2.71, 2.74, and 2.80 Å.. The Ga atoms are in puckered hexagonal layers with long distances within the layer. The short bonds are between Ga atoms of adjacent layers. Alternate Ga atoms are strongly bonded to upper or lower layers. These can be described as double layers *AA′* or *BB′*·. They are stacked in an *hcp* arrangement *AA′BB′AA′BB′*··· described as a double **P** layer (**IP**). This unusual structure is found also for black phosphorus. There is evidence for dimeric molecules in the liquid state.

The structure of manganese (α-Mn) is *bcc* (**3·2PTOT**), but with distortions because of four nonequivalent atoms (distances 2.26–2.93 Å). The high temperature structure, γ-Mn, is **3P** (*ccp*), but slightly distorted. Indium has this structure also.

4.2.2. Conversion of the *ccp–bcc* Structures

A possible mechanism for the conversion of a *ccp* structure to *bcc* for a metal involves compression. Metals are more compressible than solids such as salts, and metals are much more malleable and ductile than most other solids. In Figure 4.4, on the left side, we view a *ccp* structure parallel to the packing layers. The *ccp* structure is viewed from an angle so that *A*, *B*, and *C* positions are staggered. No attempt has been made to distinguish the distance of the atoms in each layer from the viewer. As drawn the distances from the viewer are shortest for *A* and longest for *C* (this is an arbitrary choice of sites; one could choose *C* positions closer than either *A* or *B*). In the figure the layers are compressed so that layers are converted as follows:

$\mathbf{P}_B \rightarrow \mathbf{T}_B$, $\mathbf{P}_C \rightarrow \mathbf{O}_C$, $\mathbf{P}_A \rightarrow \mathbf{T}_A$, the next $\mathbf{P}_B$ remains $\mathbf{P}_B$, $\mathbf{P}_C \rightarrow \mathbf{T}_C$, $\mathbf{P}_A \rightarrow \mathbf{O}_A$, $\mathbf{P}_B \rightarrow \mathbf{T}_B$, and the next $\mathbf{P}_C$ remains $\mathbf{P}_C$. The lowest $\mathbf{P}_A$ layer is retained as the reference layer. The **P** sites are larger than **O** and **T** sites so the structure must expand, increasing the distances within a layer and between layers for hard spheres. Metal atoms are not hard spheres so the sites can be equivalent. The figure is a correlation

TABLE 4.2. Densities of metals with *bcc* and *hcp* and/or *ccp* structures.[a]

Metal	Structure	Temp.	*a*	*c*	*c/a*	Atomic Volume (Aring^3)	Density (g/cm^3)
Li	*bcc*	78K	3.482			21.10	0.5494
	hcp	78K	3.103	5.080	1.637	21.18	0.5440
	ccp	78K	4.388			21.16	0.5445
Na	*bcc*	78K	4.235			37.98	0.9950
		5K	4.225			37.71	1.0122
	hcp	5K	3.767	6.154	1.634	37.81	1.0095
Be	*bcc*	1,250°C	2.550			8.29	1.806
	hcp	1,250°C	2.343	3.659	1.562	8.70	1.721
Ca	*bcc*	500°C	4.474			44.78	1.487
	hcp	500°C	3.97	6.49	1.635	44.29	1.503
	ccp	18°C	5.612			44.19	1.507
Sr	*bcc*	614°C	4.87			57.75	2.520
	hcp	415°C	4.33	7.05	1.628	57.23	2.543
	ccp	25°C	6.084			56.30	2.585
La	*bcc*	887°C	4.256			38.54	5.987
	hcp	293°C	3.775	12.227	2 × 1.619	37.72 37.2 (R.T.)	6.20 (R.T.)
	ccp	293°C	5.314			37.51 34.6 (23 kbar)	6.67 (23 kbar)
		887°C	5.341			38.09 37.3 (R.T.)	6.19 (R.T.)
Ce	*bcc*	757°C	4.10			34.54	6.738
	hcp	R.T.	3.677	11.862	2 × 1.613	34.72	6.704
	ccp	R.T.	5.161			34.36	6.774
						28.17 (15,000 atm)	8.26
Pr	*bcc*	821°C	4.13			35.22	6.65
	hcp	R.T.	3.67	11.83	2 × 1.611	34.50	6.78
	ccp	R.T.	5.11			33.36	7.02
Nd	*bcc*	833°C	4.13			35.22	6.80
	hcp	R.T.	3.66	11.80	2 × 1.61	34.22	7.00
	ccp	R.T.	4.80 (50 kbar)			27.65	8.67

Gd	*bcc*	1,262°C	4.06			33.46	7.807
	hcp	R.T.	3.63	5.78	1.59	32.98	7.920
Yb	*bcc*	774°C	4.44			43.76	6.569
	hcp	23°C	3.88	6.39	1.65	41.65	6.901
	ccp	R.T.	5.48			41.14	6.987
		774°C	5.60			43.90	6.548
Th	*bcc*	1,450°C	4.11			34.71	11.10
	ccp	1,450°C	5.18			34.75	11.09
Pu	*bcc* (ε)	490°C	3.634			24.00	16.51
	ccp (δ)	320°C	4.638			24.94	15.92
Ti	*bcc*	R.T.	3.284			17.71	4.491
	hcp	25°C	2.950	4.684	1.588	17.65	4.506
Zr	*bcc*	862°C	3.609			23.50	6.45
	hcp	25°C	3.23	5.15	1.59	23.26	6.51
Hf	*bcc*	2,000°C	3.610			23.52	12.61
	hcp	R.T.	3.194	5.051	1.581	22.31	13.29
Fe	α-*bcc*	20°C	2.866			11.78	7.88
	ε-*hcp*	20°C	2.705	4.37	1.616	13.85	6.70
	α-*bcc*	910°C	2.904			12.25	7.57
	γ-*ccp*	910°C	3.647			12.12	7.65
	δ-*bcc*	1,390°C	2.932			12.60	7.36
	γ-*ccp*	1,390°C	3.687			12.53	7.40
	α-*bcc*	23°C	2.805 (130 kbar)			10.43	8.89
	ε-*hcp*	23°C	2.468	3.956	1.603 (130 kbar)	11.03	8.41
Ni	*bcc*	R.T.	2.775			10.68	9.13
	hcp	R.T.	2.495	4.048	1.622	10.91	8.94
	ccp	R.T.	3.524			10.94	8.91

[a]*Source*: J. Donohue, *The Structures of the Elements*, Wiley, New York, 1974.

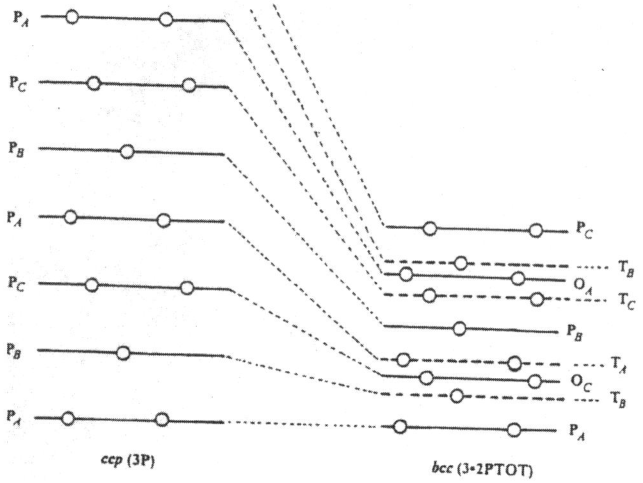

Figure 4.4. A correlation diagram for the possible conversion of *ccp* to *bcc* resulting from compression.

diagram to identify the layers in *ccp* and *bcc*. The drawing is not to scale. The $\mathbf{P}_A\mathbf{P}_B\mathbf{P}_C\mathbf{P}_A\mathbf{P}_B\mathbf{P}_C\mathbf{P}_A\mathbf{P}_B\mathbf{P}_C$ sequence for *ccp* is the same as for $\mathbf{P}_A\mathbf{T}_B\mathbf{O}_C\mathbf{T}_A\mathbf{P}_B\mathbf{T}_C\mathbf{O}_A\mathbf{T}_B\mathbf{P}_C$.

4.2.3. The Structures of Tin

Carbon is the definitive nonmetal, considering its vast variety of covalent compounds. The diamond structure (**3•2PT**, Section 4.4.3) has a network of tetrahedra. The next two elements of the carbon family, silicon and germanium, have the diamond structure, and lead is cubic close-packed. Tin shows an interesting variation. The low temperature form of tin (stable below 13.2°C) is gray tin with the diamond structure. The *CN* of Sn is 4, the Sn—Sn distance is 2.80 Å and $a_o = 6.4912$. Ordinary tin, the "high" temperature form, is white tin. In our time tin is used primarily for tin plate on iron for food containers, tin cans. In earlier periods utensils were often cast tin. In cold climates sometimes a tin utensil would crumble from "tin disease," resulting from conversion of white tin to gray tin. The conversion is slow, but once started, nucleation causes spread of the "disease."

Figure 4.5***a*** shows the projection of the cubic cell of gray tin on one face. It is identical with the structure of diamond (see Figure 4.13). The **T** sites are at heights of 25 and 75 along the vertical axis. The corner and center positions occur at 0 and 100 and centers of side faces are at 50. The crystal system of white tin is tetragonal with four atoms per cell, $\mathbf{D}_{4h}^{19}$, I4/*amd*. Figure 4.5***b*** shows the projection of the tetragonal cell of white tin along the c_o axis. This cell corresponds to the inner square of Figure 4.5***a*** with the atoms in the edges at 50 becoming those at corners at 0 and 100 for white tin. The atom at the center at 0 for gray tin remains at the center at 50 for white tin. The **T** sites are still at 25 and 75 for the new tetragonal cell. The central Sn atom in Figure 4.5***b*** in each cell is at the center of a tetrahedron formed by two Sn at 25 and

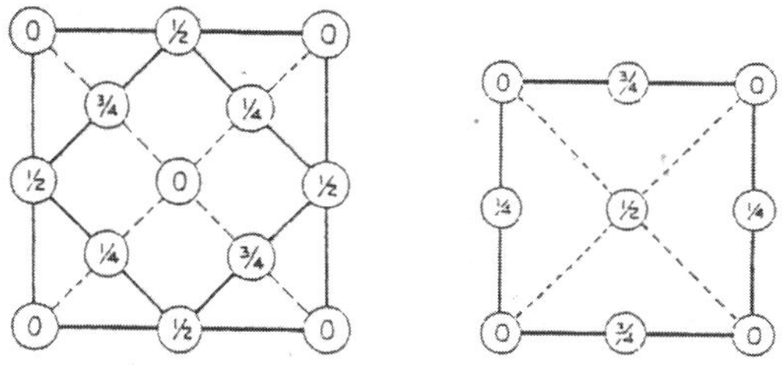

Figure 4.5. ***(a)*** A projection of the cubic cell on one face for gray tin. ***(b)*** A projection along c_o of the tetragonal cell for white tin.

two Sn at 75. Those in the edges of both cells are at the centers of tetrahedra.

The tetrahedron for white tin (high T) is compressed along the c_o direction ($c_o = 3.17488$ and $a_o = 5.8197$ at 25°C). The Sn—Sn distance within the flattened tetrahedron is 3.02 Å. The Sn atoms directly above and below the central atom (at the centers of adjacent cells) are at 3.17 Å. The ***CN*** is 6. The cell for gray tin can be represented by the same figure, but the cell is cubic. The remarkable result of the compression is that the density of white tin is 7.31 g/cm^3 and that of gray tin is 5.75 g/cm^3. Usually the higher temperature polymorph is less dense than the lower temperature form. The increase in ***CN*** to 6 is achieved by the compression causing the increase in density. Gray tin dissolves in hydrochloric acid to form $SnCl_4$ and white tin forms $SnCl_2$. In gray Sn the four valence electrons are used for bonding to the four neighbors. For white Sn the ***CN*** is 6, and reaction with HCl(aq) to form $SnCl_2$ suggests that Sn in white Sn is effectively Sn^{2+} with two electrons involved in multiple bonds or in a metallic conduction band.

Gray tin is face-centered cubic with Sn atoms in the **P** layer and one **T** layer. The cubic cell in Figure 4.5***a*** contains a smaller cell with those Sn atoms at 50 becoming the corners of the body-centered tetragonal cell. The projection of this cell is identical to Figure 4.5***b*** except the height along c is the same as a_o of the cubic cell, but a of the new cell is $1/\sqrt{2}a_o = 0.707a_o$. For white tin the body-centered cell is compressed along c until c is less than a, $c = (3.17/5.82)a = 0.54a$. The body-centered tetragonal cell for gray tin has $c = 0.707a$ and for white tin $c = 0.54a$.

4.3. The Body-Centered Cubic Structure

4.3.1. The Body-Centered Cubic Structure is Not an Anomaly

Close-packed structures (*ccp* and *hcp*) are very common for metals, noble gases, and many highly symmetrical molecules (Section 4.5). However, 18 metals have the body-centered cubic (*bcc*) structure under ordinary conditions, and at least 25 metals have the *bcc* structure at higher temperature or pressure. Hence, at least 44 metals have the *bcc* structure as the primary structure or one of the allotropic forms,

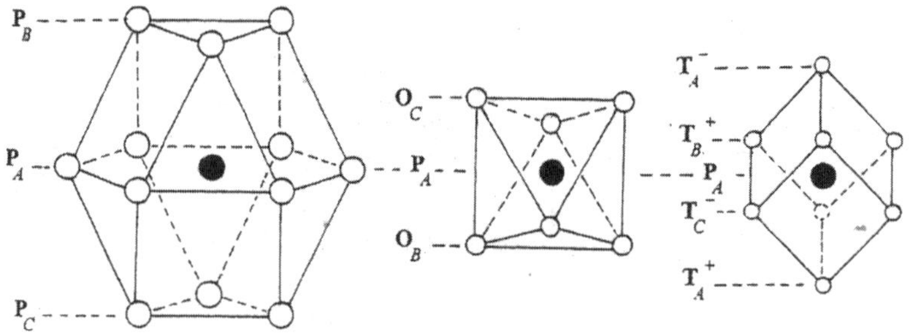

Figure 4.6. The polyhedra about a $\mathbf{P}_A$ atom in a *ccp* structure by ***(a)*** 12 **P** atoms, ***(b)*** 6 **O** sites, and ***(c)*** 8 **T** sites.

more than the numbers for *ccp* (36) and *hcp* (34). Also, it is fairly common to find the *bcc* crystal structure for large symmetrical molecules (Section 7.1). Because close-packed structures have been expected for metals and symmetrical molecules, the occurrence of the *bcc* structure has been considered as anomalous. Nevertheless, for metals, considering all polymorphs and molecular crystals, there are comparable numbers of hexagonal close-packed, cubic close-packed, and body-centered cubic structures. Let us consider why, in nature, these numbers for structures are comparable.

Figure 4.6 shows the close neighbors for an atom in $\mathbf{P}_A$, 12 **P** sites, six **O** sites, and eight **T** sites. In Figure 4.6***a***, the 12 neighbors are in a *ccp* structure with **O** and **T** sites vacant. Each **P** site is at the center of an octahedron (Figure 4.6***b***) formed by six **O** sites. Each **O** site is at the center of an octahedron formed by six **P** sites. The **P** and **O** sites are equivalent and are interchangeable. The cube in Figure 4.6***c*** consists of two tetrahedra of **T** sites, one with the base in the $\mathbf{T}_C$ layer and the apex is $\mathbf{T}_A^-$, and the other tetrahedron has the base in the $\mathbf{T}_B$ layer with the apex being $\mathbf{T}_A^+$. (The + and − labels refer to the orientation of four **P** sites around a **T** site.) The **T** sites are the closest neighbors for a *bcc* structure with ***CN*** 8. The **O** sites are 15% more distant from the central atom, giving an effective ***CN*** 14. The octahedron and cube are within the cuboctahedron (4.6***a***) giving a total ***CN*** 26 (8 + 6 + 12) for a *bcc* structure.

The remarkable increase in density of tin is caused by a minor change in structure. Gray tin, the low temperature form, is converted to white tin with considerable compression along the *c* axis. The ***CN*** of Sn increases from 4 to 6 and the density increases by 27% (see previous section). Gray Sn has the diamond structure, a *ccp* arrangement of Sn atoms with Sn also in one **T** layer, **3·2PT**. Figure 4.5 shows that the cubic cell also contains a smaller body-centered tetragonal cell. Figure 4.7***a*** shows the projection of two face-centered cubic cells without **T** sites occupied. Body-centered tetragonal cells are shown by dashed lines. The spheres at the corners of the tetragonal cell are at $\pm 1/2$, with the sphere in the center at 0. The height of the cell, *c*, is the same as a_o of the *fcc* cell. For the side of a square face of the tetragonal cell, $a = c/\sqrt{2} = a_o/\sqrt{2}$ or $0.707a_o$. Figure 4.7***a*** shows that in this projection

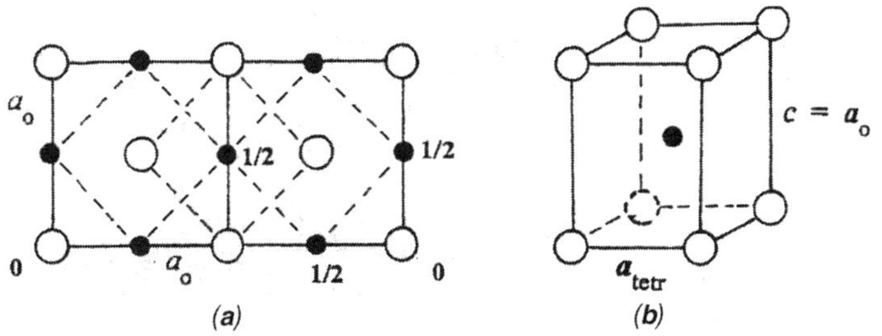

Figure 4.7. *(a)* A projection on one face of two cells of a face-centered cubic cell. The projection of the body-centered tetragonal cells is shown by dashed lines. Open and black circles show different heights in the cell. *(b)* The body-centered tetragonal cell.

there is a body-centered tetragonal cell within each cubic cell and one between each pair of adjacent cells. The same situation exists for the other two equivalent projections. Thus, *a face-centered cubic crystal can be described by body-centered tetragonal unit cells* in projection on any face. Body-centered cells and *fcc* or *ccp* structures are *quite compatible*. If the body-centered tetragonal cell is compressed until $c = a$, the new cell is body-centered cubic. The full body-centered tetragonal cell is shown in Figure 4.7*b*. We have seen that the body-centered cubic structure can be described as **3·2PTOT** with the same atoms filling **P**, **O**, and both **T** layers. Because the *bcc* cell is also obtained by compression of the body-centered tetragonal of the *ccp* structure, the *bcc* structure can be described as **3P′**, but this notation does not specify the type or extent of distortion involved. The **3·2PTOT** notation describes the relative positions of all atoms.

4.3.2. Another Hypothetical Mechanism for Conversion of the *ccp* → *bcc* Structures

If the body-centered cubic cell is elongated along what becomes the c axis, the structure becomes *fcc* (or *ccp*) when $c/a = \sqrt{2}$. In Section 4.2.2, we considered the hypothetical mechanism of conversion from *ccp* to *bcc* by compression along the packing direction $\mathbf{P}_A\mathbf{P}_B\mathbf{P}_C\mathbf{P}_A\mathbf{P}_B\mathbf{P}_C\mathbf{P}_A\ \mathbf{P}_B\mathbf{P}_C \rightarrow \mathbf{P}_A\mathbf{T}_B\mathbf{O}_C\mathbf{T}_A\mathbf{P}_B\mathbf{T}_C\mathbf{O}_A\mathbf{T}_B\mathbf{P}_C$. There is also a plausible mechanism for conversion from *ccp* to *bcc* by compression along an axial direction (becoming the c axis) until for the body-centered cell of the *ccp* structure $c = a = a_o/\sqrt{2}$.

We have noted that Pu has six allotropic forms including *ccp* (δ-Pu), body-centered tetragonal (δ′-Pu) and *bcc* (ε-Pu) structures. The tetragonal body-centered cell of δ′-Pu becomes the *bcc* cell of ε-Pu when the axial ratio is unity. There are many metals having *ccp* and *bcc* structures, but Pu is the only one of these metals that also has the intermediate body-centered tetragonal structure.

4.3.3. Densities of *hcp*, *ccp*, and *bcc* Metals

The book *The Structures of the Elements*, by Jerry Donohue, contains an abundance of data on structures and atomic volumes of metals. For a substantial number of metals, data permit comparison of atomic volumes and densities for *bcc, hcp,* and/or *ccp* structures. The data are presented in Table 4.2.

The efficiencies of packing hard spheres (made of material with density $1\,g/cm^3$) are the same for *hcp* and *ccp* structures and predict the same density, $0.7405\,g/cm^3$. The lower efficiency of packing of these hard spheres for a *bcc* structure gives a significantly lower density, $0.6802\,g/cm^3$. The density data are not consistent with this expectation. Densities for the same structures generally decrease with an increase in temperature, as expected, so we want to make comparisons under the same conditions. In some cases data are given by extrapolation. For Na, Ca, Sr, La, Pr, Nd, Gd, Th, Ti, Zr, and Hf, the densities for *bcc* are slightly less than another structure or they are about the same. For most of these the data for the *bcc* structure are for higher temperatures. We can only conclude that the densities for *bcc* are not significantly lower. For Li the density of the *bcc* structure is slightly higher than the *hcp* and *ccp* structures at the same temperatures. The density for the *bcc* structure of Be is significantly higher than that for the *hcp* structure at the same temperature. For Ce, the density for *bcc* is greater than for *hcp* and lower than for *ccp*, although the temperature is at 757°C for *bcc* and at room temperature for *hcp* and *ccp*. The density for the *bcc* structure of Yb is greater than for *ccp* at the same temperature. The density for *hcp* is less than that for *ccp* at room temperature so the density for the *hcp* structure should be less than for *bcc* at the same temperature. For Pu, the density is greater for the *bcc* structure at 490°C compared to *ccp* at 320°C. The density for *bcc* Ni is slightly, but significantly, greater than for *hcp* and *ccp* structures. Iron offers good comparisons for several structures at the same temperatures. The densities are slightly smaller for δ-*bcc* than for γ-*ccp* and for δ-*bcc* than for γ-*ccp* Fe structures. They are about the same for each pair. The density of α-*bcc* Fe is significantly greater than that of ε-*hcp* Fe at 1 atmosphere (atm.) and at 130 kbar.

The expectation of lower densities for *bcc* structures is not consistent with the data in Table 4.2. Overall comparisons indicate that the densities for *bcc, hcp,* and *ccp* structures are about the same, with the *bcc* structure slightly more likely to be more dense than *hcp* and *ccp* structures at the same temperatures. The compressibilities of metals are evident from the effects of pressure. For Nd, the *hcp* structure converts to *ccp* at 50 kbar with a 19% decrease in atomic volume.

The hard sphere model fails in predicting densities of metals with different structures because metal atoms are not hard spheres. They are very compressible. Atoms also change their shapes to accommodate the symmetry of the environment. An isolated atom is spherically symmetrical. In a unidirectional field one *p* orbital is unique. In a two-directional field with the axes at 90°, two *p* orbitals will be aligned with the field. In a tetrahedral field a carbon atom generates four equivalent sp^3 orbitals. Metal atoms adapt to the environment of *bcc, hcp,* or *ccp* structures. Compression for a metal does not mean decreasing the size of spheres. For each atom there are depressions in the electron cloud for near neighbors and bulges between neighboring atoms.

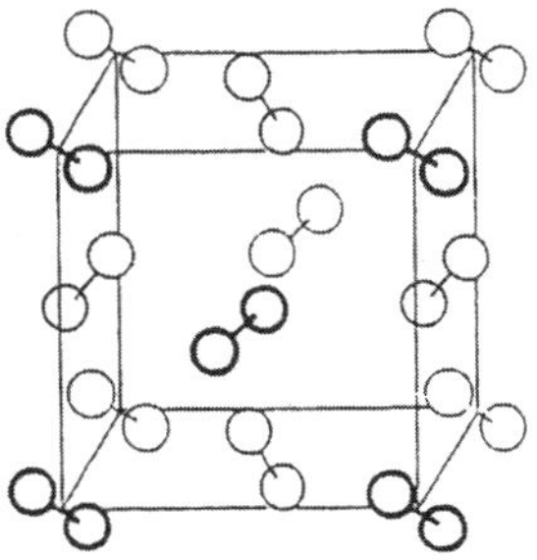

Figure 4.8. The **3P** (*ccp*) structure of α-N_2.

We have discussed the relative sizes of **P**, **O**, and **T** sites (Section 3.3), but these are based on the hard sphere model. For the *bcc* structure all sites for atoms are equivalent. We can use any atom for the origin of the cube or the center of the cube. We have used the **3·2PTOT** notation for the *bcc* structure because it describes *the relative spacings of all sites*. Figure 3.8 is the model for a *ccp* structure, with all **P**, **O**, and **T** sites shown by different balls for each type of layer. We can start a cube with any type of ball. The model is *bcc* if all balls are the same.

In molecular crystal structures we recognize short distances for covalent bonds and greater distances for nonbonded interactions (van der Waals interactions). For CCl_4, the four Cl atoms have short C—Cl distances while the Cl—Cl distances are greater. The bonded atoms move in close and nonbonding electrons buldge out between the C—Cl bonds. Metal atoms have 12 "bonded" atoms for *hcp* and *ccp* structures and for the *bcc* structure there are (8 +6) "bonded" atoms. There are 12 more neighbors in packing (**P**) layers. *One might consider an atom as a soft ball with closest neighbors pressed into the soft ball with bulges between the neighbors.*

4.4. Nonmetals

4.4.1. Diatomic Molecules

Hydrogen has one singly occupied orbital ($1s^1$) and the halogens have one singly occupied orbital, in addition to filled orbitals (ns^2np^5). Each forms single bonds so X_2 molecules are expected. H_2 is very reactive, and hydrogen forms single bonds to most elements. The halogens are also reactive, forming single bonds to most elements. The N_2 molecule has a very strong triple bond. Although we have a N_2-rich atmosphere, the N_2 molecule is very inert. The conversion of N_2 to useful compounds requires drastic conditions—except for some biological systems that achieve "fixation" of nitrogen under ambient conditions. The O_2 molecule is an unusual diradical (two unpaired electrons) and is quite reactive. It makes it possible for animals to utilize the solar energy stored by plants. It also makes it possible for us to utilize the solar energy stored in fossil fuels.

Hydrogen, H_2, at liquid helium temperature, has the **2P** (*hcp*) structure, with $c/a = 1.63$. There have been reports of a **3P** (***ccp***) form

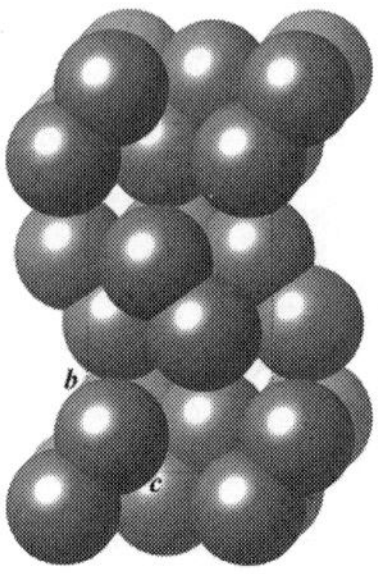

Figure 4.9. The **3P′** structure of Cl_2 (elongated face-centered cubic). A projection down the *a* axis.

obtained as thin films of H_2 or deposits on gold foil. The H_2 molecules are dumbbell shaped, not spherical. Normal close-packed structures can result from free rotation of the molecules in the crystal or by staggering the orientations of the molecules.

The structure of α-N_2 below 35.6 K is **3P** (*ccp*) as shown in Figure 4.8. The face-centered cubic structure is achieved by staggering the skewed N_2 molecules. There is a **2P** (*hcp*) structure for β-N_2 above 35.6 K. A third tetragonal structure (γ-N_2) has been reported at 4015 atm. and 20.5 K. Three crystal forms of O_2 are found also, α-O_2 (below 23.9 K), β-O_2 and γ-O_2 (above 54.4 K). The α-O_2 is monoclinic with two molecules per unit cell. The O_2 molecules are aligned, and give distorted close-packed layers parallel to the *ab* plane. The β-O_2 forms hexagonal unit cells with the long axis of O_2 molecules aligned, parallel to the *c* axis. The γ-O_2 is cubic (**3P′**, similar to α-N_2 in Figure 4.8) with disorder in the alignment of O_2 molecules.

There are two crystalline forms of F_2, α-F_2 below 45.6 K and β-F_2 above 45.6 K to the melting point (53.3 K). The α-F_2 is monoclinic but with nearly close-packed hexagonal layers of molecules in planes parallel to the *ab* plane. The F_2 molecules are aligned, but tipped by $\pm$ 18° in the *b* direction. Cubic β-F_2 is isostructural with γ-O_2. Both have disorder of the X_2 molecules.

The crystal structure of Cl_2 is orthorhombic. The molecules are aligned in alternate rows in the *bc* plane giving an elongated face-centered cubic arrangement (see Figure 4.9). For these Cl_2 molecules there is weak intermolecular bonding within the *bc* plane. Br_2 and I_2 are isostructural with Cl_2, with stronger intermolecular attraction for Br_2 and even stronger for I_2.

4.4.2. Large Discrete Molecules

Nonmetals that form more than one covalent bond can give crystals based on three-dimensional frameworks such as a diamond, or they give large discrete molecules. Boron, carbon, and sulfur are interesting examples of the latter.

The crystal structures of boron are complex. As many as 16 distinct allotropes have been reported, but some have been poorly characterized. Eight of these have been studied as single crystals and others as powders. Three of these have yielded complete structures, and for two of them unit cells and space groups were obtained. The complete characterization of the samples is complicated by the existence of many boron-rich metal borides such as $NiB_{50(?)}$ and $PuB_{100(?)}$. Trace impurities can change structures, and some of the reports are questionable, as possible boron frameworks obtained only in the presence of traces of a metal.

We consider briefly only the three for which complete structures have been reported. Rhombohedral-12 boron (R-12) is of particular interest, forming beautiful, clear red crystals. It is rhombohedral ($R\bar{3}m$, $\mathbf{D}_{3d}^5$, or alternatively, indexed as hexagonal) containing icosahedra of B atoms (Figure 4.10***a***). The B_{12} molecules are close-packed in an *ABC* sequence even though the unit is rhombohedral, not cubic. The notation is **3P(h)** to distinguish it from a *ccp* structure. The nearly perfect close-packed structure is possible because the highly symmetric icosahedra can pack as spheres. In one type of layer the six boron atoms around the waist (three in each of two planes) around a threefold axis are bonded to two neighboring icosahedra in the same layer (see Figure 4.10***b***). In a second type of layer the icosahedra are oriented with C_2 axes along the packing direction. Figure 4.10***c*** shows bonding within this layer. There is also bonding between layers. It is remarkable that different bonding patterns are accommodated with little distortion of the close-packed layers.

In tetragonal-50 boron (T-50) the B_{12} icosahedra are bonded by B—B bonds and B atom bridges. Lattice constants vary from specimen to specimen, presumably because of a large and variable degree of internal disorder. Rhombohedral-105 boron (R-105) forms black crystals with metallic luster. There are three types of icosahedra differing in B—B bond lengths. There are also fused icosahedra in trimers and other units as large as a B_{84} unit in the form of a truncated icosahedron.

In the next section we consider the diamond and graphite structures. Because these forms have been known since ancient times, it was a great surprise to find stable C_{60} molecules. This structure, called the

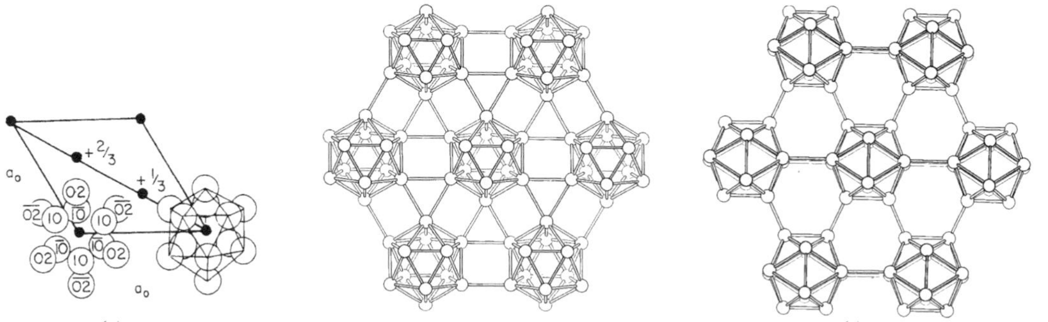

(a) *(b)* *(c)*

Figure 4.10. *(a)* The **3P(h)** structure containing B_{12} icosahedra (*Adapted from*: R.W.G. Wyckoff, *Crystal Structures*, Vol. 1, 2nd ed., Wiley, New York, 1963, p. 22.) ***(b)*** Bonding between B_{12} molecules within a close-packed layer. ***(c)*** Bonding between B_{12} molecules in another type of close-packed layer.

(*Source*: J. Donohue, the Structure of the Elements, Wiley, New York, 1974, pp. 58–59.)

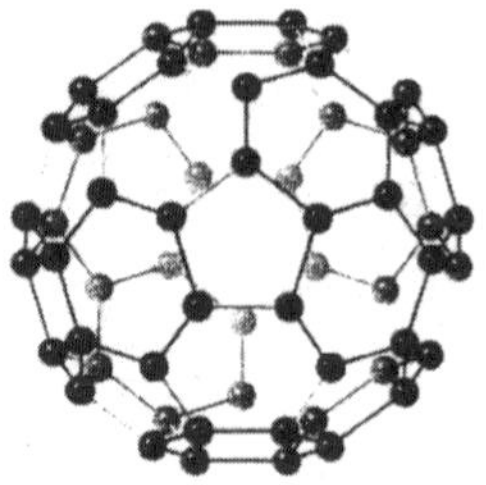

Figure 4.11. The C_{60} molecule.

buckyball, has five-membered rings sharing edges with six-membered rings. Alternate C—C edges of the six-membered rings are shared with five-membered and six-membered rings. The point group is $\mathbf{I}_h$. Because it is as close to a sphere (see Figure 4.11) as we get in molecular structures, it is not surprising that the crystal structure is **3P** (*ccp*).

Sulfur usually forms two single covalent bonds. Molecular species have been observed for S_n molecules, with $n = 1$ to 10, 12, 18, and ∞. In the crystalline state those encountered are S_6, S_7, S_8, S_9, S_{10}, S_{12}, S_{18}, and S_∞. The solids of S_6–S_{18} contain rings of single-bonded S atoms providing each S atom two bonds. S_∞ contains long chains of S atoms. When rhombic sulfur, containing S_8 molecules, is heated gently it melts to give a yellow liquid with low viscosity. Monoclinic S crystallizes from the melt (119°C) and slowly converts to the stable rhombic form containing S_8 molecules. When liquid S is strongly heated it darkens because of breaking S—S bonds in the ring to give zigzag chains with odd electrons on S atoms at the ends. It is difficult to pour from a test tube because of its high viscosity. If poured into cold water it cools to form "plastic" S. This can be stretched like rubber bands, and is dark colored. It hardens slowly and becomes yellow.

Here we consider the crystal structures of S_6 and S_8 only. The orange hexagonal prisms of rhombohedral (or a hexagonal cell can be used) sulfur contain S_6 rings. The ring can be described as an octahedron compressed along a C_3 axis. The "hole" in the ring is very small and the molecules are packed very efficiently. The density is 2.209 g/cm^3, significantly higher than other forms of S. Each S_6 molecule has short intermolecular contacts with 12 neighboring molecules. The flattened rings are close packed in layers parallel to the a_oa_o plane. The layers are packed in an *ABC* sequence even though the unit is rhombohedral, not cubic (Figure 4.12). The notation is **3P(h)** to distinguish it from a *ccp* structure. There is one molecule per rhombohedral unit cell and three molecules per hexagonal cell, $a_o = 10.818$ and $c_o = 4.280$ Å.

The familiar form of sulfur is the yellow rhombic sulfur for which there are extensive deposits. S is melted by superheated H_2O and pumped to the surface by the Frasch process. "Rhombic" sulfur has an orthorhombic structure. Each S_8 ring forms a "crown" with four S atoms above and four below an average plane. The packing of the S_8 rings is complex, with the average planes of the rings perpendicular to the *ab* plane and those in the ring planes in one layer are perpendicular to those in neighboring layers

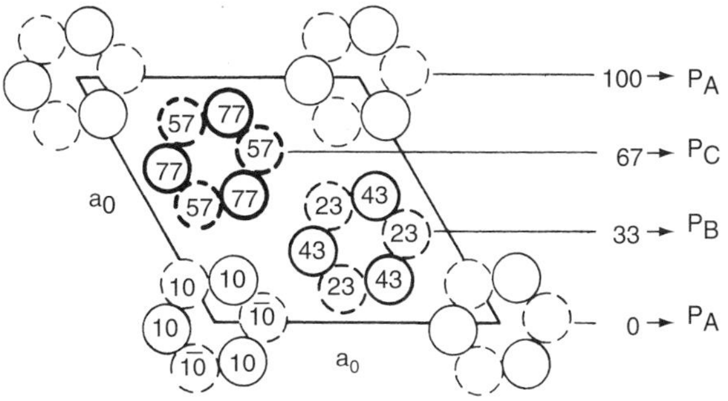

Figure 4.12. The **3P(h)** structure of the molecular crystal of S_6. The heights along the *c* direction are shown.

4.4.3. Carbon

The diamond structure was the first structure of an element determined by X-ray crystallography by W.H. and W.L. Bragg (1913). The diamond structure, **3•2PT** (*ccp*) has C in **P** layers and one of the **T** layers (Figure 4.13). The space group is $\mathbf{O}_h^7$, *Fd*3*m* with eight atoms per cell. The interatomic distance (1.54 Å) is characteristic of C—C single bonds. The structure is the same as that of zinc blende (ZnS, Figure 6.1). There is also a hexagonal diamond, **2•2PT** (*hcp*). The structure is the same as wurtzite, the *hcp* form of ZnS (see Figure 6.5). It has been found in meteorites, and has been synthesized from graphite at high pressure and temperature. Carbon is perfectly suited to these structures because the preferred bonding for C is tetrahedral with sp^3 hybridization. Carbon has four valence electrons to utilize these hybrid orbitals for four strong bonds

Graphite has layers of hexagonal networks. Each C atom is bonded to three C atoms in the same plane (sp^2 hybrids). There is one electron per C in delocalized π orbitals extending throughout the layer or, in terms of the valence bond theory, there are alternating single and double bonds. Graphite conducts electricity within the plane and can be used as an electrode.

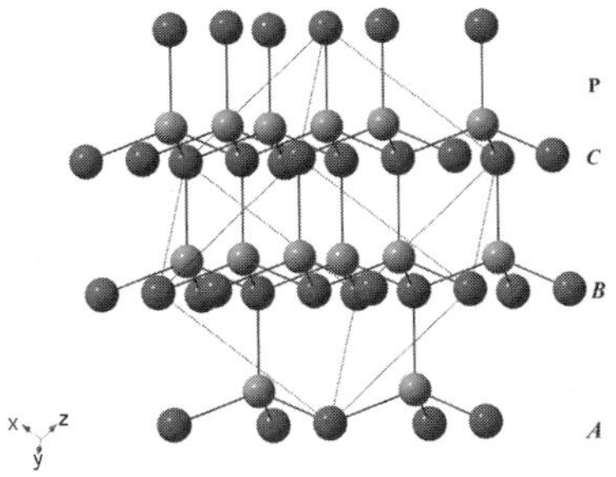

Figure 4.13. The **3•2PT** structure of diamond. The packing (**P**) layers are in a *ccp* pattern. All sites are occupied by C, but here the balls in the **T** layers are lighter. The cubic unit cell is outlined by double lines.

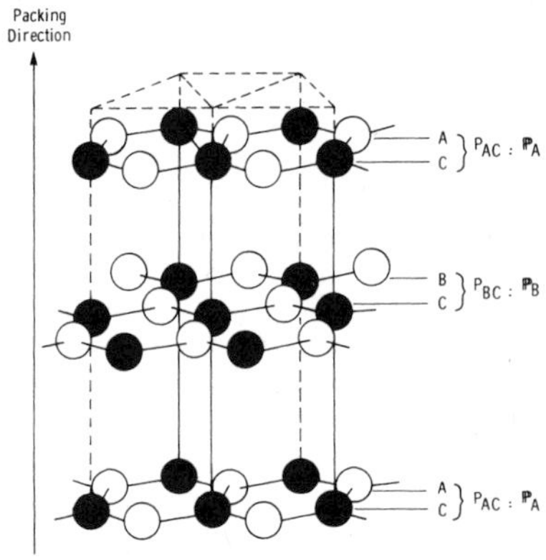

Figure 4.14. The **2$\mathbb{P}$** (double **P** layer) structure of hexagonal close-packed graphite.

There are two forms of graphite: hexagonal and rhombohedral. The individual layers are the same. They differ in the addition of a third layer. If all of the carbon atoms of the third layer are directly over those of the first layer, normal, hexagonal graphite is obtained. If all of the carbon atoms of the third layer are not directly above those of either of the first two layers, rhombohedral graphite is obtained. Sound familiar? These are the hexagonal (*AB*) or cubic (*ABC*) sequences. It is more complicated because each carbon atom in a layer has only three neighbors. However, we can get a glimpse of nature's insight by considering each layer the result of two close-packed layers forced into one plane. In the first layer, alternate carbon atoms are at *A* and *C* positions in the fused hexagons. This layer is called a double layer, labeled **$\mathbb{P}$**, in this case $\mathbb{P}_{AC}$ or $\mathbb{P}_A$. The second layer is $\mathbb{P}_{BC}$ or $\mathbb{P}_B$. In the case of hexagonal graphite the third layer is $\mathbb{P}_{AC}$ or $\mathbb{P}_A$, giving $\mathbb{P}_A$ $\mathbb{P}_B$ $\mathbb{P}_A \ldots$ corresponding to the *hcp* pattern or **2 $\mathbb{P}$** (Figure 4.14). For rhombohedral graphite the sequence is $\mathbb{P}_{AB}$ $\mathbb{P}_{BC}$ $\mathbb{P}_{CA}$ $\mathbb{P}_{AB} \ldots$, $\mathbb{P}_A$ $\mathbb{P}_B$ $\mathbb{P}_C\mathbb{P}_A \ldots$, or **3$\mathbb{P}$**, the familiar *ccp* sequence (Figure 4.15). The first layer is designated as

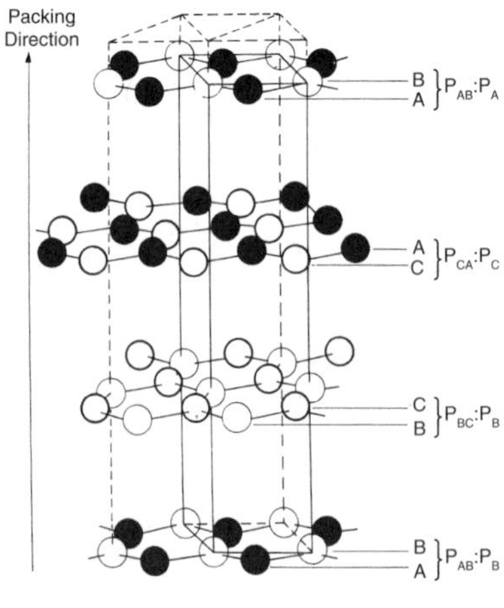

Figure 4.15. The **3$\mathbb{P}$** (double **P** layer) structure of rhombohedral graphite.

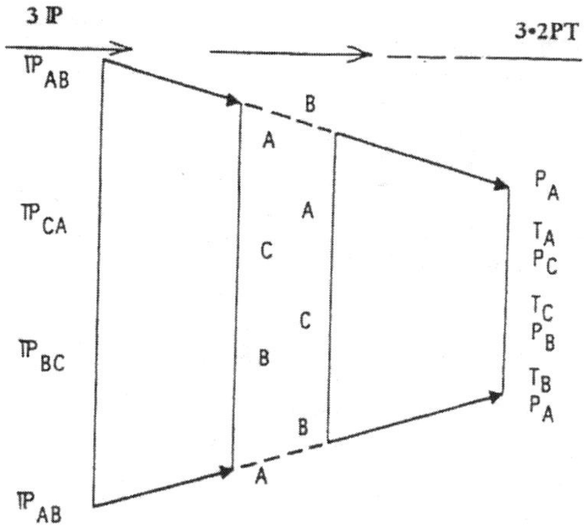

Figure 4.16. A hypothetical scheme for conversion of **3IP** graphite to diamond by compression.

$\mathbf{IP}_{AC}$ for **2IP** and $\mathbf{IP}_{AB}$ for **3IP**, to give a simple sequence for all layers. One position is in common for each pair of layers and for hexagonal graphite the *C* positions are used in all layers. The spacing between layers is much larger (nonbonding π interaction) than within a layer so the unit cell is not cubic, but rhombohedral with two atoms per unit cell.

The graphite layers show the same pattern as $\mathbf{L}_{2/3}$ partial layers (Figure 3.9*c*). However, the designation of hexagonal graphite as $\mathbf{2P}_{2/3}$ is not satisfactory because this implies a $\mathbf{P}^{A}_{2/3}\mathbf{P}^{B}_{2/3}$ sequence. Half of the carbon atoms of either form of graphite are at *same* positions as those in adjacent layers. For hexagonal graphite half of the carbon atoms are at *C* positions in each double layer. All of the atoms of a single partial layer are at the same positions and differ from those of adjacent layers. The positions of the vacancies of the double layers (the centers of the hexagons) are in *A* and *B* positions for hexagonal graphite and *A*, *B*, and *C* positions for rhombohedral graphite.

Cubic (**3•2PT**) and hexagonal (**2•2PT**) diamond have been synthesized at high temperature from graphite. We can imagine high pressure closing the big gap between layers, causing double layers to split into normal close-packed layers, each graphite layer giving a **P** layer and a **T** layer. Figure 4.16 shows a scheme one can imagine for conversion of rhombohedral graphite to cubic diamond.

4.4.4. Nitrogen Family

We have discussed the structure of crystalline N_2 in Section 4.4.1. Several allotropes of phosphorus are well known. In the gas phase there are P_4 tetrahedral molecules, and this condenses as white (also described as yellow) P containing P_4 molecules in the solid. There are two forms of white P, but the detailed structures are not known. There are inconsistencies in the structural reports. A monoclinic form of P contains cage-like P_8 and P_9 groups linked by pairs of P atoms to form

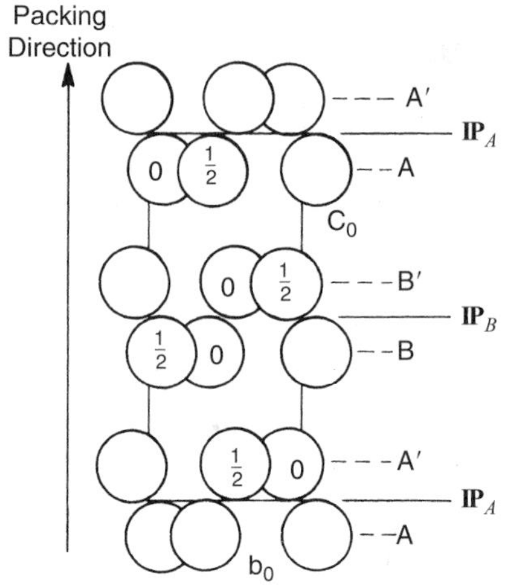

Figure 4.17. The **2IP** (double **P** layer) structure of black phosphorus.

tubes of pentagonal cross section. The tubes in different layers are perpendicular to each other. There is a metallic P described as a primitive cubic lattice related to the As structure. Red phosphorus is obtained by heating white P in an inert atmosphere or by the action of light. The structure of red P is not known. The commercial red P is amorphous. Black phosphorus is the form for which the structure is well known. It has a metallic black graphite-like appearance. The structure is orthorhombic. The P atoms form puckered sheets parallel to the *ac* plane. Each P atom is bonded to three other atoms. The sheets are so severely puckered that there are essentially two layers separated by approximately the P—P bond distance. One of these layers is slightly displaced relative to the other when viewed along the packing direction (perpendicular to the packing layers). The packing pattern of these special double layers is an *ABAB* arrangement, corresponding to the *hcp* sequence. Because the positions in a puckered sheet are only slightly displaced, we designate these positions as *AA′* and *BB′*. The full sequence is *AA′BB′*··· or $\mathbf{IP}_{AA'}\mathbf{IP}_{BB'}$ or 2 **IP** (see Figure 4.17).

There are three crystalline forms of arsenic: α-As, ε-As, and yellow As, and three amorphous forms. Stable α-arsenic is trigonal, $\mathbf{D}_{3d}^5$, $R\bar{3}m$, $a = 3.7599$ and $c = 10.4412$ Å. The structure contains puckered sheets of As atoms bonded to three others and stacked in layers perpendicular to the hexagonal *c* axis. Each sheet consists of two packing layers that can be described as a double layer (**IP**). Figure 4.18 shows the structure with the rhombohedral cell outlined. The lower layers are in an *ABC* sequence and the upper layers of double layers are in an *CAB* sequence. A projection along the packing direction (*c*) shows all atoms arranged in close-packing positions corresponding to $\mathbf{P}_A\mathbf{O}_C\mathbf{P}_B\mathbf{O}_A\mathbf{P}_C\mathbf{O}_B$. The notation is **3·2PO′(h)**. The **O** layers are not halfway between **P** layers because the As atoms in an **O** layer have three bonds to the closer **P** layer. The ε-As has been reported to be isostructural with black phosphorus, but the purity of the specimen was questionable. Yellow arsenic was reported to be cubic, consisting of As_4 molecules, but the data are incomplete because the sample is

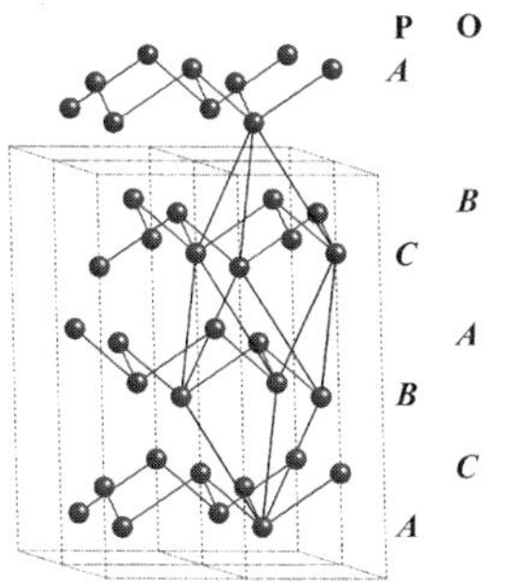

Figure 4.18. Four cells of As showing the puckered layers. Positions of close-packed layers are shown.

(*Source*: Y. Matsushita, *Chalogenide* crystal structure data Library, version 5.5B, Institute for Solid State Physics, The University of Tokyo, Tokyo, Japan, 2004; and *CrystalMaker*, by David Palmer, *CrystalMaker* Software Ltd., Begbroke Science Park, Bldg. 5, Sandy Lane, Yarnton, Oxfordshire, OX51PF, UK.)

unstable and decomposes in the X-ray beam. Only one form has been well characterized for antimony and bismuth. They are isostructural with α-As.

4.4.5. Oxygen Family

We have discussed structures of O_2, S_6, and S_8. Now we consider Se, Te, and Po. Six crystalline forms of selenium have been reported: α-Se (stable under normal conditions), α-monoclinic, and β-monoclinic Se, and three cubic forms deposited as thin films by vacuum evaporation. Metallic α-Se is trigonal, but also described as a distorted simple cubic structure, **3PO**′, similar to the structure of Te with more distortion for Se. There are infinitive helices parallel to the *c* axis. The space groups of α-Se are $\mathbf{D}_3^4$, $P3_121$ and $\mathbf{D}_3^6$, $P3_221$ for the other enantiomorph (see Section 10.1.3).

Both α- and β-monoclinic Se are deep red and revert to α-Se on heating. These forms contain Se_8 "crown" rings packed in a complicated arrangement. Cubic β-Se was reported to be simple cubic, with one atom per unit cell. Cubic α-Se, obtained by heating cubic β-Se, was reported to be face-centered cubic (*ccp* or **3P**) with four atoms in the unit cell. A third cubic Se was reported to have the diamond structure. These cubic structures have not been confirmed.

Three crystalline forms of Te have been reported: metallic α-Te, β-Te, and γ-Te. α-Tellurium is trigonal, $P3_121$, $\mathbf{D}_3^4$, $a = 4.456$, and $c = 5.921$ Å. Figure 4.19 shows two views (*a* and *c*) of the helices of Te along *c*. Figure 4.19***b*** is a projection along *c* showing the triangular positions of Te atoms at *A*, *B*, and *C*. The Te—Te bond distance is 2.835 Å, and each Te atom has two bonds at angle 103.14°, although in Figure 4.19***a*** some Te_3 groups appear as linear from this perspective. The positions of Te atoms correspond to *A*, *B*, and *C* positions in a close-packed structure. The projection of the structure in the *bc* plane (Figure 4.19***b***) show the positions in a close-packed layer. Because of bonding within the helices the *A*, *B*, and *C* planes through the helices are closer than those between the helices. Figure 4.19***c*** gives a

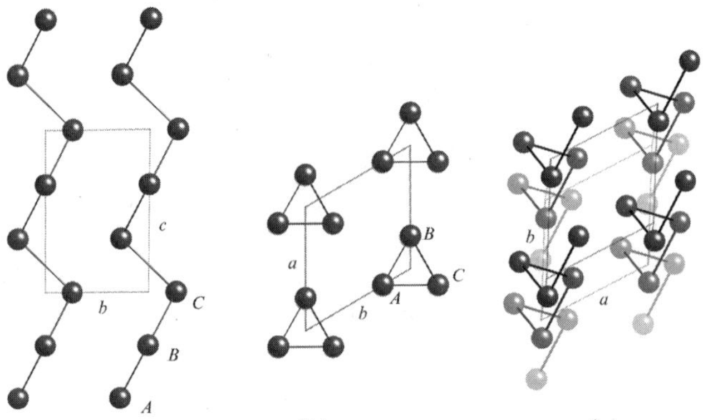

Figure 4.19. Three views of the spirals of Te.

(*Source: CrystalMaker*, by David Palmer, *CrystalMaker* Software Ltd., Begbroke Science Park, Bldg. 5, Sandy Lane, Yarnton, Oxfordshire, OX51PF, UK.)

perspective for two helices. There are two high pressure forms. From 40 to 70 kbar the structure has been indexed as an α-arsenic hexagonal unit cell. Above 70 kbar γ-Te is isostructural with β-polonium, a primitive rhombohedral lattice. Each Te atom is equidistant from six neighbors.

There are two allotropic forms of polonium, both primitive, with one atom per unit cell. α-Po is simple cubic (**3PO**) and each Po atom has six nearest neighbors (3.366 Å) at the vertices of a regular octahedron. The simple cubic structure is the same as that of NaCl (Figure 3.6) with all atoms identical. The structure of β-Po is a primitive rhombohedral lattice. Each Po atom has six nearest neighbors forming a slightly flattened trigonal antiprism (or a flattened octahedron).

4.5. Some Molecular Crystal Structures

4.5.1. The 3P Crystal Structures of CO and CO_2

Carbon monoxide, CO, has the same structure of α-N_2 (Figure 4.8) or **3P**. Carbon dioxide is also *ccp* or face-centered cubic (**3P**), (see Figure 4.20). There are four molecules in the cubic cell, $\mathbf{T}_h^6$, $Pa3$ and $a_o = 5.575$ Å at −190°C. The orientations of the linear molecules are staggered. Nitrous oxide, N_2O, has the same type of structure ($a_o = 5.656$ Å at −190°C).

4.5.2. The 3P Cubic Crystal Structure of Dibenzenechromium

Crystals of $Cr(C_6H_6)_2$ are cubic, $\mathbf{T}_h^6$, $Pa3$, with four molecules per cell and $a_o = 9.67$ Å. The $Cr(C_6H_6)_2$ molecules are *ccp* or face-centered cubic, **3P**. The Cr atoms are located at the corners and centers of the faces of the cube. The $Cr(C_6H_6)_2$ molecule has $\mathbf{D}_{6h}$ symmetry with the parallel C_6H_6 rings eclipsed.

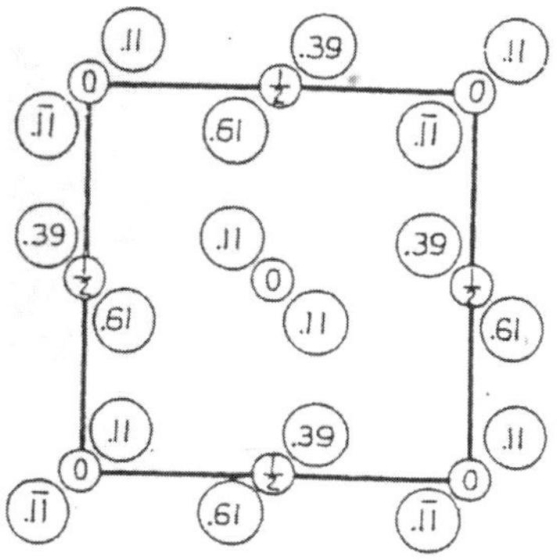

Figure 4.20. A projection of the face-centered cubic cell for CO_2.

(*Source*: R.W.G. Wyckoff, *Crystal Structure*, Vol. 1, 2nd ed., Wiley, New York, 1963, p. 368.)

4.5.3. The 3P(o) Face-Centered Crystal Structure of SO_2

Sulfur dioxide, SO_2, has an orthorhombic cell with four molecules, $\mathbf{C}_{2v}^{17}$, *Aba*2, $a_o = 6.07$, $b_o = 5.94$, and $c_o = 6.14$ Å at −130°C. The cell is face centered with S atoms at the corners and the face centers (Figure 4.21). The C_2 axes of the molecules are aligned with the c_o axis. The notation is **3P(o)**.

4.5.4. The 3P(o) Crystal Structure of B_2Cl_4

Diborontetrachloride molecules contain two bonded B atoms with two Cl atoms bonded to each B atom giving a planar ethylene-like molecule. The bond angles are close to 120°. The colorless liquid decomposes above 0°C. The orthorhombic cell contains four molecules, $\mathbf{D}_{2h}^{15}$, *Pbca*, $a_o = 11.900$, $b_o = 6.281$, and $c_o = 7.690$ Å, at −165° C. The cell is face centered (Figure 4.22) giving the simple notation **3P(o)**.

4.5.5. The 3P(o) Crystal Structure of IF_7

Iodine heptafluoride, IF_7, freezes at 6.45°C to form a cubic solid. Below −120°C it is orthorhombic with four molecules in the cell,

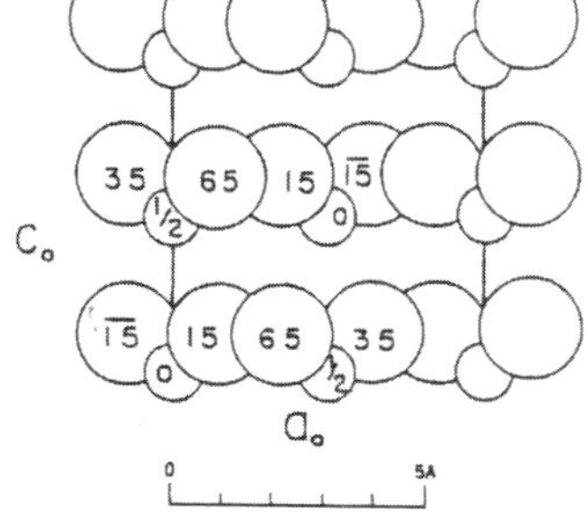

Figure 4.21. A projection along a_o of the face-centered orthorhombic cell, **3P(o)** for SO_2.

(*Source*: R.W.G. Wyckoff, *Crystal Structures*, Vol. 1, 2nd ed., Wiley, New York, 1963, p. 370.)

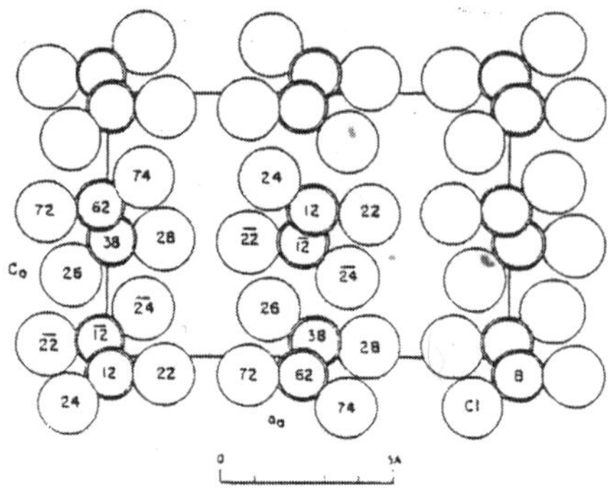

Figure 4.22. A projection along b_o of the face-centered orthorhombic cell of B_2Cl_4. The average heights of the B—B bonds are at 0 (and 100) and at 50 (side faces).

(*Source*: R.W.G. Wyckoff, Crystal Structures, Vol. 1, 2nd ed., Wiley, New York, 1963, p. 377.)

D_{2h}^{18}, *Cmca*, $a_o = 8.74$, $b_o = 8.87$, and $c_o = 6.14$ Å, at -145°C. The IF_7 molecule, with no lone pairs on I, has the structure of a badly distorted pentagonal bipyramid. In Figure 4.23, we see that I atoms are at the corners of the cell and in the center of each face giving the notation **3P(o)**.

4.5.6. The 3P(m) Crystal Structure of OsO_4

Osmium tetraoxide is tetrahedral, forming yellow crystals (m.p. 40.6°C). The crystals are monoclinic with four molecules in the cell (C_{2h}^{6}, $C2/c$, $a_o = 9.379$, $b_o = 4.515$, and $c_o = 8.632$ Å, and $\beta = 116.6°$). The structure can be described as an *ABC* arrangement of OsO_4 molecules [**3P(m)**]. This is a *ccp* sequence, but the cell is monoclinic. It can also be described as a pseudo-*ccp* arrangement of oxide ions with Os in 1/8 of the **T** sites in each **T** layer giving $\mathbf{3{\cdot}3PT_{1/8}T_{1/8}}$.

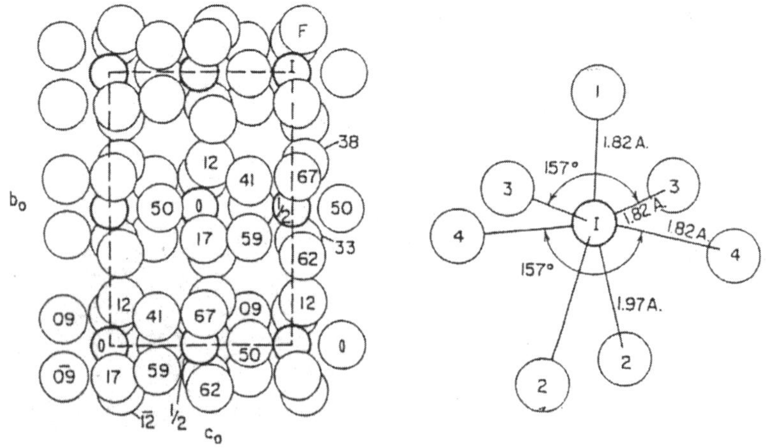

Figure 4.23. A projection along a_o of the orthorhombic cell of IF_7. The dimensions for the IF_7 molecules are shown.

(*Source*: R.W.G. Wyckoff, *Crystal Structures*, Vol. 2, 2nd ed., Wiley, New York, 1964, p. 213.)

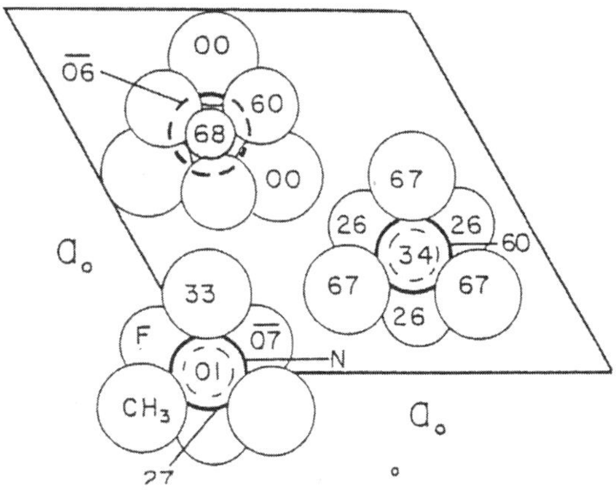

Figure 4.24. The **3P(h)** structure of $(CH_3)_3N{\cdot}BF_3$ projected along the c_o axis. Heights along c_o are shown.

(*Source*: R.W.G. Wyckoff, *Crystal Structures*, Vol. 5, 2nd ed., *The Structure of the Aliphatic Compounds*, Wiley, New York, 1965, p. 21.)

4.5.7. The 3P(h) Crystal Structure of $(CH_3)_3N{\cdot}BF_3$

The trimethylamine borontrifluoride addition compound forms hexagonal cells containing three molecules (C_{3v}^5, R3*m*, $a_o = 9.34$, $c_o = 6.10$ Å). The four atoms about N and B form tetrahedra. As seen in Figure 4.24, the $N(CH_3)_3$ and BF_3 groups are staggered viewing along the N—B bond as opposite faces of an octahedron. The cell is hexagonal but the packing arrangement is the *ccp*-pattern (*ABC*) giving the notation **3P(h)**. $(CH_3)_3N{\cdot}BH_3$ has the same structure.

4.5.8. The 3P(h) Crystal Structure of $Gd(HCO_2)_3$

Gadolinium formate [$Gd(HCO_2)_3$] forms hexagonal unit cells containing three molecules (C_{3v}^5, R3*m*, $a_o = 10.44$, $c_o = 3.98$ Å). The structure shown in Figure 4.25 has an *ABC* sequence giving the notation **3P(h)**. Gd has nine oxygen neighbors at 2.30 or 2.50 Å. The formates of Ce, Nd, Pr, and Sm have the same structure.

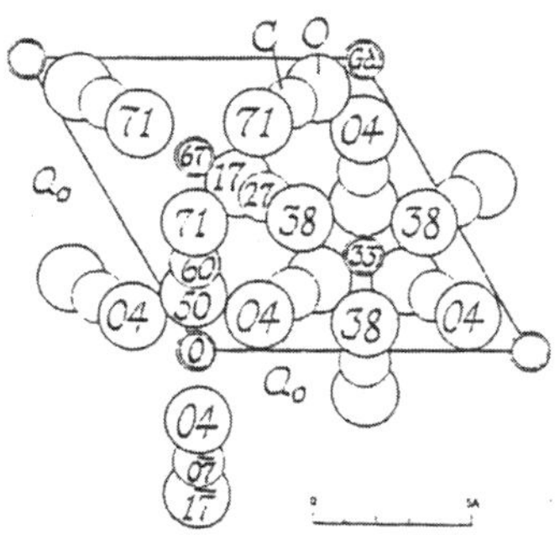

Figure 4.25. The **3P(h)** structure of $Gd(HCOO)_3$ projected along the c_o axis. Heights along c_o are shown.

(*Source*: R.W.G. Wyckoff, *Crystal Structures*, Vol. 6, 2nd ed. *The Structure of Aliphatic Compounds*, Wiley, New York, 1965, p. 135.)

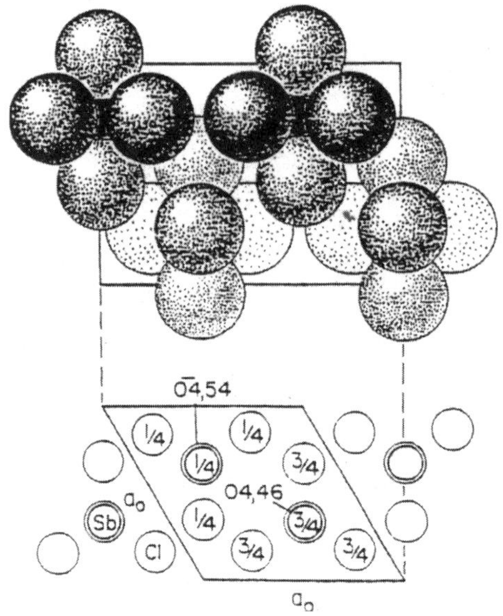

Figure 4.26. Two projections of the hexagonal structure of $SbCl_5$. The C_3 axes of $SbCl_5$ molecules are along c_o.

(*Source*: R.W.G. Wyckoff, *Crystal Structures*, Vol. 2, 2nd ed., Wiley, New York, 1964, p. 175.)

4.5.9. The 2P(h) Crystal Structure of $SbCl_5$

The crystal structure of antimony pentachloride, $SbCl_5$, is hexagonal ($\mathbf{D}_{6h}^4$, $P6_3/mmc$, $a_o = 7.49$, and $c_o = 8.01$ Å) with two molecules in the cell. The molecules are *hcp* giving the simple notation **2P(h)**. In Figure 4.26, we see that the trigonal bipyramidal molecules have their C_3 axes parallel to c_o, the packing direction. The axial Sb—Cl distance is 2.34 Å and that in the equatorial plane is 2.29 Å. The closest Cl—Cl distances (3.33 Å) are between axial Cl atoms between adjacent molecules.

4.5.10. The 2P Crystal Structure of $Sb(CH_3)_3Br_2$

Trimethylantimony dibromide, $Sb(CH_3)_3Br_2$, has trigonal bipyramidal molecules with methyl groups in equatorial positions. These large molecules are hexagonally close packed [notation **2P**] as shown in Figure 4.27***a***. The trigonal bipyramid with spacing is shown in Figure 4.27***b***. The hexagonal cell contains two molecules ($\mathbf{D}_{3h}^4$, $P\bar{6}2c$, $a_o = 7.38$, $c_o = 8.90$ Å). The other two halogen compounds, $Sb(CH_3)_3Cl_2$ and $Sb(CH_3)_3I_2$, have the same structure.

4.5.11. The 2P Crystal Structure of BCl_3

BCl_3 has a hexagonal bimolecular cell, $\mathbf{C}_6^6$, $P6_3$, with $a = 6.08$ and $c = 6.55$ Å. The trigonal planar molecules are in *A* and *B* positions as for $SbCl_5$ and $Sb(CH_3)_3Br_2$ (Figures 4.26 and 4.27). All other boron halides are also planar. Phosphorus trihalides, like NH_3 and nitrogen halides, are pyramidal because of the lone electron pair on P and N.

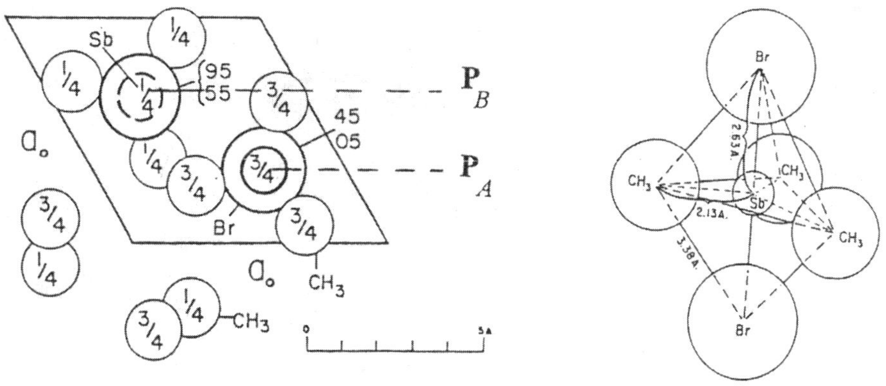

(a) (b)

Figure 4.27. ***(a)*** The $Sb(CH_3)_3\ Br_2$ structure (**2P**) projected along the c_o axis. Heights along c_o are shown. ***(b)*** The shape and dimensions of the $Sb(CH_3)_3\ Br_2$ molecule.

(*Source*: R.W.G. Wyckoff, *Crystal Structures*, Vol. 5, 2nd ed., *The Structures of Aliphatic Compounds*, Wiley, New York, 1965, p. 88.)

4.5.12. The 2P(m) Crystal Structure of Hydrazine, N_2H_4

Crystals of hydrazine, N_2H_4, are monoclinic with two molecules per cell at −15° to −40°C, $\mathbf{C}_{2h}^2$, $P2_1/m$, $a_o = 3.56$, $b_o = 5.78$, $c_o = 4.53$ Å, and $\beta = 109°30'$. Figure 4.28 shows a projection of the cell along the b_o axis. The packing direction is along b_o with $\mathbf{P}_A$ (0) and $\mathbf{P}_B$ (1/2), giving the notation **2P(m)**. The N atoms are in the packing layer with the same orientation for the N—N bonds. The positions of the H atoms were not determined.

4.5.13. 3P(o) Crystal Structure of $Ni[(CH_3)_2C_2N_2O_2H]_2$

The beautiful bright red dimethylglyoxime complex of Ni^{2+} is used for gravimetric determination of Ni because it has low solubility and is quite selective. The planar molecules are stacked along the c_o axis as shown in Figure 4.29***a***. It is a face-centered structure [**3P(o)**] deformed

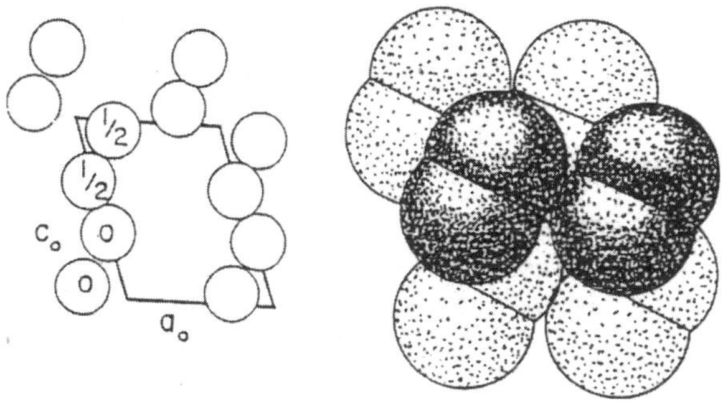

Figure 4.28. A projection along b_o for the monoclinic cell of N_2H_4. On the right the packing draw is for the same view.

(*Source*: R.W.G. Wyckoff, *Crystal Structures*, Vol. 2, 2nd ed., New York, 1964, p. 371.)

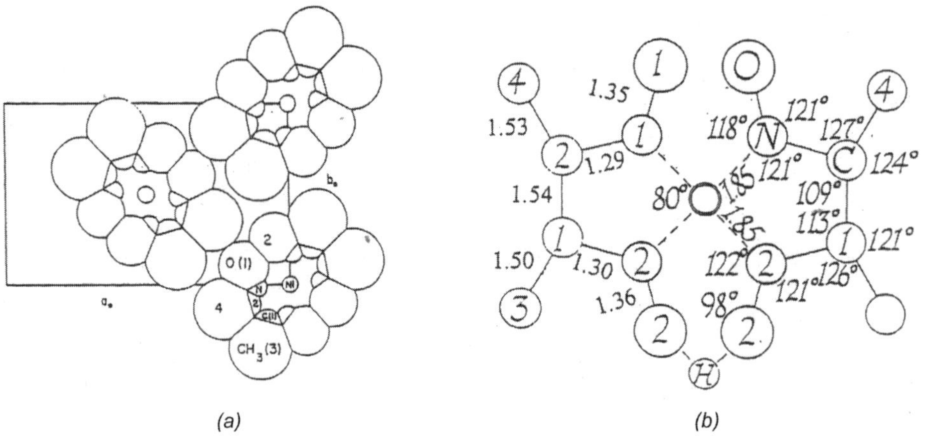

Figure 4.29. ***(a)*** The crystal structure of nickel dimethylglyoximate projected along the c_o axis. ***(b)*** The dimensions of the molecule are shown with atom numbering shown in (***a***).

(*Source*: R.W.G. Wyckoff, *Crystal structures*, Vol. 5, 2nd ed. *The Structures of Aliphatic Compounds*, Wiley, New York, 1965, p. 476.)

to form an orthorhombic cell (4 molecules/cell, $\mathbf{D}_{2h}^{26}$, *Ibam*, $a_o = 16.68$, $b_o = 10.44$, $c_o = 6.49$ Å). The planar molecules in stacking are rotated by 90° about the c_o direction as shown in the figure because the molecules are longer in one direction. The dimensions of the molecule are shown in Figure 4.29***b***. The Pd(II) complex of dimethylglyoxime has the same structure.

Chapter 5

Structures Involving P and O Layers

We begin consideration of structures of compounds as various combinations of layers. In this chapter we examine compounds that involve only **P** and **O** layers. These structures include hundreds of MX type compounds and, with partial filling of layers, compounds other than those with 1:1 atom ratios. Octahedral layers are normally halfway between **P** layers, and for a ***ccp*** arrangement the **P** and **O** layers are equivalent and can be interchanged. Ionic MX compounds are commonly encountered for **PO** structures. The configuration of neighbors of atoms in **O** sites are normally octahedral, but can be square planar or linear for partial filling of layers.

5.1. Cubic Close-Packed Structures

5.1.1. The 3·2PO Crystal Structure of NaCl

The structure of NaCl, table salt, rock salt, or halite (name of the mineral) is the most important and most common one for MX-type salts. Hundreds of salts have this structure, and many examples are listed in Table 5.1. It is characteristic of ionic MX-type compounds except for those with large differences in sizes of cations and anions. The coordination number (*CN*) is six for both ions for NaCl. Large cations, particularly with small anions, prefer a larger *CN*, commonly $CN = 8$ as in the CsCl structure. Small cations, and particularly for compounds with significant covalent character, prefer the $CN = 4$ as in ZnS.

The cubic NaCl structure, $\mathbf{O}_h^1$, $Fm3m$, $a_o = 5.64056$ Å, is designated as **3·2PO**, the index indicating that it is based on a *ccp* arrangement of anions (Cl^-) in **P** layers and that there are six (3·2) layers (**P** and **O** layers) in the repeating unit. There are four molecules (four NaCl) per cubic unit cell. The Na^+ ions fill the **O** layers. Because **P** and **O** layers are filled, the $CN = 6$ is obvious for octahedral (**O**) sites and it is also 6 for **P** sites. In the *ccp* arrangement the sequence (*ABC*...) is the same for **P** and **O** layers. They are equivalent and the designations can be reversed. Figure 5.1 shows the **3·2PO** structure with labels of the layers of NaCl. Here, a dark ball at the bottom is $\mathbf{P}_A(Cl^-)$ and at

TABLE 5.1. **Examples of compounds with the NaCl (3·2PO) structure.**

AgBr	AgCl	AgF	AmO	BaO	BaNH	BaS
BaSe	BaTe	BiSe	BiTe	CaO	CaNH	CaS
CaSe	CaTe	CdO	CeAs	CeBi	CeN	CeP
CeS	CeSb	CeSe	CeTe	CoO	CrN	CsF
CsH	DyAs	DyN	DySb	DyTe	ErAs	ErN
ErSb	ErTe	EuN	EuO	EuS	EuSe	EuTe
FeO	GdAs	GdN	GdSb	GdSe	HfC	HoAs
HoBi	HoN	HoP	HoS	HoSb	HoSe	HoTe
KCl	KCN	KF	KH	KI	KOH	KSH
KSeH	LaAs	LaBi	LaN	LaP	LaS	LaSb
LaSe	LaTe	LiBr	LiCl	LiD	LiH	LiF
LiI	LuN	MgO	MgS	MgSe	MnO	MnS
MnSe	NH_4Br	NH_4Cl	NH_4I	NaBr	NaCN	NaCl
NaF	NaH	NaI	NaSH	NaSeH	NbC	$NbN_{0.98}$
NbO	NdAs	NdBi	NdN	NdP	NdS	NdSb
NdSe	NdTe	NiO	NpN	NpO	PaO	PbS
PbSe	PbTe	PdH	PrAs	PrBi	PrN	PrP
PrS	PrSb	PrSe	PrTe	PuAs	PuB	PuC
PuN	PuO	PuP	PuS	PuTe	RbBr	RbCN
RbCl	RbF	RbH	RbI	$RbNH_2$	RbSH	RbSeH
ScAs	ScN	ScSb	SmAs	SmBi	SmN	SmO
SmP	SmS	SmSb	SmSe	SmTe	SnAs	SnSb
SnSe	SnTe	SrNH	SrO	SrS	SrSe	SrTe
TaC	TaO	TbAs	TbBi	TbN	TbP	TbS
TbSb	TbSe	TbTe	ThC	ThS	ThSb	ThSe
TiC	TiN	TiO	TmAs	TmN	TmSb	TmTe
UAs	UBi	UC	UN	UO	UP	US
USb	USe	UTe	VC	VN	VO	YAs
YN	$USb_{1.10}$	YTe	YbAs	YbN	YbO	YbSb
YbSe	YbTe	ZrB	ZrC	ZrN	ZrO	ZrP
ZrS						

the top a smaller light ball is $\mathbf{O}_A(Na^+)$. Reversing these designations is equivalent to inverting the cube. If all balls have the same color the structure is a simple (primitive) cubic structure. Polonium was mentioned as the only element with this structure. The relative heights and positions occupied by each layer are shown below for the **3·2PO** structure.

Height:	0	17	33	50	67	83	100
Position:	$\mathbf{P}_A$	$\mathbf{O}_C$	$\mathbf{P}_B$	$\mathbf{O}_A$	$\mathbf{P}_C$	$\mathbf{O}_B$	$\mathbf{P}_A$
	Cl^-	Na^+	Cl^-	Na^+	Cl^-	Na^+	Cl^-

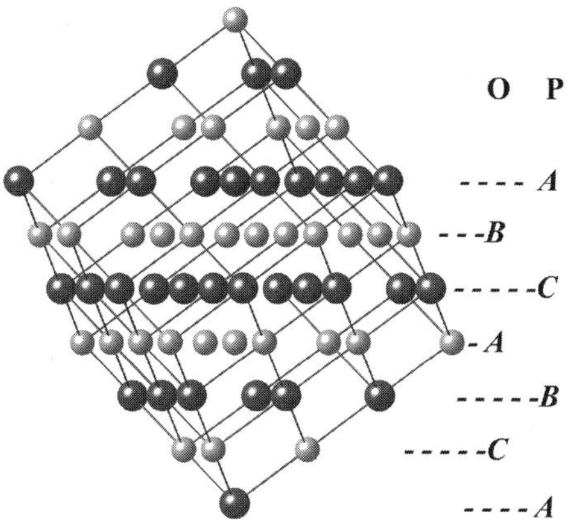

Figure 5.1. The **3•2PO** structure of NaCl. Positions of layers are shown.
(*Source*: *CrystalMaker*, by David Palmer, *CrystalMaker* Software Ltd., Begbroke Science Park, Bldg. 5, Sandy Lane, Yarnton, Oxfordshire, OX51PF, UK.)

5.1.2. The Crystal Structures of CaC_2[3•2PO(t)] and Pyrite [FeS_2] [(3•2PO)

The compound CaC_2 is known as calcium carbide, but it is actually calcium acetylide, containing the acetylide ion C_2^{2-}. It is used to react with water to produce acetylene. The structure of CaC_2 is closely related to that of NaCl. In the CaC_2 crystal all of the C_2^{2-} ions are aligned in one direction, along the c_o axis. The tetragonal unit cell is elongated in this direction as shown in Figure 5.2 ($\mathbf{D}_{4h}^{17}$, I4/*mmm*, $a_o = 3.63$, and $c_o = 6.03$ Å). The packing of layers corresponds to that of NaCl so we can consider it to be a **3•2PO(t)** structure. Some peroxides (CaO_2, SrO_2, and BaO_2) and superoxides (KO_2, RbO_2, and CsO_2) have the CaC_2 structure.

The mineral pyrite, FeS_2, has the NaCl (**3•2PO**) structure without the elongation of CaC_2. For FeS_2, there are four molecules per cubic unit cell, $\mathbf{T}_h^6$, *Pa*3, and $a_o = 5.4067$ Å. The alignments of the disulfide ions, S_2^{2-}, are staggered, avoiding elongation of the unit cell (Figure 5.3*a*). The S—S bond length in S_2^{2-} is 2.14 Å compared to 3.32 for other S—S distances. Each Fe has *CN* 6 (octahedral) and the *CN* for S is 4 (one bonded S + 3Fe). Figure 5.3*b* shows the layers of Fe and S_2. Many compounds of the type MO_2, MS_2, and MSe_2 have the pyrite structure

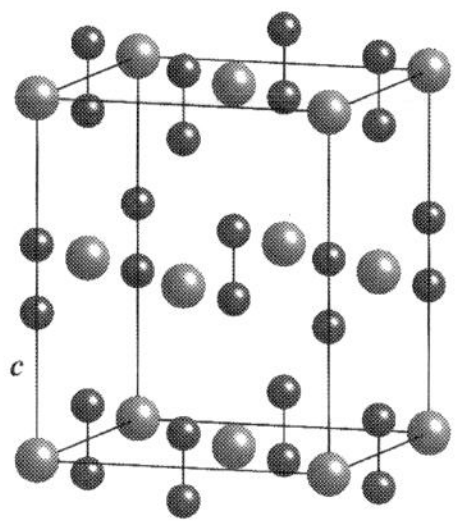

Figure 5.2. The **3•2PO(t)** structure of CaC_2.

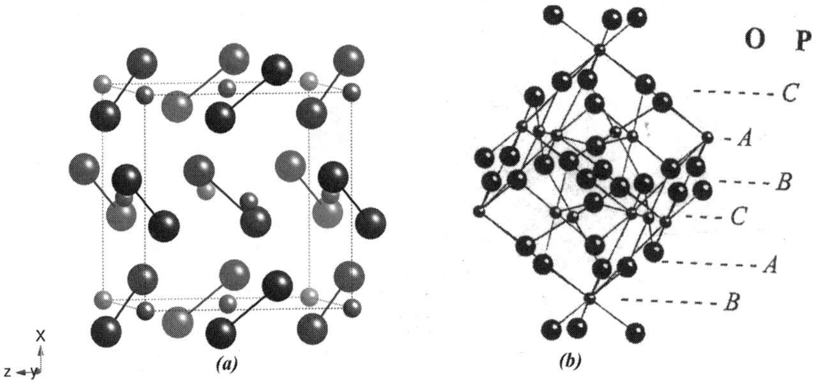

Figure 5.3. The **3·2PO** structure of pyrite (FeS_2). ***(a)*** The unit cell shows the *ccp* arrangement of Fe atoms and S—S bonding. ***(b)*** Packing layers of FeS_2 are labeled.

(*Source*: *CrystalMaker*, by David Palmer, *CrystalMaker* Software Ltd., Begbroke Science Park, Bldg. 5, Sandy Lane, Yarnton, Oxfordshire, OX51PF, UK.)

(see Table 5.2). The mineral marcasite (also FeS_2, Section 5.4.2) has a structure related to rutile (TiO_2) to be discussed later. NaCN and KCN are cubic **3·2PO** structures because of free rotation of CN^- ions. At low-temperature rotation is restricted and distortion from the cubic structure occurs.

5.1.3. The 3·2PO(o) Crystal Structure of Mercury Peroxide [HgO_2]

The **3·2PO(o)** structure of β-mercury peroxide, HgO_2, is shown in Figure 5.4. There are four molecules in the orthorhombic unit cell, $\mathbf{D}_{2h}^{15}$, *Pbca*, $a_o = 6.080$, $b_o = 6.010$, and $c_o = 4.800$ Å. The O_2^{2-} ions are centered in the center and edges of the cell. They are skewed relative to the cell edges, similar to pyrite. The O—O bond distance is 1.5 Å in the O_2^{2-} ions. The Hg^{2+} ions are in octahedral sites with four oxygen atoms in equatorial positions and O_2^{2-} ions in axial positions. Hg is bonded to both oxygen atoms of O_2^{2-} in axial positions so the *CN* of Hg^{2+} can be considered as 6 or 8. These axial O_2^{2-} ions are shared with other Hg^{2+} ions giving chains of octahedra along the *c* axis. Oxygen atoms are bonded to another oxygen and four Hg^{2+} ions in a distorted tetrahedral arrangement of Hg^{2+} ions.

Table 5.2. Compounds with the pyrite structure.

β-NaO_2	MnS_2	OsS_2	$CoSe_2$	$MnTe_2$	PtP_2	$AuSb_2$
KO_2	CoS_2	NiAsS	$NiSe_2$	$RuTe_2$	$PdAs_2$	$PdBi_2$
MgO_2	NiS_2	CoAsS	$RhSe_2$	$OsTe_2$	$PtAs_2$	$PtBi_2$
ZnO_2	RhS_2	NiSbS	$RuSe_2$	$RhTe_2$	$PdSb_2$	$RuSn_2$
CdO_2	RuS_2	$MnSe_2$	$OsSe_2$	$IrTe_2$	$PtSb_2$	

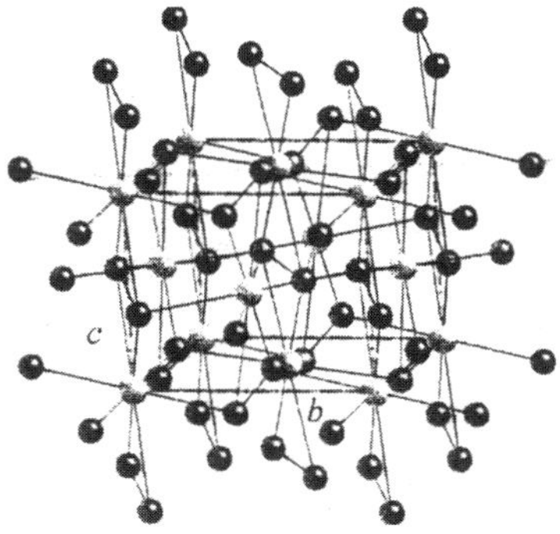

Figure 5.4. The unit cell of HgO_2. O_2^{2-} ions are in **P** sites and Hg^{2+} ions are in **O** sites.

(*Source*: Y. Matsushita, *Chalcogenide* Crystal structure data library, Version 5.5e, Institute for solid state Physics, the University of Tokyo, Tokyo, Japan, and *CrystalMaker*, by David Palmer, *CrystalMaker* Software Ltd., Begbroke Science Park, Bldg. 5, Sandy Lane, Yarnton, Oxfordshire, OX51PF, UK.)

5.1.4. The 3·2PO Crystal Structure of UB_{12}

UB_{12} is an interesting example of a compound related to NaCl even though the formulas look much different. However, with a large cation (U) and the large icosahedral B_{12} anion, a **3·2PO** cubic structure like NaCl results (see Figure 5.5) ($\mathbf{O}_h^5$, $Fm3m$, $a_o = 7.473$ Å). Each U has a cuboctahedron of 12 B atoms formed by four triangular faces of B_{12} anions approaching along the directions of the bonds of a tetrahedron. The cuboctahedra around U are clear in the figure, but the B_{12} icosahedra are not.

5.1.5. The 3·2PO Crystal Structure of $[Co(NH_3)_6][TlCl_6]$

The **3·2PO** structure of $[Co(NH_3)_6][TlCl_6]$ is an NaCl-type arrangement with the octahedral ions occupying the **P** and **O** layers (it is arbitrary to

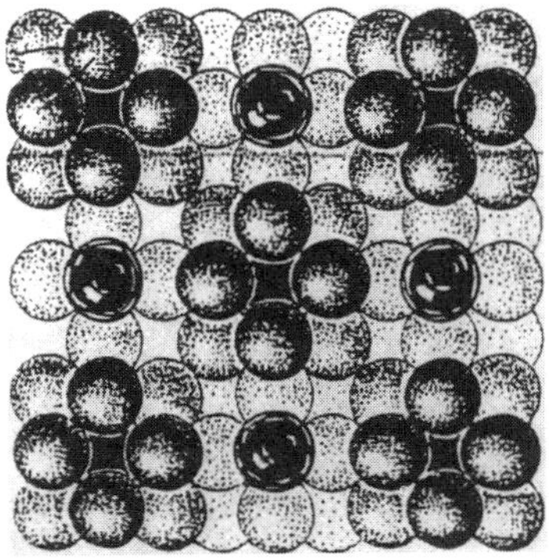

Figure 5.5. The **3·2PO** structure of UB_{12}. A packing drawing of UB_{12} viewed along a cube axis. U atoms are black.

(*Source*: R.W.G. Wyckoff, Crystal Structures Vol. 2., 2nd ed., Wiley, NewYork, 1964, p. 231.)

assign the cations or anions to **P** or **O** layers). There are four molecules in the cubic unit cell ($\mathbf{T}_h^6$, $Pa3$, $a_o = 11.42$ Å). Other complexes with this structure include $[Co(NH_3)_6][TlBr_6]$, $[Co(NH_3)_6][BiCl_6]$, $[Co(NH_3)_6][PbCl_6]$, and $[Co(NH_3)_4(H_2O)_2][TlCl_6]$. This structure is expected to be common for similar complexes. The size of the metal ion would have little effect on the size of the complex ion.

5.1.6. The 3·2PO(h) Crystal Structure of Calcite [$CaCO_3$]

The calcite form of $CaCO_3$ has a NaCl-like arrangement distorted by the spatial requirements of the trigonal planar CO_3^{2-} ion. The planes of the CO_3^{2-} ions are all parallel in the **P** layers with the Ca^{2+} ions occupying the **O** layers. The NaCl-like cubic cell can be viewed as depressed along the threefold axis perpendicular to the packing layers to give a trigonal unit, or a hexagonal unit cell can be chosen. There are six packing layers and four molecules in the rhombohedral unit cell ($\mathbf{D}_{3d}^6$, $R\bar{3}c$, $a_o = 6.361$ Å, $\alpha = 46.1°$). The extended hexagonal unit contains six molecules. Figure 5.6*a* shows a projection along *c* of the top two carbonate layers with one Ca between them. One oxygen atom of each of three carbonate ions form a face of the CaO_6 octahedron. A projection of the whole cell obscures this feature because Ca and C are eclipsed at *A*, *B*, and *C* positions. Figure 5.6*b* shows a perspective view of the hexagonal cell. The notation is **3·2PO(h)** to emphasize that the structure is not cubic. The calcite structure is found for carbonates of

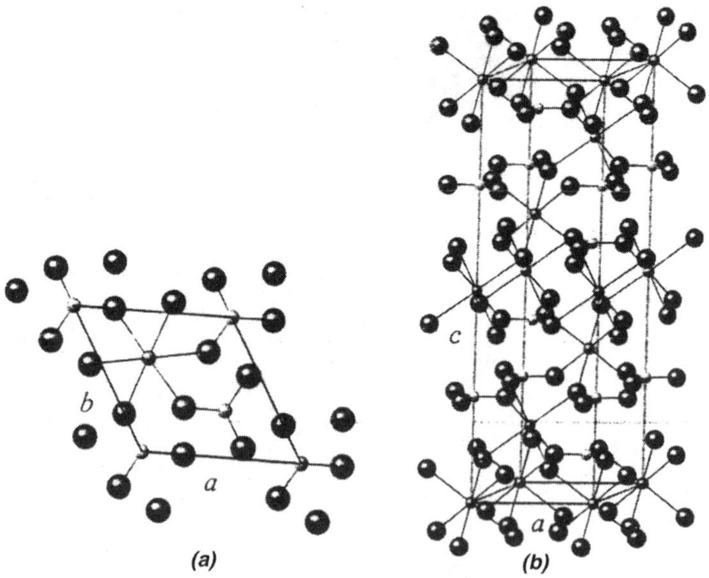

Figure 5.6. *(a)* A projection the calcite [**3·2PO(h)**] cell along the *c* axis. *(b)* A perspective view of the cell. Small dark atoms are Ca and small light atoms are carbon.

(*Source*: *CrystalMaker*, by David Palmer, *CrystalMaker* Software Ltd., Begbroke Science Park, Bldg. 5, Sandy Lane, Yarnton, Oxfordshire, OX51PF, UK.)

Mg, Mn, Fe, Co, Ni, Zn, and Cd, nitrates of Li, Na, Rb, and Ag, and borates of Sc, Lu, and In.

5.1.7. The 3·4POPO(h) Crystal Structure of $NaFeO_2$ and $AgBiSe_2$

The crystal structure of $NaFeO_2$ is rhombohedral. If the oxide ions are assigned to **P** layers, the Na^+ and Fe^{3+} ions fill alternate **O** layers; thus, all **O** layers are filled by metal ions, but because of the alternation of $\mathbf{O}_{Na}$ and $\mathbf{O}_{Fe}$ layers there are 12 layers in the repeating unit. As shown in Figure 5.7, the relative positions and heights of all packing layers are given below.

0	11	17	23	33	45	50	55	66	78	83	89	100
Na	O	Fe	O	Na	O	Fe	O	Na	O	Fe	O	Na
$\mathbf{O}_A$	$\mathbf{P}_C$,	$\mathbf{O}_B$,	$\mathbf{P}_A$	$\mathbf{O}_C$	$\mathbf{P}_B$	$\mathbf{O}_A$,	$\mathbf{P}_C$,	$\mathbf{O}_B$	$\mathbf{P}_A$,	$\mathbf{O}_C$,	$\mathbf{P}_B$	$\mathbf{O}_A$

The stronger Fe—O bonding causes closer spacing between Fe^{3+}(**O**) and O^{2-}(**P**) layers (~5–6 units) compared to spacing between Na^+(**O**) and O^{2-}(**P**) layers (10–11 units). This is a **3·4POPO(h)** structure. The sequence is *ABC* for both **P** and **O** layers corresponding to a *ccp* sequence. In the cases of CaC_2 and $CaCO_3$ the cube of NaCl is deformed to accommodate the nonspherical ions. In the case of $NaFeO_2$, the spacing of the packing layers is altered because of the great differences of Na^+ and Fe^{3+} ions in alternate **O** layers. The deviations from the normal structure are dictated by the chemistry of

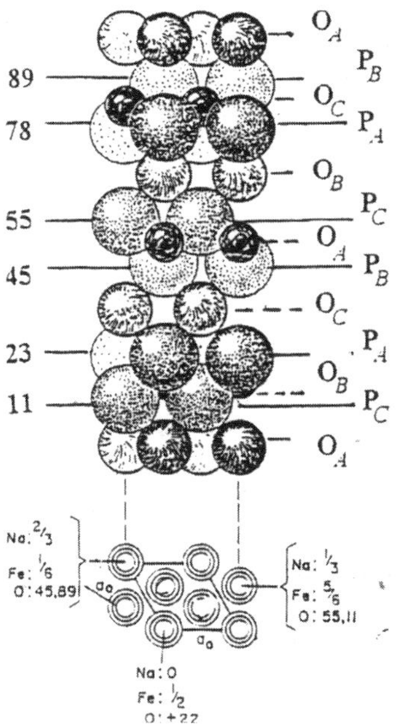

Figure 5.7. Two views of the structure of $NaFeO_2$. Oxide ions (large dotted balls) fill **P** layers. The **O** layers are labeled by position and occupancy by Na (midsize, line shaded) or Fe (black).

(*Source*: R.W.G. Wyckoff, *Crystal Structures*, Vol. 2, 2nd ed., Wiley, New York, 1964, p. 296.

TABLE 5.3. Compounds with the $NaFeO_2$ structure.

$NaTiO_2$	$LiAlO_2$	$LiCoO_2$	$LiCrO_2$	$LiNiO_2$	$LiVO_2$
$LiRhO_2$	$NaVO_2$	$NaCrO_2$	$NaFeO_2$	$NaNiO_2$	$NaInO_2$
$NaTlO_2$	$RbTlO_2$	$AgCrO_2$	$AgFeO_2$	$CuAlO_2$	$CuRhO_2$
$CuCoO_2$	$CuCrO_2$	$CuFeO_2$	$CuGaO_2$		
$NaCrS_2$	$KCrS_2$	$RbCrS_2$	$NaInS_2$		
$NaCrSe_2$	$NaInSe_2$	$RbCrSe_2$	$TlBiTe_2$	$TlSbTe_2$	

the compound. This structure is found for many $M^IM^{III}O_2$ compounds and a few corresponding compounds of S^{2-}, Se^{2-}, and Te^{2-}, as shown in Table 5.3. The compounds $LiHF_2$, $NaHF_2$, and $CaCN_2$ (calcium cyanamide) have the CsCl (**3•4PTOT**) structure with the HF_2^- and $CN_2{}^{2-}$ anions in the **P** and **O** layers with the cations in both **T** layers. The structure of $NaFeO_2$ can be described in a similar way with O^{2-} ions in **P** and **O** layers, Fe^{3+} in $\mathbf{T}^+$ layers, and Na^+ in $\mathbf{T}^-$ layers. Thus, we can see the close relationship between the CsCl (**3•4PTOT**) structure and a **3•4POPO** structure.

The **3•4POPO(h)** structure of $AgBiSe_2$ is very similar to that of $NaFeO_2$ except for more normal spacing between layers as shown in Figure 5.8. The trigonal unit cell contains one molecule, $\mathbf{D}_{3d}^3$, $P\bar{3}m$, for the hexagonal cell, $a = 4.18$ and $c = 19.67$ Å. In $AgBiSe_2$, because the Se^{2-} ions are much more polarizable than O^{2-}, both cations cause strong polarization of Se^{2-} ions. Here the chemistry causes the structure to be altered back toward to the ideal ionic model because of comparable polarizing power of the two cations. Ag and Bi fill alternate **O** layers in an *ABC* sequence.

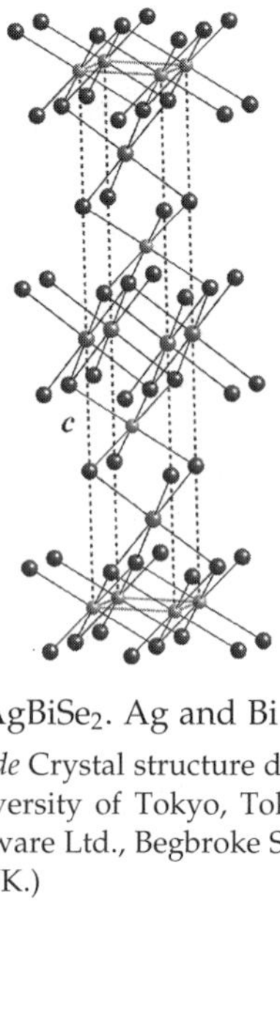

Figure 5.8. The structure of $AgBiSe_2$. Ag and Bi are in alternate **O** layers.

(*Source*: Y. Matsushita, *Chalcogenide* Crystal structure data library, verston 5.5B, Institute for Solid state Physics, The University of Tokyo, Tokyo, Japan, and *CrystalMaker*, by David Palmor, *Crystal Maker* software Ltd., Begbroke Science Park, Bldg. 5, Sandy Lane, Yarnton, Oxfordshire, OX51PF, UK.)

TABLE 5.4. **Compounds with the disordered $LiFeO_2$ structure.**

$LiCo_{0.5}Mn_{0.5}O_2$	$LiFe_{0.5}Mn_{0.5}O_2$	$LiNi_{0.5}Mn_{0.5}O_2$	$LiCo_{0.5}Ti_{0.5}O_2$
$LiFe_{0.5}Ti_{0.5}O_2$	$LiMn_{0.5}Ti_{0.5}O_2$	$LiNi_{0.5}Ti_{0.5}O_2$	$LiTiO_2$
γ-$LiTlO_2$	$NaBiS_2$	$NaBiSe_2$	$TlBiS_2$
$TlSbS_2$	$KBiS_2$	$KBiSe_2$	$LiBiS_2$
$AgBiS_2$	$AgBiSe_2$	$AgBiTe_2$	$AgSbS_2$
$AgSbSe_2$	$AgSbTe_2$	$KBiS_2$	

5.1.8. The $3{\cdot}2PO_{1/2\,1/2}$(t) and 3·4POPO(t) Crystal Structures of $LiFeO_2$

The structure of $LiFeO_2$ is similar to the cubic NaCl layer sequence of $NaFeO_2$ with ordered and disordered forms. In both forms the oxide ions are in **P** layers as for Cl^- in NaCl. Quenching $LiFeO_2$ from high temperature gives a cubic disordered structure with Li^+ and Fe^{3+} distributed randomly in **O** sites. The result is that spacing between layers is regular, giving **$3{\cdot}2PO_{1/2\,1/2}$**(t) with two molecules in the cubic unit cell. Many $M^IM^{III}X_2$ compounds have this disordered structure for oxides, sulfides, selenides, and tellurides (Table 5.4). Annealing $LiFeO_2$ at 570°C produces α-$LiFeO_2$ with an ordered distribution of metal ions and a tetragonal cell with four molecules per unit cell ($\mathbf{D}_{4h}^{19}$, $I4_1/amd$, $a_o = 4.057$, and $c_o = 8.759$ Å). The notation is **$3{\cdot}4PO^{Fe}PO^{Li}$(t)** with 12 packing layers. The spacing between layers is regular (Figure 5.9). The elongation of the cell from cubic to tetragonal results from different sizes of the cations.

5.1.9. The 3·4POPO(m) Crystal Structure of $NaNiO_2$

The structure of $NaNiO_2$ is monoclinic with two molecules per cell, C_{2h}^3, $C2/m$, $a_o = 5.33$, $b_o = 2.86$, and $c_o = 5.59$ Å. Oxygen atoms fill the

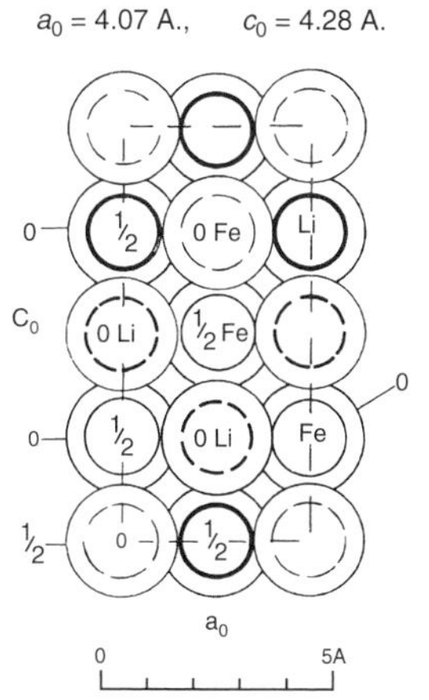

Figure 5.9. The ordered tetragonal structure of α-$LiFeO_2$ projected along the *a* axis. The Li atoms are heavily outlined and the oxygen atoms are the large, lightly outlined circles.

(*Source*: R.W.G. Wyckoff, *Crystal Structures*, Vol. 2, 2nd ed., Wiley, New York, 1964, p. 314.)

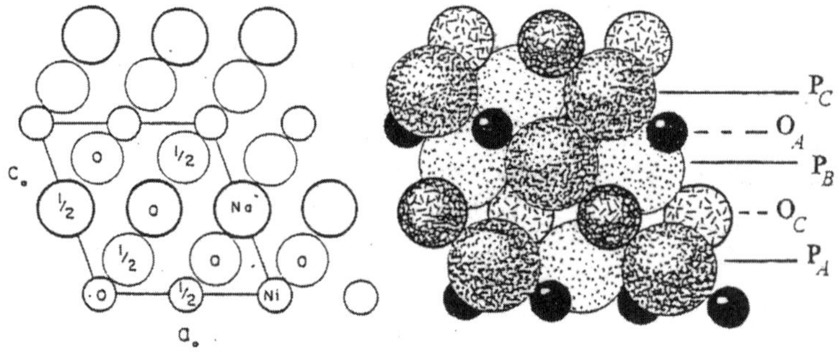

Figure 5.10. The **3·4POPO(m)** structure of low-temperature $NaNiO_2$ in projection along the *b* axis. The heavy lined large circles are Na and the small circles are Ni. In the packing drawing the layers are labeled.

(*Source*: R.W.G. Wyckoff, *Crystal Structures*, Vol. 2, 2nd ed., Wiley, New York, 1964, p. 303.)

P layers in a *ccp* pattern. The filled **O** layers of Ni and Na alternate, as seen in Figure 5.10. There are octahedra of six oxygen atoms around Ni and Na and 3Na and 3Ni atoms around each O atom forming an octahedron. For Na—O, the distances are 2.29 or 2.34 Å, and for Ni—O, the distances are 1.95 or 2.17 Å. The full notation is $\mathbf{3{\cdot}4P_{\mathit{A}}O^{Na}P_{\mathit{B}}O^{Ni}P_{\mathit{C}}O^{Na}P_{\mathit{A}}O^{Ni}P_{\mathit{B}}O^{Na}P_{\mathit{C}}O^{Ni}(m)}$. It can be simplified as **3·4POPO(m)** (Ni and Na layers alternate).

5.2. Hexagonal Close-Packed Structures

5.2.1. The 2·2PO Crystal Structure of NiAs

Following the pattern of NaCl (**3·2PO**) using an *hcp* arrangement we have the structure (**2·2PO**) for NiAs (the mineral niccolite, see Figure 5.11). The hexagonal cell contains two molecules, $\mathbf{C}_{6v}^4$, $C6mc$, $a_o = 3.602$, and $c_o = 5.009$ Å. The As atoms (dark balls) occupy *A* and *B* positions in **P** layers, an *hcp* pattern. The Ni atoms (small light balls) occupy **O** layers. Octahedral sites occur at positions *different* than those of the neighboring **P** layers. Because **P** sites occur only at *A* and *B* positions, all **O** sites must occur at *C* positions only for an *hcp*

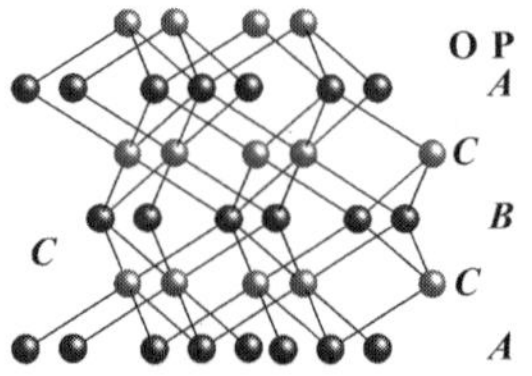

Figure 5.11. The **2·2PO** structure of NiAs. The As atoms (dark) occupy $\mathbf{P}_A$ and $\mathbf{P}_B$ layers with Ni atoms (light) in $\mathbf{O}_C$ layers.

(*Source*: *CrystalMaker*, by David Palmer, *CrystalMaker* Software Ltd., Begbroke Science Park, Bldg. 5, Sandy Lane, Yarnton, Oxfordshire, OX51PF, UK.)

TABLE 5.5 Compounds with the NiAs structure.

TiS	TiSe	TiTe	MnTe	VP	CoSb	PtB	RhSn
VS	VSe	VTe	ZrTe	MnAs	NiSb	FeSn	AuSn
CrS	CrSe	CrTe	PdTe	NiAs	PdSb	NiSn	MnBi
FeS	FeSe	FeTe	RhTe	MnSb	PtSb	CuSn	RhBi
CoS	CoSe	CoTe	IrTe	CrSb	IrSb	PdSn	
NiS	NiSe	NiTe		FeSb		PtSn	
NbS							

structure. The Ni atoms in **O** sites have an octahedral arrangement (*CN* 6) of As atoms. Each As atom is at the center of a trigonal prism formed by six Ni atoms. The positions and heights are given below.

Height:	0	25	50	75	100
Position:	$\mathbf{P}_A$	$\mathbf{O}_C$	$\mathbf{P}_B$	$\mathbf{O}_C$	$\mathbf{P}_A$
Atom:	As	Ni	As	Ni	As

The notation for this sequence is **2·2PO** indicating an *hcp* arrangement with 4 (**2·2**) layers in the repeating unit. There are about 50 compounds with the **2·2PO** structure given in Table 5.5.

There are important consequences of the features of the **2·2PO** structure, and there are severe limitations of the kinds of compounds found with this structure. In the **3·2PO** (NaCl) structure each ion is surrounded with six octahedrally arranged ions of opposite charge. Both ions are well shielded by ions of opposite charge, maximizing the attraction of ions of opposite charge and minimizing the repulsion between ions of the same charge. In NiAs, all Ni atoms occur at *C* positions with little shielding from neighboring Ni atoms. In fact, the Ni atoms form chains (or "wires") along the C positions. Because of the strong Ni–Ni interaction, NiAs conducts electricity well along this one direction. This structure is very unfavorable for typical ionic salts such as NaCl. The **2·2PO** structure is limited to compounds with appreciable covalency, or highly polarizable anions and strongly polarizing cations. The structure is not found for halides or oxides or for alkali or alkaline earth metals that form ionic compounds.

5.2.2. The $\mathbf{2{\cdot}2PO_{1/2}PO_{1/2}}$ Structures of Cuproscheelite [$CuWO_4$] and Sanmartinite [$ZnWO_4$]

Cuproscheelite, $CuWO_4$, is triclinic, $\mathbf{C}_i^1$, $P\bar{1}$, $a = 4.7026$, $b = 5.8389$, $c = 4.8784$ Å, $\alpha = 91.677°$, $\beta = 92.469°$, and $\gamma = 82.805°$, with two molecules in the unit cell. In Figure 5.12***a***, the oxygen atoms are in *AB* close-packed layers with the packing direction along *b*. Cu and W occupy half of alternate **O** layers. **O** sites occur only at *C* layers, but the metal ions occupy two sites alternating with two vacancies along *b*. The occupied sites for W are above and below the vacancies in the Cu layer. The CuO_6 and WO_6 octahedra share edges in

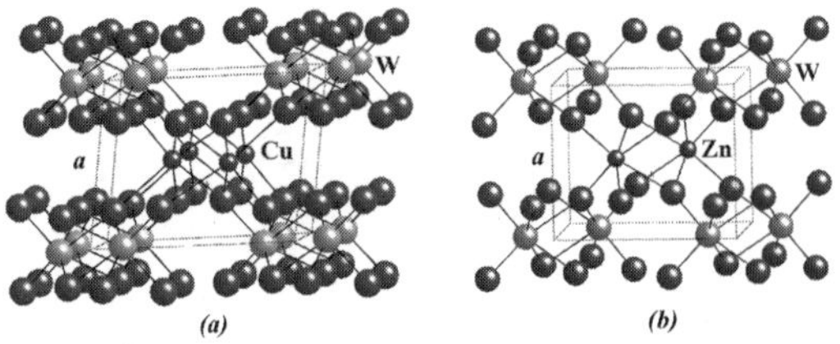

Figure 5.12. The $\mathbf{2{\cdot}2PO_{1/2}PO_{1/2}}$ structures of $CuWO_4$ and $ZnWO_4$. ***(a)*** The triclinic of $CuWO_4$. Oxygen atoms are in *AB* **P** layers. Cu and W are in 1/2 of **O** sites of alternate layers. ***(b)*** The monoclinic cell of $ZnWO_4$.

(*Source*: *CrystalMaker*, by David Palmer, *CrystalMaker* Software Ltd., Begbroke Science Park, Bldg. 5, Sandy Lane, Yarnton, Oxfordshire, OX51PF, UK.)

pairs with apices shared with adjacent **O** layers. The notation is $\mathbf{2{\cdot}2PO^{Cu}_{1/2}PO^{W}_{1/2}}$**(triclinic)**.

Sanmartinite, $ZnWO_4$, also has the $\mathbf{2{\cdot}2PO^{Zn}_{1/2}PO^{W}_{1/2}}$**(m)** structure. The structure is very similar to $CuWO_4$, with Zn and W occupying half of **O** sites in alternate **O** layers (Figure 5.12*b*). The octahedra share edges in pairs along *b* and are aligned with vacancies in adjacent **O** layers, as for $CuWO_4$. Crystals of $ZnWO_4$ are much more symmetrical than $CuWO_4$; they are monoclinic, $\mathbf{C}^4_{2h}$, $P2/c$, $a = 4.72$, $b = 5.70$, $c = 4.95$ Å, $\beta = 90.15^\circ$, with two molecules in the unit cell. The oxygen atoms are in *AB* close-packed layers stacked along the *a* axis with Zn and in half of alternate **O** layers.

$CuWO_4$, $ZnWO_4$, and $CaWO_4$ (scheelite) provide interesting comparisons. $CaWO_4$ is totally different from $CuWO_4$ and $ZnWO_4$. In $CaWO_4$ the ***CN*** of Ca is 8 and that of W is 4. The description of the structure of $CaWO_4$ is in terms of close-packed layers of Ca and W with oxygen atoms in all **T** sites ($\mathbf{3{\cdot}6P_{1/2\ 1/2}TT}$, Section 6.3.8). For $CuWO_4$ and $ZnWO_4$, the oxygen atoms are in close-packed **P** layers with the metal ions in octahedral sites staggered to avoid repulsion between close cations, all at the same *C* sites. The crystallographic descriptions are much different; $CuWO_4$ is triclinic and $ZnWO_4$ is monoclinic. The space groups give no clue that the environments of ions are very similar. That close similarity is clear from the notation $\mathbf{2{\cdot}2PO_{1/2}PO_{1/2}}$ for both compounds.

5.2.3. The $\mathbf{2{\cdot}2PO_{1/2}O_{1/2}}$(o) Structure of Aragonite [$CaCO_3$]

We have considered the **3·2PO(h)** structure of calcite ($CaCO_3$). Aragonite is another modification of $CaCO_3$. It occurs as blue colored crystals, but also colors range from colorless, white, yellow, and green to pink. Many colorless minerals, such as calcite and aragonite, and gems show beautiful colors because of traces of transition metal ions. Colors can change, with changes in the amounts of impurities. Pearls consist of a composite material called nacre, which consists of tiny crystals of aragonite crystals held together by an organic binder. The orthorhombic unit cell of aragonite contains four molecules ($\mathbf{D}^{16}_{2h}$, *Pnma*, $a_o = 7.968$, $b_o = 5.741$, and $c_o = 4.959$ Å at 26°C). Arago-

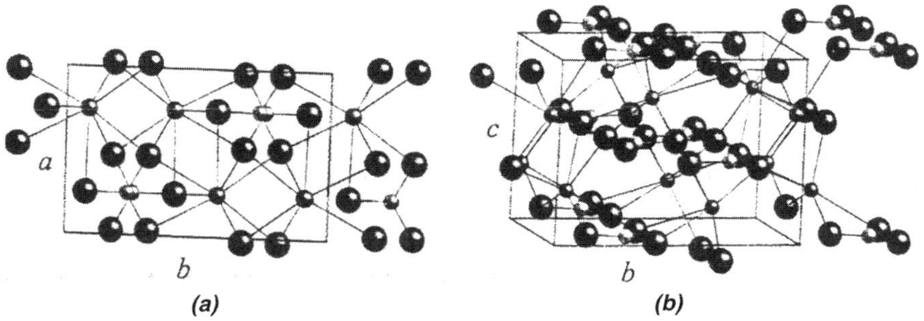

Figure 5.13. The **2·2P$_{1/2}$O$_{1/2}$(o)** structure of aragonite ($CaCO_3$). *(a)* A projection of the orthorhombic cell along the *c* axis. Ca atoms are dark small atoms. The triangular CO_3^{2-} ions are apparent with C atoms small light atoms. *(b)* A perspective view of the cell.

(*Source: CrystalMaker*, by David Palmer, *CrystalMaker* Software Ltd., Begbroke Science Park, Bldg. 5, Sandy Lane, Yarnton, Oxfordshire, OX51PF, UK.)

nite is similar to calcite as a 1:1 compound, but with Ca^{2+} occupying **P** layers with the planes of CO_3^{2-} ions parallel in **O** layers. Figure 5.13***a*** is a projection along *c* showing the staggered CO_3^{2-} ions in adjacent layers. Figure 5.13***b*** shows a perspective view of the cell. The packing sequence is *hcp* with four layers repeating, suggesting **2·2PO(o)** with Ca^{2+} ions in $\mathbf{P}_A$ and $\mathbf{P}_B$ positions. All CO_3^{2-} ions are at $\mathbf{O}_C$ positions, but they are staggered on opposite faces of an octahedron. (For calcite, we assigned the carbonate ions to **P** layers because anions are usually larger than cations and occupy **P** layers. The assignment to **P** and **O** layers is arbitrary for *ccp*. For aragonite, the carbonate ions are in $\mathbf{O}_C$ positions for this *hcp* pattern.) The orientations of CO_3^{2-} ions relative to Ca^{2+} ions in calcite and aragonite are shown in Figure 5.14. In calcite each oxygen is directed between two Ca^{2+} ions: one above and one below. In aragonite, each oxygen of CO_3^{2-} is directed toward a Ca^{2+} below in the figure and one Ca^{2+} ion in a layer above is on a line between each pair of oxygen atoms forming the edge of the triangular ion. The spacing of layers for aragonite is given below:

Height:	8	25	42	58	75	92
Ion:	CO_3^{2-}	Ca^{2+}	CO_3^{2-}	CO_3^{2-}	Ca^{2+}	CO_3^{2-}
Position:	$\mathbf{O}_C$	$\mathbf{P}_A$	$\mathbf{O}_C$	$\mathbf{O}_C$	$\mathbf{P}_B$	$\mathbf{O}_C$

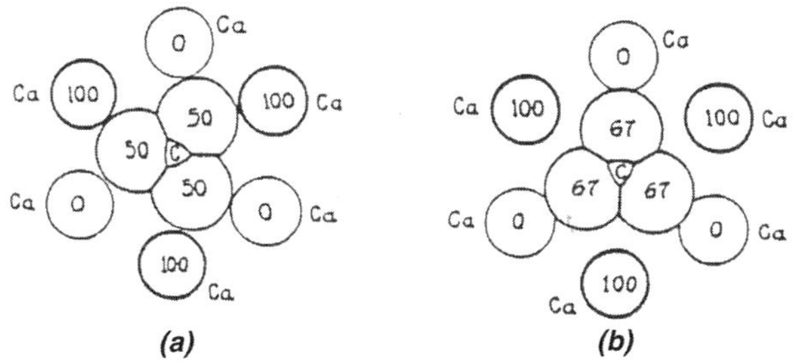

Figure 5.14. A comparison of the orientations of CO_3^{2-} ions in *(a)* calcite and *(b)* aragonite.

(*Source*: L. Bragg, *The Crystalline States*, C. Bell and Sons, London, 1949, p. 131).

The average height for the CO_3^{2-} layers between $\mathbf{P}_A$ and $\mathbf{P}_B$ is 50 as expected for an **O** layer. However, as for NiAs, there is a problem for **O** layers in an *hcp* structure because all sites are at *C* positions. In NiAs, a strong Ni–Ni interaction occurs. For aragonite, the CO_3^{2-} ions are triangles with the extension in the plane of the layer. The spacing from Ca^{2+} layers to the adjacent CO_3^{2-} layers is 17 in each case. Between two Ca^{2+} layers there are two CO_3^{2-} layers. We might consider these CO_3^{2-} layers as a double **O** layer (**OO**), resulting from crowding of CO_3^{2-} ions in a *C* layer and attraction between CO_3^{2-} ions and Ca^{2+} ions. This results in splitting the **O** layer into two sublayers, alternating the heights of adjacent CO_3^{2-} ions. Instead of writing the notation as **2·2PO** (this neglects the separate **O** layers) we give it as $\mathbf{2{\cdot}2PO_{1/2}O_{1/2}}$(**o**) or as $\mathbf{2{\cdot}2POO_{1/2\ 1/2}}$(**o**). The aragonite structure is found for KNO_3, $SrCO_3$, YBO_3, and $LaBO_3$. If aragonite is heated above 400°C it is transformed to calcite.

5.2.4. The 2·2PO Crystal Structure of $NaPO_3NH_3$

$NaPO_3NH_3$ can be considered as a 1:1 compound having two molecules per hexagonal unit cell ($\mathbf{C}_6^6$, $C6_3$, $a_o = 5.773$, and $c_o = 6.031$ Å). The $PO_3NH_3^-$ zwitterion is shown as $PO_3NH_2^{2-}$ in Figure 5.15***a***. The unit cell is shown in Figure 5.15***b***. The anions are in **P** layers of the *hcp* arrangement with Na^+ ions in all **O** layers at *C* positions with four layers in the repeating unit and the notation is **2·2PO**.

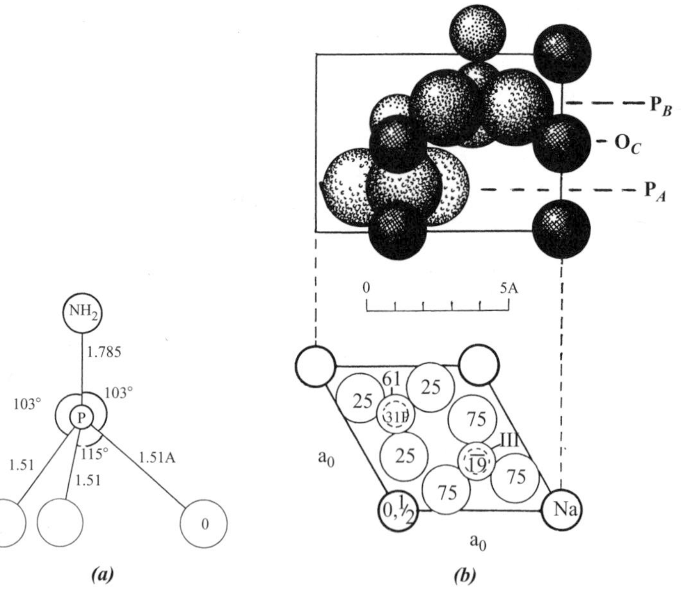

Figure 5.15. The **2·2PO** structure of $NaH(PO_3NH_2)$. ***(a)*** The dimensions of the $PO_3NH_2^{2-}$ ion. ***(b)*** Two projections of the hexagonal structure. The Na^+ ions are heavy lined circles and the O atoms are largest circles. In the packing drawing the $PO_3NH_2^{2-}$ ions are orientated as in ***(a)*** and layers are labeled with Na^+ in $\mathbf{O}_C$ layers. H atoms are not shown.

(*Source*: R.W.G. Wyckoff, *Crystal Structures*, Vol. 3. 2nd ed., Wiley, New York, 1965, p. 129.)

The oxygen atoms extend between Na^+ ions, providing some shielding. The N atoms (of NH_3^+) extend upward in the figure forming three equivalent hydrogen bonds with oxygen atoms of three other anions.

5.2.5. The 4•2PO Crystal Structure of γ′-MoC and TiP

The structure of γ′-MoC has similarities to both NiAs and NaCl. There are eight layers in the repeating unit as shown in Figure 5.16. Carbon atoms are in **P** layers and Mo atoms are in **O** layers, and the sequence of layers for **4•2PO** is $|\mathbf{P}_A\mathbf{O}_C\mathbf{P}_B\mathbf{O}_C\mathbf{P}_A\mathbf{O}_B\mathbf{P}_C\mathbf{O}_B|\mathbf{P}_A\ldots$. The *ABAC* sequence is a combination of the NiAs ($\mathbf{P}_A\mathbf{P}_B$) and the NaCl ($\mathbf{P}_B\mathbf{P}_A\mathbf{P}_C$) sequence. Why did nature choose this odd sequence? For the **2•2PO** (NiAs) structure all Mo would be in *C* positions with Mo–Mo interaction throughout. For the **3•2PO** (NaCl) structure Mo would be in *A*, *B*, and *C* positions with *no* interaction between Mo atoms. The actual sequence in **4•2PO** provides for Mo–Mo bonding in pairs. The hexagonal cell contains four molecules, $\mathbf{D}_{6h}^4$, $P6_3/mmc$, $a_o = 2.932$, and $c_o = 10.97$ Å. The compounds α-TiAs, β-ZrP, ZrAs, HfP, and ε-NbN have this structure.

Essentially, the same **4•2PO** structure occurs for TiP with P in **P** layers as shown in Figure 5.17. The figure identifies the NiAs ($\mathbf{P}_A\mathbf{P}_B$) and NaCl ($\mathbf{P}_B\mathbf{P}_A\mathbf{P}_C$) sequences. We see the trigonal prismatic arrangement of Ti around P in the NiAs sequence region and an octahedral arrangement of Ti around P in the NaCl sequence region. As for MoC, there is close interaction between pairs of Ti atoms in *C–C* and *B–B* layers. The hexagonal cell of TiP contains four molecules, $\mathbf{D}_{6h}^4$, $P6_3/mmc$, $a_o = 3.513$, and $c_o = 11.75$ Å.

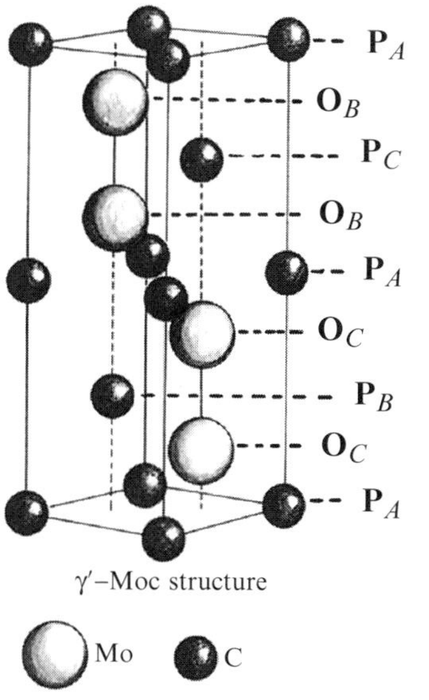

Figure 5.16. The **4•2PO** structure of γ′-MoC. The C atoms are in **P** layers.
(*Source*: F.S. Galasso, *Structure and Properties of Inorganic Solids*, Pergamon, Oxford, 1970, p. 150.)

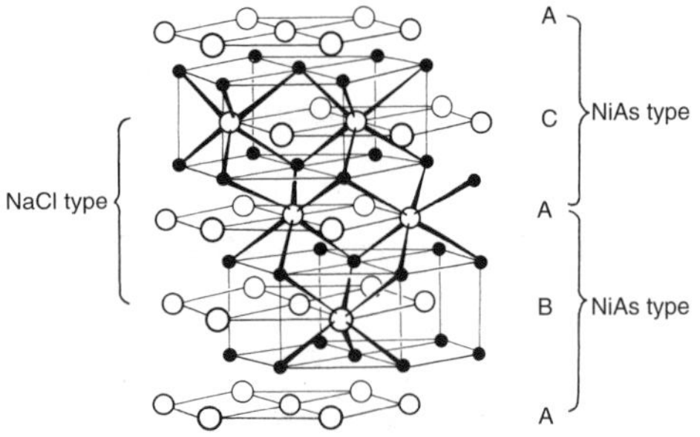

Figure 5.17. The trigonal prismatic and octahedral coordination of phosphorus atoms in TiP

(*Source*: D.M. Adams, *Inorganic Solids*, Wiley, New York, 1974, p. 313.)

5.3. *ccp* PO Crystal Structures with Partial Filling of Layers

5.3.1. The $3{\cdot}4PO_{2/3}PO_{1/3}$(t) Crystal Structure of Anatase [TiO_2]

There are three allotropic forms of TiO_2: anatase, rutile, and brookite. Anatase has a *ccp* pattern and rutile has an *hcp* pattern. The structure of brookite is more complex. Anatase has an elongated tetragonal cell with four molecules per cell, $\mathbf{D}_{4h}^{19}$, $I4_1/amd$, $a_o = 3.785$, and $b_o = 9.514$ Å. In Figure 5.18*a*, instead of all **O** layers being half-filled, they alternate as $\mathbf{PO}_{2/3}\mathbf{PO}_{1/3}$. There are 12 packing layers repeating, giving the notation $\mathbf{3{\cdot}4PO_{2/3}PO_{1/2}(t)}$. The octahedra are irregular, Ti—O distances are 1.40 Å for equatorial positions and 1.97 Å for axial positions. The axial O—Ti—O bonds are linear with 85° and 95° bond angles in the equatorial plane. This plane of each octahedron is tipped by 17° relative to the axial direction. There are two orientations of the octahedra in the $\mathbf{O}_{2/3}$ layer. Figure 5.18***b*** shows how edges are shared between octahedra. There are chains of octahedra sharing edges along *a* and *b*.

5.3.2. The $3{\cdot}4P_{3/16}O_{1/16}$(h) Crystal Structure of $AlCl_3{\cdot}6H_2O$

The structure of $AlCl_3{\cdot}6H_2O$ is similar to that of NaCl with Cl^- ions in **P** layers and octahedral $[Al(H_2O)_6]^{3+}$ ions in 1/16 of **O** sites. Each $[Al(H_2O)_6]^{3+}$ ion is surrounded octahedrally by six Cl^- ions. The small fractions of layers occupied results from treating large $[Al(H_2O)_6]^{3+}$ ions as units. The hexagonal cells contain two molecules, $\mathbf{D}_{3d}^{6}$, $R\bar{3}c$, $a_o = 11.76$, and $c_o = 11.824$ Å. There are 12 packing layers giving the notation $\mathbf{3{\cdot}4P_{3/16}O_{1/16}(h)}$. The spacings of the **O** layers (*ABCABC*) along c_o are 0, 1/6, 1/3, 1/2, 2/3, and 5/6 (see Figure 5.19).

5.3.3. The $3{\cdot}2P_{1/4\ 1/4}O_{1/4\ 1/4}$ Crystal Structure of Martensite [FeC]

Martensite, FeC, is an important phase in the Fe–C system. It is a supersaturated solution of carbon in α-Fe (*bcc*). It is extremely hard. In Figure 5.20, we see the *bcc* structure of Fe with carbon in vertical

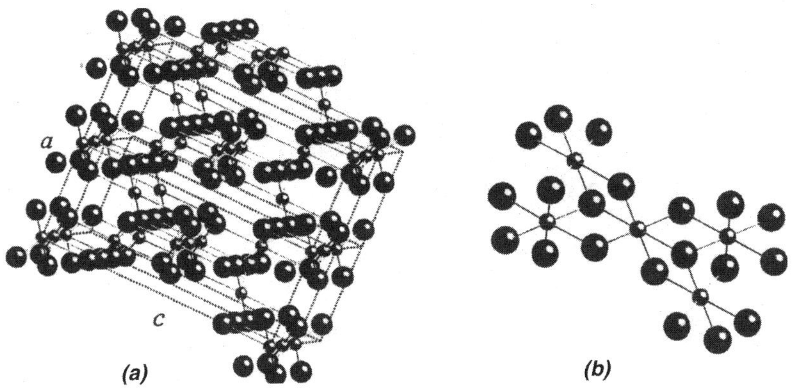

Figure 5.18. *(a)* A perspective view of the anatase (TiO_2) cell showing the **O** layers 2/3 and 1/3 occupied, ***(b)*** a view showing the edge sharing TiO_6 octahedra.

(*Source*: *CrystalMaker*, by David Palmer, *CrystalMaker* Software Ltd., Begbroke Science Park, Bldg. 5, Sandy Lane, Yarnton, Oxfordshire, OX51PF, UK.)

edges and the top and bottom faces. A range of positions of Fe atoms is shown by the cylinders. The structure has been described as half of the NaCl pattern. It differs from the NaCl structure in having one-quarter of sites occupied by Fe and C each in **P** and **O** layers. NaCl has four molecules in the unit cell and FeC has only two molecules. The notation is $\mathbf{3{\cdot}2P^{Fe}_{1/4}{}^{C}_{1/4}O^{Fe}_{1/4}{}^{C}_{1/4}}$.

5.3.4. The $3{\cdot}2P_{1/4\;3/4}O_{1/4}$ Crystal Structure of Perovskite [$CaTiO_3$]

The well-known structure of the mineral perovskite, $CaTiO_3$, is more simple than might be expected. Ti^{4+} ions fill one-quarter of **O** sites and

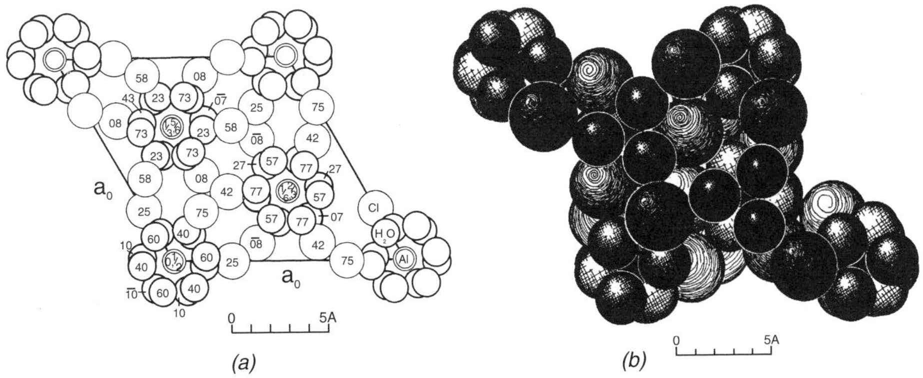

Figure 5.19. The structure of $AlCl_3{\cdot}6H_2O$. ***(a)*** A projection on the hexagonal base of the cell with heights of atoms shown. ***(b)*** A packing drawing viewed along the *c* axis. The Cl atoms are largest spheres. Al atoms (small, black) are covered by the triangular arrangements of oxygens (octahedral faces).

(*Source*: R.W.G. Wyckoff, *Crystal Structures*, Vol. 3, 2nd ed., Wiley, New York, 1963, pp. 792–793.)

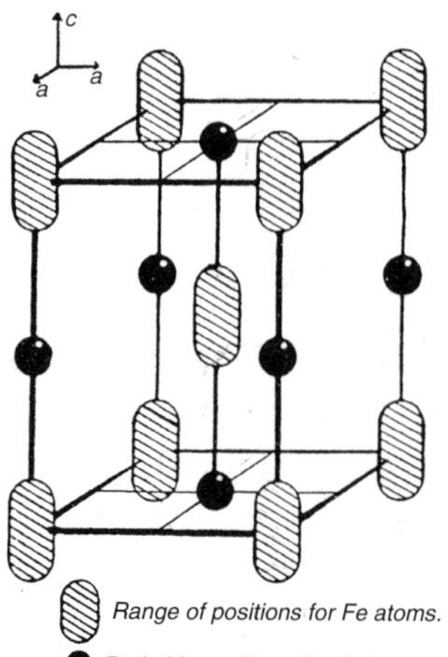

Figure 5.20. The structure of martensite (FeC).

(*Source*: H. Lipson and A.M.B. Parker, *J. Iron Steel Inst.*, Vol. 149 p. 123, 1944.)

Ca^{2+} and O^{2-} ions together fill **P** layers in a *ccp* pattern. The stoichiometry requires that one-quarter of the **P** sites be filled by Ca^{2+}. The Ti^{4+} ions occupy only one-quarter of the **O** sites, only those surrounded by six O^{2-} ions. The sequence of the packing layers (*A*, *B*, and *C*) are shown in Figure 5.21*a* for a $\mathbf{3{\cdot}2P_{1/4\ 3/4}O_{1/4}}$ structure with six repeating packing layers. Figure 5.21*b* shows a cube formed by Ca^{2+} ions with Ti^{4+} at the center of an octahedron formed by six O^{2-} in the faces of the cube. The stoichiometry is correct: 1 Ti, $6 \times 1/2 = 3$ O and $8 \times 1/8 = 1$ Ca. Another representation showing the beautiful simplicity of the structure is given in Figure 5.21*c*, where TiO_6 octahedra form a cube with Ca^{2+} at the center. Again, the stoichiometry is correct (counting only O in the edges of the cube; others are outside the cell): 1 Ca, $8 \times 1/8 = 1$ Ti and $12 \times 1/4 = 3$ O. Each O^{2-} ion is shared by two Ti^{4+} ions. Ca^{2+} in a **P** site is surrounded by 12 O^{2-} ions as part of each packing layer without any Ca^{2+} neighbors. The space group for $CaTiO_3$ is $\mathbf{O}_h^1$, $Pm3m$, $a_o = 3.84$ Å.

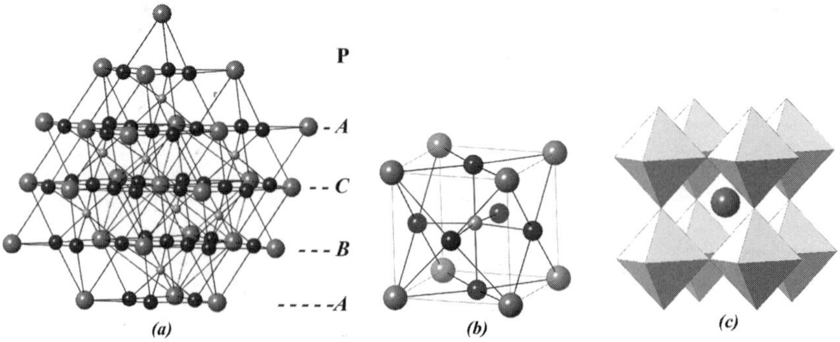

Figure 5.21. The structure of perovskite ($CaTiO_3$). *(a)* The packing layers are identified (Ti, small; O, larger dark; Ca, large light). *(b)* A cube showing the cubic unit cell showing the TiO_6 octahedron and the cube formed by Ca atoms. *(c)* A cubic arrangement of TiO_6 octahedra with Ca^{2+} at the center.

(*Source*: *CrystalMaker*, by David Palmer, *CrystalMaker* Software Ltd., Begbroke Science Park, Bldg. 5, Sandy Lane, Yarnton, Oxfordshire, OX51PF, UK.)

The highly symmetrical perovskite structure is very common for $MM'X_3$ type compounds. Some of the many hundreds of such compounds are listed in Table 5.6. Because of their important ferroelectric, ferromagnetic, and superconducting properties, many compounds have been synthesized varying the ratios of metals to optimize desired properties. New compounds are reported frequently. We will discuss structures related to perovskite in Sections 5.3.5, 5.3.6, 5.3.7, 5.4.11, 5.4.12, and 5.4.13.

5.3.4.1. Superconductors

Kamerlingh Onnes, at the University of Leiden, discovered superconductivity in 1911. He found that the resistance of some metallic wires became zero at very low temperature; it did not just *approach* zero, there was *no* dissipation of heat. At that time his laboratory was the only one equipped for studies at the temperature of liquid He (*bp* 4.1 K). Theoretical explanations of the phenomenon did not appear until the work of John Bardeen, Leon Cooper, and Robert Schrieffer in 1957. They received the Nobel Prize in Physics in 1972. The expense and difficulty of applying superconductivity to practical problems limits the applications. Nevertheless, superconductor magnets of very high field are now widely used in NMR in chemistry and the medical diagnostic applications of NMR called MRI (magnetic resonance imaging—they wanted to avoid the word "nuclear").

There was great excitement when "high-temperature" superconductors (T_C 30–40 K) were reported in 1986 by George Bednorz and Alex Müller, and in 1987, by Paul Chu and Maw-Kuen Wu ($T_C \sim 90$ K). Then there was great activity to find a superconductor above 77.2 K, the boiling point of N_2. With liquid N_2 as the refrigerant, many large-scale applications are viable. It not practicable to maintain long-distance power lines at liquid He temperature, but the enormous saving in power loss is practicable at liquid N_2 temperature ($T_C > 77.2$ K). Maglev trains become practicable. The train is elevated by magnetic levitation avoiding friction with the tracks. Because there are no wheels on the track, the train is propelled by timed magnetic pulses pulling and pushing the train. A third class of superconductors was discovered in 1988 by teams at the University of Arkansas and the National Metal Research Institute in Tsukuba, Japan. These materials have T_C about 125 K, well above the goal of 77.2 K.

The first class of superconductors (T_C 30–40 K) contains La, Ba, Cu, and O. The second class ($T_C \sim 90$ K) contains Y, Ba, Cu, and O in the ratios 1:2:3:7 or 1–2–3 for short. These compounds have the perovskite structure (Figure 5.21), Cu replaces Ti of perovskite, and Y and Ba replace Ca. The proportions of elements have been varied extensively so these conductors are called YBCO (for the elements). There is a problem in producing ceramic wires of $YBa_2Cu_3O_7$ because of poor alignment at grain boundaries. Doping the grain boundaries with Ca^{2+} (Ca^{2+} replacing some Y^{3+}) improved the current-carrying capacity at 77 K. The third class compounds contain Bi, Sr, Ca, Cu, and O or Tl, Ba, Ca, Cu, and O. These materials have perovskite-type structures, but they are not as well ordered. All three classes of superconductors

TABLE 5.6 Some compounds with the perovskite structure.

$AgZnF_3$	$AlBiO_3$	$BaCeO_3$	$BaFeO_3$	$BaMoO_3$	$BaPbO_3$	$BaPrO_3$
$BaPuO_3$	$BaSnO_3$	$BaThO_3$	$BaTiO_3$	$BaUO_3$	$BaZrO_3$	$BaZrS_3$
$CaCeO_3$	$CaMnO_3$	$CaMoO_3$	$CaSnO_3$	$CaThO_3$	$CaTiO_3$	$CaVO_3$
$CaZrO_3$	$CdCeO_3$	$CdSnO_3$	$CdThO_3$	$CdTiO_3$	$CeAlO_3$	$CeCrO_3$
$CeFeO_3$	$CeGaO_3$	$CeVO_3$	$CrBiO_3$	$CsCaF_3$	$CsCdBr_3$	$CsCdCl_3$
$CsHgBr_3$	$CsHgCl_3$	$CsIO_3$	$CsMgF_3$	$CsPbBr_3$	$CsPbCl_3$	$CsPbF_3$
$CsZnF_3$	$DyAlO_3$	$DyFeO_3$	$DyMnO_3$	$EuAlO_3$	$EuCrO_3$	$EuFeO_3$
$EuTiO_3$	$FeBiO_3$	$GdAlO_3$	$GdCoO_3$	$GdCrO_3$	$GdFeO_3$	$GdMnO_3$
$HgNiF_3$	$KCaF_3$	$KCdF_3$	$KCoF_3$	$KCrF_3$	$KCuF_3$	$KFeF_3$
KIO_3	$KMgF_3$	$KMnF_3$	$KNbO_3$	$KNiF_3$	$KTaO_3$	$KZnF_3$
$LaAlO_3$	$LaCoO_3$	$LaCrO_3$	$LaFeO_3$	$LaGaO_3$	$LaMnO_3$	$LaNiO_3$
$LaRhO_3$	$LaTiO_3$	$LaVO_3$	$LiBaF_3$	$LiUO_3$	Li_xWO_3	$MgCeO_3$
NH_4CoF_3	NH_4MnF_3	NH_4NiO_3	$NaAlO_3$	$NaMgF_3$	$NaMnF_3$	$NaNbO_3$
$NaTaO_3$	$NaWO_3$	$NaZnF_3$	$NdAlO_3$	$NdCoO_3$	$NdCrO_3$	$NdFeO_3$
$NdGaO_3$	$NdMnO_3$	$NdVO_3$	$PbCeO_3$	$PbSnO_3$	$PbThO_3$	α-$PbTiO_3$
β-$PbTiO_3$	$PbZrO_3$	$PrAlO_3$	$PrCoO_3$	$PrCrO_3$	$PrFeO_3$	$PrGaO_3$
$PrMnO_3$	$PrVO_3$	$PuAlO_3$	$PuMnO_3$	$RbCaF_3$	$RbCoF_3$	$RbIO_3$
$RbMgF_3$	$RbMnF_3$	$RbZnF_3$	$SmAlO_3$	$SmCoO_3$	$SmCrO_3$	$SmFeO_3$
$SmVO_3$	$SrCeO_3$	$SrCoO_3$	$SrFeO_3$	$SrHfO_3$	$SrMoO_3$	$SrPbO_3$
$SrSnO_3$	$SrThO_3$	$SrTiO_3$	$SrZrO_3$	$TaSnO_3$	$TlCoF_3$	$TlIO_3$
$YAlO_3$	$YCrO_3$	$YFeO_3$				

$CsCd(NO_2)_3$	$CsHg(NO_2)_3$	$KCd(NO_2)_3$	$NH_4Cd(NO_2)_3$	$RbCd(NO_2)_3$	$RbHg(NO_2)_3$
$TlCd(NO_2)_3$	$TlHg(NO_2)_3$	$BaNi_{0.33}Nb_{0.67}O_3$	$BaCa_{0.5}W_{0.5}O_3$	$BaCe_{0.5}Nb_{0.5}O_3$	$BaCo_{0.33}Ta_{0.67}O_3$
$BaCo_{0.5}W_{0.5}O_3$	$BaDy_{0.5}Nb_{0.5}O_3$	$BaEr_{0.5}Nb_{0.5}O_3$	$BaEu_{0.5}Nb_{0.5}O_3$	$BaFe_{0.5}Nb_{0.5}O_3$	$BaFe_{0.5}Nb_{0.5}O_3$
$BaFe_{0.5}W_{0.5}O_3$	$BaGd_{0.5}Nb_{0.5}O_3$	$BaHo_{0.5}Nb_{0.5}O_3$	$BaIn_{0.5}Nb_{0.5}O_3$	$BaLi_{0.5}Re_{0.5}O_3$	$BaLu_{0.5}Nb_{0.5}O_3$
$BaMg_{0.5}W_{0.5}O_3$	$BaMo_{0.5}Co_{0.5}O_3$	$BaMn_{0.5}Ni_{0.5}O_3$	$BaNa_{0.5}Re_{0.5}O_3$	$BaNd_{0.5}Nb_{0.5}O_3$	$BaNi_{0.33}Ta_{0.67}O_3$
$BaNi_{0.5}W_{0.5}O_3$	$BaPr_{0.5}Nb_{0.5}O_3$	$BaSc_{0.5}Nb_{0.5}O_3$	$BaSc_{0.5}Ta_{0.5}O_3$	$BaSm_{0.5}Nb_{0.5}O_3$	$BaTb_{0.5}Nb_{0.5}O_3$
$BaTm_{0.5}Nb_{0.5}O_3$	$BaY_{0.5}Nb_{0.5}O_3$	$BaYb_{0.5}Nb_{0.5}O_3$	$BaYb_{0.5}Ta_{0.5}O_3$	$BaZn_{0.33}Nb_{0.67}O_3$	$BaZn_{0.5}W_{0.5}O_3$
$BaCaZrGeO_3$	$BaLa_{0.5}Nb_{0.5}O_3$	$CaNi_{0.33}Nb_{0.67}O_3$	$CaNi_{0.33}Yb_{0.67}O_3$	$(Ca, Na)(Ti, Nb)O_3$	$K_{0.5}Bi_{0.5}TiO_3$
$K_{0.5}Ce_{0.5}TiO_3$	$K_{0.5}La_{0.5}TiO_3$	$K_{0.5}Nd_{0.5}TiO_3$	$KBaTiNbO_6$	$KBaCaTiZrNbO_9$	$K_2CeLaTi_4O_{12}$
$La_{0.6}Ba_{0.4}MnO_3$	$La_{0.6}Ca_{0.4}MnO_3$	$La_{0.6}Sr_{0.4}MnO_3$	$LaMg_{0.5}Ge_{0.5}O_3$	$LaMg_{0.5}Ti_{0.5}O_3$	$LaNi_{0.5}Ti_{0.5}O_3$
$LaZr_{0.5}Ca_{0.5}O_3$	$LaMg_{0.5}Ge_{0.5}O_3$	$LaZr_{0.5}Mg_{0.5}O_3$	$LaMnO_{3.07\text{–}3.15}$	$LaMnO_{3.10\text{–}3.23}$	$Na_{0.5}Bi_{0.5}TiO_3$
$Na_{0.5}La_{0.5}TiO_3$	$NaBaTiNbO_6$	$NdMg_{0.5}Ti_{0.5}O_3$	$PbMg_{0.33}Nb_{0.67}O_3$		

TABLE 5.7. Ferromagnetic compounds with the ordered perovskite structure.

Ba_2FeMoO_6	Sr_2FeMoO_6	Ca_2FeMoO_6	Sr_2CrMoO_6
Ca_2CrMoO_6	Sr_2CrWO_6	Ca_2CrWO_6	Ba_2FeReO_6
Sr_2FeReO_6	Ca_2FeReO_6	Sr_2CrReO_6	Ca_2CrReO_6

contain Cu, O, at least one alkaline earth metal (Ba, Sr, and/or Ca) and Y, La, Bi, or Tl. They are all layered structures and they share Cu–O planes. These Cu–O planes are parallel to cell faces and are not in close-packed layers.

5.3.5. The $3{\cdot}4\mathbf{P}_{1/4\ 3/4}\mathbf{O}_{1/4}$ Crystal Structure of Ba_2FeMoO_6

The structure of Ba_2FeMoO_6 [or $Ba(Fe_{0.5}Mo_{0.5})O_3$ to show the similarity to perovskite] is known as an ordered perovskite structure. Perovskite structures containing more than one transition metal are described as ordered or disordered with respect to the arrangement of the metals in the **O** layers. For Ba_2FeMoO_6, Fe and Mo are in alternate **O** layers. The notation $3{\cdot}2\mathbf{P}_{1/4\ 3/4}\mathbf{O}_{1/4}$ describes a perovskite structure, but for the ordered Ba_2FeMoO_6 structure the unit cell is increased to include 12 packing layers and eight molecules, giving the notation $3{\cdot}4\mathbf{P}_{1/4\ 3/4}\mathbf{O}_{1/4}$. These compounds are important ferromagnetic materials; therefore, many combinations of different transition metals occupying the **O** layers have been studied. Some important ferromagnetic compounds with the ordered perovskite structure are shown in Table 5.7.

5.3.6. The $3{\cdot}2\mathbf{P}_{1/4\ 3/4}\mathbf{O}_{1/4}$ Crystal Structure of $Ba_3SrTa_2O_9$

Another ordered perovskite structure is encountered for $Ba_3SrTa_2O_9$ with Sr in one-quarter of the **O** sites in one layer and Ta in one-quarter of the **O** sites in the two adjacent **O** layers as shown in Figure 5.22. Because Sr atoms occupy one-quarter of the **O** layers in a *ccp* pattern (*A*, *B*, *C*), there are only six layers in the repeating unit giving $3{\cdot}2\mathbf{P}_{1/4\ 3/4}\mathbf{O}_{1/4}$:

$\mathbf{P}_{1/4\ 3/4}$	$\mathbf{O}_{1/4}$	$\mathbf{P}_{1/4\ 3/4}$	$\mathbf{O}_{1/4}$	$\mathbf{P}_{1/4\ 3/4}$	$\mathbf{O}_{1/4}$
A (Ba,O)	*C* (Ta)	*B*(Ba,O)	*A*(Sr)	*C*(Ba,O)	*B*(Ta)

5.3.7. The $3{\cdot}2\mathbf{P}_{3/4}\mathbf{O}_{1/4}$ Crystal Structure of ReO_3 and Cu_3N

The structure of ReO_3 is strictly cubic, $\mathbf{O}_h^1$, $Pm\bar{3}m$, $a_o = 3.751$ Å. The O^{2-} ions are in a *ccp* pattern with one-quarter of sites vacant. This is the perovskite structure without Ca^{2+}, giving a network of ReO_6 octahedra sharing vertices (see Figure 5.23). The $3{\cdot}2\mathbf{P}_{3/4}\mathbf{O}_{1/4}$ structure of ReO_3 is found for UO_3, MoF_3, NbF_3, TaF_3, NbO_2F, TaO_2F, and TiO_2F. In the oxofluorides, the oxygen and fluorine atoms are randomly distributed among the **P** sites occupied.

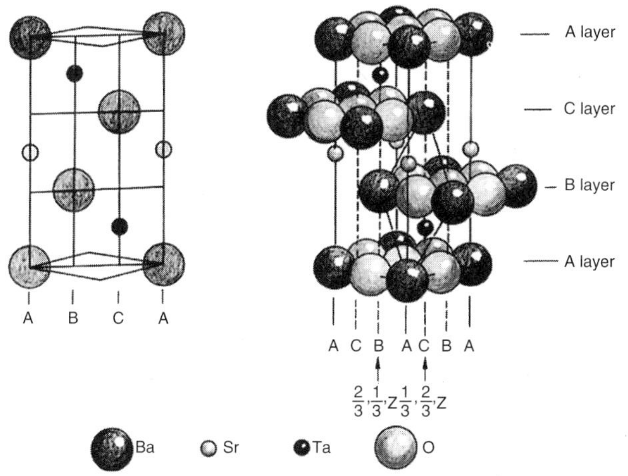

Figure 5.22. The structure of $Ba_3SrTa_2O_9$. The packing layers are identified. The drawing at the left shows the positions of atoms in the vertical plane at the center of the drawing to the right.

(*Source*: F.S. Galasso, *Structure and Properties of Inorganic Solids*, Pergamen, Oxford, 1970, p. 200.)

Copper(I) nitride, Cu_3N, crystals are also cubic, $\mathbf{O}_h^1$, $Pm\bar{3}m$, $a_o = 3.814$ Å, with one molecule in the unit cell. Each N has an octahedral arrangement of six Cu atoms. Cu atoms have linear bonds to two N atoms and eight nearest Cu neighbors (see Figure 5.24). These are two sets of four planar Cu atoms bonded to the N atoms. The N atoms occupy sites for a simple cubic cell. To accommodate Cu and N in the **P** system, Cu atoms occupy **P** sites in an *ABC* sequence with N atoms in **O** sites. The **P** sites are three-quarters occupied and **O** sites are one-quarter occupied to give the notation $\mathbf{3{\cdot}2P_{3/4}O_{1/4}}$.

5.3.8. The $\mathbf{3{\cdot}2P_{3/4}O_{3/4}}$ Crystal Structure of NbO

NbO has a highly ordered *ccp* structure, $\mathbf{O}_h^1$, $Pm\bar{3}m$, $a_o = 4.210$ Å. Nb atoms are in corners of the cell and in the center of side faces. Oxygen atoms are in the center and centers of top and bottom edges

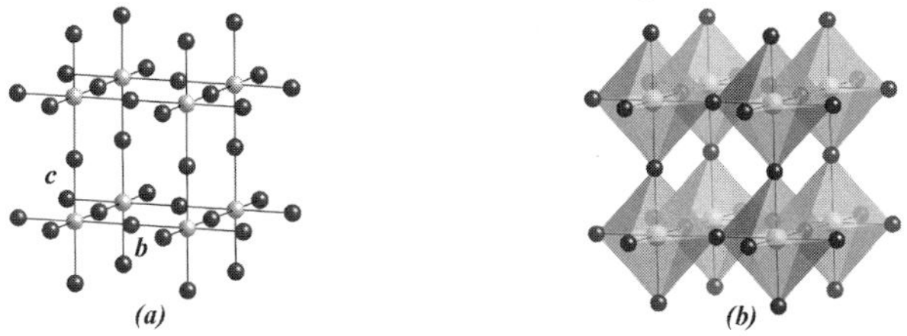

Figure 5.23. *(a)* The unit cell of ReO_3. *(b)* A cube formed by ReO_6 octahedra sharing all oxygens.

(*Source*: *CrystalMaker*, by David Palmer, *CrystalMaker* Software Ltd., Begbroke Science Park, Bldg. 5, Sandy Lane, Yarnton, Oxfordshire, OX51PF, UK.)

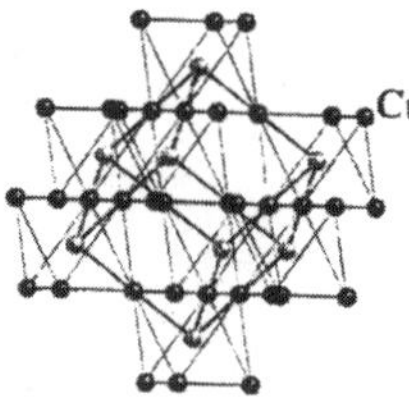

Figure 5.24. The $\mathbf{3{\cdot}2P_{3/4}O_{1/4}}$ structure of Cu_3N.

(*Source:* Y. Matsushita, *Chalcogenide* crystal structure data library, Version 5.5B, Institute for Solid State Physics, The University of Tokyo, Tokyo, Japan, 2004; and *CrystalMaker*, by David Palmer, *CrystalMaker* Software Ltd., Begbroke Science Park, Bldg. 5, Sandy Lanc, Yarnton, Oxfordshire, OX51PF, UK.)

(Figure 5.25). These occupancies correspond to $\mathbf{P_{3/4}}$ and $\mathbf{O_{3/4}}$. Either atom can be assigned to **P** or **O** sites. Nb has *CN* 4 with square planar NbO_4 units in planes parallel to faces. The notation is $\mathbf{3{\cdot}2P_{3/4}O_{3/4}}$.

5.3.9. The $3{\cdot}2P_{1/4\ 3/4}O_{1/2}(o)$ Crystal Structure of Atacamite $[Cu_2Cl(OH)_3]$

The mineral atacamite, $Cu_2Cl(OH)_3$, is an ore for Cu, occurring only in arid regions because it dissolves in ground water. Beautiful crystals are translucent and bright emerald green to nearly black. The orthorhombic unit cell contains two molecules, $\mathbf{D}_{2h}^{16}$, *Pnma*, $a_o = 6.01$, $b_o = 9.13$, $c_o = 6.84$ Å. Hydroxohalides, $M_2X(OH)_3$, or M_2XY_3 compounds often have the same structures as dihalides or dioxides, MX_2. The usual structure involves X atoms in close-packed **P** layers with M atoms in **O** layers. For $Cu_2Cl(OH)_3$ the Cu^{2+} ions fill 1/2 of **O** layers, OH^- ions fill three-quarter of **P** layers and Cl^- ions fill one-quarter of these **P** layers. The pattern is *ccp*, giving the notation $\mathbf{3{\cdot}2P_{1/4\ 3/4}O_{1/2}(o)}$. We do not need to specify which anions fill one-quarter or three-quarter of the **P** layers because it is determined by the stoichiometry. Wells gave a complex figure that shows the vacancy of **O** sites in a NaCl-type structure. All Cu^{2+} ions have a square planar $Cu(OH)_4^{2-}$ configuration. Half of them also have two Cl^- ions, giving an octahedral arrangement. The other $Cu(OH)_4^{2-}$ units have a fifth OH^- and one Cl^-, completing a distorted octahedral arrangement. Figure 5.26*a* shows the network of $Cu(OH)_4^{2-}$ ions, and

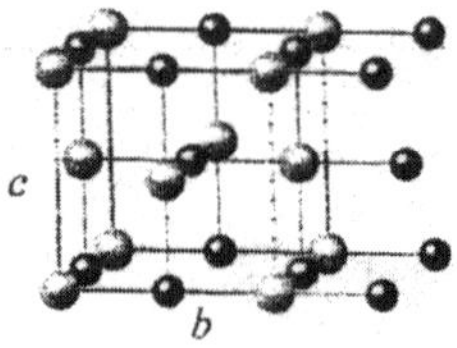

Figure 5.25. The structure of NbO. Nb atoms are larger atoms.

(*Source*: *CrystalMaker*, by David Palmer, *CrystalMaker* Software Ltd., Begbroke Science Park, Bldg. 5, Sandy Lane, Yarnton, Oxfordshire, OX51PF, UK.)

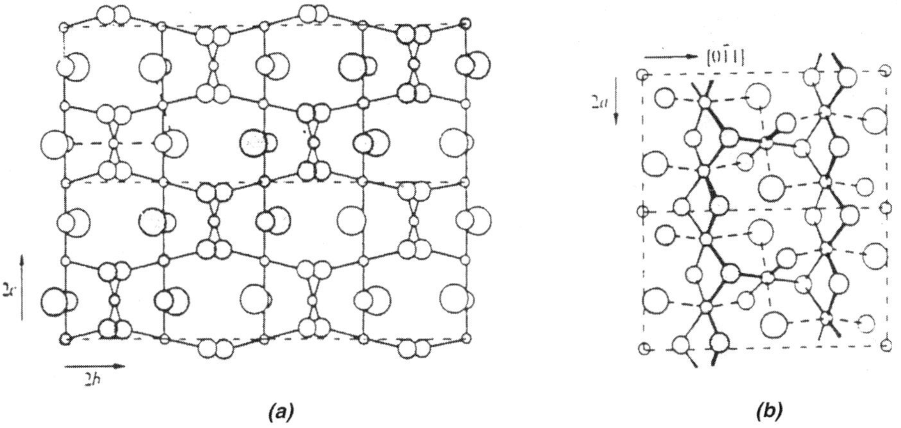

Figure 5.26. The structure of atacamite $[Cu_2Cl(OH)_3]$ ***(a)*** The network of $Cu(OH)_4^{2-}$ ions. ***(b)*** In the center $Cu(OH)_2Cl_2$ units are aligned vertically. On each side there are more distorted $Cu(OH)_4(OH)Cl$ units.

(*Source*: A.F. Wells, *Structural Inorganic Chemistry*, 3rd ed., Oxford UP, Oxford, 1962, p. 170.)

Figure 5.26***b*** shows the $Cu(OH)_4Cl_2$ units aligned vertically in the center with more distorted $Cu(OH)_5Cl$ units on each side.

5.3.10. The $2{\cdot}2P_{1/4\ 3/4}O_{1/2}$(o) Crystal Structure of Cu_2CsCl_3

The notation for Cu_2CsCl_3 is $\mathbf{2{\cdot}2P_{1/4\ 3/4}O_{1/2}}$(**o**) or $\mathbf{2{\cdot}2P_{1/4\ 3/4}T_{1/2}}(C_2)$ (**o**) as for Ag_2CsI_3 (Section 5.4.24), but the structures differ significantly. In both cases Cs and the anion (here Cl) are in **P** layers (1:3) with the smaller Cu atoms in **O** (or **T**) layers (see Figure 5.27). The Cu atoms are halfway between **P** layers as expected for an **O** layer, but the Cu are at the centers of distorted tetrahedra with a C_2 axis aligned with the packing direction, $\mathbf{T}(C_2)$. Each Cs atom has 12 Cl neighbor atoms in a distorted arrangement corresponding to that for an *hcp* structure. There is substantial deformation of the **P** layers occupied by Cs and Cl atoms. Cu atoms are at the centers of very distorted tetrahedra with two Cu–Cl distances 2.16 Å and two 2.53 Å. The $CuCl_4$ tetrahedra form double chains along c_o. Edges are shared along the length of the chain and between the individual chains. The short distances between pairs of Cu atoms (~ 3.1 Å) through the long tetrahedral edge indicates weak Cu—Cu bonding. The Cu—Cu distance in the metal is 2.56 Å. The structure is orthorhombic with four molecules per unit cell, $\mathbf{D}_{2h}^{17}$, *Cmcm*, $a_o = 9.49$, $b_o = 11.88$, and $c_o = 5.61$ Å.

5.3.11. The $3{\cdot}2PO_{1/5}$(t) Crystal Structure of Bismuth Pentafluoride

The crystal structure of BiF_5 is tetragonal, $\mathbf{C}_{4h}^5$, I4/m, a = 6.581, and $c = 4.229$ Å. Figure 5.28 shows the Bi atoms in a body-centered cell. It is not clear from this cell, but examination of the arrangements of the

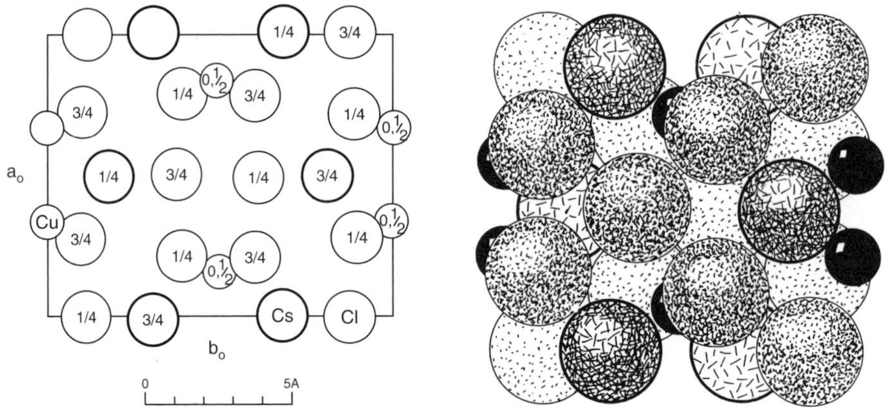

Figure 5.27. The structure of Cu_2CsCl_3. A projection along the *c* axis and a packing drawing of the same view. Cu atoms are black, Cl atoms are largest dotted spheres and Cs atoms are line shaded.

(*Source*: R.W.G. Wyckoff, *Crystal Structures*, Vol. 2, 2nd ed., Wiley, New York, 1964, p. 502.)

F^- ions using *CrystalMaker* shows that they are in close-packed layers in an *ABC* sequence. Bi atoms occupy one-fifth of octahedral sites. The BiF_6 octahedra share F^- ions along the *c* direction. The notation is $\mathbf{3{\cdot}2PO_{1/5}(t)}$.

5.3.12. The $3{\cdot}2P^{K}{}_{1/4}{}^{F}{}_{1/2}O_{1/8}(t)$ Crystal Structure of K_2NiF_4

Because Ni readily forms square planar complexes, K_2NiF_4 might be expected to contain planar NiF_4 units. The tetragonal crystals ($\mathbf{D}_{4h}^{17}$, $I4/mmm$, $a = 4.043$, and $c = 13.088$ Å) contains regular NiF_6 octahedra (Ni—F = 2.006 and 2.007 Å). These octahedra are centered in cubes of eight K^+ ions (Figure 5.29). F^- ions are in the centers of the faces of the cubes. The K^+ and F^- ions are in close-packed layers distorted by differences of sizes of the ions and vacancy of 1/4 of the sites. Ni^{2+} ions are in one-eighth of octahedral sites in each **O** layer.

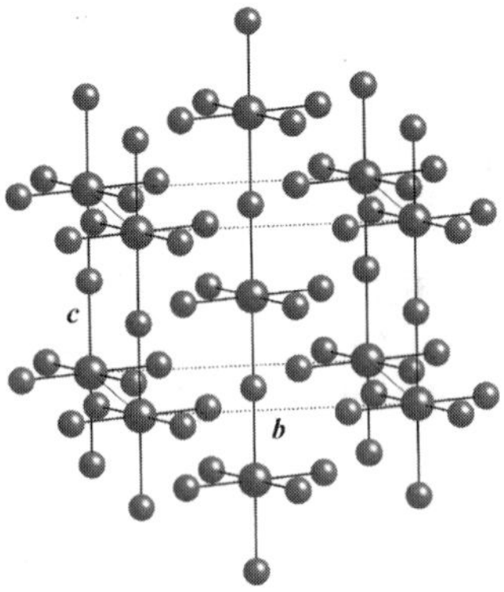

Figure 5.28. The unit cell of BiF_5. Bi^{5+} ions occupy 1/5 of **O** sites.

(*Source*: Y. Matsushita, *Chalcogenide* crystal structure data library, Version 5.5B, Institute for Solid State Physics, The University of Tokyo, Tokyo, Japan, 2004; and *CrystalMaker*, by David Palmer, *CrystalMaker* Software Ltd., Begbroke Science Park, Bldg. 5, Sandy Lane, Yarnton, Oxfordshire, OX51PF, UK.)

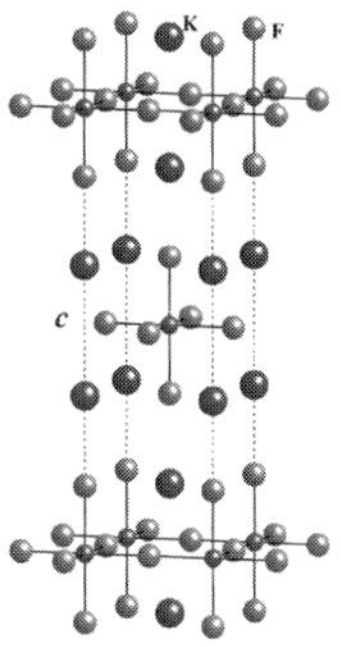

Figure 5.29. The crystal cell of K_2NiF_4 with the NiF_6^{2-} octahedron at the center of the figure.

(*Source*: Y. Matsushita, *Chalcogenide* crystal structure data library, Version 5.5B, Institute for Solid State Physics, The University of Tokyo, Tokyo, Japan, 2004; and *CrystalMaker*, by David Palmer, *CrystalMaker* Software Ltd., Begbroke Science Park, Bldg. 5, Sandy Lane, Yarnton, Oxfordshire, OX51PF, UK.)

The close-packed layers are in an *ABC* sequence. The notation is $\mathbf{3{\cdot}2P^K{}_{1/4}{}^F{}_{1/2}O_{1/8}(t)}$. In the *ab* plane each F is bonded to two Ni atoms, $\cdots$Ni—F—Ni—F$\cdots$ sequence along *a* or *b*. Along *c*, F^- ions are not shared, the sequence is $\cdots$F—Ni—F—K—K$\cdots$.

5.4. *hcp* PO Crystal Structures with Partial Filling of Layers

5.4.1. The $\mathbf{2{\cdot}2PO_{1/2}(t)}$ Crystal Structure of Rutile [TiO_2]

Rutile is a common Ti mineral. We have already considered another TiO_2 mineral, anatase ($\mathbf{3{\cdot}2PO_{1/2}(t)}$, Section 5.3.1). SnO_2 (cassiterite) has the rutile structure, and sometimes the structure is called the cassiterite structure. It is a common structure for M^{IV} oxides and fluorides of small M^{II} cations. It is also common for double oxides, $M^{III}M^{V}O_4$, such as $FeTaO_4$. Compounds with the rutile structure, including fluorides, oxides, and double oxides, are given in Table 5.8. The flattened unit cell of rutile is tetragonal with two molecules as shown in Figure 5.30*a*, $\mathbf{D}_{4h}^{14}$, $P4/mnm$, $a_o = 4.59373$, and $c_o = 2.95812$ Å, at 25°C. Figure 5.30*b* shows the octahedra of other Ti atoms. The oxygen atoms are in **P** layers in an *hcp* pattern with Ti atoms in half of the **O** sites. Because

TABLE 5.8. **Compounds with the Tetragonal TiO_2 Structure.**

CoF_2	FeF_2	MgF_2	MnF_2	NiF_2	PdF_2
PdF_2	CrO_2	GeO_2	IrO_2	MnO_2	MoO_2
NbO_2	OsO_2	PbO_2	RuO_2	SnO_2	TaO_2
TeO_2	WO_2				
$AlSbO_4$	$CrNbO_4$	$CrSbO_4$	$CrTaO_4$	$FeNbO_4$	$FeSbO_4$
$FeTaO_4$	$GaSbO_4$	$RhNbO_4$	$RhSbO_4$	$RhTaO_4$	$RhVO_4$

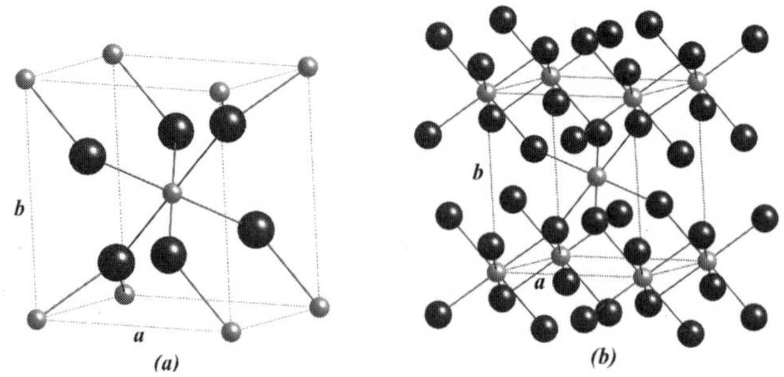

Figure 5.30. The **2·2PO$_{1/2}$(t)** structure of rutile (TiO_2) *(a)* A perspective view of the cell showing the TiO_6 octahedra in the cell. The smaller atoms are Ti. *(b)* A perspective view of the rutile cell showing the octahedra around each Ti.

(*Source*: *CrystalMaker*, by David Palmer, *CrystalMaker* Software Ltd., Begbroke Science Park, Bldg. 5, Sandy Lane, Yarnton, Oxfordshire, OX51PF, UK.)

the **P** sites are at *A* and *B* positions, all Ti atoms are at *C* positions giving chains of octahedra sharing edges. **O** sites occupied are staggered to avoid sharing octahedral faces. Each oxygen atom is bonded to three Ti atoms. The octahedra are slightly distorted, with four closer equatorial oxygen atoms at 1.95 Å with axial bonds at 1.98 Å.

5.4.2. The 2·2PO$_{1/2}$(o) Crystal Structure of Marcasite [FeS_2]

Pyrite (Section 5.1.2) is a FeS_2 mineral with a crystal structure closely related to the NaCl structure. Marcasite is another FeS_2 mineral closely related to the NiAs structure (Section 5.2.1). Pyrite has a cubic structure achieved by staggering the $S_2{}^{2-}$ ions. The crystal structure of marcasite is orthorhombic, $\mathbf{D}_{2h}^{12}$, *Pnnm*, $a = 4.436$, $b = 5.414$, and $c = 3.381$ Å, and can be described as a hexagonal sequence of S layers with Fe in half of the octahedral sites. The marcasite cell is shown in Figure 5.31***a***. There are chains of octahedra sharing edges along *c*. Along *a* and *b* directions octahedra are joined by vertex sharing.

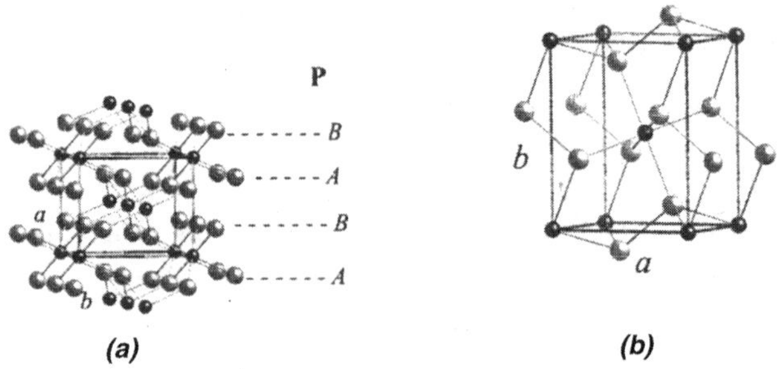

Figure 5.31. The **2·2PO$_{1/2}$(o)** structure of marcasite, FeS_2. *(a)* A perspective view of the cell with the **P** layers occupied by S labeled. The FeS_6 octahedra share edges along *c*. *(b)* The cell showing the S—S bonds in cell edges.

(*Source*: *CrystalMaker*, by David Palmer, *CrystalMaker* Software Ltd., Begbroke Science Park, Bldg. 5, Sandy Lane, Yarnton, Oxfordshire, OX51PF, UK.)

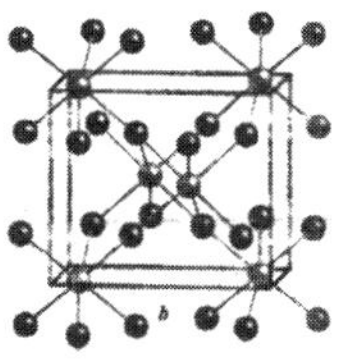

Figure 5.32. The **2·2PO$_{1/2}$(o)** structure of ReO_2.

(*Source*: Y. Matsushita, *Chalcogenide* crystal structure data library, Version 5.5B, Institute for Solid State Physics, The University of Tokyo, Tokyo, Japan, 2004; and *CrystalMaker*, by David Palmer, *CrystalMaker* Software Ltd., Begbroke Science Park, Bldg. 5, Sandy Lane, Yarnton, Oxfordshire, OX51PF, UK.)

Only half of the octahedral sites are occupied and the chains of octahedra along *c* are staggered to avoid close Fe neighbors at *C* positions. In Figure 5.31***b***, a projection of the cell along *c* shows the bending of octahedra to provide close S_2 units in the edges. The bond length (broken line) for the S_2 units is 2.21 Å compared to 3.22 Å for other S—S distances. The notation is **2·2PO$_{1/2}$(o)**.

5.4.3. The 2·2PO$_{1/2}$(o) Crystal Structure of ReO$_2$

The crystal structure of rhenium dioxide, ReO_2, is orthorhombic, $\mathbf{D}_{2h}^{14}$, *Pbcn*, $a = 4.8094$, $b = 5.6433$, and $c = 4.6007$ Å, with four molecules in the unit cell. There are chains of ReO_6 octahedra sharing edges along the *c* direction. Figure 5.32 shows the cell. Oxygen atoms are in *A* and *B* positions, with Re in *C* positions. The Re atoms are staggered along the packing direction (*a*) to avoid close Re—Re neighbors. The notation is **2·2PO$_{1/2}$(o)**.

5.4.4. The 2·2PO$_{1/2}$(o) Crystal Structure of Diaspore [AlO(OH)]

Diaspore, AlO(OH), is orthorhombic, *Pbnm*, $\mathbf{D}_{2h}^{16}$, $a = 4.4$, $b = 9.43$, and $c = 2.84$ Å. Figure 5.33***a*** is a projection of the unit cell showing AlO_6 octahedra sharing edges in pairs along *b*. The central unit is aligned along the *a* direction with vacancies on the *b* edges. Figure 5.33***b*** provides a perspective view of the cell. Oxygen atoms occur only at *A* and *B* positions so all octahedral sites are at *C* positions. Staggering of the rows of AlO_6 sharing edges along *c* avoids close neighbors for Al along the packing direction, *a*. Each oxygen atom is shared by two Al in equatorial positions and one Al in an apical position.

5.4.5. The 2·2PO$_{1/2}$(o) Crystal Structure of Cu(OH)$_2$

The crystal structure of $Cu(OH)_2$ is orthorhombic, $\mathbf{C}_{2v}^{12}$, $Cmc2_1$, $a = 2.947$, $b = 10.593$, and $c = 5.256$. Figure 5.34***a*** shows the unit cell. The CuO_6 octahedra share edges along *a*, and along *c*, an apical oxygen atom of one CuO_6 is an equatorial oxygen atom of a CuO_6 octahedron above or below. The Cu—O bond lengths are shorter for oxygens shared in edges. Figure 5.34***b*** shows packing layers with position. The packing direction is vertical. Oxygen atoms are in *A* and *B* positions and Cu atoms are only in *C* positions. To avoid close spacing

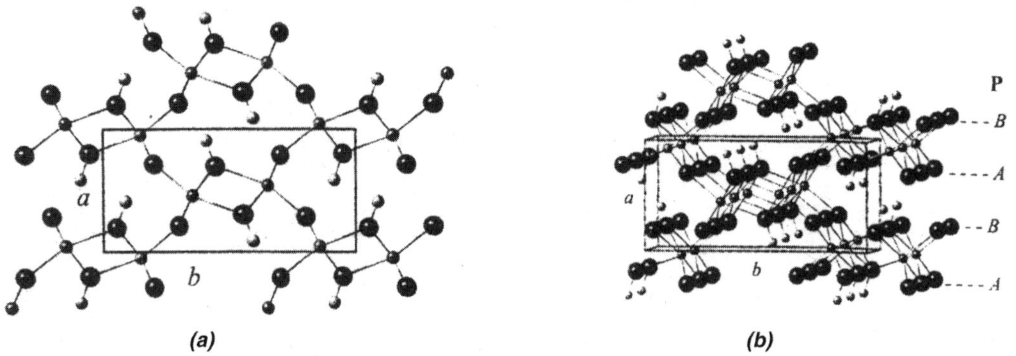

Figure 5.33. (*a*) A projection of the diaspore, AlO(OH), cell showing the AlO_6 octahedra sharing edges in pairs to form chains along *b*. (*b*) A perspective view of the cell showing additional sharing of oxygen atoms between AlO_6 octahedra.

(*Source*: *CrystalMaker*, by David Palmer, *CrystalMaker* Software Ltd., Begbroke Science Park, Bldg. 5, Sandy Lane, Yarnton, Oxfordshire, OX51PF, UK.)

along the packing positions, rows of the octahedral sites are staggered in a different pattern from that for AlO(OH) (previous section).

5.4.6. The $2{\cdot}3PO_{1/2}PO_{1/2}PO_{1/2}(t)$ Crystal Structure of $ZnSb_2O_6$

The structure of $ZnSb_2O_6$ has been described as a trirutile structure, considering it as three rutile-type cells stacked, one on another. Zn atoms fill one-half of sites in one **O** layer and Sb atoms fill one-half of sites in the next two **O** layers. The structure shown in Figure 5.35 shows partial occupancy by Zn and Sb in all **O** layers. Oxygen atoms fill all **P** layers (*A* and *B* positions) in an *hcp* pattern with all **O** sites at *C* positions. There are six repeating layers giving $\mathbf{2{\cdot}3PO_{1/2}PO_{1/2}PO_{1/2}}$(t). This structure is found for MSb_2O_6 type compounds of Mg, Fe, Co, Ni, and Zn and MTa_2O_6 type compounds of Mg, Co, and Ni. The tetragonal unit cell contains two molecules, $\mathbf{D}_{4h}^{14}$, $P4/mnm$, $a_o = 4.66$, and $c_o = 9.24$ Å.

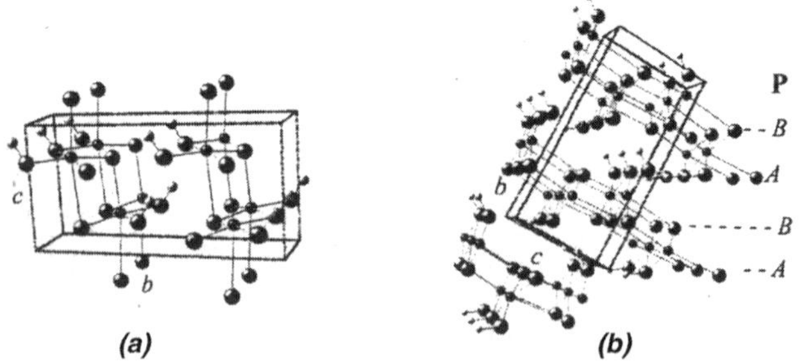

Figure 5.34. (*a*) A perspective view of the $Cu(OH)_2$ cell showing oxygen atoms forming elongated CuO_6 octahedra. (*b*) Another perspective view of the cell with **P** layers of oxygen atoms labeled.

(*Source*: *CrystalMaker*, by David Palmer, *CrystalMaker* Software Ltd., Begbroke Science Park, Bldg. 5, Sandy Lane, Yarnton, Oxfordshire, OX51PF, UK.)

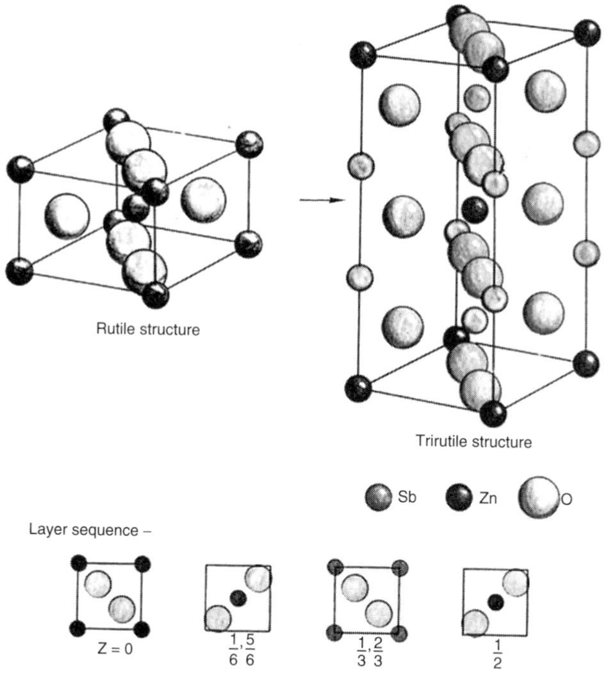

Figure 5.35. The structure of $ZnSb_2O_6$, shown as trirutile based on the rutile (TiO_2) structure.

(*Source*: F.S. Galasso, *Structure and Properties of Inorganic Solids*, Pergamon, Oxford, 1970, p. 47.)

5.4.7. The 2·6PO$_{2/3}$(h) Crystal Structure of Corundum [α-Al_2O_3]

The mineral corundum (α-Al_2O_3 or alumina) is important in many applications. Because of its great hardness it is used for watch bearings and abrasives. Its high melting point makes it important for refractories and its high dielectric constant makes it a useful insulator. The red color of rubies is caused by small amounts of Cr^{3+} substituted for Al^{3+} in corundum. Sapphires are corundum with small amounts of Fe, Ti, or other impurities giving pink, green, yellow, or, for those most prized, blue crystals. Star sapphires contain small amounts of TiO_2 crystals. Because of its importance the structure of corundum and other forms of Al_2O_3 and its hydrates have been investigated extensively. The corundum structure is found for oxides of Ti, V, Cr, Fe (α-Fe_2O_3), Ga, and Rh, γ-Al_2S_3 and Co_2As_3.

Corundum has oxide ions in **P** layers in an *hcp* pattern as shown in the projection along *c* in Figure 5.36***a***. Oxygen atoms are at *A* and *B* positions with Al in *C* positions. Al^{3+} ions fill two-thirds of the sites in **O** layers. The pattern for two-thirds occupancy of a layer is shown in Figure 3.9***c***. This is a hexagonal network with the vacancies at the centers of hexagons. The vacancies are staggered for each **O** layer added, requiring six **O** layers and a total of 12 **P** and **O** layers in the repeating unit. The result is that pairs of Al^{3+} ions are aligned in position in adjacent layers, but three Al^{3+} ions are never aligned. All Al^{3+} ions are at *C* positions for an *hcp* pattern. The corundum structure is described by the **2·6PO$_{2/3}$(h)** notation. The full pattern for layers is

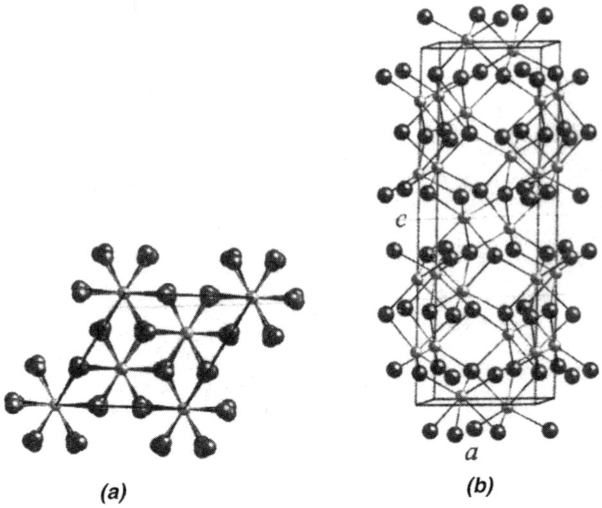

Figure 5.36. The $\mathbf{2{\cdot}6PO_{2/3}}$ structure of corundum (Al_2O_3). (***a***) A projection of the cell along *c* showing oxygen atoms at *A* and *B* positions and Al at *C* positions. (***b***) A perspective view of the cell showing the layers. Al atoms are not halfway between **P** layers.

(*Source*: *CrystalMaker*, by David Palmer, *CrystalMaker* Software Ltd., Begbroke Science Park, Bldg. 5, Sandy Lane, Yarnton, Oxfordshire, OX51PF, UK.)

shown in Figure 5.36***b***. Al atoms are not halfway between oxygen layers. They are shifted away from the neighbor in the adjacent layer and toward vacancies above or below. The rhombohedral unit cell contains two molecules, $\mathbf{D}^6_{3d}$, $R\bar{3}c$, $a_o = 5.128$ Å, and $\alpha = 55.33°$.

5.4.8. The $2{\cdot}2PO_{2/3}$(h) Crystal Structure of Hematite [α-Fe_2O_3]

The crystal structure of hematite, $\alpha - Fe_2O_3$, is trigonal, $\mathbf{D}^6_{3d}$, $R\bar{3}c$, $a = 5.038$, and $c = 13.772$ Å. In Figure 5.37***a*** the cell shows that oxygen atoms are in *A* and *B* positions. Fe are in octahedral sites at *C* positions. Pairs of octahedra along *c* share faces as seen in the side faces of the cell. In other directions edges are shared. Because the Fe atoms of face-bridged octahedra are at *C* positions they are shifted away from one another and toward vacant *C* sites above and below. Thus, the Fe atoms are displaced from the usual spacing of octahedral sites halfway between **P** layers. The long Fe—O distances are 2.12 Å and the short distances are 1.95 Å. Figure 5.37***b*** shows a projection along *c*, the packing direction, verifying that oxygen atoms are in *A* and *B* positions with Fe in *C* positions. The octahedral sites are two-thirds filled giving the notation $\mathbf{2{\cdot}2PO_{2/3}}$(**h**).

5.4.9. The $2{\cdot}6PO^{Ti}_{2/3}PO^{Fe}_{2/3}$(h) Crystal Structure of Ilmenite [$FeTiO_3$]

The structure of ilmenite, $FeTiO_3$, is essentially the same as that of corundum (α-Al_2O_3). The symmetry of the trigonal cell is lower

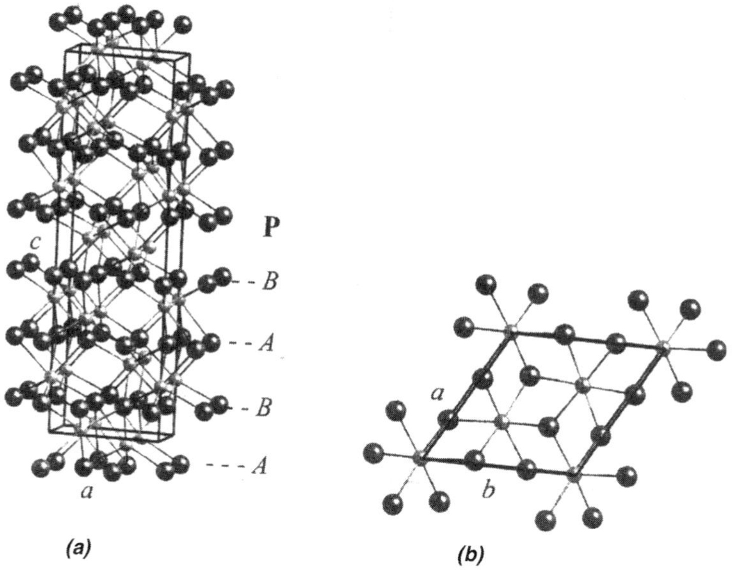

Figure 5.37. The **2·2PO$_{2/3}$(o)** structure of hematite, α-Fe_2O_3. **(*a*)** A perspective view of the cell with oxygen **P** layers labeled. Fe atoms are above or below the mid-portion between oxygen layers. **(*b*)** A projection of the cell along *c* showing the oxygen atoms in *A* and *B* positions with Fe atoms in *C* positions.

(*Source*: *CrystalMaker*, by David Palmer, *CrystalMaker* Software Ltd., Begbroke Science Park, Bldg. 5, Sandy Lane, Yarnton, Oxfordshire, OX51PF, UK.)

because there are two cations, C_{3i}^2, $R\bar{3}$, $a_o = 5.0884$, and $c = 14.0885$ Å. It is the most common Ti ore. The oxide ions fill **P** layers in an *hcp* pattern. The Fe^{2+} and Ti^{4+} ions each occupy two-thirds of sites in alternate **O** layers, all at *C* positions. There are Fe–Ti pairs in adjacent **O** layers, but there are vacant sites aligned with a Fe–Ti pair in the next adjacent **O** layers. The **2·6PO$_{2/3}$PO$_{2/3}$(h)** structure is shown in Figure 5.38*a*. The **O** layers are slightly puckered because some sites of an **O** layer have a metal ion above and some below at the same position. The average heights for the **O** layers along the *c* axis are:

Fe	−0.03-.03		0.31-.36		0.64-.69	
Ti		0.14-.19		0.47-.53		0.81-.86

The projection along *c* (Figure 5.38***b***) corresponds to that for α-Al_2O_3 (Figure 5.36***a***).

The ilmenite structure is found for $MMnO_3$ (M = Ni and Co), $MTiO_3$ (M = Mg, Mn, Ni, Co and Cd), $CrRhO_3$, $FeRhO_3$, $FeVO_3$, $LiNbO_3$, $MgGeO_3$, $NaSbO_3$, and $NaBiO_3$. For $MM'O_3$ compounds with large M cations the perovskite structure is encountered.

5.4.10. The 2·2PO$_{1/3\ 1/3}$(o) Crystal Structure of $LiSbO_3$

The structure of $LiSbO_3$ is similar to that of α-Al_2O_3, with one-third of sites occupied by Li and Sb each in all **O** layers. The oxide ions fill

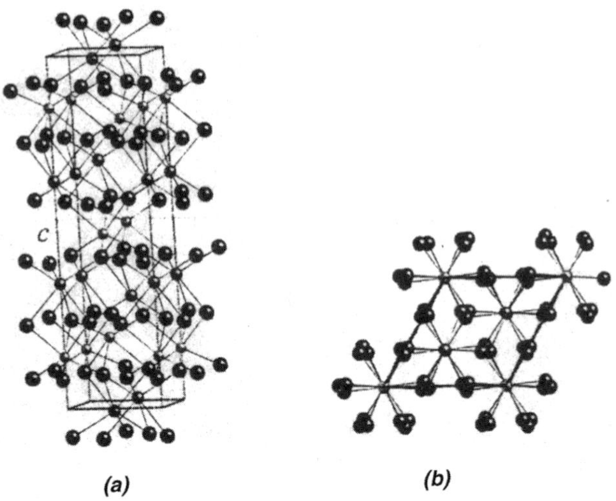

Figure 5.38. **(a)** A perspective view of Ilmenite, $FeTiO_3$, with Ti (smallest light atoms) and Fe (mid-size atoms) in alternating octahedral layers. ***(b)*** A projection of the cell along *c*. Fe and Ti are eclipsed.

(*Source*: *CrystalMaker*, by David Palmer, *CrystalMaker* Software Ltd., Begbroke Science Park, Bldg. 5, Sandy Lane, Yarnton, Oxfordshire, OX51PF, UK.)

P layers with an *hcp* pattern and all **O** sites are at *C* positions. The **O** sites occupied have the same hexagonal pattern (Figure 5.39***a***) found for α-Al_2O_3. The octahedra occupied by Sb atoms form chains with edges shared. LiO_6 octahedra are adjacent to SbO_6 octahedra, alternating from one side to the other (Figure 5.39***b***). There is a string of vacant **O** sites between each of the strings of SbO_6 and LiO_6 octahedra. For $LiSbO_3$ there are four molecules in the orthorhombic cell, $\mathbf{D}_{2h}^{6}$, *Pnna*, $a_o = 4.893$, $b_o = 8.491$, and $c_o = 5.183$Å, with four repeating packing layers, giving $\mathbf{2{\cdot}2PO_{1/3\ 1/3}(o)}$.

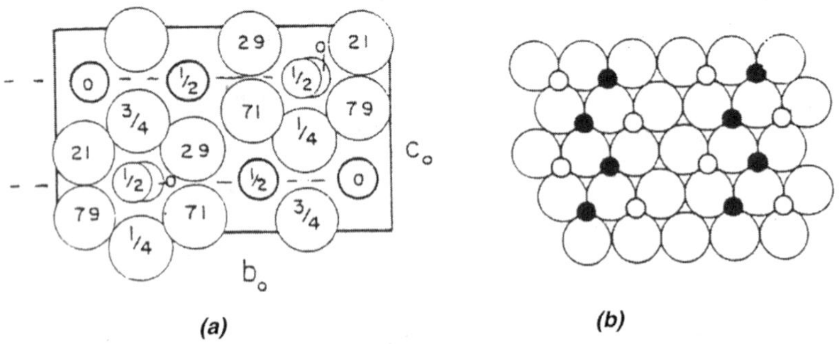

Figure 5.39. ***(a)*** A projection of the $LiSbO_3$ cell along the *a* axis. Sb atoms have heavy lines and Li are small circles. ***(b)*** A **P** layer of oxide atoms with chains of Sb (black circles) and Li (open circles).

(*Source*: R.W.G. Wyckoff, *Crystal Structures*, Vol. 2, 2nd ed., Wileys, New York, 1964, p. 420.)

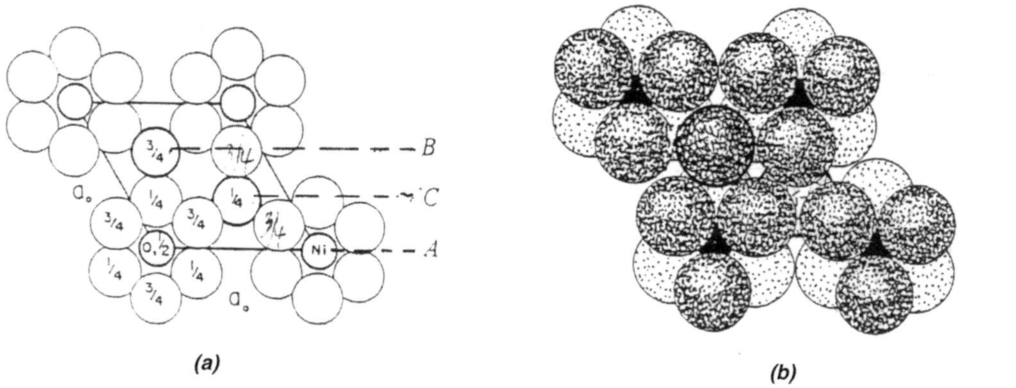

Figure 5.40. *(a)* A projection of the $BaNiO_3$ hexagonal cell on the base. The large heavy circles are Ba atoms. *(b)* A packing drawing along the *c* axis. Ni atoms are small and black and Ba atoms are line shaded.

(*Source*: R.W.G. Wyckoff, *Crystal Structures*, Vol. 2, 2nd ed., Wiley, New York, 1964, p. 418–419.)

5.4.11. The $2 \cdot 2\mathbf{P}_{1/4\ 3/4}\mathbf{O}_{1/4}$ Crystal Structure of $BaNiO_3$

The structure of $BaNiO_3$ is closely related to perovskite ($CaTiO_3$, $3 \cdot 2\mathbf{P}_{1/4\ 3/4}\mathbf{O}_{1/4}$), but the packing sequence is *hcp* rather than *ccp* as for perovskite. The hexagonal cell contains two molecules, C_{6v}^4, $C6_3mc$, $a_o = 5.580$ and $c_o = 4.832$ Å. As shown in Figure 5.40, each Ni atom is at the center of an octahedron formed by six O atoms. With **P** layers at *B* and *C* positions all Ni atoms in **O** layers are at *A* positions, sharing three O atoms with another Ni atom. The Ba atom is at the center of a trigonal prism formed by six Ni atoms. The packing sequence of this structure is shown below:

Height:	0	25	50	75
Position:	*A*	C	A	B
Atom:	Ni	Ba,O	Ni	Ba,O
Layer:	$\mathbf{O}_{1/4}$	$\mathbf{P}_{1/4\ 3/4}$	$\mathbf{O}_{1/4}$	$\mathbf{P}_{1/4\ 3/4}$

This structure is not common, but it is found for $BaTiS_3$, $SrTiS_3$, and the low-temperature form of $BaMnO_3$.

5.4.12. The $6 \cdot 2\mathbf{P}_{1/4/\ 3/4}\mathbf{O}_{1/4}$ Crystal Structure of $BaTiO_3$

Like γ′-MoC and TiP, $BaTiO_3$ has overlaping sequences corresponding to NiAs ($\mathbf{P}_A\mathbf{P}_B$ or here $\mathbf{P}_C\mathbf{P}_A$) and NaCl ($\mathbf{P}_A\mathbf{P}_B\mathbf{P}_C$ or here $\mathbf{P}_A\mathbf{P}_C\mathbf{P}_B$). As for $CaTiO_3$ (perovskite), Ba and O fill **P** layers with Ti occupying one-quarter of **O** sites. The sequence is a combination of that for perovskite ($3 \cdot 2\mathbf{P}_{1/4\ 3/4}\mathbf{O}_{1/4}$) and that for $BaNiO_3$($2 \cdot 2\mathbf{P}_{1/4\ 3/4}\mathbf{O}_{1/4}$). The full sequence as seen in Figure 5.41 is:

$$\mathbf{P}_C\ \mathbf{O}_B\ \mathbf{P}_A\ \mathbf{O}_B\ \mathbf{P}_C\ \mathbf{O}_A\ \mathbf{P}_B\ \mathbf{O}_C\ \mathbf{P}_A\ \mathbf{O}_C\ \mathbf{P}_B\ \mathbf{O}_A$$

hcp (from $\mathbf{P}_C$ to $\mathbf{P}_C$)   *hcp* (from $\mathbf{P}_B$ to $\mathbf{P}_B$)

ccp (from $\mathbf{P}_A$ to $\mathbf{P}_A$)

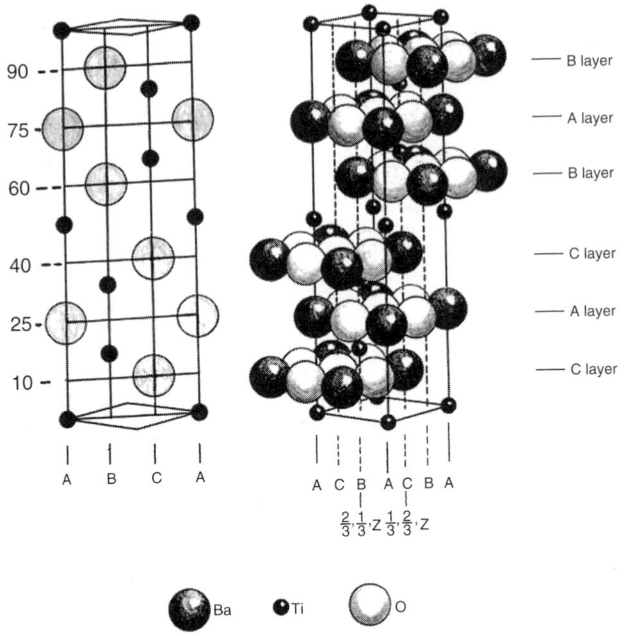

Figure 5.41. The structure of $BaTiO_3$. Note the *CACBAB* sequence of **P** layers. The spacing of layers is shown on the left in the vertical plane through the center of the cell.

(*Source*: F.S. Galasso, *Structure and Properties of Inorganic Solids*, Pergamon, Oxford, 1970, p. 203.)

There are six **P** layers and a total of 12 (**6•2**) repeating layers total, giving the notation $\mathbf{6{\cdot}2P}_{1/4\ 3/4}\mathbf{O}_{1/4}$. The spacing between **P** layers is 15 units for the NiAs regions ($\mathbf{P}_C\mathbf{O}_B\mathbf{P}_A\mathbf{O}_B\mathbf{P}_C$) with close pairs of Ti atoms (both *B* or *C* positions), while spacing is 20 units between **P** layers ($\mathbf{P}_C\mathbf{O}_A\mathbf{P}_B$) and these Ti atoms have no other close Ti atom. Other compounds with the $BaTiO_3$ structure are given in Table 5.9.

5.4.13. The $\mathbf{4{\cdot}2P}_{1/4\ 3/4}\mathbf{O}_{1/4}$ Crystal Structure of $BaMnO_3$

The structure of $BaMnO_3$ is the perovskite type ($CaTiO_3$, $\mathbf{3{\cdot}2P}_{1/4\ 3/4}\mathbf{O}_{1/4}$), but there are four **P** layers repeating. Figure 5.42 shows the structure and the heights of layers are given below.

Height:	0	11	25	39	50	61	75	89	100
Layer:	**P**	**O**	**P**	**O**	**P**	**O**	**P**	**O**	**P**
Position:	*A*	*B*	*C*	*B*	*A*	*C*	*B*	*C*	*A*

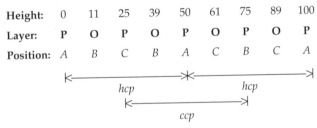

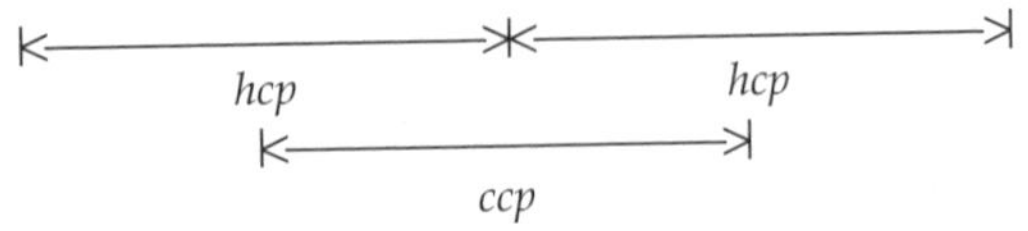

Like perovskite, Ba and O share **P** layers with Mn occupying one-quarter of each **O** layer. There are two sequences of **P** layers, *ACA* and *ABA*, of the *hcp* pattern, but they share one *A* layer, giving a *CAB* or *ccp* pattern. This results in four **P** layers (*ACAB*) in the repeating unit. This pattern puts Mn atoms in pairs at the same positions (*BB* and *CC*) in

TABLE 5.9. **Compounds with the $BaTiO_3$ structure.**

Ba_2CoOsO_6	Ba_2CrTaO_6	Ba_2FeSbO_6	Ba_2CrOsO_6	Ba_2InIrO_6	Ba_2MnOsO_6
Ba_2NiOsO_6	Ba_2ErIrO_6	Ba_2FeOsO_6	Ba_2RhUO_6	Ba_2ScIrO_6	
$Ba_3CoTi_2O_9$	$Ba_3FeTi_2O_9$	$Ba_3MoCr_2O_9$	$Ba_3IrTi_2O_9$	$Ba_3ReCr_2O_9$	$Ba_3MnTi_2O_9$
$Ba_3UCr_2O_9$	$Ba_3WCr_2O_9$	$Ba_3OsTi_2O_9$	$Ba_3RuTi_2O_9$	$Ba_3ReFe_2O_9$	
Ba(Fe, Ir)O_3	Ba(Cr, Ti)O_3	Ba(Rh, Ti)O_3	$Ba(Pt_{0.1}Ti_{0.9})O_3$		

adjacent **O** layers. The short Mn–Mn distance is 2.62 Å, compared to 2.7–2.8 Å in the metal. There are four molecules in the hexagonal cell. The space group is $\mathbf{D}_{6h}^4$, $P6_3/mmc$, $a_o = 5.669$, and $c_o = 9.375$ Å. There is a total of eight packing layers in the repeating unit, giving the notation $\mathbf{4{\cdot}2P_{1/4\,3/4}O_{1/4}}$.

5.4.14. The $2{\cdot}2PO_{1/3}$ Crystal Structure of Cs_3O

Cs_3O is a suboxide forming hexagonal crystals with two molecules per cell, $\mathbf{D}_{6h}^3$, $C6/mcm$, $a_o = 8.78$, and $c_o = 7.52$ Å. This is essentially an *hcp* structure, with Cs in **P** layers and O in one-third of the **O** sites. The notation is $\mathbf{2{\cdot}2PO_{1/3}}$. There are columns of Cs_6O octahedra, each sharing a pair of opposite faces. The Cs–O distance is 2.98 Å, similar to that in Cs_2O. The Cs–Cs distances in the strings of octahedra vary from 3.80–4.34 Å. The Cs–Cs distances between chains is ~ 5.8 Å. The structure is shown in Figure 5.43. The normal oxide, Cs_2O, has the $CdCl_2$ layer structure (Section 5.5.1). Stacking

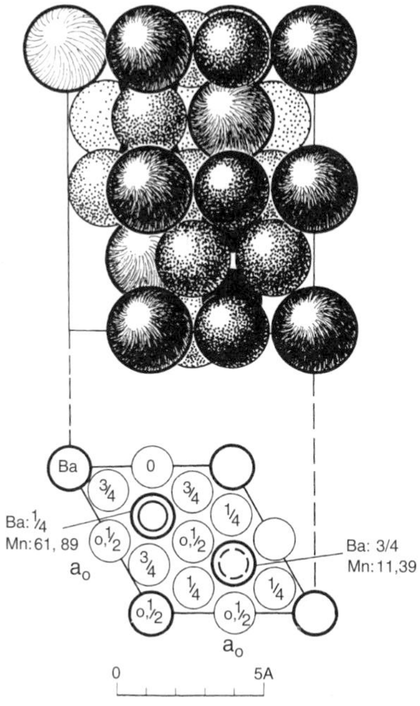

Figure 5.42. A projection on the base of the $BaMnO_3$ hexagonal cell. The small heavy circles are Mn and the larger ones are Ba. Above there is a packing drawing along the *c* axis. The Mn atoms in **O** layers are black. Oxygen atoms are dot shaded and smaller than the Ba atoms.

(*Source*: R.W.G. Wyckoff, *Crystal Structures*, Vol. 2, 2nd ed., Wiley, New York, 1964, p. 417.)

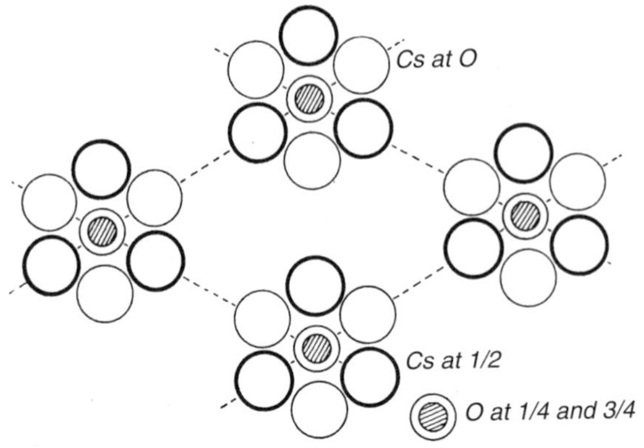

Figure 5.43. A projection on the base of the Cs_3O hexagonal cell.
(*Source*: A. F. Wells, *Structural Inorganic Chemistry*, 3rd ed., Oxford UP, Oxford, 1962, p. 536.)

filled Cs layers without O atoms between them indicates the polarizability of Cs atoms.

5.4.15. The $2{\cdot}2PO_{2/9}PO_{1/9}$ Crystal Structure of UCl_6

The structure of UCl_6 is similar to that of Cs_3O (previous section), but the roles of ions are reversed. The space group is $\mathbf{D}_{3d}^3$, $P\bar{3}m1$, $a_o = 10.97$, and $c_o = 6.04$ Å. For UCl_6, the structure corresponds to the more common situation with Cl atoms in **P** layers (A and B) with U atoms in **O**

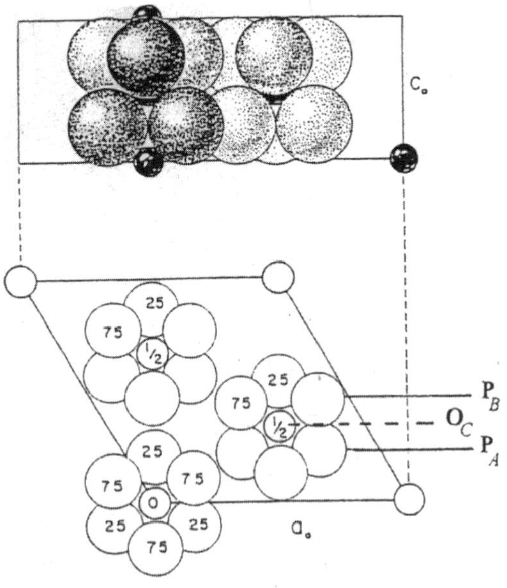

Figure 5.44. A projection on the base of the UCl_6 hexagonal cell. U atoms are smaller. Above is a packing drawing along the c axis.
(*Source*: R.W.G. Wyckoff, *Crystal Structures*, Vol. 2, 2nd ed., Wiley, New York, 1964, p. 205.)

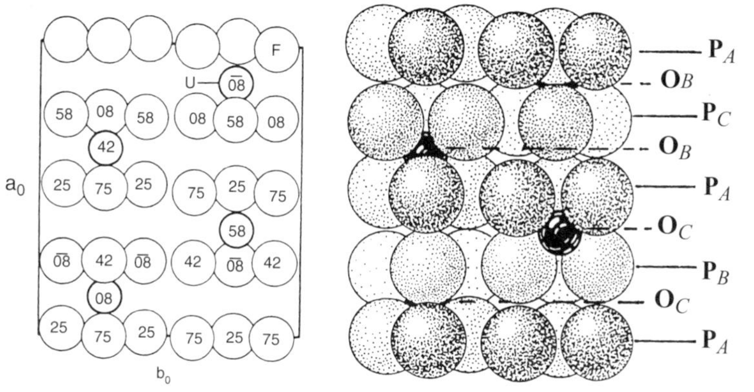

Figure 5.45. A projection of the structure of UF_6 along the *c* axis and a packing drawing with layers identified. Note the *ABAC* sequence for **P** layers.
(*Source*: R.W.G. Wyckoff, *Crystal Structures*, Vol. 2, 2nd ed., Wiley, New York 1964, p. 207.)

layers (*C* positions). From Figure 5.44 we see that there are two U atoms inside the cell at 1/2 and there are eight U atoms at the corners at 0 and 100, giving $8 \times 1/8 = 1$ U. Thus, one **O** layer has twice as many U atoms as the other, giving four layers in the repeating unit or $\mathbf{2{\cdot}2PO_{2/9}PO_{1/9}}$. The two **P** layers are filled with Cl atoms (position *A* at 25 and position *B* at 75). There are three UCl_6 molecules in the cell.

5.4.16. The $4{\cdot}2PO_{1/6}(o)$ Crystal Structure of UF_6

UF_6 is orthorhombic with four molecules per cell, $\mathbf{D}_{2h}^{16}$, *Pnma*, $a_o = 9.900$, $b_o = 8.962$, and $c_o = 5.207$ Å. The F^- ions are close packed with four **P** layers repeating, $\mathbf{P}_A\ \mathbf{P}_B\ \mathbf{P}_A\ \mathbf{P}_C \cdots$. This double hexagonal close-packing pattern is shown in Figure 5.45. The U^{VI} ions occupy one-sixth of each **O** layer. The **O** sites occur in pairs at the same positions (*B,B* and *C,C*) as shown. This staggering and partial filling avoids U^{VI} ions aligned in adjacent layers, giving $\mathbf{4{\cdot}2PO_{1/6}(o)}$. The octahedra are somewhat distorted with one U—F bond longer. The compound $OsOF_5$ has the same structure.

5.4.17. The $2{\cdot}2PO_{1/3}$ Crystal Structure of Zirconium Trihalides

The compounds $ZrCl_3$, $ZrBr_3$, and ZrI_3 have structures similar to that of NiAs (**2·2PO**) except that only one-third of octahedral sites are occupied in each **O** layer by the Zr^{3+} ions. The notation is $\mathbf{2{\cdot}2PO_{1/3}}$. The ZrX_6 octahedra are all at *C* positions forming chains of octahedra sharing faces. As shown in Figure 5.46, the *C* positions adjacent to each Zr^{3+} ion in the same layer are vacant. The Zr–Zr distance for ZrI_3 is comparable to that in the metal. The layers are somewhat distorted because there is Zr–Zr interaction in occupied **O** sites and none for empty sites. The Zr halides are hexagonal with two molecules in the cell, $\mathbf{D}_{6h}^{3}$, $P6_3/mcm$, for $ZrCl_3$, $a_o = 6.36$ and $c_o = 6.14$ Å, for $ZrBr_3$, $a_o = 6.75$ and $c_o = 6.315$ Å, and for ZrI_3, $a_o = 7.25$, and $c_o = 6.64$ Å. HfI_3 has the same structure.

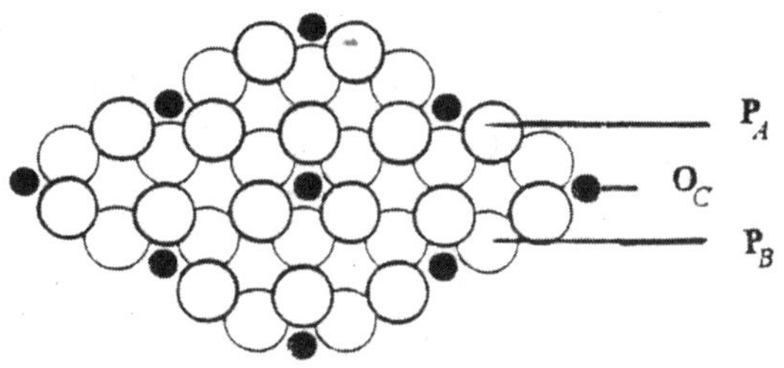

Figure 5.46. The **2·2PO$_{1/3}$** structure of $ZrCl_3$. Two superimposed close-packed layers are shown. Zr atoms (black) occupy chains of octahedral sites running parallel to the packing direction (c).

5.4.18. The 2·2PO$_{1/3}$ Crystal Structures of VF_3 and AlF_3

The VF_3 structure is found for trifluorides of Ti, Cr, Fe, Co, Ga, Mo, Ru, Rh, Ir, and Pd. The metal ions occupy one-third of **O** sites with F^- ions in the filled **P** sites in an *hcp* pattern as shown in Figure 5.47. The hexagonal cell contains six molecules, $\mathbf{D}_{3d}^6$, $R\bar{3}c$, $a_o = 5.170$, and $c_o = 13.402$ Å. The planes of V^{3+} and F^- ions are equally spaced and stacked normal to the principal axis. V^{3+} ions, at *C* positions, are at the center of an almost regular octahedron of six F^- ions. The heights along the c_o axis and positions are:

Height:	0	8	17	25	33	42	49	58	66	75	83	92
Position:	O_C	P_A	O_C	P_B	O_C	P_A	O_C	P_B	O_C	P_A	O_C	P_B

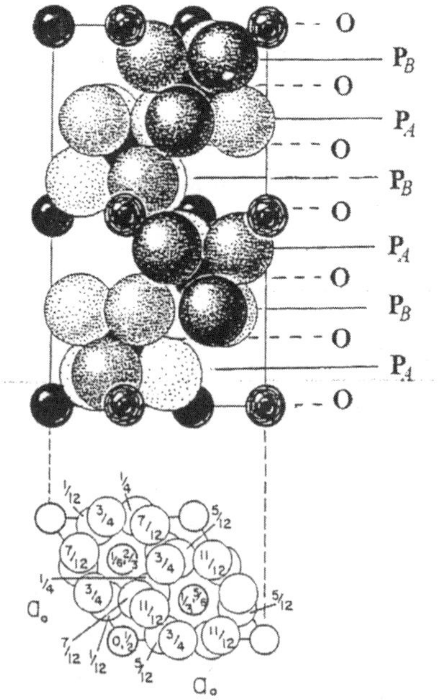

Figure 5.47. A projection on the base of the VF_3 hexagonal cell and a packing drawing with layers labeled (V atoms are small and black).

(*Source:* R.W.G. Wyckoff, *Crystal Structures*, Vol. 2, 2nd ed., Wiley, New York, 1964, p. 49.)

TABLE 5.10. **Compounds in the Cr–S solid phase system.**

Compound	System	Notation
CrS	Monoclinic	**2·2PO**
Cr_7S_8	Hexagonal	**2·2POPO$_{3/4}$**
Cr_5S_6	Hexagonal	**2·4POPO$_{2/3}$**
Cr_3S_4	Monoclinic	**2·4POPO$_{1/2}$**
α-Cr_2S_3	Hexagonal	**2·6POPO$_{1/3}$**
β-Cr_2S_3	Rhombohedral	**2·6POPO$_{1/3}$**

The **P** positions are slightly shifted among layers. The notation for VF_3 is **2·2PO$_{1/3}$**.

The structure of AlF_3 can be described by the same notation as VF_3, **2·2PO$_{1/3}$**, but there is more distortion ($\mathbf{D}_3^7$, R32; R.W.G. Wycoff, *Crystal Structures*, vol. 2. 2nd ed., p. 47, Wiley, New York, 1964). The octahedron around each Al^{3+} ion is distorted with three F^- ions at greater distance (1.89 Å) than the other three F^- (1.70Å).

5.4.19. The PO-type Crystal Structures of Chromium Sulfides

Studies of solid phases of the Cr–S system reveal six compounds of composition from CrS to Cr_2S_3. Except for CrS, the structures are similar with S atoms close packed (*hcp*) and with Cr atoms filling the **O** layers incompletely (see Table 5.10). The structure of CrS is intermediate between that of NiAs (**2·2PO**) and PtS (**3·3P T$_{1/2}$T$_{1/2}$**, Section 6.3.6). The notation for Cr_7S_8 is given as **2·2POPO$_{3/4}$**, but there are vacancies in one of the two **O** layers as required by the stoichiometry. Here we consider only Cr_5S_6 and β-Cr_2S_3 in more detail.

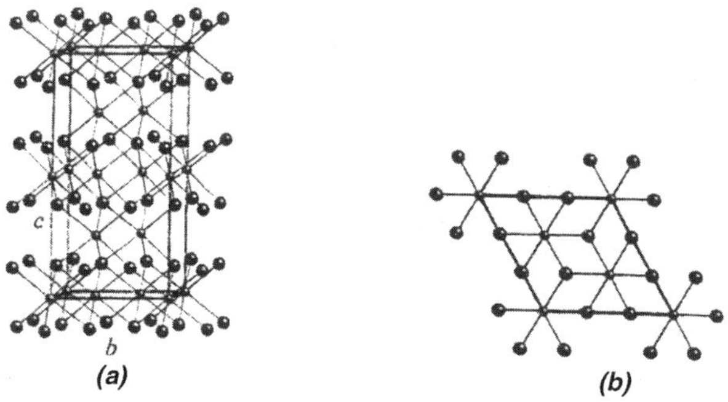

Figure 5.48. *(a)* The Cr_5S_6 hexagonal cell. *(b)* A projection of the cell along the *c* axis.

(*Source*: Y. Matsushita, *Chalcogenide* crystal structure data library, Version 5.5B, Institute for Solid State Physics, The University of Tokyo, Tokyo, Japan, 2004; and *CrystalMaker*, by David Palmer, *CrystalMaker* Software Ltd., Begbroke Science Park, Bldg. 5, Sandy Lane, Yarnton, Oxfordshire, OX51PF, UK.)

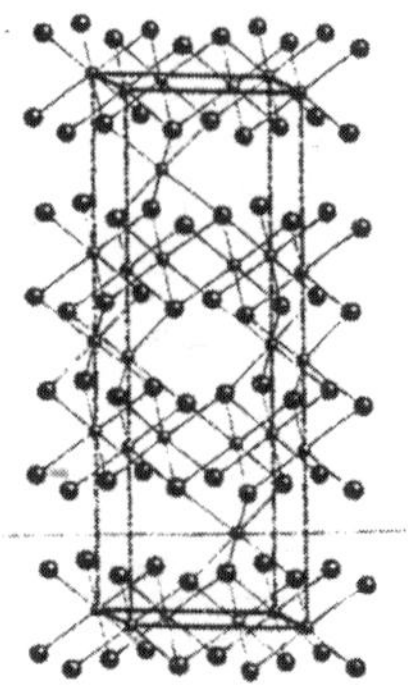

Figure 5.49. The β-Cr_2S_3 hexagonal cell.

(*Source*: Y. Matsushita, *Chalcogenide* crystal structure data library, Version 5.5B, Institute for Solid State Physics, The University of Tokyo, Tokyo, Japan, 2004; and *CrystalMaker*, by David Palmer, *CrystalMaker* Software Ltd, Begbroke Science Park, Bldg. 5, Sandy Lane, Yarnton, Oxfordshire, OX51PF, UK.)

5.4.19.1. The 2·4POPO$_{2/3}$ Crystal Structure of Cr_5S_6

The structure of Cr_5S_6, as shown in Figure 5.48, has a regular *hcp* pattern for S in **P** layers. Cr atoms are in **O** layers with alternate layers full and two-thirds filled. All **O** sites are at C positions. The hexagonal cell has two molecules, $\mathbf{D}^4_{3d}$, P$\bar{3}$1*c*, $a_o = 5.939$, and $c_o = 11.192$ Å. There are eight layers in the repeating unit giving **2·4POPO$_{2/3}$**.

5.4.19.2. The 2·6POPO$_{1/3}$ Crystal Structure of β-Cr_2S_3

There are two structures of Cr_2S_3: β-Cr_2S_3 is trigonal. The α-Cr_2S_3 has the same space group as Cr_5S_6 with alternate **O** layers only one-third filled. There are 12 layers repeating. The notation is the same (**2·6POPO$_{1/3}$**) for α-Cr_2S_3 and β-Cr_2S_3 structures. The structure for β-Cr_2S_3 is shown in Figure 5.49, $\mathbf{C}^2_{3i}$, R$\bar{3}$, and for the hexagonal cell, $a = 5.937$ and $c = 16.698$ Å.

5.4.20. The 2·2PO$_{1/3\ 1/3}$ Crystal Structure of $LiIO_3$

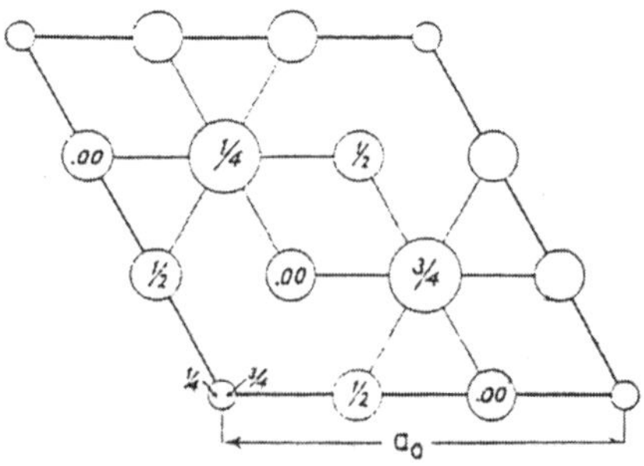

Figure 5.50. A projection on the base of the $LiIO_3$ hexagonal cell. The largest atoms are I and the smallest ones are Li.

(*Source*: R.W.G. Wyckoff, *Crystal Structures*, Vol. 2, 2nd ed., Wiley, New York, 1964, p. 387.)

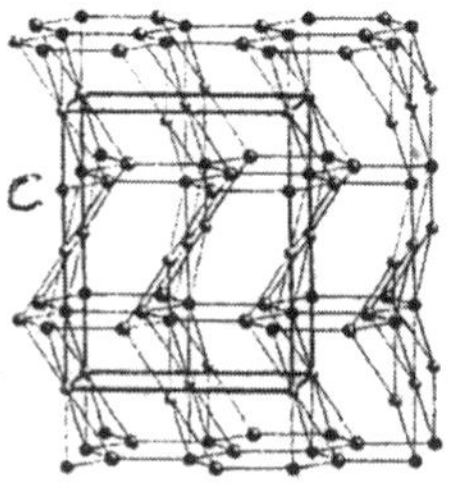

Figure 5.51. The orthohrombic cell of AgCuS. S atoms are largest in **P** layers with Cu (dark). Ag atoms are between **P** layers.

(*Source*: Y. Matsushita, *Chalcogenide* crystal structure data library, Version 5.5B, Institute for Solid State Physics, The University of Tokyo, Tokyo, Japan, 2004; and *CrystalMaker*, by David Palmer, *CrystalMaker* Software Ltd., Begbroke Science Park, Bldg. 5, Sandy Lane, Yarnton, Oxfordshire, OX51PF, UK.)

The metal iodates, and halates generally, contain discrete XO_3^- ions. The structure of $LiIO_3$ is very unusual in that there are no discrete IO_3^- ions. The structure and heights of ions are shown in Figure 5.50. The oxygen atoms fill the *hcp* **P** layers. Both iodine and Li atoms each occupy one-third of sites of the **O** layers (at *C* positions) and both iodine and Li atoms are at centers of six oxygen atoms forming octahedra. Each oxygen atom shared by two I atoms is equidistant from the I atoms. The crystal structure is hexagonal with two molecules per cell, $\mathbf{D}_6^6$, C6$_3$2, $a_o = 5.469$, and $c_o = 5.155$ Å. The notation is $\mathbf{2{\cdot}2PO_{1/3\ 1/3}(h)}$ and the stoichiometry makes it clear that oxygen atoms occupy the **P** layer. The packing sequences are shown below:

Height:	0	25	50	75
Atom:	O	Li, I	O	Li, I
Layer:	P_A	$O_{1/3\ 1/3}$	P_B	$O_{1/3\ 1/3}$

5.4.21. The $2{\cdot}2P_{1/3\ 1/3}O_{1/3}(o)$ Crystal Structure of Stromeyerite, AgCuS

The mineral stromeyerite, AgCuS, has four molecules in the orthorhombic cell, $\mathbf{D}_{2h}^{17}$, *Cmcm*, $a_o = 4.06$, $b_o = 6.66$, and $c_o = 7.99$ Å. The structure is simple but very curious as shown in Figure 5.51. The **P** layer is only partly filled, one-third by Cu and one-third by S. The graphite-like layer consists of hexagons formed by three Cu and three S. Each Cu has three close S neighbors in the same plane. The Ag atoms occupy one-third of the octahedral sites, those having six S neighbors forming very skewed octahedra. The packing sequence is *AB* (*hcp*) with Ag atoms at *C* positions. There are zigzag S–Ag–S chains along the packing direction. There are four layers in the repeating unit giving the notation $\mathbf{2{\cdot}2P_{1/3\ 1/3}O_{1/3}(o)}$.

5.4.22. The $2{\cdot}2PO_{2/3}PO_{1/3}$ Crystal Structure of $CoI_2{\cdot}6H_2O$

The structure of cobalt(II) iodide hexahydrate is hexagonal with one molecule per cell, $\mathbf{D}_{3d}^1$, $P\bar{3}1m$, $a_o = 7.26$, and $c_o = 4.79$ Å. Figure 5.52 does not show H atoms. The positions for layers of atoms are:

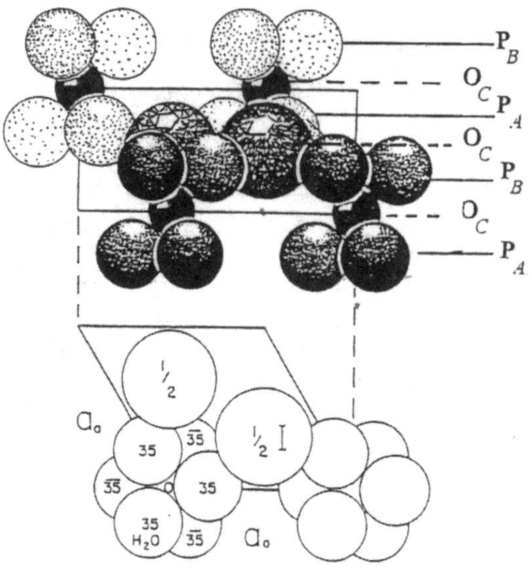

Figure 5.52. A projection on the base of the hexagonal cell of $CoI_2{\cdot}6H_2O$ and a packing drawing along the *c* direction. Co atoms are small and black, I atoms are largest. H atoms are not shown. See text for occupancies of layers.

(*Source*: R.W.G. Wyckoff, *Crystal Structures*, Vol. 3, 2nd ed., Wiley, New York, 1965, p. 791.)

Height:	0	35	50	65	100
Atoms:	Co	O	I	O	Co
Layers:	$\mathbf{O}_C$	$\mathbf{P}_A$	$\mathbf{O}_C$	$\mathbf{P}_B$	$\mathbf{O}_C$

The oxygen atoms are in $\mathbf{P}_A$ and $\mathbf{P}_B$ layers. The compound is $[Co(H_2O)_6]I_2$; as expected, the Co atoms occupy one-third of $\mathbf{O}_C$ layers. The layer of I atoms are halfway between Co layers at *C* positions so it is an **O** layer also with two-thirds of the sites filled. The sequence of layers repeating is $|\mathbf{P}_A\mathbf{O}_C^{I}\mathbf{P}_B\mathbf{O}_C^{Co}|$ so the notation is $\mathbf{2{\cdot}2PO_{2/3}PO_{1/3}}$.

5.4.23. The $2{\cdot}2PO_{2/3}PO_{1/3}$ Crystal Structure of Sb_2PbO_6

The structure of Sb_2PbO_6 is a typical *hcp* pattern of oxygen atoms with antimony and lead atoms in octahedral sites at C positions (**O** layers). The antimony occupies one-third of one **O** layer and lead fills two-thirds of the next one. The heights of layers are:

Height:	29	50	71	100
Atom:	O	Sb	O	Pb
Layer:	$\mathbf{P}_A$	$\mathbf{O}_{2/3}$	$\mathbf{P}_B$	$\mathbf{O}_{1/3}$

The **O** layers are halfway between the **P** layers, but the spacing is greater for the larger Pb atoms (see Figure 5.53). The hexagonal unit cell has one molecule, $\mathbf{D}_3^1$, P312, $a_o = 5.287$, and $c_o = 5.364$ Å. There are four packing units in the repeating unit giving the notation

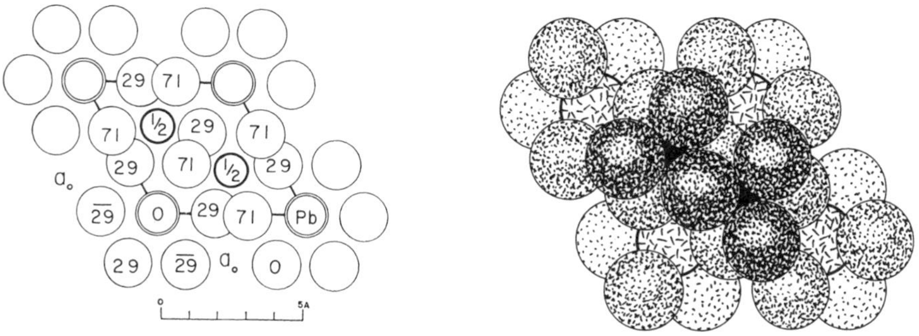

Figure 5.53. A projection on the hexagonal base of the Sb_2PbO_6 cell (Sb atoms are the smallest circles) and a packing drawing along the *c* axis. Pb atoms are doubly outlined and Sb atoms are heavily outlined or black.

(*Source*: R.W.G. Wyckoff, *Crystal Structures*, Vol. 3., 2nd ed., Wiley, New York, 1965, p. 365.)

2·2PO$_{2/3}$PO$_{1/3}$. Compounds having this structure include those of the As_2MO_6 type (M = Cd, Co, Hg, and Sr) and those of the Sb_2MO_6 type (M = Ca, Cd, Hg, and Sr).

5.4.24. The 2·2P$_{1/4\ 3/4}$O$_{1/2}$(o) Crystal Structure of Ag_2CsI_3

The structure of Ag_2CsI_3 is orthorhombic with four molecules per cell, $\mathbf{D}_{2h}^{16}$, *Pnma*, $a_o = 11.08$, $b_o = 13.74$, and $c_o = 6.23$ Å. As shown in Figure 5.54, each **P** layer is shared by the large Cs and I atoms in the ratio 1:3 (**P**$_{1/4\ 3/4}$). Each Cs has the usual arrangement of 12 Cl for an *hcp* structure, six in a hexagon in the same plane as Cs with three Cl above and below. The small silver atoms occupy one-half of each **O** layer (**O**$_{1/2}$), but each Ag atom is at the center of a tetrahedron formed by two I atoms from **P** layers above and below. This unusual tetrahedral arrangement results from the distortion of the **P** layers because of packing of atoms of different size. The notation is **2·2P$_{1/4\ 3/4}$O$_{1/2}$(o)**. The spacing of the layers (Ag at 0 and 50) identifies these as **O** layers, not **T** layers. The tetrahedra in **T** layers in the **PTOT** scheme are

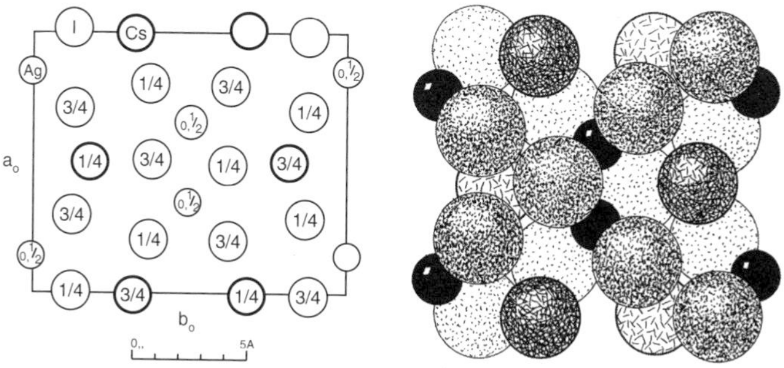

Figure 5.54. The **3·2P$_{1/4\ 3/4}$O$_{1/2}$(o)** structure of Ag_2CsI_3. A projection along the *c* axis and a packing drawing of the same view. Ag atoms are black and I atoms are largest dotted spheres.

(*Source*: R.W.G. Wyckoff, *Crystal Structures*, Vol. 2., 2nd ed., Wiley, New York, 1965, p. 499.)

formed by three atoms in one **P** layer and one in the next **P** layer. This structure could be described as a $\mathbf{2{\cdot}2P_{1/4\ 3/4}T_{1/2}}(C_2)(\mathbf{o})$ structure, with a C_2 axis of the tetrahedra aligned with the packing direction. This arrangement requires the Ag sites to be halfway between **P** layers.

5.5. Layer Structures (POP)

5.5.1. The 3·3POP(h) Layer Crystal Structure of $CdCl_2$

The structure of $CdCl_2$ is hexagonal with three molecules per cell ($\mathbf{D}_{3d}^5$, $R\bar{3}m$, $a = 3.85$, and $c = 17.46$ Å). There is a *ccp* pattern of **P** layers filled by Cl^- ions. The Cd^{2+} ions fill **O** layers in only *alternate* **O** layers (see Figure 5.55*a*). The **P–O–P** (Cl–Cd–Cl) "sandwiches" are stacked without cations between sandwiches. For $CdCl_2$, the negative charges of Cl^- layers are polarized inward toward the Cd^{2+} layer to such an extent that the interaction between the Cl^- layer of one sandwich and the Cl^- layer of the next sandwich is comparable to the van der Waals interaction between CCl_4 molecules. The repeating sequence is given below:

$$|\mathbf{P}_A\mathbf{O}_C\mathbf{P}_B\ \mathbf{P}_C\mathbf{O}_B\mathbf{P}_A\ \mathbf{P}_B\mathbf{O}_A\mathbf{P}_C|$$

The **O** sites are at different positions in the repeating unit for *ccp*. The sandwiches consist of $CdCl_6$ octahedra sharing six edges to form sheets. The top and bottom edges of the octahedra are not shared. The notation is **3·3POP(h)**, showing nine repeating layers and a

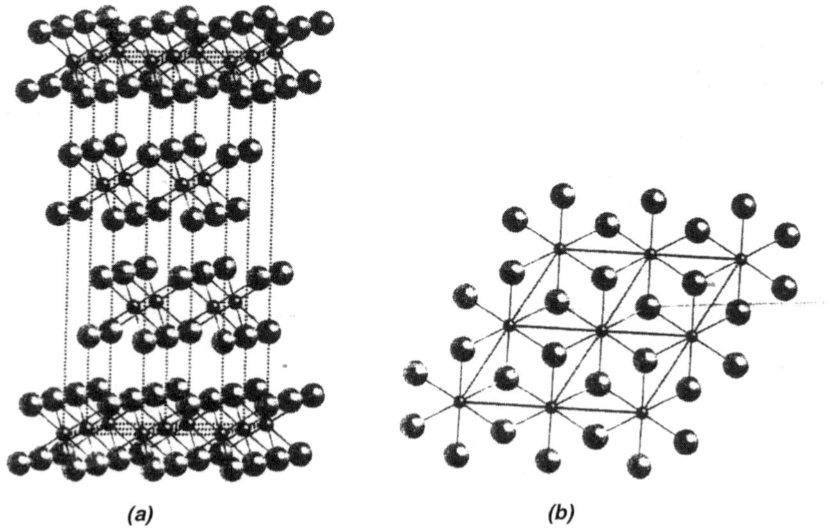

Figure 5.55. ***(a)*** A perspective view showing the $CdCl_2$ cell with alternate **O** layers filled by Cd. ***(b)*** A projection of the cell of one Co–Cd–Cl sandwich showing the regular octahedra.

(*Source*: *CrystalMaker*, by David Palmer, *CrystalMaker* Software Ltd., Begbroke Science Park, Bldg. 5, Sandy Lane, Yarnton, Oxfordshire, OX51PF, UK.)

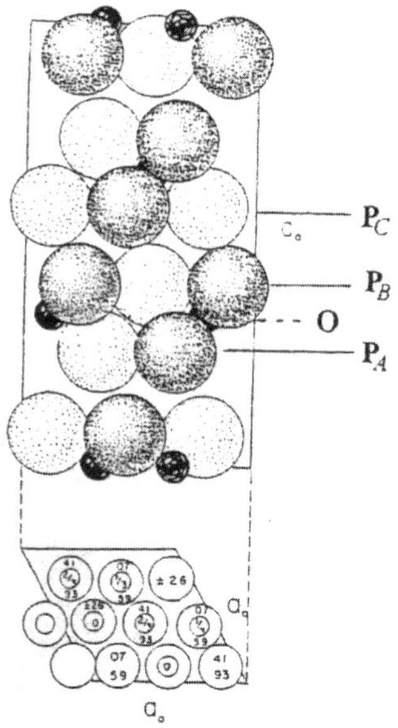

Figure 5.56. A projection on the base of the hexagonal cell of $CrCl_3$ and a packing drawing along the *c* axis (Cr atoms are black). Alternate **O** layers are two-thirds filled by Cr.

(*Source*: R.W.G. Wyekoff, *Crystal Structures*, Vol. 2, 2nd ed., Wiley, New York, 1964, p. 54.)

rhombohedral cell. Figure 5.55***b*** is a projection of one Cl–Cd–Cl sandwich. The notation for a layer structure can be given as **POP** or **PPO**. Repetition of either of these gives the full sequence given above. **POP** emphasizes the sandwich packing.

The $CdCl_2$ structure is found for the chlorides of Mg, Mn, Co, Ni, and Zn; the bromides of Ni and Zn; the iodides of Ni, Zn and Pb. It is interesting that Cs_2O has the anti-$CdCl_2$ structure, with the roles of the ions reversed. Here the polarization of the large Cs^+ ions by O^{2-} ions permits packing of adjacent Cs layers without anions between them.

5.5.2. The $3{\cdot}3PO_{2/3}P(h)$ Layer Crystal Structure of $CrCl_3$

The structure of $CrCl_3$ is hexagonal with six molecules per cell, D_3^5, $P3_212$, $a_o = 6.00$, and $c_o = 17.3$ Å. The packing of layers is the same as for $CdCl_2$ except the alternate **O** layers are occupied by Cr^{3+} in two-thirds of the sites (see Figure 5.56). Each octahedron shares three edges in the plane of the **POP** sandwich. The **P** layers are in an *ABC* sequence with nine layers in the repeating unit, giving $\mathbf{3{\cdot}3PO_{2/3}P(h)}$ for the notation. The $CrCl_3$ structure is found for CrI_3, $RuCl_3$ and $TiCl_3$. Ordinary chromium chloride is a hydrate. The solid is green, and gives green solution containing Cr complexes with water and one or two Cl^- ions coordinated. The solution slowly becomes violet, containing $[Cr(H_2O)_6]^{3+}$. Heating hydrated $CrCl_3$ gives the oxide and HCl. Anhydrous $CrCl_3$ is pink and feels slippery, like graphite.

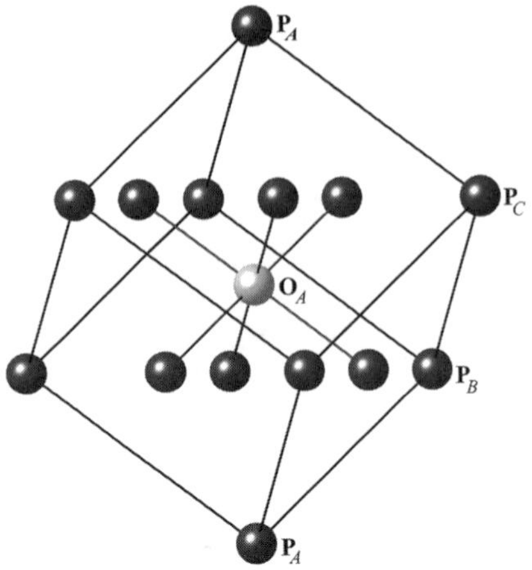

Figure 5.57. The cubic cell for Fe_4N, a face-centered cube (*ccp*) of Fe with N at the center.

5.5.3. The 3·3$PO_{1/2}$P Layer Crystal Structure of Fe_4N

Iron nitride, Fe_4N, has a *ccp* arrangement of Fe atoms with an N atom at the center of the face-centered cube (see Figure 5.57). The cubic cell contains one molecule, $a_o = 3.7885$ Å. This places N in an octahedral layer. It fills one-half of sites in alternate **O** layers. The packing layer sequence is more complicated than the simple figure:

$$|\mathbf{P}_A\mathbf{P}_B\mathbf{O}_A\mathbf{P}_C\mathbf{P}_A\mathbf{O}_C\mathbf{P}_B\mathbf{P}_C\mathbf{O}_B|$$

The notation is **3·3$PO_{1/2}$P**, showing that there is a *ccp* sequence, nine layers repeating, and alternate **O** layers are one-half filled by N. This structure is found for Mn_4N, Ni_4N, and Fe_3NiN.

5.5.4. The 2·3/2POP Layer Crystal Structure of CdI_2

The hexagonal cell of CdI_2 contains one molecule, C^1_{3v}, P3*m*1, $a_o = 4.24$, and $c_o = 6.84$ Å. The I–Cd–I "sandwich" layers or sheets of CdI_6 octahedra are arranged in the same way as those of $CdCl_2$. For CdI_2, the sandwiches are stacked with an *hcp* pattern for **P** layers (Figure 5.58) rather than *ccp* as for $CdCl_2$. The repeating unit is

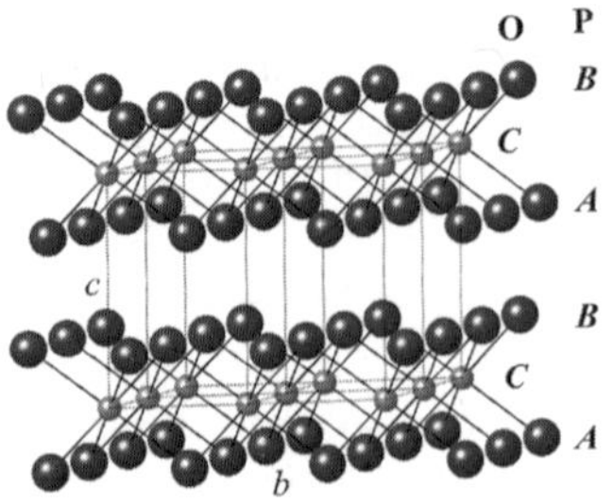

Figure 5.58. The **2·3/2POP** layer structure of CdI_2.

(*Source*: *CrystalMaker*, by David Palmer, *CrystalMaker* Software Ltd., Begbroke Science Park, Bldg. 5, Sandy Lanc, Yarnton, Oxfordshire, OX51PF, UK.)

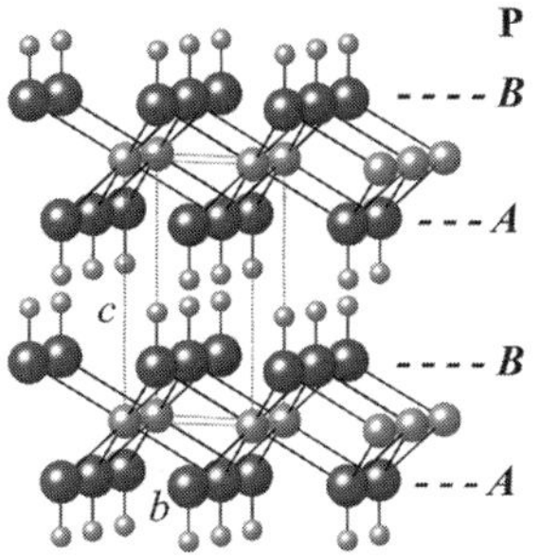

Figure 5.59. The **2·3/2POP** structure of brucite, $Mg(OH)_2$. A perspective view showing the layers. Mg atoms are dark small balls. H atoms are small light balls.

(*Source*: *CrystalMaker*, by David Palmer, *CrystalMaker* Software Ltd., Begbroke Science Park, Bldg. 5, Sandy Lane, Yarnton, Oxfordshire, OX51PF, UK.)

$\mathbf{P}_A\mathbf{O}_C\mathbf{P}_B$. Cd atoms are only at *C* positions. The notation is **2·3/2POP**; the 2 of the index indicates an *hcp* pattern for **P** layers, and the product **2·3/2** indicates that there are three repeating layers. The CdI_2 structure is found for bromides of Cd, Fe, Co, and Ni; iodides of Mg, Ca, Mn, Fe, Co, Zn, and Pb; the hydroxides of Ca, Cd, Mn, Fe, Co, and Ni, and sulfides of Ti, Pt, and Sn.

5.5.5. The 2·3/2POP Layer Crystal Structure of Brucite [$Mg(OH)_2$]

Brucite, $Mg(OH)_2$, has a structure very similar to that of CdI_2. The trigonal unit cell contains one molecule, $\mathbf{D}_{3d}^3$, $P\bar{3}m1$, $a_o = 3.142$, and $c_o = 4.766$ Å. Figure 5.59 shows the layers. Hydrogen atoms are on the outside of the **POP** sandwiches. Mg^{2+} ions are only at *C* positions, but alternate **O** layers are vacant.

5.5.6. The 2·3/2POP Layer Crystal Structure of $(NH_4)_2SiF_6$

The hexagonal unit cell of $(NH_4)_2SiF_6$ contains one molecule, $\mathbf{D}_{3d}^3$, $P\bar{3}m1$, $a_o = 5.76$, and $c_o = 4.77$ Å. The space group is the same as that of $Mg(OH)_2$. The structure, as shown in Figure 5.60, corresponds to those and of CdI_2 and $Mg(OH)_2$ with NH_4^+ ions replacing I or

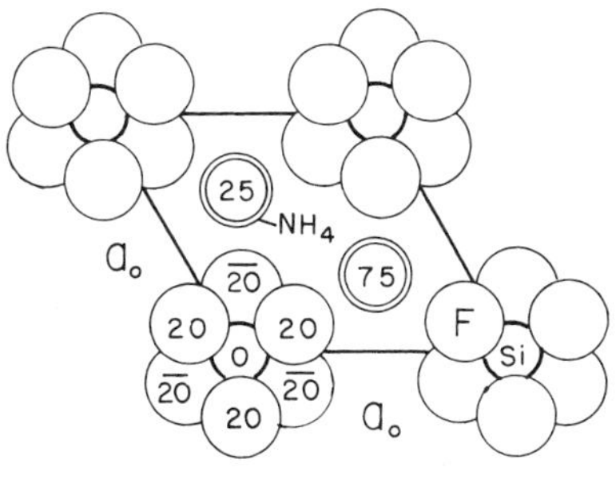

Figure 5.60. A projection on the base of the hexagonal cell of $(NH_4)_2SiF_6$.

(*Source*: R.W.G. Wyckoff, *Crystal Structures*, Vol. 3, 2nd ed., Wiley, New York, 1965, p. 351.)

TABLE 5.11. Compounds with the $(NH_4)_2SiF_6$ structure.

$(NH_4)_2TiF_6$	K_2RuF_6	Rb_2ReF_6	Cs_2ReF_6
$(NH_4)_2GeF_6$	K_2PtF_6	Rb_2PtF_6	Cs_2PtF_6
Na_2PuF_6	K_2GeF_6	Rb_2GeF_6	Cs_2RuF_6
K_2TiF_6	Rb_2TiF_6	Cs_2TiF_6	Cs_2ThF_6
K_2MnF_6	Rb_2ZrF_6	Cs_2ZrF_6	Cs_2UF_6
K_2ReF_6	Rb_2HfF_6	Cs_2HfF_6	Cs_2PuF_6
Tl_2TiF_6			

OH and SiF_6^{2-} replacing Cd or Mg. The roles of cations and anions are reversed. The **O** sites in alternate layers are filled by SiF_6^{2-} and each SiF_6^{2-} ion is surrounded by six NH_4^+ ions. There are many favorable situations for strong hydrogen bonds. Other compounds with this structure are listed in Table 5.11. In Section 8.2.9, the structure is considered in terms of individual Si, F, and NH_4^+ units, giving puckered double layers of NH_4^+ and F^-.

5.5.7. The 2·3/2POP(m) Layer Crystal Structure of Calaverite [$AuTe_2$]

$AuTe_2$ occurs as the mineral calaverite. The monoclinic cell contains two molecules, C_{2h}^3, $C2/m$, $a_o = 7.18$, $b_o = 4.40$, and $c_o = 5.07$ Å and $\beta = 90^\circ$. The sequence of packing is the same as CdI_2, but the cell is twice as large. Each Au is in an octahedral arrangement of Te atoms, but two Te atoms are closer giving linear Te–Au–Te molecules in the *ac* planes at the same heights as seen in Figure 5.61. There are six layers in the repeating unit giving the notation **2·3/2POP(m)** for the monoclinic structure.

5.5.8. The 2·3/2POP(m) Layer Crystal Structure of $CuBr_2$

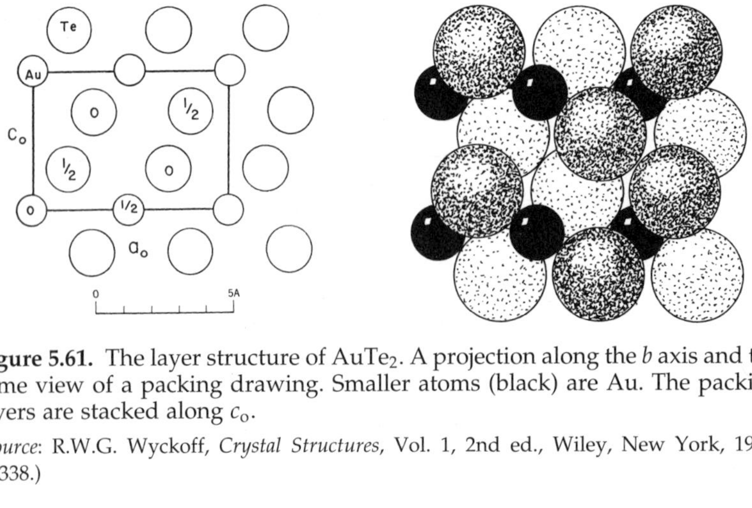

Figure 5.61. The layer structure of $AuTe_2$. A projection along the *b* axis and the same view of a packing drawing. Smaller atoms (black) are Au. The packing layers are stacked along c_o.

(*Source*: R.W.G. Wyckoff, *Crystal Structures*, Vol. 1, 2nd ed., Wiley, New York, 1963, p. 338.)

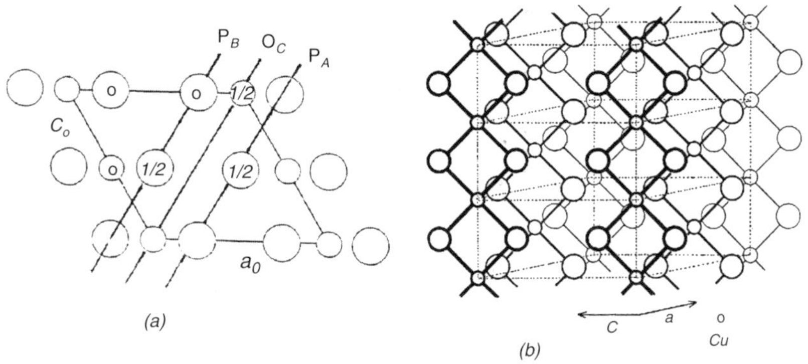

Figure 5.62. ***(a)*** A projection on the base of the $CuBr_2$ hexagonal cell. The larger atoms are Br. ***(b)*** A view of $CuBr_2$ showing the chains of square planar $CuBr_4$ units.

(*Source*: R.W.G. Wyckoff, *Crystal Structures*, Vol. 1, 2nd ed., Wiley, New York, 1963, p. 346.)

The structure of $CuBr_2$ (also $CuCl_2$) is monoclinic with two molecules per cell, $\mathbf{C}_{2h}^3$, B2/m, $a_o = 7.14$, $b_o = 3.46$, $c_o = 7.18$ Å, and $\beta = 121.25^\circ$. Four Br atoms of the six Br neighbors give planar [$CuBr_4$] units in the *ab* planes as seen in Figure 5.62***a***. Each Cu of a planar [$CuBr_4$] has one more distant Br atom of another [$CuBr_4$] unit above and one below completing a distorted octahedron around Cu. In this way the [$CuBr_4$] units form staggered chains, as seen in Figure 5.62***b***. The packing layers are shown in the figure. The **P** layers are at *A* and *B* positions and Cu fills $\mathbf{O}_C$ layers giving the notation is **2·3/2POP(m)**.

5.5.9. The 2·3/2POP(m) Layer Crystal Structure of Tl_2O

Thallium(I) oxide, Tl_2O, has a monoclinic modification, $\mathbf{C}_{2h}^3$, $C2/m$, $a = 6.082$, $b = 3.52$, $c = 13.24$ Å, and $\beta = 108.2^\circ$, with four molecules in the unit cell. Figure 5.63 shows the Tl atoms in close-packed layers in an *AB* sequence. Alternate **O** layers are filled by oxygen atoms giving the notation **2·3/2POP(m)**. The octahedra share edges. This is an unusual structure with oxygen atoms in octahedra. The Tl–O distances are 2.511–2.531 Å. The Tl–Tl distances within a plane are 3.514 Å, within

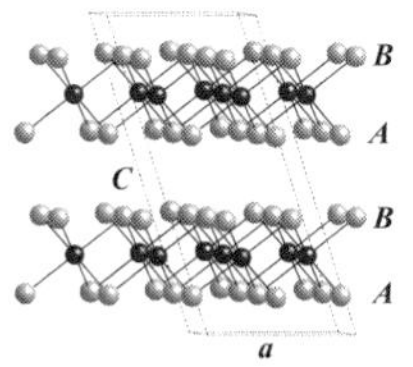

Figure 5.63. The monoclinic cell of Tl_2O.

(*Source*: Y. Matsushita, *Chalcogenide* crystal structure data library, Version 5.5B, Institute for Solid State Physics, The University of Tokyo, Tokyo, Japan, 2004; and *CrystalMaker*, by David Palmer, *CrystalMaker* Software Ltd., Begbroke Science Park, Bldg. 5, Sandy Lane, Yarnton, Oxfordshire, OX51PF, UK.)

the octahedra 3.622 and 3.641 Å and between adjacent Tl layers without oxygen atoms between are 3.862 Å. Tl–Tl distances in hexagonal elemental Tl are 3.410 and 3.456 Å and in the body-centered cubic form 3.352 and 3.871 Å.

There is a trigonal form of Tl_2O, $\mathbf{D}_{3d}^5$, $R\bar{3}m$, $a = 3.516$, and $c = 37.84$ Å. The large c results from a complex sequence of positions of Tl atoms. It is *ABAB CACA BCBC*. It is a **POP** sandwich structure with the notation **12·3/2POP(h)**.

5.5.10. The 3·5/3POPPO(h) Crystal Structure of La_2O_3

The lanthanum oxide crystal structure is known as the A-type M_2O_3 or A-type rare earth structure. The crystals are hexagonal with one molecule per cell, $\mathbf{D}_{3d}^3$, Cm, $a_o = 3.9373$, and $c_o = 6.1299$ Å. Although the unit cell is hexagonal, this is not an *hcp* sequence because oxygen atoms occur at *A*, *B*, and *C* positions. Figure 5.64***a*** shows that there are triple layers in the sequence with the following heights along c_o:

Heights:	0	24	37	63	76	100
Atoms:	O	La	O	O	La	O
Positions(1):	$\mathbf{P}_A$	$\mathbf{O}_C$	$\mathbf{P}_B$	$\mathbf{P}_C$	$\mathbf{O}_B$	$\mathbf{P}_A$
Positions(2):	$\mathbf{O}_C$	$\mathbf{P}_A$	$\mathbf{T}_B$	$\mathbf{T}_A$	$\mathbf{P}_B$	$\mathbf{O}_C$

For positions (1), there are adjacent oxygen layers only for $\mathbf{P}_B$ (at 37) and $\mathbf{P}_C$ (at 63). If this were to be designated as a **ℙ** structure it would require pairs of adjacent oxygen layers on each side of the La layers. There would be six repeating layers, but there are only five repeating layers. All layers are filled. There are two tripled **POP** layers with one **P** layer shared. The notation is **3·5/3POPPO(h)**, indicating an *ABC* sequence but only five layers repeating. La atoms are in distorted octahedral sites because of the unequal spacings of near oxygen layers. La atoms are only in *B* and *C* positions with the $\mathbf{O}_A$ layers empty. The *CN* of La is seven with a seventh oxygen at the same position as La

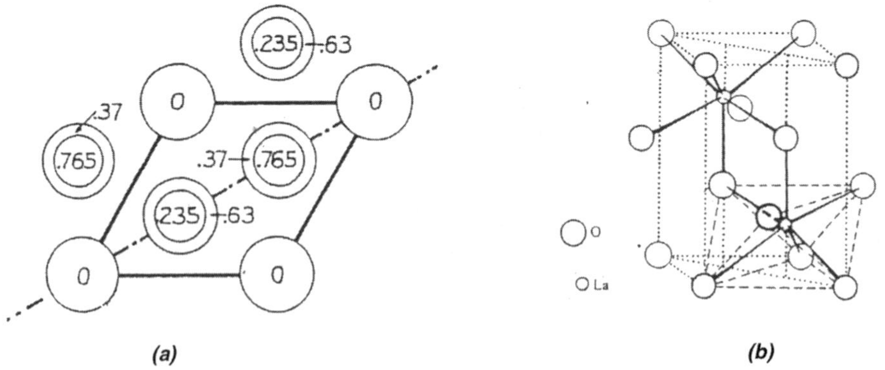

Figure 5.64. ***(a)*** A projection on the base of the hexagonal unit of La_2O_3. La atoms are the smaller circles and heights are along *c*. ***(b)*** A pictorial view of the structure showing ***CN*** 7 for La.

(*Source*: R.W.G. Wyckoff, *Crystal Structures*, Vol. 2, 2nd ed., Wiley, New York, 1964, p. 2.)
(*Source*: A.F. Wells, *Structural In Organic Chemistry*, 3rd ed., Oxford UP, Oxford, 1962, p. 546.)

TABLE 5.12. **Compounds with the hexagonal La_2O_3 structure.**

Ac_2O_3	Pr_2O_3	La_2O_2Se	Sm_2O_2Se	β-Al_2S_3	Mg_3Sb_2
Ce_2O_3	β-Am_2O_3	Pr_2O_2Se	Ho_2O_2Se	Th_2N_3	
Nd_2O_3	Pu_2O_3	Nd_2O_2Se	Yb_2O_2Se	Mg_3Bi_2	

above or below an octahedral face, (Figure 5.64*b*). The oxygen atoms in layers $\mathbf{P}_A$ layers have six La atoms in an octahedral arrangement while those in $\mathbf{P}_B$ and $\mathbf{P}_C$ layers have a regular tetrahedral arrangement of four La atoms. Each oxygen of $\mathbf{P}_B$ and $\mathbf{P}_C$ layers caps an LaO_6 octahedron.

This layered structure is different from $CdCl_2$, CdI_2, and $CrCl_3$ because the oxygen atoms of adjacent **P** layers are bonded to La atoms giving *CN* 7. Many oxides and other compounds (see Table 5.12) have this structure. Y_2O_2S, Pu_2O_2S, and all Ln_2O_2S compounds (Ln represents a lanthanide ion) except Pm_2O_2S have the La_2O_3 structure. Smaller lanthanide metals are found in the C–M_2O_3 structure with $CN = 6$ for the metal ion (Section 6.3.7).

Another description of the La_2O_3 structure places La in $\mathbf{P}_A$ and $\mathbf{P}_B$ layers (Positions 2) with oxygen atoms in both **T** layers between $\mathbf{P}_A$ and $\mathbf{P}_B$ layers and oxygen filling $\mathbf{O}_C$ layers between $\mathbf{P}_B$ and $\mathbf{P}_A$ layers. There are five layers repeating, $\mathbf{P}_A^{La}$ $\mathbf{T}_B$ $\mathbf{T}_A$ $\mathbf{P}_B^{La}\mathbf{O}_C$, giving the notation $\mathbf{2{\cdot}5/2P}_A^{La}\mathbf{TTP}_B^{La}\mathbf{O}_C$. This describes the two roles of oxygen atoms and this description requires that LaO_6 octahedra are capped to give *CN* 7 for La. The spacings of layers for La in **P** layers are expected for the

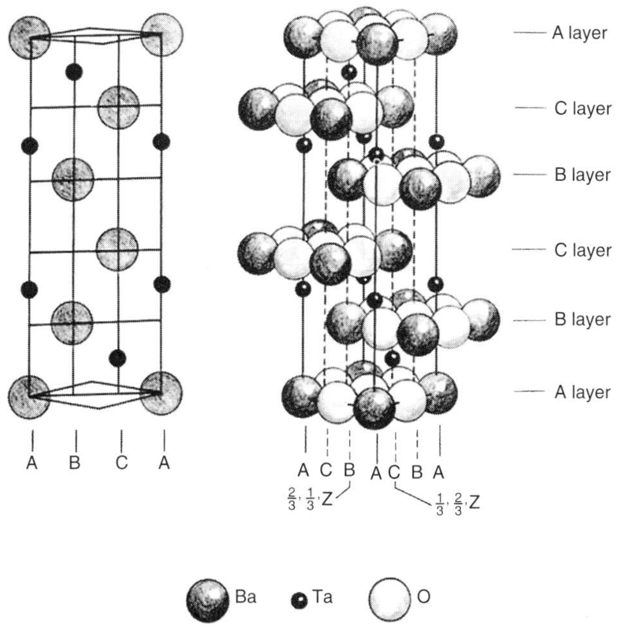

Figure 5.65. The layer structure of $Ba_5Ta_4O_{15}$ with the **P** layers labeled. On the left there is a plane through the center of the cell.

(*Source*: F.S. Galasso, *Structure and Properties of Inorganic Solids*, Pergamon, Oxford, 1970, p. 201.)

PTOT scheme. $\mathbf{O}_C$ layers are exactly halfway between **P** layers and the distance of each **T** layer from the nearest **P** layer is one-quarter of the distance between **P** layers.

5.5.11. The Layered Crystal Structure of $Ba_5Ta_4O_{15}$

The structure of $Ba_5Ta_4O_{15}$ is similar to that of perovskite ($CaTiO_3$, $\mathbf{3{\cdot}2P_{1/4\ 3/4}O_{1/4}}$; Section 5.3.4), but it is a double-decker sandwich or double-layered structure. The hexagonal cell is unimolecular, $\mathbf{D}^3_{3d}$, $P\bar{3}m1$, $a_o = 5.79$, and $c_o = 11.75$ Å. The Ba and oxygen atoms fill **P** layers in a 1:3 ratio. There are five **P** layers in the sequence, $\mathbf{P}_A\ \mathbf{P}_B\ \mathbf{P}_C\ \mathbf{P}_B\ \mathbf{P}_C$. Ta atoms occupy one-quarter of sites in four of the five **O** layers in the repeating unit with one **O** layer empty. This produces two double deckers $|\mathbf{P}_A\ \mathbf{O}_C\ \mathbf{P}_B\ \mathbf{O}_A\ \mathbf{P}_C|$ and $|\mathbf{P}_B\ \mathbf{O}_A\ \mathbf{P}_C\ \mathbf{O}_B\ \mathbf{P}_A|$ giving $|\mathbf{P}_A\ \mathbf{O}_C\ \mathbf{P}_B\ \mathbf{O}_A\ \mathbf{P}_C\ \mathbf{P}_B\ \mathbf{O}_A\ \mathbf{P}_C\ \mathbf{O}_B|$ with the **O** layer between the double deckers unoccupied (see Figure 5.65). The Ta atoms are at centers of octahedral arrangements of six oxygen atoms. The index is **5·9/5**, indicating that there are 5 **P** layers and a total of nine layers in the repeating unit. The notation $\mathbf{5{\cdot}9/5P_{1/4\ 3/4}\ O_{1/4}P_{1/4\ 3/4}\ O_{1/4}\ P_{1/4\ 3/4}(h)}$ agrees with the stoichiometry for the nine layers, giving as sums 5/4 Ba, 4/4 Ta, and 15/4 O; multiplying by 4 to convert to integers, we get $Ba_5Ta_4O_{15}$.

Chapter 6

Crystal Structures Involving P and T Layers

Because interstitial **T** sites are smaller than **O** sites, structures with only **P** and **T** layers, compared to **PO** structures, are usually encountered for smaller cations and those with more covalent character and more highly directional bonding patterns.

6.1. Cubic Close-Packed PT Structures

6.1.1. The 3·2PT Crystal Structure of Zinc Blende [ZnS]

There are two polymorphic structures of ZnS, zinc blende (or sphalerite) (**3·2PT**) and wurtzite (**2·2PT**). In zinc blende there is a *ccp* arrangement of S atoms with Zn atoms filling one of the two **T** layers as shown in Figure 6.1. The diamond has the same structure, with the sites of **P** and one **T** layer filled by C atoms (Section 4.3.3). The structure of zinc blende has six (**3·2**) layers in the repeating unit. This structure is encountered for many binary compounds with significant covalent character as shown in Table 6.1. The space group for zinc blende is $\mathbf{T}_d^2$, F$\bar{4}$3m, and $a_o = 5.4093$ Å, for the cubic unit cell .

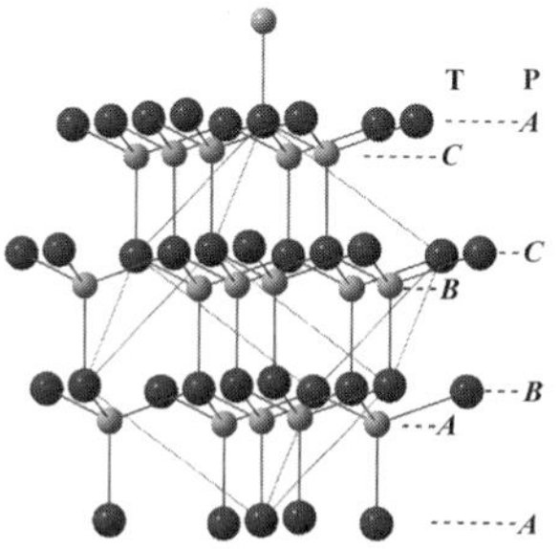

Figure 6.1. The **3·2PT** structure of zinc blende (ZnS). The **P** layers (S) are dark balls.

(*Source*: *CrystalMaker*, by David Palmer, *CrystalMaker* Software Ltd., Begbroke Science Park, Bldg. 5, Sandy Lanc, Yarnton, Oxfordshire, OX51PF, UK.)

6.1.2. The 3·2PT Crystal Structure of I_c Ice

Ice I_c is cubic, $\mathbf{O}_h^7$, $Fd3m$, $a = 6.35$ Å with eight molecules in the unit cell. It is a **3·2PT** structure like that of the zinc blende with oxygen atoms replacing S and Zn in both **P** and **T** layers. It can also be described as related to β-cristobalite (Section 10.1.2). I_c ice crystallizes from vapor at −120 to −140°C. In Figure 6.2, the hydrogen atoms are at midpoints between oxygen atoms.

6.1.3. The $3{\cdot}2PT_{2/3}$ Crystal Structure of γ-Ga_2S_3

The low-temperature form of Ga_2S_3 is γ-Ga_2S_3, with the $\mathbf{3{\cdot}2PT_{2/3}}$ structure. It is cubic ($a_o = 5.441$ Å) and similar to the zinc blende with one **T** layer only two-thirds filled. This deficit cubic structure is encountered for Ga_2Se_3, Ga_2Te_3, α-In_2S_3, and α-In_2Te_3. Cubic γ-Ga_2S_3, after sintering at 1,000°C for 12 days, changes to α-Ga_2S_3 with the *hcp* wurtzite structure. The space group for the hexagonal cell of α-Ga_2S_3 (with six molecules) is $\mathbf{C}_6^3$, $P6_5$, $a_o = 6.389$, and $c_o = 18.086$ Å. For metals, transformation from *ccp* to *bcc* structures is common, and transformation from *ccp* to *hcp* is less common. For a **PT** structure, **P** and **T** layers involve *A*, *B*, and *C* positions for a *ccp* pattern, but only *A* and *B* positions for *hcp*. Perhaps the transformation from *ccp* to *hcp* is facilitated with **T** vacancies.

6.1.4. The Crystal Structures of Metal Sulfides Related to the Zinc Blende, $3{\cdot}4PT_{1/2\ 1/2}(t)$ [$CuFeS_2$] and $3{\cdot}4PT_{1/2\ 1/4\ 1/4}(t)$ [Cu_2FeSnS_4]

The mineral chalcopyrite ($CuFeS_2$) has the $\mathbf{3{\cdot}4PT_{1/2\ 1/2}(t)}$ structure. Chalcopyrite (*Gr*. Copper pyrite) is the most important ore for Cu. Like pyrite, it is known as fool's gold because of its yellow color and bright luster. It has the zinc blende structure, with Cu and Fe atoms each filling one-half of the **T** layer as shown in Figure 6.3***a***. The mineral stannite (Cu_2FeSnS_4) has the related structure shown in the figure. For Cu_2FeSnS_4 (Figure 6.3***b***) there is only one **T** layer filled between *ccp* **P** layers of S atoms. In the figure alternate **T** layers are filled by Cu atoms and by Fe and Sn atoms, but these are not packing layers: packing layers are along body diagonal directions. All **T** packing layers are mixed, giving the notation $\mathbf{3{\cdot}4PT_{1/2\ 1/4\ 1/4}(t)}$. In both minerals all S and metal atoms have tetrahedral configurations. For both

TABLE 6.1. **Compounds with the zinc blende (3·2PT) structure.**

AgI	BP	CuBr	GaSb	InSb
AlAs	BeS	CuCl	HgS	MnS
AlP	BeSe	CuF	HgSe	MnSe
AlSb	BeTe	CuI	HgTe	SiC
BAs	CdS	GaAs	InAs	ZnSe
BN	CdTe	GaP	InP	ZnTe

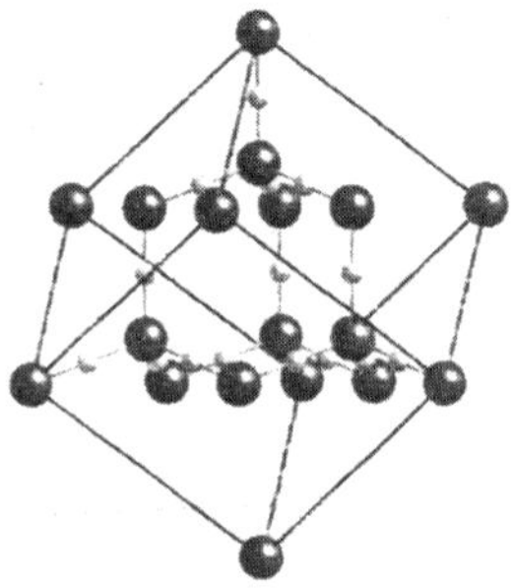

Figure 6.2. The **3•2PT** structure of $\mathbf{I}_c$ ice. Oxygen atoms are in **P** and **T** sites. H atoms are shown at mid-points between O atoms.

(*Source*: *CrystalMaker*, by David Palmer, *CrystalMaker* Software Ltd., Begbroke Science Park, Bldg. 5, Sandy Lane, Yarnton, Oxfordshire, OX51PF, UK.)

structures the space group of the tetragonal unit cell (with four molecules) is $\mathbf{D}_{2d}^{12}$, $I\bar{4}2d$, $a_o = 5.24$, and $c_o = 10.30$ Å.

6.1.5. The $PT_{2/4\ 1/4}$(t) Crystal Structures of $M_2M'X_4$ Compounds

The $M_2M'X_4$ compounds considered here are related to that of the zinc blende as shown in Figure 6.4. All of these have *ccp* anions (Se^{2-} or I^-). In the In_2CdSe_4 structure the Cd atoms are at the corners of the cell ($1/8 \times 8 = 1$Cd), the In atoms are in the centers of four faces ($1/2 \times 4 = 2$ In), and there are four Se atoms within the cell giving In_2CdSe_4. Two **T** sites in the top and bottom faces are vacant. One **T** layer between **P** layers is occupied, one-half filled by In, one-quarter filled by Cd, and one-quarter empty. The notation is **3•2PT**$_{2/4\ 1/4}$(**t**). The

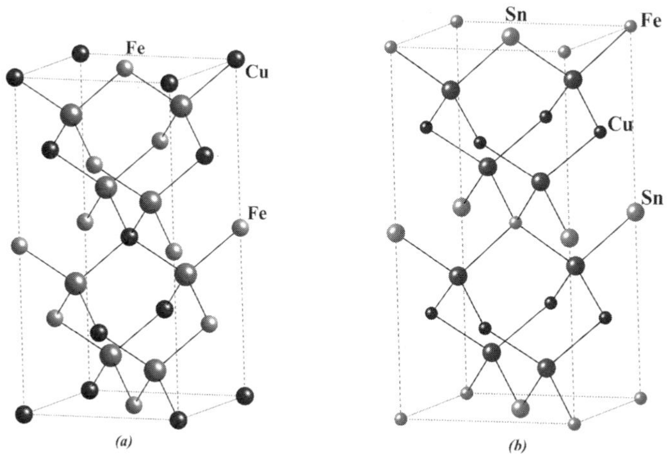

Figure 6.3. *(a)* The chalcopyrite ($CuFeS_2$) structure with Fe and Cu sharing **T** layers. S atoms are larger balls. *(b)* The stannite (Cu_2FeSnS_4) structure.

(*Source*: *CrystalMaker*, by David Palmer, *CrystalMaker* Software Ltd., Begbroke Science Park, Bldg. 5, Sandy Lane, Yarnton, Oxfordshire, OX51PF, UK.)

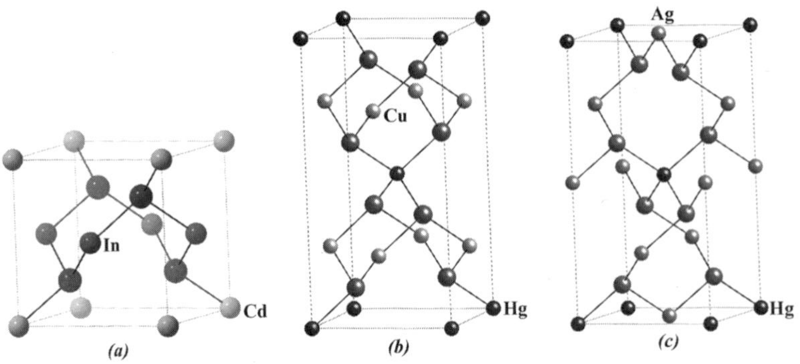

Figure 6.4. The structures of *(a)* In_2CdSe_4, *(b)* β-Cu_2HgI_4, and *(c)* β-Ag_2HgI_4.
(*Source*: *CrystalMaker*, by David Palmer, *CrystalMaker* Software Ltd., Begbroke Science Park, Bldg. 5, Sandy Lane, Yarnton, Oxfordshire, OX51PF, UK.)

cell for β-Cu_2HgI_4 in the figure corresponds to two cells for In_2CdSe_4 with the upper cell inverted and the cells joined by a layer of one Hg atom within the cell rather than four Hg atoms at the corners. Each **T** layer occupied is one-half filled by Cu, one-quarter filled by Hg, and one-quarter empty. The notation is **3·4PT**$_{2/4\ 1/4}$**(t)**. The structure of β-Ag_2HgI_4 is closely related to that of β-Cu_2HgI_4. The empty **T** sites are arranged differently, but they are still in the **T** layers with Ag and Hg. The notation is the same as for β-Cu_2HgI_4, **3·4PT**$_{2/4\ 1/4}$**(t)**. There are 12 layers repeating for β-Cu_2HgI_4 and β-Ag_2HgI_4. Both structures are tetragonal, for β-Cu_2HgI_4, $\mathbf{D}_{2d}^{11}$, $I\bar{4}2m$, $a_o = 6.092$, and $c_o = 12.242$ Å, and for β-Ag_2HgI_4, $\mathbf{S}_4^2$, $I\bar{4}$, $a_o = 6.317$, and $c_o = 12.633$ Å. For α-Ag_2HgI_4, there is a structure with a random distribution of Ag^+ and Hg^{2+}.

6.2. Hexagonal Close-Packed PT Structures

6.2.1. The 2·2PT Crystal Structure of Wurtzite [ZnS]

In addition to the zinc blende (**3·2PT**), ZnS has the wurtzite structure based on an *hcp* pattern of S atoms. The space group of the hexagonal cell with two molecules is $\mathbf{C}_{6v}^4$, $P6_3mc$, $a_o = 3.811$, and $c_o = 6.234$ Å. This is a beautiful, simple structure as shown in Figure 6.5***a***, with one of two **T** layers between **P** layers filled by Zn atoms (**2·2PT**). The figure shows the close-packed **P** and **T** layers. The ZnS_4 tetrahedra are symmetrical. **P** and **T** layers involve only *A* and *B* positions, with **O** layers at *C* positions vacant and the empty *C* channels, which are clear in Figure 6.5***b***. This is a view along the packing direction from the top of Figure 6.5***a***. The hexagonal pattern is like honey comb.

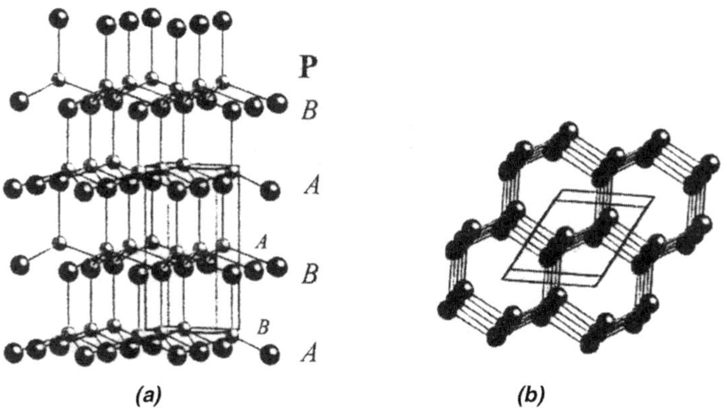

Figure 6.5. *(a)* The **2·2PT** structure of wurtzite (ZnS). The **P** layers (S) are large dark balls. *(b)* Open channels at *C* positions in wurtzite.

(*Source*: *CrystalMaker*, by David Palmer, *CrystalMaker* Software Ltd., Begbroke Science Park, Bldg. 5, Sandy Lane, Yarnton, Oxfordshire, OX51PF, UK.)

There are our layers in the packing unit with two molecules per cell. The heights of layers are given below:

Height:	0	37	50	87	100
Atom:	S	Zn	S	Zn	S
Position:	$\mathbf{P}_A$	$\mathbf{T}_B$	$\mathbf{P}_B$	$\mathbf{T}_A$	$\mathbf{P}_A$

Compounds with the wurtzite structure are shown in Table 6.2.

6.2.2. The 2·2PT Crystal Structure of I_h Ice

Ice I_h has the **2·2PT** structure like that of wurtzite, with oxygen atoms replacing S and Zn in **P** and **T** layers. It is also similar to β-tridymite (Section 10.1.1). The space group is $\mathbf{D}^4_{6h}$, $P6_3/mmc$ for the hexagonal unit with four molecules in the cell, $a_o = 4.5227$ and $c_o = 7.3671$ Å at 0°C. The H atoms form hydrogen bonds between O atoms, four H bonds for each O. The structure is shown in Figure 6.6. Ice I_h shares the open *C* channels with wurtzite. This open structure explains the lower density of ice compared to liquid H_2O. Two H atoms are shown between oxygen atoms. These represent the disordered H sites, but only half of these are occupied. The pairs of close oxygen atoms at the same *A* or *B* positions represent **P** and **T** sites.

Table 6.2. Compounds with the wurtzite (2·2PT) structure.

AgI	CuBr	InN	NH_4F	ZnO
AlN	CuCl	MgTe	NbN	ZnSe
BeO	CuH	MnS	SiC	ZnTe
CdS	CuI	MnSe		
CdSe	GaN	MnTe		

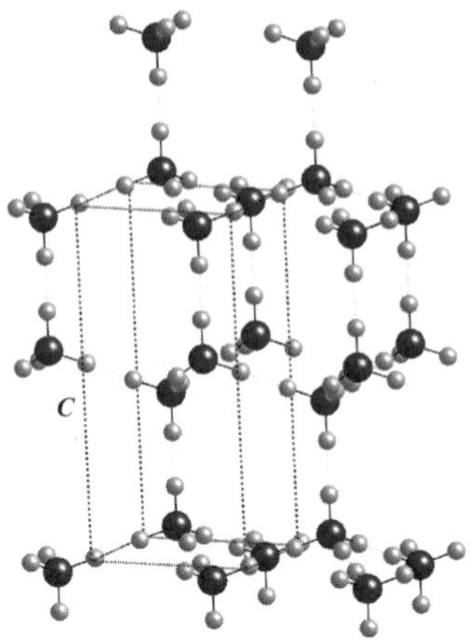

Figure 6.6. The $\mathbf{I}_h$ ice structure. Oxygen atoms are in **P** and **T** sites of the wurtzite structure. Both of the disordered H positions between oxygen atoms are shown.

(*Source*: *CrystalMaker*, by David Palmer, *CrystalMaker* Software Ltd., Begbroke Science Park, Bldg. 5, Sandy Lanc, Yarnton, Oxfordshire, OX51PF, UK.)

6.2.3. The 2·2PT$_{1/2}$(t) Crystal Structure of Germanium Sulfide [GeS_2]

Germanium sulfide, GeS_2, has a tetragonal crystal structure, $\mathbf{D}_{2h}^{12}$, I$\bar{4}$2*d*, $a = 5.48$, and $c = 9.143$ Å with four molecules in the unit cell. Pairs of GeS_4 tetrahedra, aligned along the *c* axis, share edges (see Figure 6.7***a***). The other two S atoms are shared with other GeS_4 tetrahedra. For each row along *c* **T** sites are occupied in pairs alternating with empty pairs, The S atoms are in close-packed **P** layers in an *AB* sequence (see Figure 6.7***b***). There is slight displacement of **P** layers causing the apical bonds for the GeS_4 tetrahedra, normally perpendicular to the packing layers, to be slightly inclined. The notation is **2·2PT$_{1/2}$(t)**.

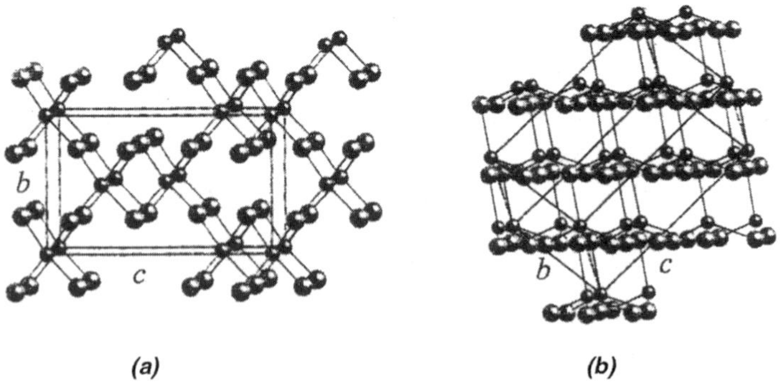

Figure 6.7. The structure of GeS_2. Ge atoms are dark smaller balls. ***(a)*** GeS_4 tetrahedra share edges in pairs along *c*. ***(b)*** **P** (S) and **T** (Ge) layers are shown.

(*Source*: *CrystalMaker*, by David Palmer, *CrystalMaker* Software Ltd., Begbroke Science Park, Bldg. 5, Sandy Lanc, Yarnton, Oxfordshire, OX51PF, UK.)

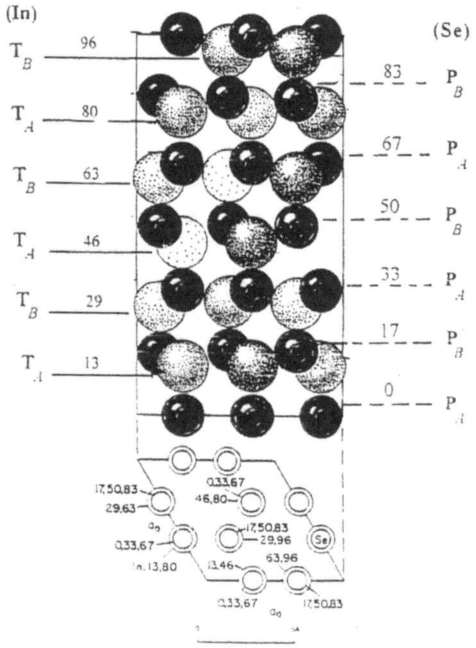

Figure 6.8. The structure of In_2Se_3. The Se atoms (black balls) are in $\mathbf{P}_A$ and $\mathbf{P}_B$ layers.

(*Source*: R.W.G. Wyckoff, *Crystal Structures*, Vol. 2, 2nd ed., Wiley, New York, 1964, p. 23.)

6.2.4. The 2•6PT$_{2/3}$ Crystal Structure of In_2Se_3

Introducing vacancies into the simple **2•2PT** structure causes more complexity. For In_2Se_3, there are one-third vacancies in the **T** layer occupied, giving **2•6PT$_{2/3}$** with 12 layers in the repeating unit. The Se atoms are in an *hcp* pattern with In atoms in two-thirds of one **T** layer as shown in Figure 6.8. There are six molecules in the hexagonal cell, $\mathbf{C}_6^3$, P6$_5$, $a_o = 7.11$, and $c_o = 19.3$ Å.

6.2.5. The 2•2PT$_{3/4}$ $_{1/4}$(o) Crystal Structure of Enargite [Cu_3AsS_4]

The mineral enargite (Cu_3AsS_4), an important ore for Cu and As, has a **2•2PT$_{3/4}$ $_{1/4}$(o)** structure. It is similar to wurtzite except that each **T** layer occupied is one-quarter filled by As and three-quarters by Cu. The stoichiometry indicates the occupancies of **T** sites. The structure is shown in Figure 6.9. Because of the simple ordered pattern of partial filling by As and Cu there are four packing layers in the packing unit and two molecules per cell. It also has open *C* channels. The orthorhombic cell contains two molecules, $\mathbf{C}_{2v}^7$, *Pmn*2$_1$, $a_o = 6.46$, $b_o = 7.43$, and $c_o = 6.18$ Å.

6.2.6. The 2•2PT$_{2/4}$ $_{1/4}$ Crystal Structure of Al_2ZnS_4

The structure of the high-temperature form of Al_2ZnS_4 has an *hcp* pattern of S atoms like wurtzite. The occupied **T** layer is randomly

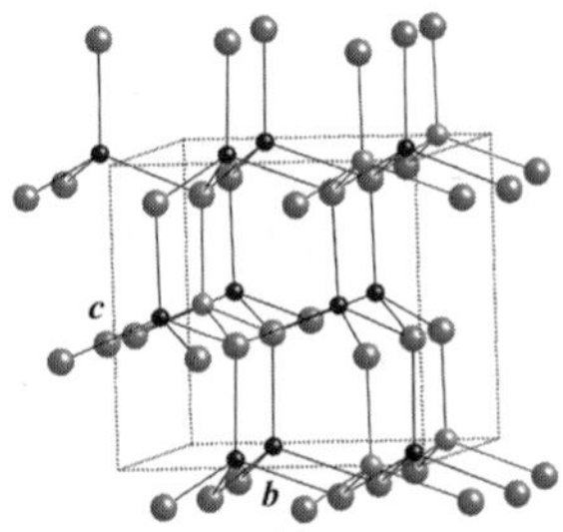

Figure 6.9. The structure of Cu_3AsS_4. The S atoms (large atoms) are in $\mathbf{P}_A$ and $\mathbf{P}_B$ layers. **T** layers are occupied by Cu (small dark atoms) and As.

(*Source*: Y. Matsushita, *Chalcogenide* crystal structure data library, Version 5.5B, Institute for Solid State Physics, The University of Tokyo, Tokyo, Japan, 2004; and *CrystalMaker*, by David Palmer, *CrystalMaker* Software Ltd., Begbroke Science Park, Bldg. 5, Sandy Lane, Yarnton, Oxfordshire, OX51PF, UK.)

three-quarters occupied by Al and Zn in the ratio 2:1, giving the notation $\mathbf{2{\cdot}2PT}_{2/4\ 1/4}$. Below 26°C this structure is transformed to a *ccp* form similar to that of spinel (see Section 7.3.3).

6.3. Structures Involving P and Both T Layers

There are many examples of structures based on the *ccp* **PO** type, but few based on the *hcp* **PO** type. We noted that the **2·2PO** structure of NiAs is limited by the fact that for an *hcp* pattern all **O** sites, and only **O** sites, occur at *C* positions. Screening between these **O** sites is poor. NiAs is unusual, and the structure is limited to compounds where the interaction of atoms in *C* positions is tolerable. The interaction can be avoided in cases of partial filling of layers. In a *ccp* structure all **P**, **O**, and **T** sites are staggered (*A*, *B*, and *C*) positions. For *hcp* structures, **P** and **T** sites are limited to *A* and *B* positions. There are pairs of **T** sites at the same *A* or *B* position: one above and one below each **P** layer. These are very close together so that there would be strong repulsion if both sites of a pair are occupied. In Figure 6.10 we see that the **T** sites ($\mathbf{T}^+$ and $\mathbf{T}^-$) just above and below the central **P** layer are very close without screening. The repulsion can be avoided if these **T** layers are partially occupied. The **O** sites are only at *C* positions without screening.

6.3.1. The 3·3PTT Crystal Structure of Li_2O and Fluorite [CaF_2]

In the structure of lithium oxide, Li_2O, as shown in Figure 6.11, there is a *ccp* (**3P**) framework of oxide ions with Li^+ ions filling both **T** layers ($\mathbf{T}^+$ and $\mathbf{T}^-$) between **P** layers. The very close $\mathbf{T}^+$ and $\mathbf{T}^-$ layers (above and below each **P** layer) are clearly shown. There is no problem with all **T** sites occupied for a *ccp* arrangement because the **T** sites are staggered (*A*, *B*, and *C*). Each oxide ion is at the center of a cube formed by eight Li^+ ions. This structure is the reference structure for many related tetrahedral structure involving partial occupancy of **T** and/or **P** layers.

The mineral fluorite (CaF_2) has the same **3·3PTT** structure but with the roles of cations and anions reversed. Because CaF_2 is more com-

Figure 6.10. A model of the *hcp* (**2P**) structure with **T** and **O** (white) layers filled. All **P** and **T** layers are at *A* and *B* positions. $\mathbf{T}^+$ and $\mathbf{T}^-$ sites are very close.

mon, this structure is commonly known as the fluorite structure and the Li_2O structure is usually known as the antifluorite structure. The fluorite structure is cubic with four molecules in the cell, $\mathbf{O}_h^5$, $Fm3m$, $a_o = 5.46295$ Å at 28°C. In the **3·3PTT** structure of CaF_2 the Ca^{2+} ions fill the **P** layers in a *ccp* arrangement with F^- ions filling all **T** layers. The fluorite structure is shown in Figure 6.12 in comparison with that of the zinc blende (**3·2PT**). The figure clearly shows that for CaF_2 a second tetrahedron of F^- ions is added within the cell to the one tetrahedron of Zn atoms in ZnS, thus forming a cube within the cell. The cube around each Ca^{2+} ion can be visualized for two adjacent cells. For a Ca^{2+} ion in the center of a shared face of two adjacent cells there is a square group of four F^- within each cell (a face of the internal cube) forming a cube with Ca^{2+} at the center.

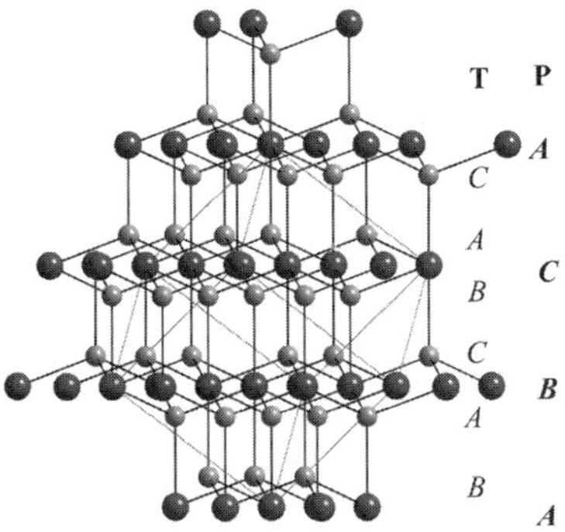

Figure 6.11. The **3·3PTT** structure of Li_2O. Dark large balls are oxygen atoms. (*Source*: *CrystalMaker*, by David Palmer, *CrystalMaker* Software Ltd., Begbroke Science Park, Bldg. 5, Sandy Lane, Yarnton, Oxfordshire, OX51PF, UK.)

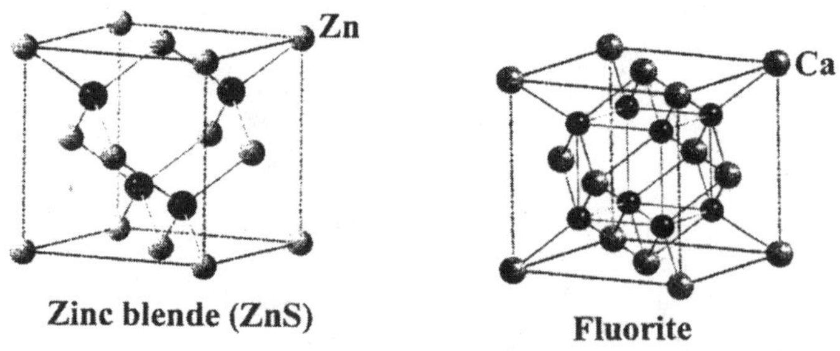

Figure 6.12. The **3·3PTT** structure of fluorite (CaF_2) with all **T** layers filled by F atoms (dark atoms) and Ca atoms in a *ccp* arrangement. The zinc blende (ZnS) structure with one **T** layer filled by S is shown for comparison.

(*Source*: *CrystalMaker*, by David Palmer, *CrystalMaker* Software Ltd., Begbroke Science Park, Bldg. 5, Sandy Lane, Yarnton, Oxfordshire, OX51PF, UK.)

The usual situation for a close-packed model for ionic compounds involves a close-packed arrangement of the anions with cations in **T** and **O** layers. The basis for this is that the common case involves anions that are larger than cations. However, the **PTOT** pattern applies even though the relative sizes of cations and anions can vary greatly. For a *ccp* arrangement **P** and **O** layers are equivalent and the labels can be interchanged. Also **P** and **O** layers together are equivalent to the pair of **T** layers and they can be interchanged. The CaF_2 structure can be described as a *ccp* arrangement of F^- filling **P** and **O** layers, giving the notation **3·3POT** (or **3·3PTO**, Section 7.4.1).

Compounds with the Li_2O structure are shown in Table 6.3. Those with the CaF_2 structure are shown in Table 6.4.

6.3.2. The 3·3PTT Crystal Structure of MgAgAs

The structure of MgAgAs is an ordered *ccp* structure with Ag in T^+ layers and Mg in T^- layers. The notation is $\mathbf{3{\cdot}3P^{As}T^{Ag}T^{Mg}}$. Figure 6.13 shows the MgAgAs structure in relationship to that of the zinc blende (**3·2PT**). The As atoms are in **P** layers in a *ccp* arrangement and Ag atoms replace Zn atoms (of ZnS). The tetrahedral arrangement of Mg atoms is added to the AgAs structure (same as that of ZnS) to give the cubic arrangement of Mg_4Ag_4 within the *fcc* cell of As atoms. The structure corresponds to that of CaF_2 shown in Figure 6.12.

TABLE 6.3. **Compounds with the Li_2O (antifluorite) (3·3PTT) structure.**

Be_2C	K_2Se	Li_2Se	Na_2Se
Ir_2P	K_2Te	Li_2Te	Na_2Te
K_2O	Li_2NH	Na_2O	Rb_2O
K_2S	Li_2S	Na_2S	Rb_2S
RhP_2			

TABLE 6.4. **Compounds with the fluorite (CaF_2, 3·3PTT) structure.**

Al_2Au	LiMgP	CdF_2	CeH_2	$GeLi_5N_3$
Al_2Pt	Li_5P_3Si	CeOF	DyH_2	Li_5SiN_3
As_3GeLi_5	Li_5P_3Ti	EuF_2	ErH_2	Li_5TiN_3
As_3Li_5Si	Mg_2Pb	HgF_2	GdH_2	UN_2
As_3Li_5Ti	$MgPu_2$	HoOF	HoH_2	AmO_2
AsLiZn	Mg_2Si	LaOF	LuH_2	CeO_2
$AuGa_2$	Mg_2Sn	NdOF	NbH_2	CmO_2
$AuIn_2$	$NiSi_2$	β-PbF_2	NdH_2	NpO_2
$CoSi_2$	PRh_2	PrOF	PrH_2	PaO_2
Ga_2Pt	$PtSn_2$	PuOF	ScH_2	PoO_2
$GeLi_5P_3$	Be_2B	RaF_2	SmH_2	PuO_2
$GeMg_2$	Be_2C	SmOF	TbH_2	TbO_2
In_2Pt	AcOF	$SrCl_2$	TmH_2	ThO_2
Ir_2P	BaF_2	SrF_2	YH_2	UO_2
$IrSn_2$	CaF_2	β-YOF		ZrO_2(H.T.)

There are four molecules in the unit cell, $\mathbf{T}_d^2$, $F\bar{4}3m$. Compounds with the MgAgAs structure are given in Table 6.5.

6.3.3. The 3·3PTT Crystal Structure of K_2PtCl_6

If we consider K_2PtCl_6 as an M_2X compound it has the simple Li_2O (antifluorite) structure! The octahedral $PtCl_6^{2-}$ are stacked in a *ccp* arrangement with K^+ ions filling all **T** layers. The *ccp* arrangement of $PtCl_6^{2-}$ octahedra and the cubic arrangement of K^+ in both **T** layers are shown in Figure 6.14*a*. The notation is **3·3PTT**. The structure corresponds to that of Li_2O and that of CaF_2 (Figure 6.12) reversed. The space group is $\mathbf{O}_h^5$, $Fm3m$, $a_o = 9.76$ Å.

It is much more complicated, but we can consider the structure of K_2PtCl_6 by treating the atoms individually. The K and Cl atoms fill the **P** layers in a *ccp* arrangement in the ratio 1:3 ($\mathbf{P}^{K}_{1/4}{}^{Cl}_{3/4}$). The Pt atoms

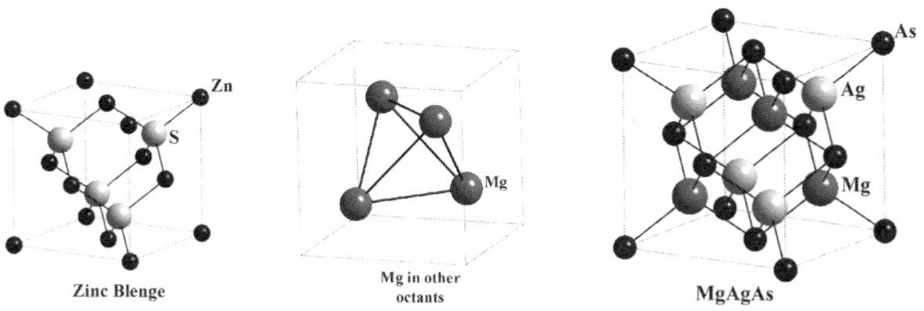

Figure 6.13. The zinc blende (ZnS) structure is shown with the Mg atoms in the other **T** layer. They are combined to give a fluorite-like structure for MgAgAs.

(*Source*: *CrystalMaker*, by David Palmer, *CrystalMaker* Software Ltd., Begbroke Science Park, Bldg. 5, Sandy Lane, Yarnton, Oxfordshire, OX51PF, UK.)

TABLE 6.5. **Compounds with the MgAgAs (3·3PTT) structure.**

AgZnAs	LiMgBi	CuMgSb	LiZnP
MgLiAs	NiMgBi	CuMgSn	MgNiSb
NaZnAs	CdCuSb	CuMnSb	LiMgN
CuMgBi	CoMnSb	LiMgSb	LiZnN

occupy one-quarter of alternate **O** layers as shown in Figure 6.14*b*; these are the **O** sites surrounded by six Cl atoms. One $PtCl_6^{2-}$ octahedron is outlined. The notation is $\mathbf{3{\cdot}3P}_{1/4\ 3/4}\mathbf{P}_{1/4\ 3/4}\mathbf{O}_{1/4}$. This structure is similar to that of perovskite ($CaTiO_3$, $\mathbf{3{\cdot}3P}_{1/4\ 3/4}\mathbf{O}_{1/4}$), except for perovskite all **O** layers are one-quarter occupied. Many complexes with the K_2PtCl_6 structure are given in Table 6.6.

6.3.4. The 3·3PTT Crystal Structure of $[N(CH_3)_4]_2SnCl_6$

The structure of $[N(CH_3)_4]_2SnCl_6$ is similar to that of K_2PtCl_6 (Figure 6.14) with $N(CH_3)_4^+$ ions replacing K^+. The cubic cell contains four molecules with $a_o = 12.87$ Å($\mathbf{O}_h^5$, $Fm3m$). The $[SnCl_6]^{2-}$ anions are cubic close packed with $N(CH_3)_4^+$ ions filling both **T** layers giving **3·3PTT**. This is the undistorted Li_2O or CaF_2-type structure.

6.3.5. The 3·3PTT(h) Crystal Structure of $CaSi_2$

Silicon is usually found in oxides and silicates. With very electropositive metals it forms silicides. The compound $CaSi_2$ can be viewed as containing Ca^{2+} and Si^- ions. The crystal structure is trigonal, $\mathbf{D}_{3h}^5$, $R\bar{3}m$, $a = 3.8259$, and $c = 15.904$ Å with three molecules in the unit cell. The Si^- ion has the same valence shell as As, and layers of

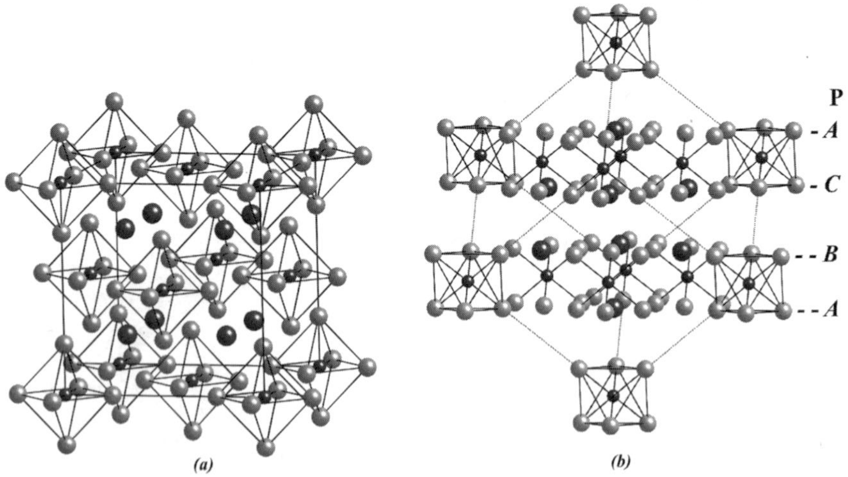

Figure 6.14. *(a)* The **3·3PTT** structure of K_2PtCl_6. The $[PtCl_6]^{2-}$ octahedra are *ccp* with K^+ in all **T** layers. *(b)* The structure of K_2PtCl_6 with K (larger and dark) and Cl (light color) atoms in **P** layers and Pt atoms in alternate **O** layers.

(*Source*: *CrystalMaker*, by David Palmer, *CrystalMaker* Software Ltd., Begbroke Science Park, Bldg. 5, Sandy Lane, Yarnton, Oxfordshire, OX51PF, UK.)

TABLE 6.6. **Complexes with the K_2PtCl_6(3·3PTT) structure.**

Cs_2CoF_6	Cs_2CrF_6	Cs_2GeCl_6	Cs_2GeF_6	Cs_2MnF_6	Cs_2MoBr_6
Cs_2MoCl_6	Cs_2NiF_6	Cs_2PbCl_6	Cs_2PdBr_6	Cs_2PdCl_6	Cs_2PdF_6
Cs_2PoBr_6	Cs_2PoI_6	Cs_2PtBr_6	Cs_2PtCl_6	Cs_2SeCl_6	Cs_2SiF_6
Cs_2SnBr_6	Cs_2SnCl_6	Cs_2SnI_6	Cs_2TeBr_6	Cs_2TeCl_6	Cs_2TeI_6
Cs_2TiCl_6	Cs_2TiF_6	Cs_2CrCl_6	$K_2CrF_6K_2$	K_2MnF_6	K_2MoCl_6
K_2NiF_6	K_2OsBr_6	K_2OsCl_6	K_2PdBr_6	K_2PdCl_6	K_2PtBr_6
K_2ReBr_6	K_2ReCl_6	K_2RuCl_6	K_2SeBr_6	K_2SiF_6	K_2SnBr_6
K_2SnCl_6	K_2TeCl_6	K_2TiCl_6	$(NH_4)_2GeF_6$	$(NH_4)_2IrCl_6$	$(NH_4)_2OsBr_6$
$(NH_4)_2PbCl_6$	$(NH_4)_2PdBr_6$	$(NH_4)_2PdCl_6$	$(NH_4)_2PoBr_6$	$(NH_4)_2PoCl_6$	$(NH_4)_2PtBr_6$
$(NH_4)_2PtCl_6$	$(NH_4)_2SeBr_6$	$(NH_4)_2SeCl_6$	$(NH_4)_2SiF_6$	$(NH_4)_2SnBr_6$	$(NH_4)_2SnCl_6$
$(NH_4)_2TeBr_6$	$(NH_4)_2TeCl_6$	$(NH_4)_2TiBr_6$	$(NH_4)_2TiCl_6$	Na_2MoF_6	Rb_2CrF_6
Rb_2MnF_6	Rb_2MoBr_6	Rb_2MoCl_6	Rb_2NiF_6	Rb_2PbCl_6	Rb_2PdBr_6
Rb_2PdCl_6	Rb_2PtBr_6	Rb_2PtCl_6	Rb_2RuF_6	Rb_2SbCl_6	Rb_2SeCl_6
Rb_2SiF_6	Rb_2SnBr_6	Rb_2SnCl_6	Rb_2SnI_6	Rb_2TeCl_6	Rb_2TiCl_6
Rb_2TiF_6	Rb_2ZrCl_6	Tl_2MoCl_6	Tl_2PtCl_6	Tl_2SiF_6	Tl_2SnCl_6
Tl_2TeCl_6	Tl_2TiF_6				

Si atoms could be considered as double layers similar to those in elemental arsenic (Section 4.4.4). The layers of Ca atoms are in an *ABC* sequence and are aligned with Si atoms above and below (see Figure 6.15). Each Si has a tetrahedral arrangement of three Ca at 3.093 Å and the axial Ca at 3.138 Å. Thus, the puckered Si layer can be considered as two **T** layers. In the figure a $\mathbf{T}^+$ site is labeled *C* (its position) and a $\mathbf{T}^-$ site is labeled *B*. In addition to the four Ca neighbors, each Si atom is bonded to three Si atoms at 2.417 Å. The three

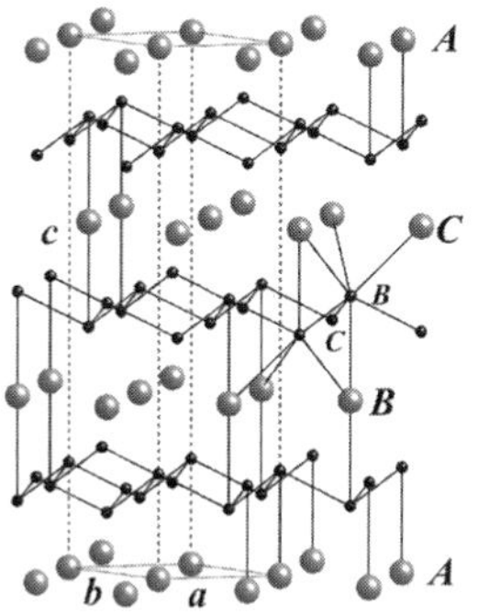

Figure 6.15. On the right of the $CaSi_2$ cell small Si atoms can be seen in $\mathbf{T}^+$ and $\mathbf{T}^-$ sites, respectively

(*Source*: Y. Matsushita, *Chalcogenide* crystal structure data library, Version 5.5B, Institute for Solid State Physics, The University of Tokyo, Tokyo, Japan, 2004; and *CrystalMaker*, by David Palmer, *CrystalMaker* Software Ltd., Begbroke Science Park, Bldg. 5, Sandy Lane, Yarnton, Oxfordshire, OX51PF, UK.)

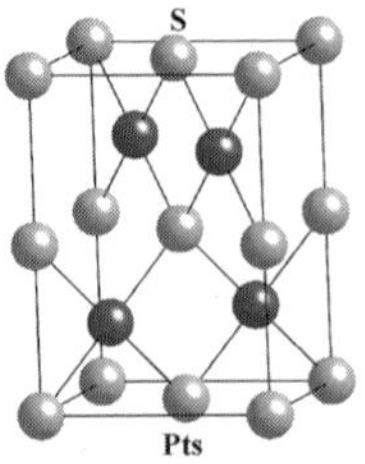

Figure 6.16. The structure of PtS showing the square planar coordination of Pt by four S and the tetrahedral coordination of S.

(*Source*: *CrystalMaker*, by David Palmer, *CrystalMaker* Software Ltd., Begbroke Science Park, Bldg. 5, Sandy Lane, Yarnton, Oxfordshire, OX51PF, UK.)

Ca atoms forming the base of a tetrahedron and three Si atoms form a trigonal prism with the Si_3 face capped by the axial Ca atom. The notation is **3·3PTT(h)**.

6.3.6. The Crystal Structures of PtS and PbO

The structures of PtS (cooperite) and PbO are derived from that of fluorite (CaF_2, **3·3PTT**). The same notation, $\mathbf{3{\cdot}3PT_{1/2}T_{1/2}}$ with the metal in the **T** layers, applies to PtS and PbO, although the patterns of MX_4 units differ. Figure 6.16 shows the PtS structure with two chains of square planar PtS_4 units. S atoms are in **T** sites. The tetragonal unit cell contains two molecules, $\mathbf{D}_{4h}^{9}$, $P4_2/mmc$, $a = 3.41$, and $c = 6.10$ Å. The drawing contains four PtS. It is larger than the unit cell to show the PtS_4 chains. PtO and PdS have the PtS structure.

Red PbO has a bimolecular tetragonal unit cell shown in Figure 6.17*a*, $\mathbf{D}_{4h}^{7}$, $P4/nmm$, $a = 3.975$, and $c_o = 5.023$ Å at 27°C. Four **T** sites about each Pb are occupied, but these are in a plane parallel to the base of the cell giving square pyramids as shown in Figure 6.17*b*. Such "one-sided" coordination is common for compounds of metals having an unshared electron pair (a "lone" pair) in an orbital directed away from the ligands (here 4S in the base of the pyramid). SnO, InBi, and LiOH have the PbO structure.

Another approach to the structures of PtS and PbO is to examine the **3·3PTT** structure of CaF_2. There is a *ccp* arrangement of Ca^{2+} ions in CaF_2 with a cubic arrangement of $8F^-$ ions in all **T** sites, Figure 6.18*a*. For ZnS (zinc blende, **3·2PT**) one **T** layer is removed leaving a tetrahedral arrangement of S^{2-} ions (Figure 6.18*b*) around Zn^{2+}. Removal of two S^{2-} ions from each **T** layer as shown in Figure 6.18*c* gives square planar PtS_4 units. For PbO removal of a different set of two anions from each **T** layer gives the square pyramidal PbO_4 unit (see Figure 6.18*d*). For CaF_2, ZnS, PtS, and PbO the metal ions are in packing (**P**) *ccp* positions.

6.3.7. The $\mathbf{3{\cdot}3PT_{3/4}T_{3/4}}$ Crystal Structure of Tl_2O_3

There are three structures (A, B, and C) encountered for metal sesquioxides (M_2O_3). Tl_2O_3 (the mineral avicennite) represents the C-type

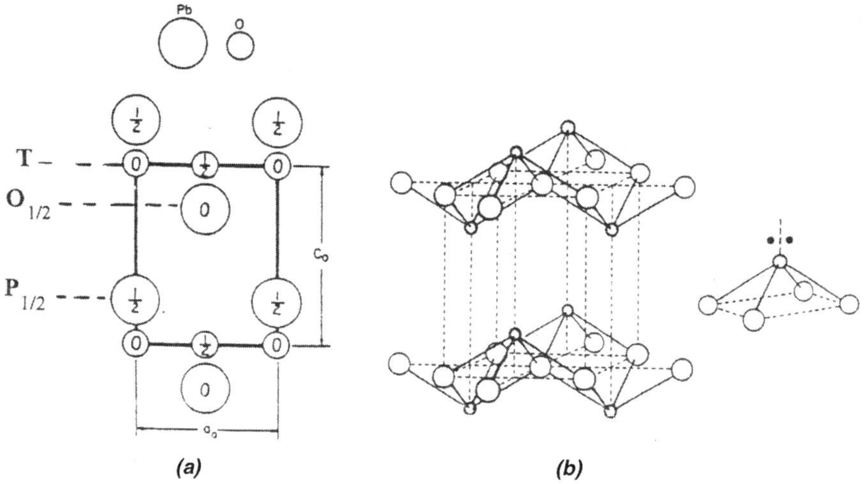

Figure 6.17. *(a)* A projection of the tetragonal PbO cell on a face. Large circles are Pb. *(b)* A view of the PbO structure showing the square pyramidal PbO_4 units.

(*Source*: R.W.G. Wyckoff, *Crystal Structures*, Vol. 1, 2nd ed., Wiley, New York, 1963, p. 136.)
(*Source:* A. F. Wells, *Structural Inorganic Chemistry*, 3rd ed., Oxford UP, Oxford, 1962.)

structure. La_2O_3 (Section. 5.5.10) is an example of an A-type M_2O_3. Tl_2O_3 has a cubic crystal structure, T_h^7, $Ia\bar{3}$, and $a = 10.5344$ Å. It is a fluorite-type structure with vacancies in oxygen layers (Figure 6.19). The Tl atoms are in close-packed layers stacked in an *ABC* sequence along the body diagonal of the cell. Oxygen ions occupy 3/4 of sites in **T** layers. Tl ions have ***CN*** 6 in distorted octahedra. The Tl sites correspond to those of Ca in fluorite with two oxide ions missing from the cube. The cube consists of three atoms of each closest oxygen layer and one of the next closer oxygen layer. One TlO_6 unit results from two oxide ions removed along one face diagonal of the cube and the other TlO_6 unit results from two oxide ions removed along a body diagonal of the cube. The notation based on these figures is $\mathbf{3{\cdot}3PT_{3/4}T_{3/4}}$ with Tl^{3+} ions in *ccp* **P** positions. Because of the equivalence of the

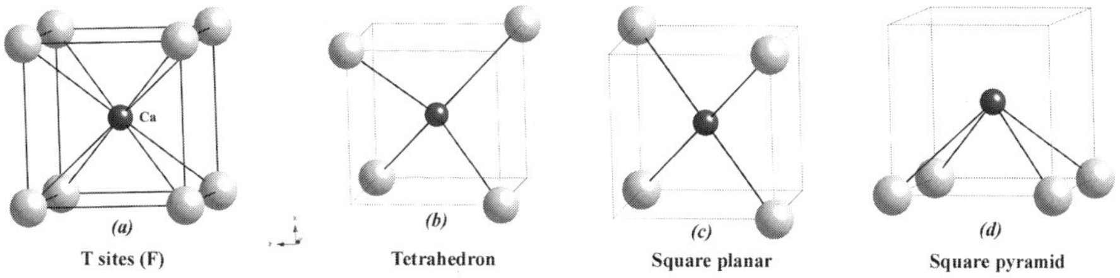

Figure 6.18. Square planar, square pyramidal, and tetrahedral coordination in structures related to CaF_2. *(a)* The eight **T** sites in the cell. The cube consists of two tetrahedra. *(b)* Omission of four **T** sites of one layer leaving a tetrahedron. *(c)* A square planar arrangement obtained on omission of two **T** sites of each **T** layer. *(d)* A square pyramid obtained on omission of two different **T** sites of each **T** layer.

(*Source*: *CrystalMaker*, by David Palmer, *CrystalMaker* Software Ltd., Begbroke Science Park, Bldg. 5, Sandy Lane, Yarnton, Oxfordshire, OX51PF, UK.)

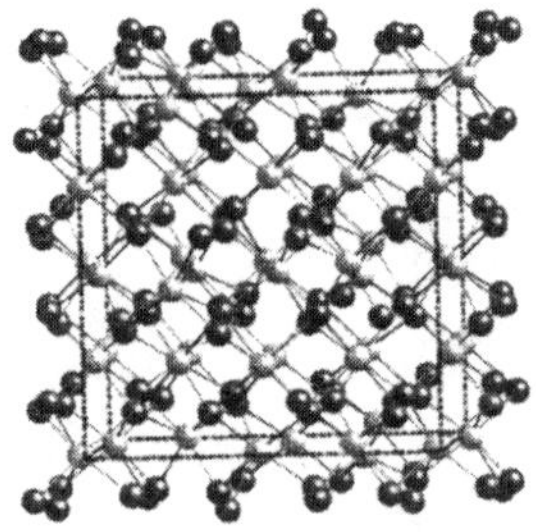

Figure 6.19. The cubic cell of Tl_2O_3. Packing layers are stacked along the body diagonal. Oxygen atoms (dark balls) are in **T** layers.

(*Source*: Y. Matsushita, *Chalcogenide* crystal structure data library, Version 5.5B, Institute for Solid State Physics, The University of Tokyo, Tokyo, Japan, 2004; and *CrystalMaker*, by David Palmer, *CrystalMaker* Software Ltd., Begbroke Science Park, Bldg. 5, Sandy Lane, Yarnton, Oxfordshire, OX51PF, UK.)

combination of **P** and **O** layers with the two **T** layers, the notation can be given as $\mathbf{3{\cdot}3P_{3/4}O_{3/4}T}$ with Tl in **T** sites. Other oxides with the C-type M_2O_3 structures are shown in Table 6.7.

6.3.8. The $3{\cdot}6P_{1/2\ 1/2}TT(t)$ Crystal Structure of Scheelite [$CaWO_4$]

The structure of the mineral scheelite ($CaWO_4$) is closely related to the fluorite (CaF_2, **3·3PTT**) structure. Complex structures with some deformations can be visualized more readily from descriptions of the ideal model in terms of well-known structures. The scheelite structure can be considered as the result of stacking CaF_2 cubes. There is a *ccp* arrangement of **P** layers filled by Ca and W ($\mathbf{P_{1/2\ 1/2}}$). The oxygen atoms fill both **T** layers. Figure 6.20 shows one CaF_2 unit on the left, and two of these units with deformations for scheelite are shown on the right. Each oxygen atom has two Ca atoms and two W atoms as neighbors forming a tetrahedron. The distortions of the tetrahedra result from displacements of oxygen atoms. The framework of **P** layers (Ca and W atoms) are not distorted significantly. The scheelite structure involves eight cubes and 18 packing layers in the repeating unit. The *ccp* pattern requires **3·3** (9) layers, and the pattern of Ca and W positions doubles this number (18) of layers giving the notation $\mathbf{3{\cdot}6P_{1/2\ 1/2}TT(t)}$. There are four molecules in the tetragonal unit cell, C_{4h}^6, $I4_1/a$, $a_o = 5.24$, and $c_o = 11.38$ Å. The *CN* of Ca is 8 and W has *CN* 4, a tetrahedron flattened along a C_2 axis. The compounds $CuWO_4$

TABLE 6.7. **Metal oxides with the C-type M_2O_3 structure.**

α-Am_2O_3	Cm_2O_3	Dy_2O_3	Er_2O_3	Eu_2O_3	β-Fe_2O_3
Gd_2O_3	Ho_2O_3	In_2O_3	Lu_2O_3	β-Mn_2O_3	Nd_2O_3
Pr_2O_3	Pu_2O_3	Sc_2O_3	Sm_2O_3	Tb_2O_3	Tl_2O_3
Tm_2O_3	Y_2O_3	Yb_2O_3			

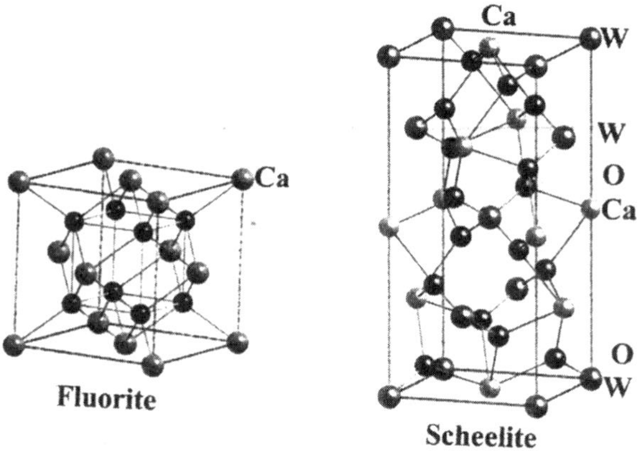

Figure 6.20. The unit cell for fluorite and the unit cell for scheelite ($CaWO_4$), a combination of two units of fluorite.

(*Source*: *CrystalMaker*, by David Palmer, *CrystalMaker* Software Ltd., Begbroke Science Park, Bldg. 5, Sandy Lane, Yarnton, Oxfordshire, OX51PF, UK.)

and $ZnWO_4$ (Section 5.2.2) have much different structures. In both compounds the metal ions are in octahedral sites.

Because the mineral scheelite and the related pyrochlore (Section 6.3.9) are important laser host materials for rare earth metal ions, many such compounds and their doped crystals have been investigated intensively. Compounds with the scheelite structure are given in Table 6.8.

6.3.9. The $3{\cdot}6P_{1/2}\ {}_{1/2}T_{3/4}T$ Crystal Structure of Pyrochlore [$Ca_2Ta_2O_7$]

The idealized composition of pyrochlore is $Ca_2Ta_2O_7$. The structure is adopted by minerals of the general formula $(NaCa)_2(Nb,Ta)_2\,O_6(OH,F)$.

TABLE 6.8. **Compounds with the scheelite [$CaWO_4$, $3{\cdot}6P_{1/2}\ {}_{1/2}TT(t)$] structure.**

$LiDyF_4$	$AgIO_4$	NH_4IO_4	$TlReO_4$	$Na_{0.5}La_{0.5}MoO_4$	$Ho(Ti_{0.5}W_{0.5})O_4$
$LiErF_4$	$AgReO_4$	NH_4ReO_4	$UGeO_4$	$Na_{0.5}La_{0.5}WO_4$	$Lu(Ti_{0.5}Mo_{0.5})O_4$
$LiEuF_4$	$BaMoO_4$	$NaIO_4$	$YNbO_4$	$Ce(Ti_{0.5}Mo_{0.5})O_4$	$Nd(Ti_{0.5}Mo_{0.5})O_4$
$LiGdF_4$	$BaWO_4$	$NaReO_4$	$ZrGeO_4$	$Dy(Ti_{0.5}Mo_{0.5})O_4$	$Nd(Ti_{0.5}W_{0.5})O_4$
$LiHoF_4$	$BiAsO_4$	$NaTcO_4$	$K_{0.5}Bi_{0.5}MoO_4$	$Dy(Ti_{0.5}W_{0.5})O_4$	$Sm(Ti_{0.5}Mo_{0.5})O_4$
$LiLuF_4$	$CaMoO_4$	$PbMoO_4$	$K_{0.5}Ce_{0.5}WO_4$	$Er(Ti_{0.5}Mo_{0.5})O_4$	$Sm(Ti_{0.5}W_{0.5})$
$LiTbF_4$	$CaWO_4$	$PbWO_4$	$K_{0.5}La_{0.5}WO_4$	$Er(Ti_{0.5}W_{0.5})O$	$Pr(Ti_{0.5}Mo_{0.5})O_4$
$LiTmF_4$	$CdMoO_4$	$RbIO_4$	$Li_{0.5}Bi_{0.5}MoO_4$	$Eu(Ti_{0.5}Mo_{0.5})O_4$	$Tb(Ti_{0.5}Mo_{0.5})O_4$
$LiYF_4$	$CeGeO_4$	$RbReO_4$	$Li_{0.5}La_{0.5}MoO_4$	$Eu(Ti_{0.5}W_{0.5})O_4$	$Tb(Ti_{0.5}W_{0.5})O_4$
$LiYbF_4$	$HfGeO_4$	$SrMoO_4$	$Li_{0.5}La_{0.5}WO_4$	$Gd(Ti_{0.5}Mo_{0.5})O_4$	$Tm(Ti_{0.5}Mo_{0.5})O_4$
$CsCrO_3F$	$KReO_4$	$SrWO_4$	$Na_{0.5}Bi_{0.5}MoO_4$	$Gd(Ti_{0.5}Mo_{0.5})O_4$	$Y(Ti_{0.5}Mo_{0.5})O_4$
$KCrO_3F$	$KRuO_4$	$ThGeO_4$	$Na_{0.5}Ce_{0.5}WO_4$	$Ho(Ti_{0.5}Mo_{0.5})O_4$	$Yb(Ti_{0.5}Mo_{0.5})O_4$

TABLE 6.9. Compounds with the pyrochlore ($Ca_2Ta_2O_7$) structure.

$NaCaNb_2O_6F$	$Dy_2Sn_2O_7$	$Eu_2Sn_2O_7$	$Lu_2Ru_2O_7$	$Sm_2Ru_2O_7$	$Y_2Ru_2O_7$
$BaSrNb_2O_6F$	$Dy_2Tc_2O_7$	$Gd_2Ru_2O_7$	$Lu_2Sn_2O_7$	$Sm_2Sn_2O_7$	$Y_2Sn_2O_7$
$Ca_2Sb_2O_7$	$Dy_2Ti_2O_7$	$Gd_2Sn_2O_7$	$Nd_2Hf_2O_7$	$Sm_2Tc_2O_7$	$Y_2Ti_2O_7$
$Ca_2Ta_2O_7$	$Er_2Ru_2O_7$	$Gd_2Ti_2O_7$	$Nd_2Ru_2O_7$	$Sn_2Ti_2O_7$	$Y_2Zr_2O_7$
$Cd_2Nb_2O_7$	$Er_2Sn_2O_7$	$Ho_2Ru_2O_7$	$Nd_2Sn_2O_7$	$Tb_2Ru_2O_7$	$Yb_2Ru_2O_7$
$Cd_2Sb_2O_7$	$Er_2Tc_2O_7$	$Ho_2Sn_2O_7$	$Nd_2Zr_2O_7$	$Tb_2Sn_2O_7$	$Yb_2Sn_2O_7$
$Cd_2Ta_2O_7$	$Er_2Ti_2O_7$	$La_2Hf_2O_7$	$Pr_2Ru_2O_7$	$Tm_2Ru_2O_7$	$Yb_2Ti_2O_7$
$Dy_2Ru_2O_7$	$Eu_2Ru_2O_7$	$La_2Sn_2O_7$	$Pr_2Sn_2O_7$	$Tm_2Sn_2O_7$	$Zr_2Ce_2O_8$

Compounds with this structure are listed in Table 6.9. $Ca_2Ta_2O_7$ can be regarded as an anion-deficient fluorite (CaF_2, **3·3PTT**) structure retaining the *ccp* arrangement of O^{2-} and Ca^{2+}. For the $A_2B_2O_7$ formula type the larger A ions retain eightfold coordination but the B ions have *CN* 6.

The unit cell of pyrochlore can be considered as eight CaF_2-type cells stacked as octants of a cube. There are two types of cells shown in Figure 6.21. They differ in the position of the oxygen vacancy and the relative positions of atoms A and B. Only two of the eight octants are shown to make it easier to visualize the relative positions of the three types of atoms. The octant on the lower left is type **I** and the other one is type **II**. There are four of each type, and they are not in adjacent cells. Atoms A and B are in *ccp* layers (a face-centered cubic structure) with oxygens in **T** layers. The type **I** cube has A ions located on face

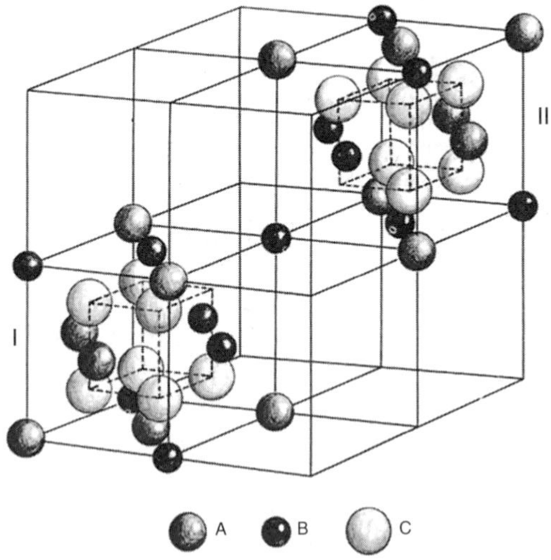

Figure 6.21. Only two of the octants of pyrochlore ($Ca_2Ta_2O_7$) are shown. Each octant has the fluorite structure with one oxygen missing. This simplified structure shows the orientation of the incomplete inner cube.

(*Source*: F. S. Galasso, *Structure and Properties of Inorganic Solids*, Pergamon, Oxford, 1970, p. 106.)

diagonals from the lower left corner with the oxygen vacancy opposite this corner. The diagonally opposite type **II** cube shown is the mirror image of the type **I** cube. The two vacancies decrease the *CN* of ion B. Pyrochlore is similar to scheelite (previous section) except for one-eighth deficiency of oxygen. The large cubic unit cell has 18 packing layers with 16 molecules, O_h^7, $Fd3m$, $a_o = 10.420$ Å. The notation is $\mathbf{3{\cdot}6P_{1/2\,1/2}T_{3/4}T}$, with half occupancy of **P** layers by Ca and Ta, and oxygen atoms filling one **T** layer and three-quarter of the other **T** layer.

6.3.10. The $3{\cdot}3P^{Cu}T_{1/4}T_{1/4}$ Crystal Structure of Cuprite [Cu_2O]

The structure of the mineral cuprite (Cu_2O) is related to that of fluorite. The cubic cell contains two molecules, O_h^4, $Pn3m$, and $a_o = 4.27$ Å. It also corresponds to two interpenetrating β-cristobalite or ice networks (Sections 10.1.2 and 6.1.2, Figures 10.3 and 6.2). These interpenetrating networks are not connected by Cu—O bonds. In this respect it is unique among inorganic M_2X or MX_2 compounds. The structure of Cu_2O is shown in Figure 6.22*a* as a model (small balls for Cu with layers labeled), and Figure 6.22*b* shows the relationships for Cu and oxygen. In Figure 6.22*b*, one of the networks is shown with heavy lines and solid circles for oxygen and the other is shown with lighter lines and open circles. The structure was studied by the father and son team, W.H. and W.L. Bragg in 1916. There is a body-centered framework of oxygen atoms with the Cu atoms in a face-centered (*ccp*) arrangement, filling the **P** layers. Oxygen atoms occupy one-quarter of the sites in each **T** layer. Cu atoms are equidistant from two oxygen atoms (*CN* 2). The cubic cell has nine layers repeating, giving $\mathbf{3{\cdot}3P^{Cu}T_{1/4}T_{1/4}}$. The same structure occurs for Ag_2O, Pb_2O, $Cd(CN)_2$, and $Zn(CN)_2$. The cyanides have reversed roles of the ions (an antistructure).

6.3.11. The $3{\cdot}4PT_{1/6}T_{1/6}(m)$ Crystal Structure of Al_2Br_6

The structure of $AlCl_3$ is similar to that of $CrCl_3$ (Section 5.5.2) for which the metal ion is in **O** layers with *CN* 6. $AlBr_3$ occurs as dimers,

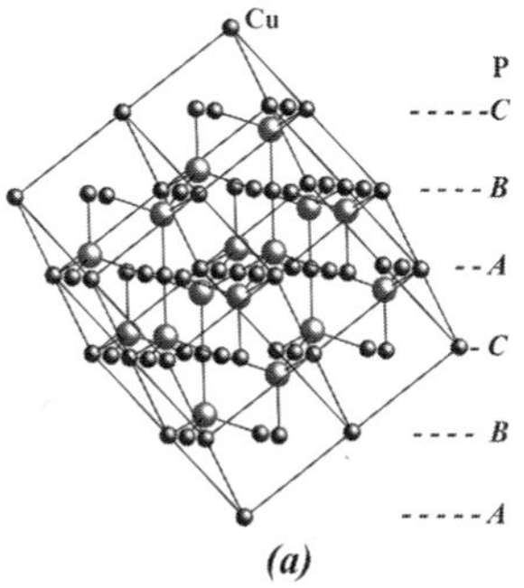

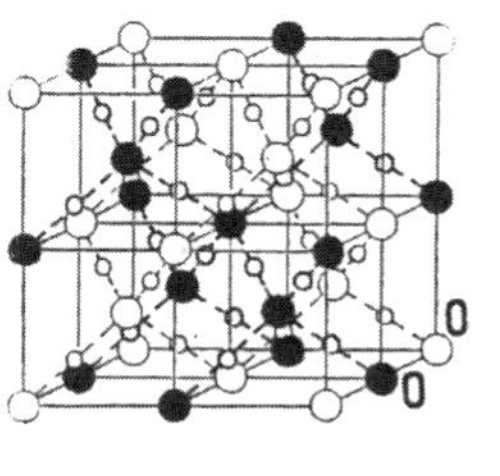

Figure 6.22. *(a)* A model of Cu_2O with layers labeled. *(b)* A drawing of the cubic unit cell for Cu_2O. Oxygen atoms for one of the independent interpenetrating Cu–O networks are blackened and lines are heavier.

(*Source*: *CrystalMaker*, by David Palmer, *CrystalMaker* Software Ltd., Begbroke Science Park, Bldg. 5, Sandy Lane, Yarnton, Oxfordshire, OX51PF, UK.)
(*Source*: L. Bragg, *The Crystalline State*, G. Bell and Sons, London, 1949, p. 93.)

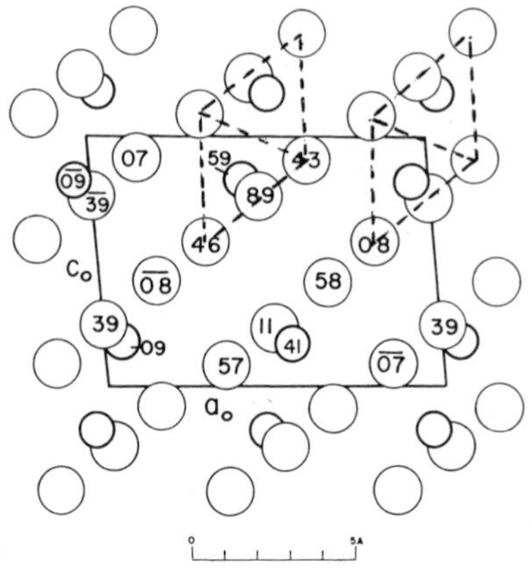

Figure 6.23. View of the structure of Al_2Br_6 along the *b* axis with heights shown. The Br atoms (large circles) are in a *ccp* arrangement.

(*Source*: R.W.G. Wyckoff, *Crystal Structures* Vol. 2, 2nd ed., Wiley, New York, 1964, p. 58.)

Al_2Br_6, in the vapor and in nonpolar solvents. In the solid there is an almost perfect *ccp* arrangement of Br^- ions with Al^{3+} ions in one-sixth of both **T** layers between **P** layers (see Figure 6.23). The $AlBr_4$ tetrahedra occur in pairs, sharing an edge forming Al_2Br_6 molecules. The crystal is usually described as having a molecular structure. With the large Br^- ions in **P** layers the Al^{3+} ions fit into pairs of **T** sites to form Al_2Br_6 units without deforming the *ccp* arrangement of Br^-. There are two Al_2Br_6 molecules per monoclinic unit cell, C_{2h}^5, $P2_1/a$, $a_o = 10.20$, $b_o = 7.09$, $c_o = 7.48$ Å, and $\beta = 96°$. There are 12 layers in the packing unit giving the notation $\mathbf{3{\cdot}4PT_{1/6}T_{1/6}(m)}$.

6.3.12. The $3{\cdot}3PT_{1/8}T_{1/8}$ Crystal Structure of SnI_4

The structure of SnI_4 has a *ccp* arrangement of I atoms with Sn atoms in one-eighth of the sites in each of the **T** layers. The space group is T_h^6, $Pa3$, and $a_o = 12.273$ at 26°C. The tetrahedral SnI_4 units (molecules) are isolated without sharing I atoms (Figure 6.24). The pattern of

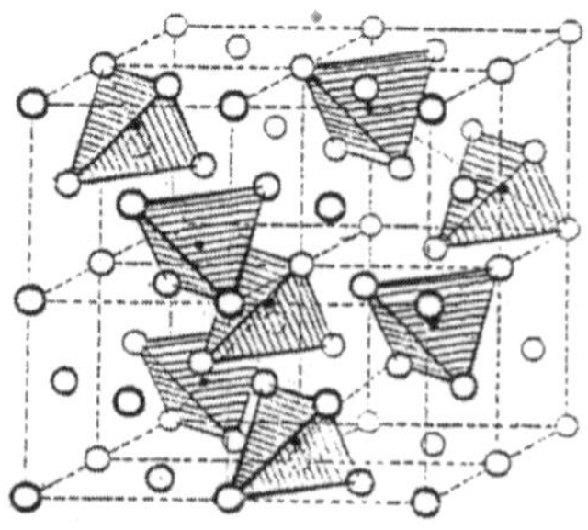

Figure 6.24. The structure of SnI_4. The I atoms (open circles) are in a *ccp* (face-centered cubic) arrangement.

(*Source*: A.F. Wells, Structural Inorganic Chemistry, 3rd ed., Oxford UP, Oxford, 1962.)

I atoms is very simple, but the pattern of occupancy of T sites requires nine layers in the repeating unit. There are eight molecules in the large cell. The SnI_4 structure occurs for tetrachlorides of Zr and Pt, tetrabromides of Ti and Zr, and tetraiodides of Ti, Ge, and Si.

6.3.13. The Crystal Structure of Cubanite [$CuFe_2S_3$]

The mineral cubanite, $CuFe_2S_3$, has a very unusual structure. The orthorhombic cell contains four molecules, $\mathbf{D}_{2h}^{16}$, *Pnma*, $a_o = 6.46$, $b_o = 11.117$, and $c_o = 6.233$ Å. It is based on an *hcp* structure with two **T** layers filled between **P** layers. It has been studied intensively because it is a good ferromagnetic material. The structure consists of slices of the wurtzite structure (**2·2PT**) joined such that pairs of FeS_4 tetrahedra share edges. These are joined by CuS_4 tetrahedra sharing three apices with FeS_4 tetrahedra (see Figure 6.25*a*). The tetrahedra point up ($\mathbf{T}^+$) and down ($\mathbf{T}^-$). The FeS_4 tetrahedra, sharing an edge, point in opposite directions. The CuS_4 tetrahedra share an apex (one S) on each side of the double chain of FeS_4 tetrahedra and point in the same direction as the attached FeS_4 ($\mathbf{T}^+$ on one side and $\mathbf{T}^-$ on the other). The ferromagnetism of cubanite is probably the result of the unusual structure involving double chains of FeS_4 tetrahedra insulated by CuS_4 tetrahedra.

We noted that an *hcp* arrangement with all **T** layers filled is unlikely because of repulsion of atoms in the very close **T** sites at the same positions (*A* or *B*) just above and below a **P** layer. For cubanite, these filled **T** layers are *between* **P** layers. In Figure 6.26*b*, the $\mathbf{P}_A$ layer (large atoms), is taken as lying in the plane of the paper. The $\mathbf{T}^+$ tetrahedra are shown as triangles and the $\mathbf{T}^-$ tetrahedra are not outlined as triangles. For cubanite, the sequence is $|\mathbf{P}_A\mathbf{T}_B^-\mathbf{T}_A^+\mathbf{P}_B\mathbf{T}_A^-\mathbf{T}_B^+|$, giving the notation $\mathbf{2{\cdot}3PT}_{1/6\ 2/6}\mathbf{T}_{1/6\ 2/6}$(o) (the + and − signs are omitted as they

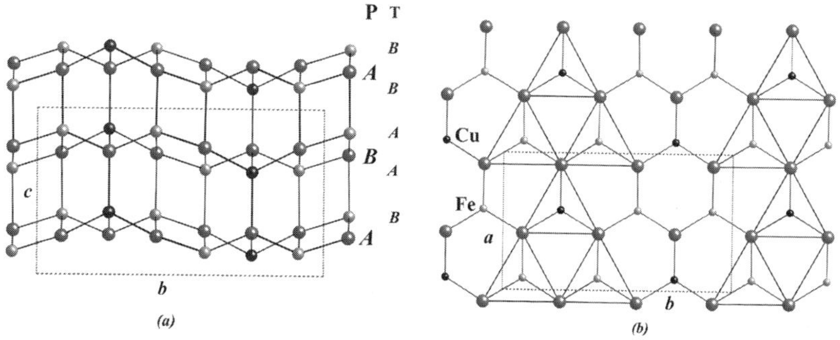

Figure 6.25. ***(a)*** A projection of cubanite ($CuFe_2S_3$) showing S in **P** positions (large atoms) with Cu (dark small atoms) and Fe (light small atoms) in **T** sites. ***(b)*** A projection of one layer of S in **P** positions and Cu and Fe in **T** sites.

(*Source*: Y. Matsushita, *Chalcogenide* crystal structure data library, Version 5.5B, Institute for Solid State Physics. The University of Tokyo, Tokyo, Japan, 2004; and *CrystalMaker*, by David Palmer, *CrystalMaker* Software Ltd., Begbroke Science Park, Bldg. 5, Sandy Lane, Yarnton, Oxfordshire, OX51PF, UK.)

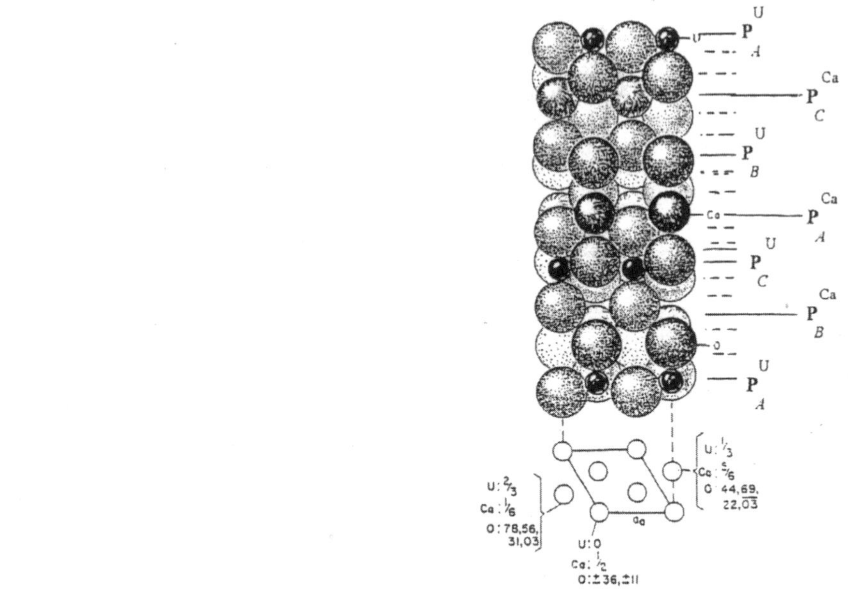

Figure 6.26. Two projections of the hexagonal $Ca(UO_2)O_2$ structure.
(*Source*: R.W.G. Wyckoff, *Crystal Structures*, Vol. 2, 2nd ed., Wiley, New York, 1964, p. 324.)

describe the normal feature). The troublesome **T** layer pairs between **P** layers are occupied, but only half of the sites are occupied in each **T** layer and they are staggered.

6.3.14. The 3·6P^U^TTP^Ca^TT(h) Structure of $Ca(UO_2)O_2$

The structure of calcium uranyl dioxide, $Ca(UO_2)O_2$, is related to the CaF_2 structure with Ca and U atoms replacing Ca of CaF_2 in alternate **P** layers. Oxygen atoms replace F of CaF_2 in both **T** layers. The UO_2^{2+} ion is treated as separate atoms because the linear UO_2^{2+} ion is aligned with the packing direction. U atoms are in **P** layers with oxygen atoms in **T** sites directly above and below at the same *A*, *B*, or *C* positions. The hexagonal structure is shown in Figure 6.26. This is based on a *ccp* sequence so each atom occurs at the *A*, *B*, and *C* positions. The heights of the atoms and their positions from the figure are listed below.

Height:	0	3	11	17	22	31	33	36	44	50	56	64	66	69	78	83	89	97	0
Atom:	U	O	O	Ca	O	O	U	O	O	Ca	O	O	U	O	O	Ca	O	O	U
Position:	*A*	*B*	*A*	*B*	*C*	*B*	*C*	*A*	*C*	*A*	*B*	*A*	*B*	*C*	*B*	*C*	*A*	*C*	*A*
Layer :	**P**	**T**	**T**	**P**	**T**	**T**	**P**	**T**	**T**	**P**	**T**	**T**	**P**	**T**	**T**	**P**	**T**	**T**	**P**

For UO_2^{2+} ions, the positions and heights of the atoms are: U (0) O (±11) at position *A*, U (33) O (22, 44) at position *C*, and U (66) O (56, 78) at position *B*. The second closest oxygen atoms are those at the same positions and bonded to U. The notation is **3·6P^U^TTP^Ca^TT(h)**. The $Ca(UO_2)O_2$ structure is based on an *ABC* sequence, but it is hexagonal,

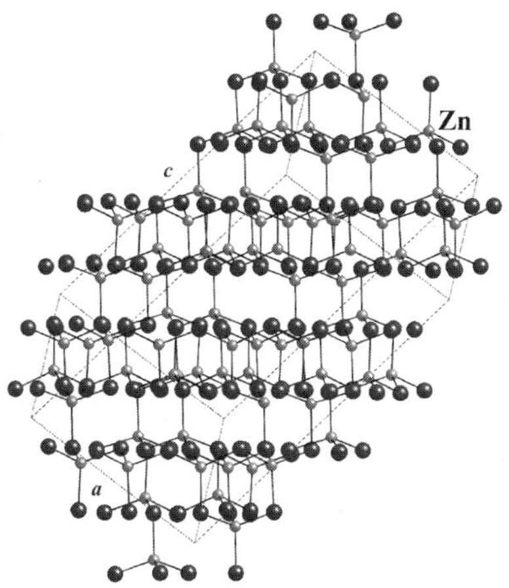

Figure 6.27. The layers of the zinc iodide structure.
(*Source*: *CrystalMaker*, by David Palmer, *CrystalMaker* Software Ltd., Begbroke Science Park, Bldg. 5, Sandy Lane, Yarnton, Oxfordshire, OX51PF, UK.)

not cubic. The interaction between Ca^{2+} and oxide ions is nondirectional, but there is covalent bonding in the linear UO_2^{2+} ions. The relationships between **P** and **T** sites accommodates the UO_2^{2+} ions very well.

6.3.15. The $3{\cdot}6PT_{3/8}T_{3/8}PT_{1/8}T_{1/8}(t)$ Crystal Structure of Zinc Iodide

Zinc iodide, ZnI_2, crystals are tetragonal, $\mathbf{D}_{4h}^{20}$, $I4_1/acd$, $a = 12.284$, and $b = 23.582$ Å, with 32 molecules in the unit cell. The **P** layers are filled by I atoms in an *ABC* sequence. All **T** layers are partially filled by Zn atoms with T^+ and T^- layer occupancies alternating as **3/8 3/8** and **1/8 1/8**. Because the **P** layers occur in groups of three (*ABC*) and **T** occupancies alternate, there are 18 layers repeating causing the very large unit cell (Figure 6.27). The notation is $\mathbf{3{\cdot}6PT_{3/8}T_{3/8}PT_{1/8}T_{1/8}(t)}$. The complex pattern of occupancies of **T** layers was revealed using *CrystalMaker* to expand layers. Projections of **P** layers with adjacent **T** layers resolved the problem. There are continuous chains of tetrahedra sharing apices along *c* for $\mathbf{T}_{3/8}$ layers and chains of tetrahedra without sharing apices along *c* for $\mathbf{T}_{1/8}$ layers.

6.3.16. The 2·3TPT Layer Crystal Structure of Molybdenite [MoS_2]

The hexagonal cell of MoS_2 contains two molecules, $\mathbf{D}_{6h}^{4}$, $P6_3/mmc$, $a_o = 3.1604$, and $c_o = 12.295$ Å, at 26°C. The structure of the mineral molybdenite, MoS_2, contains S–Mo–S "sandwiches" stacked as in CdI_2 with important differences. For each sandwich the layers of S atoms are in the *same* positions (*AA* and *BB*) as seen in Figure 6.28. The Mo atoms follow an *hcp* pattern *ABAB*. Each Mo is at the center of a trigonal prism formed by six S atoms of two adjacent layers. We saw that the octahedron of an **O** site can be distorted to give planar [$CuBr_4$] units (Section 5.5.8, $CuBr_2$) or linear [Te–Au–Te] units (Section 5.5.7,

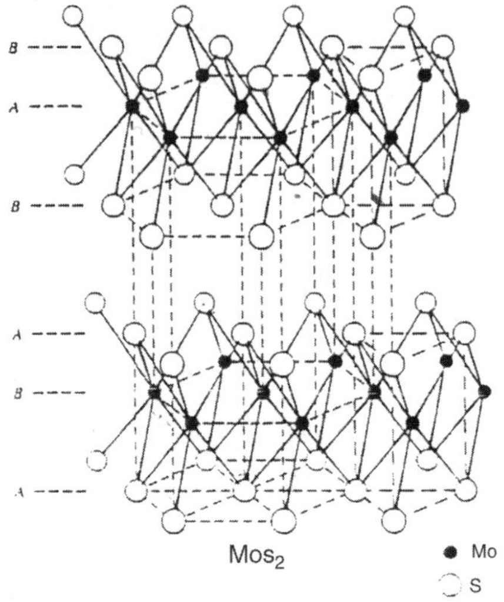

Figure 6.28. The **2·3TPT** structure of molybdenite (MoS_2). The Mo atoms are in **P** layers for an ***hcp*** arrangement with S filling **T** layers above and below each **P** layer. The **P** layer between the sandwiches is empty.

$AuTe_2$). Octahedral sites are not normally distorted to give a trigonal prism. For MoS_2 all atoms in **P** layers occur at A or B positions. This eliminates any *ccp*-type arrangement and it eliminates occupancy of **O** sites for any arrangement (**O** sites are only at C positions for an *hcp* arrangement). The basis for this structure must involve an *hcp* arrangement without **O** layers occupied. An arrangement $\mathbf{P}_A\mathbf{P}_A$ or $\mathbf{P}_B\mathbf{P}_B$ is not compatible with an *hcp* arrangement and, in fact, it is not a close-packed structure. Normally we begin by assigning large anions to **P** sites, but here it does not work. The structure is a **PTOT** arrangement if we assign Mo to **P** sites ($\mathbf{P}_B\mathbf{P}_A$) and S to sites in both **T** layers. Between two **P** layers there are two **T** layers in the sequence $|\mathbf{P}_B\mathbf{T}_A\mathbf{T}_B\mathbf{P}_A\mathbf{T}_B\mathbf{T}_A|$. This is the sequence seen in the figure. The **T** layers above and below a **P** layer are at the same positions ($\mathbf{T}_B\mathbf{P}_A\mathbf{T}_B$ or $\mathbf{T}_A\mathbf{P}_B\mathbf{T}_A$) so six S atoms form a trigonal prism around each Mo in a **P** site. Adjacent **T** layers are staggered ($\mathbf{T}_A\mathbf{T}_B$ or $\mathbf{T}_B\mathbf{T}_A$). The notation is **2·3TPT**. Within the sandwich Mo is strongly bonded to the S atoms. There are three Mo atoms sharing each S atom forming the base of a tetrahedron. The fourth Mo atom of the elongated tetrahedron belongs to the next layer. It is farther away because of the repulsion between S layers without cations between them.

Sandwich-type **TPT** layer structures are not expected for an *hcp* framework with filled **T** layers because the very close **T** sites at the same A or B positions. In MoS_2 the S—S distance between sandwiches is 3.66 Å, approximately the sum of the radii of S^{2-} ions, as expected with no cations between the S layers. The Mo—S distance is 2.35 Å and the S—S distance within the prism is 2.98 Å. The S—S bond length in S_8 molecule is 2.04 Å. The relatively short S—S distance between S atoms in adjacent **T** sites (2.98 Å) indicates some S—S bonding.

S—S bonding is not unusual in S compounds. The S—S bonding could be an important factor in placing bonded S atoms at the same *A* or *B* positions. The spacings along c_o of layers for MoS_2 are:

Height:	13	25	37	63	75	87
Atom:	S	Mo	S	S	Mo	S
Layer:	$\mathbf{T}_A$	$\mathbf{P}_B$	$\mathbf{T}_A$	$\mathbf{T}_B$	$\mathbf{P}_A$	$\mathbf{T}_B$
Spacing:		12	12	26	12	12

Because of the weak interaction between the sandwiches, MoS_2 is a useful solid lubricant. It feels slippery like graphite. The MoS_2 structure is found for WS_2, $MoSe_2$, WSe_2, and $MoTe_2$. Because of the greater polarizabilities of Se and Te compared with S, selenides and tellurides should be better solid lubricants than sulfides but the toxicity and unpleasant odors of selenides and tellurides make them poor choices.

6.3.17. The 2·3TPT Layer Crystal Structure of ReB_2

The ReB_2 crystal structure is hexagonal, $\mathbf{D}_{6h}^4$, $P6_3/mmc$, $a_o = 2.900$, and $c_o = 7.478$ Å, with two molecules per cell. This is a sandwich structure similar to that of MoS_2. Re atoms fill $\mathbf{P}_A$ and $\mathbf{P}_B$ layers with boron atoms filling both **T** layers (Figure 6.29) giving the notation **2·3TPT.** The **T** layers are at the same positions on each side of Re layers, as required in the **PTOT** system. Each boron atom has four Re atoms, forming a distorted tetrahedron and three close boron atoms as neighbors. Each Re atom has six neighbor boron atoms, forming a trigonal prism with each triangular face capped, giving *CN* 8. The spacing of layers are as follows:

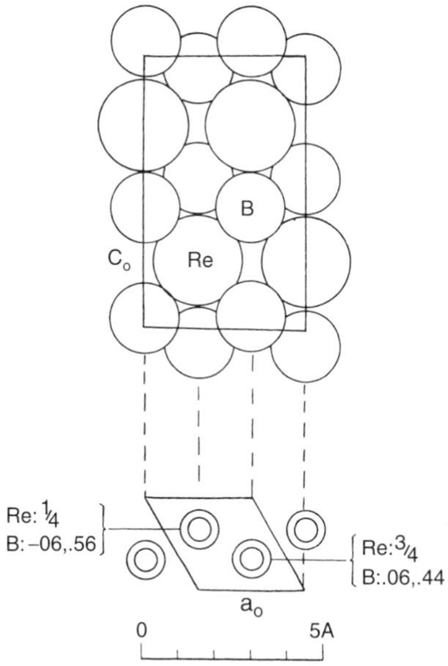

Figure 6.29. Two projections of the structure of ReB_2.
(*Source*: R.W.G. Wyckoff, *Crystal Structures*, Vol. 1, 2nd ed., Wiley, New York, 1963, p. 366.)

Height:	06		25		44	56		75		94	106		
Atom:	B		Re		B	B		Re		B	B		
Layer:	$\mathbf{T}_B$		$\mathbf{P}_A$		$\mathbf{T}_B$	$\mathbf{T}_A$		$\mathbf{P}_B$		$\mathbf{T}_A$	$\mathbf{T}_B$		
Spacing:		19		19	12		19		19	12			

The spacing along c_o between bonded boron layers in the sandwiches is 19, greater than for S layers of MoS_2, and the spacing between nonbonded layers (between sandwiches) is 12, much less than that for nonbonded S layers of MoS_2. The differences in spacing suggest that there is no B—B bonding within the sandwiches. There is much less repulsion between nonbonded boron layers because of the small atoms with few electrons and the positions of the **T** layers are shifted.

ReB_3 has a very similar hexagonal structure, with boron atoms added at the corners of the cell at 0 and 50. The added boron atoms are at the centers of octahedral sites. We can treat this as a **PTOT** structure or consider the three closely spaced boron layers (95, 0, 5 and 45, 50, 55) as a triple layer (Section 8.3.1).

6.3.18. The 4·2$\mathbf{P}_A$TT$\mathbf{P}_A$ Quadruple-Layered Crystal Structure of GaS

Gallium monosulfide, GaS, gives hexagonal crystals with four molecules per cell, $\mathbf{D}_{6h}^4$, $P6_3/mmc$, $a_o = 3.547$, and $c_o = 14.74$ Å. The structure is quadruple layered, S—Ga—Ga—S, with Ga—Ga bonding. The cages formed by the four layers are shown in Figure 6.30; atoms occur only at A and B positions and the sequence. The heights along c_o are:

Height:	10	17	33	40	60	67	83	90
Atom:	S	Ga	Ga	S	S	Ga	Ga	S
Layer:	$\mathbf{P}_A$	$\mathbf{T}_B$	$\mathbf{T}_B$	$\mathbf{P}_A$	$\mathbf{P}_B$	$\mathbf{T}_A$	$\mathbf{T}_A$	$\mathbf{P}_B$

The Ga atoms in **T** sites within the quadruple layer are at the same positions, but different from those for S. All atoms are at A and B positions. Each Ga has three S atoms at the base of a tetrahedron with

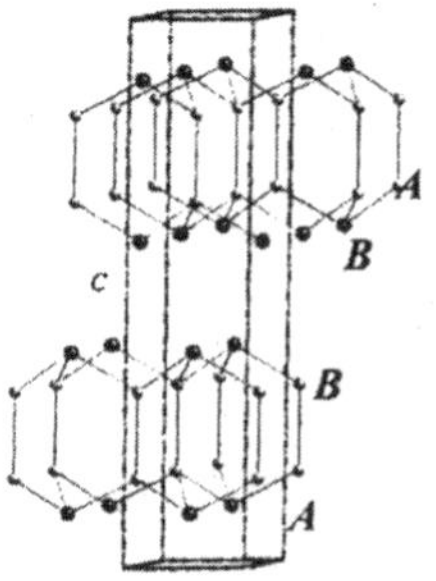

Figure 6.30. The structure of GaS.

(*Source*: Y. Matsushita, *Chalcogenide* crystal structure data library, Version 5.5B, Institute for Solid State Physics. The University of Tokyo, Tokyo, Japan, 2004; and *CrystalMaker*, by David Palmer, *CrystalMaker* Software Ltd., Begbroke Science Park, Bldg. 5, Sandy Lane, Yarnton, Oxfordshire, OX51PF, UK.)

another Ga at the apex, joining the two tetrahedra. The Ga layer is closer to the S layer because this is the center of the **T** site to the base of the tetrahedron. The bond distance for Ga—Ga is 2.52 Å and for Ga—S it is 2.30 Å. In the metal, Ga atoms occur in pairs with the shortest distance of 2.47 Å. Each layer is filled, giving the notation 4•2$\mathbf{P}_A\mathbf{TTP}_A$. InSe has the same structure.

6.3.19. The 4•4$\mathbf{P}_A\mathbf{TTP}_A$ Quadruple-Layered Crystal Structure of α-GaSe

The crystal structure of α-GaSe is also hexagonal and similar to that of GaS. The cages (Figure 6.31*a*) are the same as those for GaS (Figure 6.30). The hexagonal cell of α-GaSe is $\mathbf{C}_{4v}^4$, P6$_3$*mc*, $a = 3.755$, and $c = 31.99$ Å, containing six molecules. The hexagonal cell is very long because of the complex sequence of positions, $\mathbf{P}_A\mathbf{T}_B\mathbf{T}_B\mathbf{P}_A$ $\mathbf{P}_B\mathbf{T}_C\mathbf{T}_C\mathbf{P}_B$ $\mathbf{P}_C\mathbf{T}_B\mathbf{T}_B\mathbf{P}_C$ $\mathbf{P}_B\mathbf{T}_A\mathbf{T}_A\mathbf{P}_B$. Each layer is filled giving the notation 4•4$\mathbf{P}_A\mathbf{TTP}_A$.

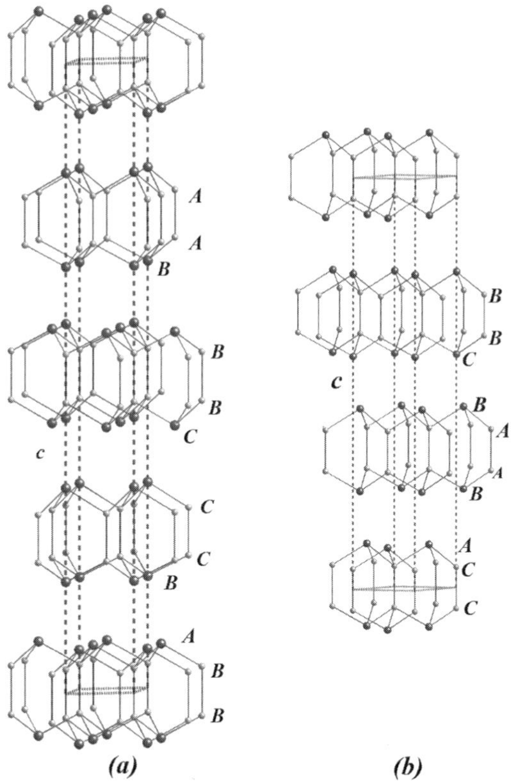

Figure 6.31. *(a)* Sequence of layers for α-GaSe. *(b)* Sequence of layers for β-GaSe.

(*Source*: Y. Matsushita, *Chalcogenide* crystal structure data library, Version 5.5B, Institute for Solid State Physics. The University of Tokyo, Tokyo, Japan, 2004; and *CrystalMaker*, by David Palmer, *CrystalMaker* Software Ltd., Begbroke Science Park, Bldg. 5, Sandy Lane, Yarnton, Oxfordshire, OX51PF, UK.)

6.3.20. The 6·2P_ATTP_A Quadruple-Layered Crystal Structure of β-GaSe

GaS and α-GaSe have two repeating quadruple layers following an *hcp* pattern. β-GaSe has a trigonal structure, C_{3v}^5, R3*m*, $a = 3.747$, $c = 23.91$ Å, with six molecules per cell and three repeating quadruple layers. There is a *ccp* pattern using all three positions for all atoms as seen in Figure 6.31*b*. The heights along c_o and positions are:

Height:	05	10	24	28	38	44	57	62	72	77	90	95	05	10
Atom:	Ga	Se	Se	Ga	Ga	Se	Se	Ga	Ga	Se	Se	Ga	Ga	Se
Layer:	T_C	P_A	P_B	T_A	T_A	P_B	P_C	T_B	T_B	P_C	P_A	T_C	T_C	P_A

The notation is **6·2PTTP** for $\mathbf{P}_A\mathbf{T}_C\mathbf{T}_C\mathbf{P}_A\mathbf{P}_B\mathbf{T}_A\mathbf{T}_A\mathbf{P}_B\mathbf{P}_C\mathbf{T}_B\mathbf{T}_B\mathbf{P}_C$. Both α-GaSe and β-GaSe have the same cages and differ in the stacking of the quadruple layers.

6.4. Structures of SiC

SiC was made accidentally by E.G. Acheson in 1891. He recognized its abrasive power and named it carborundum [carbo(n) and (co)rundum (Al_2O_3)]. SiC is formed from coke and SiO_2 by sublimation in an electric furnace. Its hardness on the Mohs scale is 9.5, next to that of diamond (10). Carborundum has excellent abrasive power because of its hardness and the tendency to fracture to give sharp cutting edges. SiC is chemically stable, and is oxidized in air only above 1,000°C. β-SiC is useful as a high-temperature semiconductor. SiC has been found as the mineral moissanite and in a meteorite found in Colorado.

Because Si and C form tetrahedral sp^3 bonds, one might expect the diamond structure, or for a more direct comparison, the **3·2PT** ZnS structure, but there are many other forms as well. All of the structures have tetrahedral SiC_4 and CSi_4 units joined. Each atom has a tetrahedral arrangement of four atoms of the other kind and 12 second-nearest neighbors of the same kind. These are the six atoms in the same layer and three in each of the adjacent layers. The different forms of SiC are not polymorphic forms. The hexagonal unit cells of different forms have the same dimensions $a_o = b_o = 3.078$ Å with the height of the cell, c_o, differing by an integral multiple of 2.518 Å. Structures of this type differing in layering are known as *polytypes*. For SiC, there are more than 70 polytypes ranging from $c_o = 5$ Å (2 layers) to $c_o = 1{,}500$ Å (594 layers), and there appears to be no limit to the number of layers. Only three simple close-packed structures are described below.

6.4.1. The 3·2PT Crystal Structure of β-SiC

The structure of β-SiC is cubic, similar to that of the zinc blende or cubic diamond (**3·2PT**), as shown in Figure 6.32. For the cubic cell $a = 4.439$ Å and for a hexagonal cell (with $c \perp$ to packing layers) $a_o = 3.073$ and $c_o = 7.57$ Å. This is probably the simplest SiC with the

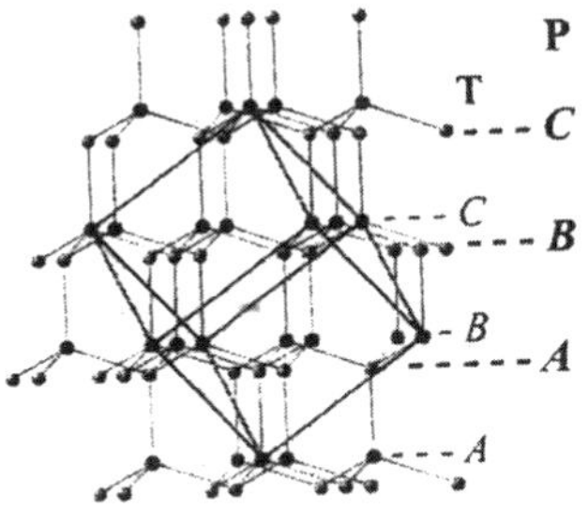

Figure 6.32. The **3·2PT** structure of β-SiC with **P**(C) and **T**(Si) layers labeled.

(*Source*: *CrystalMaker*, by David Palmer, *CrystalMaker* Software Ltd., Begbroke Science Park, Bldg. 5, Sandy Lane, Yarnton, Oxfordshire, OX51PF, UK.)

ABC sequence of **P** layers and **T** layers. Those related to the wurtzite structure (**2·2PT**) have more complicated sequences of layers.

6.4.2. The 6·2PT Crystal Structure of α-SiC

α-SiC is the most commonly used form in industry and the one most extensively studied. There are six packing layers in the sequence *ABCACB*, as shown in Figure 6.33. The unit cell contains six molecules and the notation is **6·2PT**, $C_{6\nu}^{6}$, P6$_3$, $a_o = 3.073$, and $c_o = 15.08$ Å. For the hexagonal forms of SiC the value of c_o is about 2.51 Å times the numbers of **P** layers, the same as the number of molecules per cell.

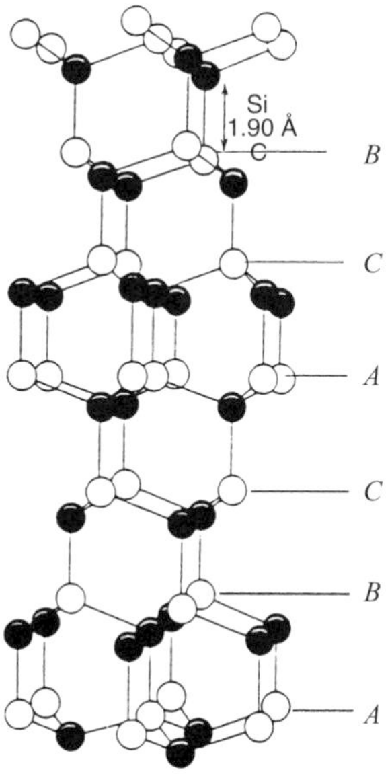

Figure 6.33. The **6·2PT** structure of hexagonal α-SiC showing the full 6 layer sequence (carbon open circles).

(*Source*: R. Verma and P. Krishna, *Polymorphism and Polytypism Crystals*, Wiley, New York, 1966, pp. 82, 87.)

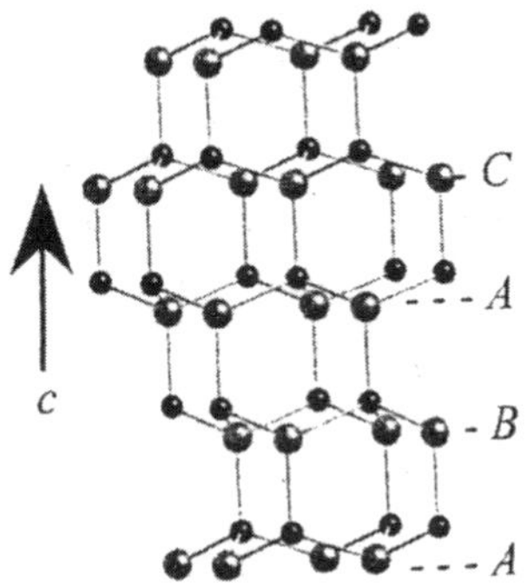

Figure 6.34. The **4·2PT** structure of hexagonal SiC with an *ABAC* sequence of layers.

(*Source*: *CrystalMaker*, by David Palmer, *CrystalMaker* Software Ltd., Begbroke Science Park, Bldg. 5, Sandy Lane, Yarnton, Oxfordshire, OX51PF, UK.)

6.4.3. The 4·2PT Crystal Structure of the *ABAC* Sequence for SiC

The hexagonal *ABAC* sequence SiC is shown in Figure 6.34. The unit cell ($\mathbf{C}_6^6$, P6$_3$) has $a_o = 3.078$ and $c_o = 10.046$ Å. Positions of **T** sites are the same as that of the **P** atom at the apex of the tetrahedron.

Chapter 7

Crystal Structures Involving P, T, and O Layers

We began with simple close-packed structures (*ccp* and *hcp*) encountered for metals and simple molecular elements and compounds. Because many metals have the body-centered cubic (*bcc*) structure under suitable conditions, we had to include this structure with the **3·2PTOT** notation. Next we considered about 60 structures involving **P** and **O** layers. There were about 40 structures involving **P** and **T** layers, some of which are more complex because there are two **T** layers for each **P** layer. Now we consider structures involving **P**, **T**, and **O** layers. As earlier, we begin with simple cases. Some molecular species give structures based on a *bcc* pattern involving all layers occupied by the individual molecules. Then we consider compounds with structures similar to that of CsCl. If atoms of a CsCl-type structure were the same it would be *bcc*. These are **3·2PTOT** structures with cations and anions occupying different layers. Next, we consider the more complex structures involving combinations of **P**, **T**, and **O** layers, some of which are partially filled. Finally, we consider structures involving **P**, **O**, and one **T** layer, including cases that have partially filled layers.

In Figure 3.12 we examined the addition of atoms in **O** or **T** layers to the face-centered cube (*ccp*) to give common structures such as NaCl, ZnS, Li_2O, etc. We might consider the fully occupied **PTOT** structure, such as CsCl (**3·2PTOT**), the general case from which the various combinations of occupancies can be derived by omitting some layers as shown in Scheme 7.1. This scheme is for the *ccp* pattern, and there are also many structures involving partial filling of layers. Here we use the *ccp* pattern because there are significant omissions of examples involving combinations of layers for the *hcp* pattern. In fact, there are no examples of the *hcp* **2·2PTOT** or **2·2PTT** structures with both **T** layers filled without chemical interaction. We have noted that there is strong repulsion if both **T** layers are filled because the **T** layers are very close for $\mathbf{T}_B\mathbf{P}_A\mathbf{T}_B$ and $\mathbf{T}_A\mathbf{P}_B\mathbf{T}_A$. These cases are encountered for partially filled **T** layers, and there are examples of the other combinations of layers for the *hcp* pattern. ReB_3 is the only example of a **2·2PTOT** structure with filled layers (Section 7.3.1). This structure is unusual, with long distances from $\mathbf{P}^{Re}$ layers to **T** layers and close spacings within the **TOT** layers suggest boron–boron bonding.

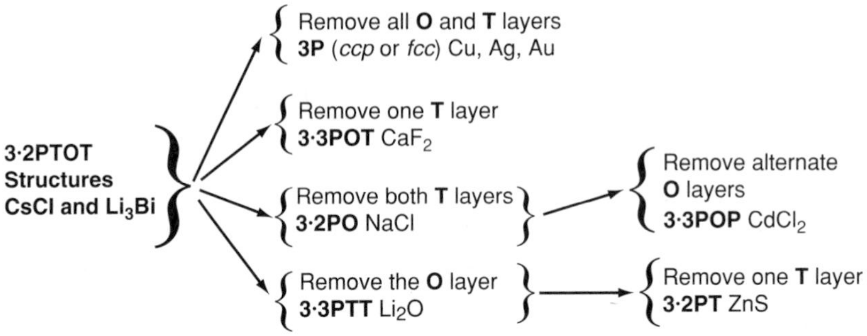

Scheme 7.1

It is important to recognize the arrangements of atoms for **P**, **O**, and **T** layers. Figure 7.1 shows these relationships for the *ccp* pattern. Figure 7.1***a*** shows the **P** positions for the *ccp* or face-centered cubic structure and Figure 7.1***b*** shows this framework of the NaCl structure with Na^+ ions added in **O** positions. For a *ccp* structure the **P** and **O** layers are identical and interchangeable. Figure 7.1***c*** shows one atom in each octant of the *ccp* unit cell. The atoms are in the **T** sites for both **T** layers. Figure 7.1***d*** shows the tetrahedral arrangement of atoms in one **T** layer in a *ccp* framework. The zinc blende (**3•2PT**) has this structure. Figure 7.1***e*** shows the Li_2O (**3•3PTT**) structure. The oxygen atoms are in **P** positions with Li in all **T** sites. The internal cube of **T** sites represents a combination of the tetrahedra of four atoms in each of the two **T** layers. Here the assignments of atoms to sites are in accord with the figures. Normally we assign larger ions to **P** sites (Cl^- for

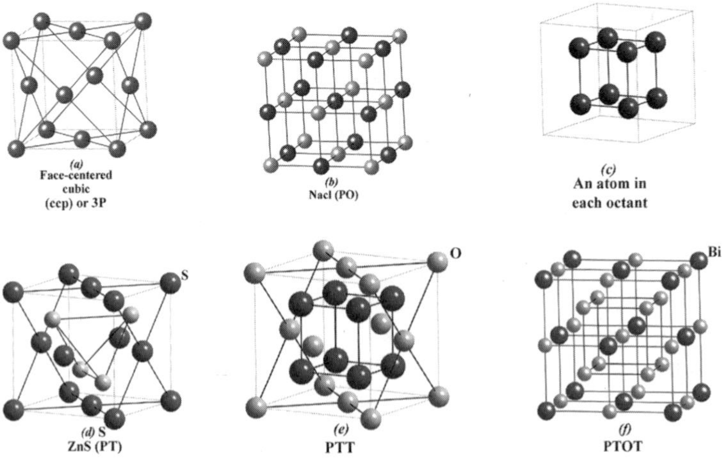

Figure 7.1. Relationships are shown among the different sites for *ccp* structures. (*Source*: *CrystalMaker*, by David Palmer, *CrystalMaker* Software Ltd., Begbroke Science Park, Bldg. 5, Sandy Lane, Yarnton, Oxfordshire, OX51PF, UK.)

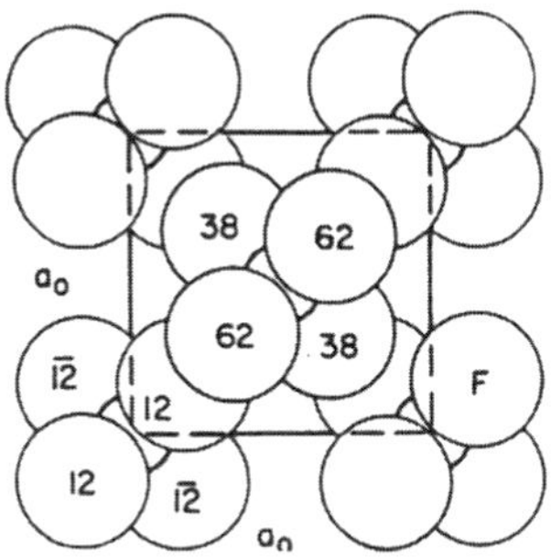

Figure 7.2. The cubic unit cell of SiF_4 projected along an axis of the cube. (*Source*: R.W.G. Wyckoff, *Crystal Structures*, Vol. 2, 2nd ed., Wiley, New York, 1964, p. 127.)

NaCl, S^{2-} for ZnS and O^{2-} for Li_2O). Figure 7.1*f* shows the BiF_3 structure (**3•4PTOT**) with Bi atoms in **P** layers (*ccp*) and F in the **O** layer and both **T** layers. The two **T** layers are equivalent. The combination of **P** and **O** layers are equivalent to the two **T** layers and the pairs of layers are interchangeable. For BiF_3 we could assign Bi to the **O** layer or to either of the **T** layers.

7.1. Molecular 3•2PTOT (*bcc*) Crystal Structures

In Chapter 4 we considered 12 molecular crystal structures based on *hcp* (**2P**) and *ccp* (**3P**) structures. As for metals, many molecular crystal structures are based on *bcc* (**3•2PTOT**) structures.

7.1.1. The 3•2PTOT Crystal Structure of SiF_4

SiF_4 is a gas with a short liquidus range, *b.p.* – 86°C and *m.p.* – 90.2°C. It crystallizes as the simple molecular *bcc* (**3•2PTOT**) structure as shown in Figure 7.2. At −145°C the bimolecular unit cell has $a_0 = 5.41$ Å with the Si—F distance 1.56 Å, T_d^3, $I\bar{4}3m$.

7.1.2. The 3•2PTOT(t) Crystal Structure of B_4Cl_4

The B_4Cl_4 molecule has a simple structure: a tetrahedron of four B atoms surrounded by a larger tetrahedron of four Cl atoms. The bonding within the B_4 tetrahedron involves three-center bonds, but the B—Cl bonds are single (electron pair) bonds. In Figure 7.3 the corner tetrahedra are centered at heights 50 and 150 with the one at the center of the cell centered at 100, forming a body-centered or **3•2PTOT(t)** structure. The tetrahedral edges are aligned with the edges of the unit cell. The crystals of B_4Cl_4 are tetragonal with two molecules in the cell, $\mathbf{D}_{4h}^{15}$, $P4_2/nmc$, $a_o = 8.09$, and $c_o = 5.45$ Å.

7.1.3. The 3•2PTOT(t) Crystal Structure of Pentaborane [B_5H_9]

The B_5H_9 molecules are square pyramidal. Each boron has one terminal H and there are four bridging H atoms in the edges of the square

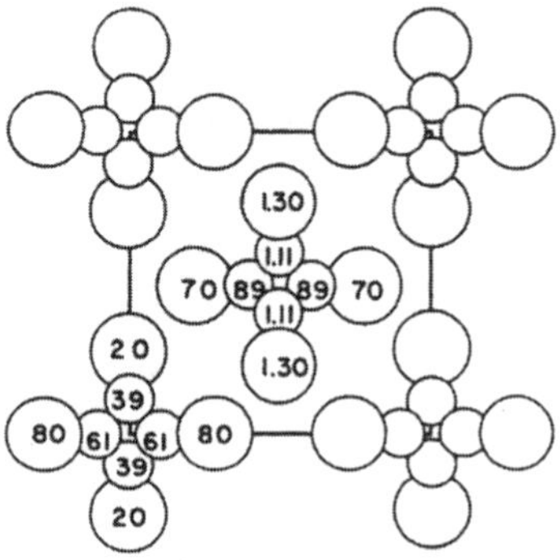

Figure 7.3. A projection of the tetragonal body-centered cell of B_4Cl_4.
(*Source*: R.W.G. Wyckoff, *Crystal Structures*, Vol. 1, 2nd ed., Wiley, New York, 1963, p. 180.)

base of the pyramid. The apical boron is bonded to the four boron atoms in the base by multiple-center bonds. The crystals melt at −47°C. The tetragonal cell contains two molecules, C_{4v}^9, I4*mm*, $a_o = 7.16$, and $c_o = 5.38$ Å. The cell is body centered (Figure 7.4) giving the notation **3·2PTOT(t)**. The C_{4v} axes of the molecules are aligned with the c_o axis.

7.1.4. The 3·2PTOT(m) Crystal Structure of S_4N_4

The molecular structure of S_4N_4 is unusual, having a tetrahedral arrangement of S atoms with the N atoms at the corners of a square.

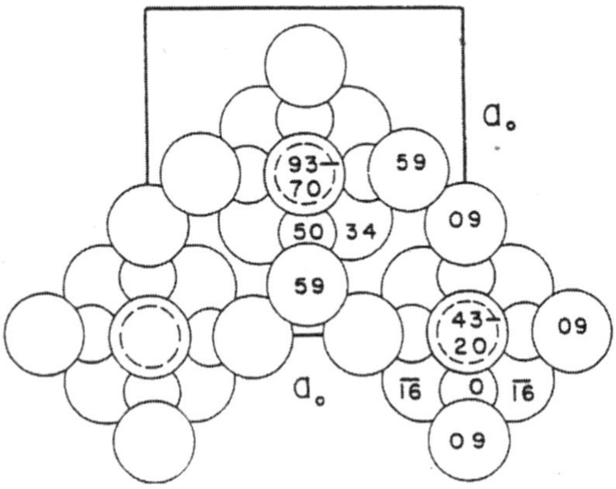

Figure 7.4. A projection along c_o of the atoms in the tetragonal cell of B_5H_9. Boron atoms are the smaller circles. The centers of the four B atoms in the square base of the pyramidal molecules are at 0 (and 100) and 50 giving a body-centered cell.
(*Source*: R.W.G. Wyckoff, *Crystal Structures*, Vol. 2, 2nd ed., Wiley, New York, 1964, p. 238.)

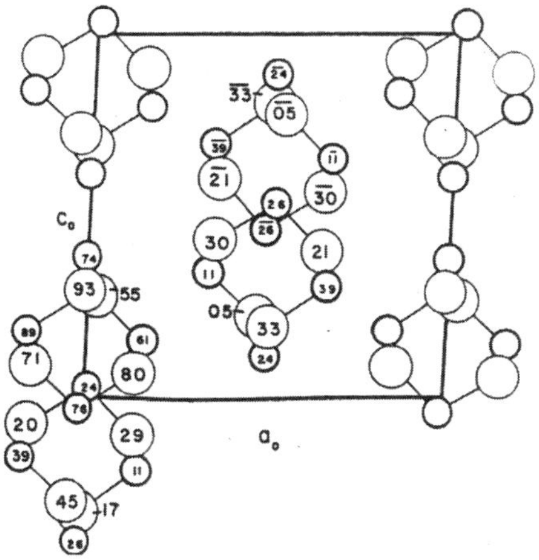

Figure 7.5. A projection along b_o of the monoclinic structure of S_4N_4. The larger circles are S atoms. Heights are shown.

(*Source*: R.W.G. Wyckoff, *Crystal Structures*, Vol. 1, 2nd ed., Wiley, New York, 1963, p. 178.)

Two S atoms of one tetrahedral edge span one edge of the square of N atoms and the two S atoms of the opposite tetrahedral edge span the opposite edge of the N_4 square. The S—S lengths for S atoms not linked by N is 2.58 Å. The S—S length between S_4N_4 molecules is 3.7 Å. The S_4N_4 molecule can be viewed as an eight-membered S—N ring in an extreme boat conformation giving a tetrahedral arrangement of S atoms (see Figure 7.5). The crystal structure is distorted, but it is a **3•2PTOT(m)** body-centered structure. The S_4N_4 molecules are not centered at the corners and center of the cell, but an N atom is approximately at these special positions. The monoclinic cell of S_4N_4 has four molecules, $\mathbf{C}_{2h}^5$, $P2_1/b$, $a_o = 8.75$, $b_o = 7.16$, $c_o = 8.65$ Å, and $\beta = 87.5°$.

7.1.5. The 3•2PTOT(t) Crystal Structure of $(NSF)_4$

The $(NSF)_4$ molecule is related to S_4N_4. Both have eight-membered S—N rings with an F atom attached to each S in $(NSF)_4$. The conformation of the S—N ring of $(NSF)_4$ is changed compared to S_4N_4 such that the N atoms are in a tetrahedral arrangement in $(NSF)_4$ and the S atoms form a flattened tetrahedron. Essentially the N—S ring forms an inner N_4 tetrahedron, a larger flattened tetrahedral arrangement of four S atoms, and there is an outer flattened tetrahedron of four F atoms—a tetrahedron within a tetrahedron within another tetrahedron! The crystal structure is tetragonal with two molecules per unit cell, $\mathbf{D}_{2d}^4$, $P\bar{4}2_1c$, $a_o = 9.193$, and $c_o = 4.299$ Å. It is a body-centered tetragonal or **3•2PTOT(t)** structure (see Figure 7.6). Even though the large molecule is complex, it has high symmetry and packs efficiently in the body-centered structure.

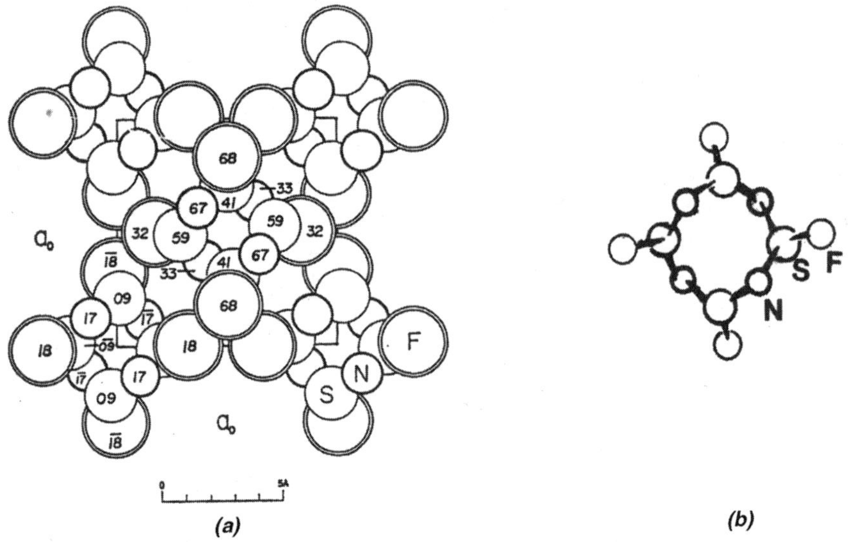

Figure 7.6. (*a*) The tetragonal structure of $(NSF)_4$ projected along the c_o axis. (*b*) The $(NSF)_4$ molecule.

(*Source*: R.W.G. Wyckoff, *Crystal Structures*, Vol. 4, 2nd ed., Wiley, New York, 1965, p. 105.)

7.1.6. The 3·2PTOT(t) Crystal Structures of XeF_2 and XeF_4

The inert gases form few compounds, but Xe reacts with F_2 to form XeF_2, XeF_4, and XeF_6. XeF_2 molecules are linear, with three equatorial unshared electron pairs (lone pairs) and two linear bonds. XeF_2 crystals (*m.p.* 129°C) are tetragonal ($a_o = 4.32$ and $c_o = 6.99$ Å). The body-centered cell is tetragonal because the linear molecules are aligned along the c_o direction (see Figure 7.7). The notation is **3·2PTOT(t)**.

XeF_4 is square planar as expected for a molecule with four electron bonding pairs and two unshared electron pairs on Xe. Crystals of XeF_4 are monoclinic ($P2_1/n$) with two molecules per cells ($a = 5.05$, $b = 5.92$, $c = 5.77$ Å, and $\beta = 99.6°$). It is a body-centered cell and the notation is **3·2PTOT(t)**.

7.1.7. The Body-Centered Crystal Structure of Bis(salicylaldehydato)copper(II)

Crystals of bis(salicylaldehydato)copper(II), $Cu[C_6H_4O(CHO)]_2$, are monoclinic with two molecules per cell, $\mathbf{C}_{2h}^5$, $P2_1b$, $a_o = 8.72$, $b_o = 6.19$, $c_o = 11.26$ Å, and $\beta = 104.8°$. Figure 7.8 shows that the Cu atoms are at the corners and in the center of the cell [**3·2PTOT(m)**]. The long direction of the molecules is aligned along b_o. The four oxygen atoms coordinated to Cu are nearly square planar, as seen in the structural formula.

7.1.8. The Body-Centered Crystal Structure of Pd_4Se

Tetrapalladium selenide, Pd_4Se, is tetragonal with two molecules per cell, $\mathbf{D}_{2d}^4$, $P\bar{4}2_1c$, $a_o = 5.2324$, and $c_o = 5.6470$ Å. At the center of the cell

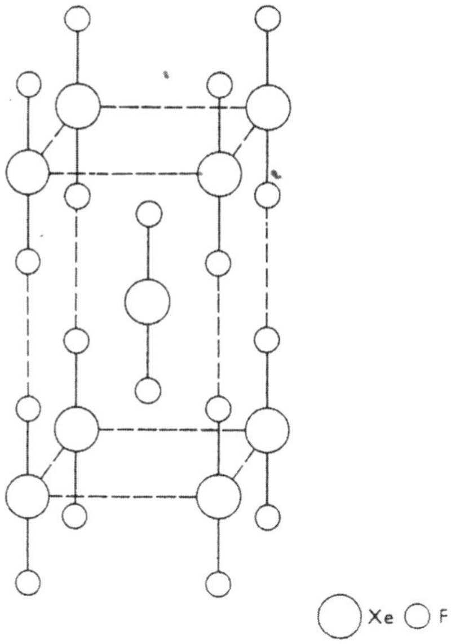

Figure 7.7. The body-centered molecular structure of XeF_2.

in Figure 7.9 there is a tetrahedral Pd_4Se molecule (Se at 1/2, two Pd at 15, and two Pd at 85) and there is one Pd_4Se at each corner. Four Se atoms are shown at zero, there are four more at 100. Only Pd atoms in the cell are shown. The body-centered tetragonal structure of Pd_4Se is described as **3·2PTOT(t)**. Pd_4S has the same structure.

7.2. The 3·2PTOT Crystal Structures of the MX Type

7.2.1. The 3·2PTOT Crystal Structure of CsCl

The CsCl structure is well known. The unit cell reminds us of the *bcc* structure of many elements, but the ion at the center of the unit cell is

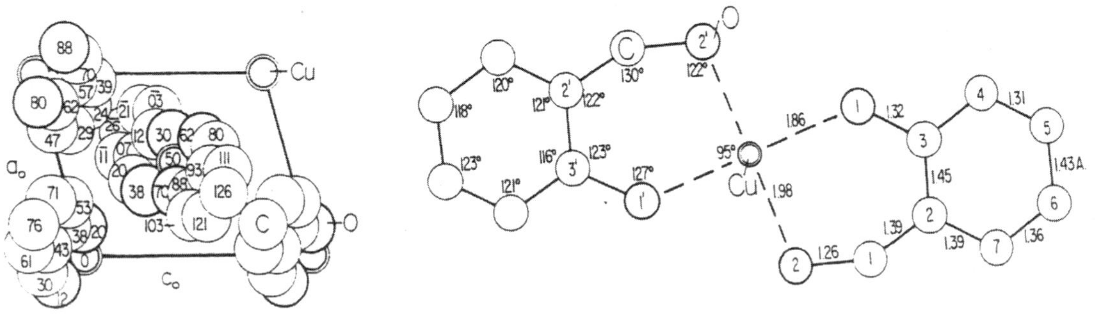

Figure 7.8. A projection along b_o for one monoclinic form of bis(salicylaldehidato)copper(II).
(*Source*: R.W.G. Wyckoff, Vol. 6, Part 1, *Crystal Structures, The Structures of Aromatic Compounds*, Wiley, New York, 1966, p. 229.)

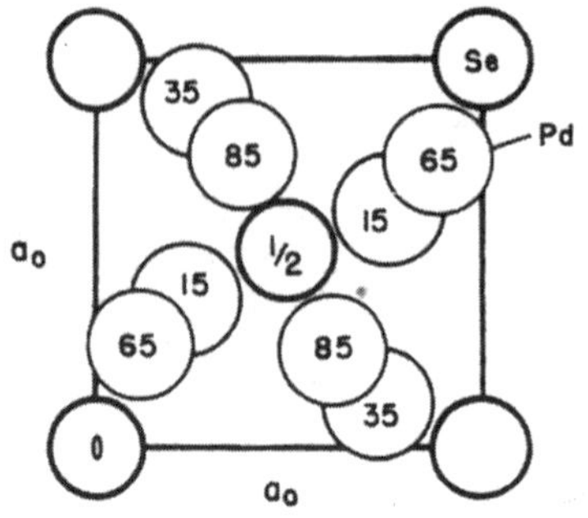

Figure 7.9. A projection along the c_o axis of the tetragonal structure of Pd_4Se. The Se atoms are smaller heavy line circles.

(*Source*: R.W.G. Wyckoff, *Crystal Structures*, Vol. 2, 2nd ed., Wiley, New York, 1964, p. 137.)

different from those at the corners of the cube. Usually models have Cs^+ at the center, emphasizing the *CN* 8 for Cs^+. This is an important distinction from the NaCl structure (*CN* 6, **3·2PO**) for the larger Cs^+ ion. However, the roles of the cations and anions can be reversed for CsCl, as for NaCl. In the **PTOT** system one ion (here Cl^- is chosen) fills the **P** and **O** layers with the other ion (here Cs^+) fills both **T** layers in a *ccp* pattern (**3·2PTOT**). In Figure 7.10 the layers are labeled. This figure emphasizes the close-packed layers. The cell is cubic and the packing direction (vertical in the figure) is along the body diagonal of the cell. The unit cell is a cube of eight gray balls (Cl^-) with one black ball (Cs^+) at the center. The cell contains one molecule of CsCl ($1Cs^+ + 8 \times 1/8\ Cl^-$). The sequence of layers, their positions and heights are:

Height:	0	17	30	56	67	83	|	0
Packing layer:	P(O)	T	O(P)	T	P(O)	T	|	O(P)
CsCl ions:	Cl	Cs	Cl	Cs	Cl	Cs		Cl
Position:	A	B	C	A	B	C		A

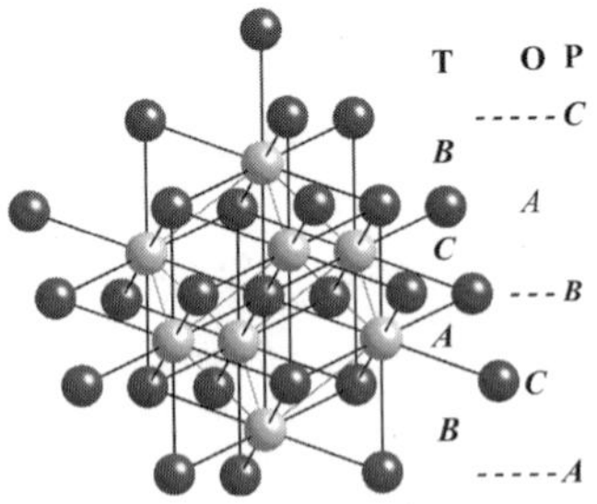

Figure 7.10. A model of the CsCl cell. Cl (black balls) are in **P** and **O** layers with Cs (gray balls) in both **T** layers.

(*Source*: *CrystalMaker*, by David Palmer, *CrystalMaker* Software Ltd., Begbroke Science Park, Bldg. 5, Sandy Lane, Yarnton, Oxfordshire, OX51PF, UK.)

TABLE 7.1. **Compounds with the CsCl structure.**

CsBr	CsCl	CsCN	CsI	$CsNH_2$	CsSH
CsSeH	RbCl	TlBr	TlCl	TlCN	TlI
ThTe	ND_4Br	ND_4Cl	NH_4Br	NH_4Cl	
NH_4I	AgCd	AgCe	AgLa	AgMg	AgZn
AlNd	AlNi	AuCd	AuMg	AuZn	BeCo
BeCu	BePd	CaTl	CdCe	CdLa	CdPr
CdPd	CuZn	LiAg	LiHg	LiTl	MgCe
MgHg	MgLa	MgPr	MgSr	MgTl	SrTl
TlBi	TlSb	ZnCe	ZnLa	ZnPr	

There are six (**3·2**) layers in the repeating unit because **P** and **O** positions are equivalent in a *ccp* structure. The space group for the cubic is $\mathbf{O}_h^1$, Pm3m, $a_o = 4.123$ Å. More than 50 compounds are known with the CsCl structure (Table 7.1), including pseudobinary compounds, $K(NF_2)$, $N(CH_3)_4(ICl_2)$, $PCl_4(ICl_2)$, and $NH_4(ClO_2)$.

7.2.2. The 3·2PTOT Crystal Structures of CaB_6 and LaB_6

In CaB_6 the B_6^{2-} anion is a perfect octahedron. For CaB_6, the crystals are cubic and have the CsCl or **3·2PTOT** structure. There is one molecule per cell, Pm3m, $\mathbf{O}_h^1$, $a_o = 4.1450$Å. Figure 7.11***a*** shows the unit cell of CaB_6 with heights and a packing diagram. Figure 7.11***b*** shows the corresponding cell for LaB_6 with La in the center. Many compounds have this structure as shown in Table 7.2.

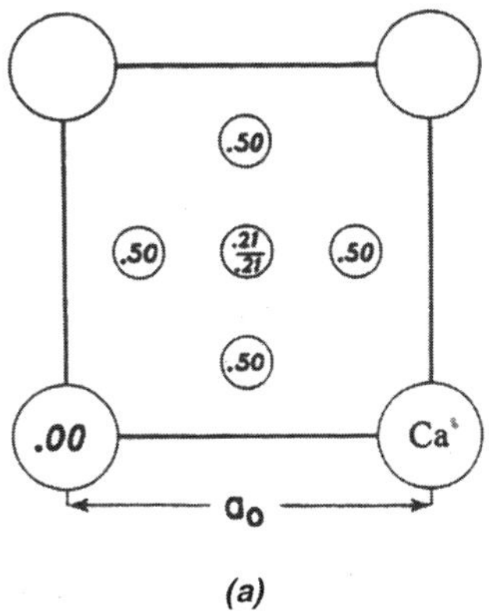

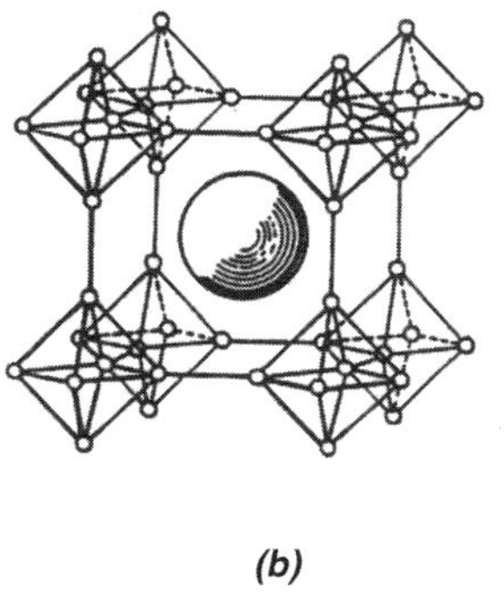

Figure 7.11. (*a*) The cubic unit ceil of CaB_6. Ca atoms are at the corners with B_6 in the center. (*b*) The structure of LaB_6 with La in the center with B_6 octahedra at the corners of the cube.

(*Source*: R.W.G. Wyckoff, *Crystal Structures*, Vol. 2, 2nd ed., Wiley, New York, 1964, p. 202.)

TABLE 7.2. **Compounds with the CaB_6 structure.**

BaB_6	CeB_6	DyB_6	ErB_6	EuB_6
GdB_6	HoB_6	LaB_6	LuB_6	NdB_6
PrB_6	PuB_6	SiB_6	SmB_6	SrB_6
TbB_6	ThB_6	TmB_6	YB_6	YbB_6

7.2.3. The 3·4PTOT Crystal Structure of BiF_3

The crystal structure of BiF_3 is cubic, $\mathbf{O}_h^5$, $Fm3m$ with $a = 5.853$Å. The cell has Bi^{3+} ions in a cubic face-centered arrangement, **P** positions, with F^- ions in all **O** and **T** sites. Bi atoms are at the center of a cubic arrangement of eight F at 2.438 Å with F atoms capping four faces at 2.815 Å. Those in the cube are in **T** sites, and those capping faces are in **O** sites. The cell is shown in Figure 7.1*f* and it is the same at that of $BiLi_3$ in Figure 9.28*a*.

7.2.4. The 3·2PTOT(t) Crystal Structure of $TlAlF_4$

Surprisingly, $TlAlF_4$ is a pseudobinary compound containing octahedral AlF_6 units rather than the expected tetrahedral AlF_4^- units. The AlF_6 octahedron is at the center of a tetragonal cell with Tl atoms at the corners. Two F atoms along the *c* direction are not bonded to another Al, but they are bonded to four Tl atoms in top and bottom faces. The F atoms in the tetragonal faces are shared with another Al atom (see Figure 7.12). Each Tl atom is surrounded by a cube of eight F atoms. The following compounds have the $TlAlF_4$ structure: $LiAlF_4$, $NaAlF_4$, KVF_4, $KFeF_4$, $TlCrF_4$, $TlGaF_4$, and β-$RbFeF_4$.

7.2.5. The 3·2PTOT Crystal Structure of $PbMo_6S_8$

There is a family of compounds known as Chevrel phases (Roger Chevrel, 1971) with the general formula $M_xMo_6X_8$, where M represents a rare earth metal, a transition or main group metal, and possibly a vacancy; X can be S, Se, or Te. The idealized structure of $PbMo_6S_8$ is

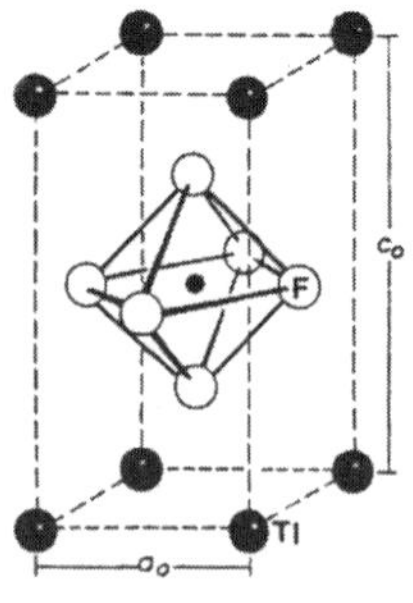

Figure 7.12. The tetragonal cell of $TlAlF_4$.

(*Source*: R. Roy, Ed., *The Major Ternary Structural Families*, Springer-Verlag, New York, 1974, p. 105.)

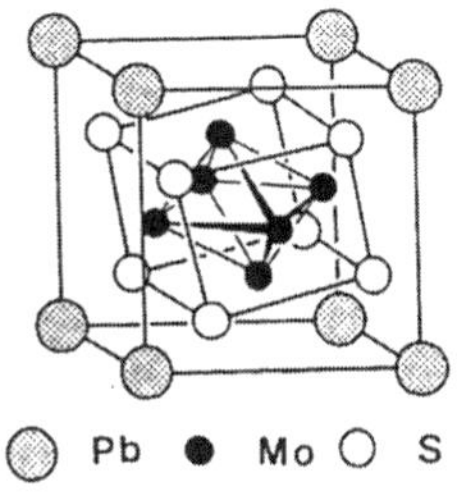

Figure 7.13. The $PbMo_6S_8$ cell. The Mo atoms are in the centers of the faces of the S_8 cube.

shown in Figure 7.13. It is a CsCl-like **3·2PTOT** structure containing the Mo_6X_8 cluster at the center of the cube formed by eight Pb atoms. The Mo_6 octahedron is elongated along a C_3 axis to form a trigonal antiprism inside a cube formed by eight S atoms. Compounds of metal clusters such as $Mo_6Cl_8^{4+}$ and $Mo_6S_8^{4-}$ ions are well known. The Chevrel phases are of interest as superconductors for which the superconductivity persists in the presence of high magnetic fields. As for the perovskite-type superconductors, many variations in composition of Chevrel phases have been investigated.

7.2.6. The 3·2PTOT Crystal Structure of NH_4Cl

NH_4Cl has the **3·2PTOT** or CsCl structure below 184.3°C. This structure is found for NH_4Br below 137.8°C and for NH_4I below −17°C. Above these temperatures these halides have the NaCl (**3·2PO**) structure. These ammonium halides have been studied extensively by X-ray, electron, and neutron diffraction techniques. Although halogens larger than F form weak hydrogen bonds, the **3·2PTOT** structure allows the most favorable orientation for hydrogen bonding for NH_4^+ ions aligned with the tetrahedral edges aligned with the cubic face diagonals (Figure 7.14).

7.2.7. The 3·2PTOT(o) Crystal Structure of NH_4HF_2

NH_4HF_2 forms the **3·2PTOT(o)** or CsCl-type structure with distortions to give an orthorhombic unit cell (Figure 7.15), $\mathbf{D}_{2h}^7$, *Pmna*, $a_o = 8.408$, $b_o = 8.163$ and $c_o = 3.670$Å. The cell in the figure corresponds to the upper half ot the **PTOT** cell in Figure 7.1f. The NH_4^+ ions are slightly above or below the centers of the octants. The H atoms are in **P**

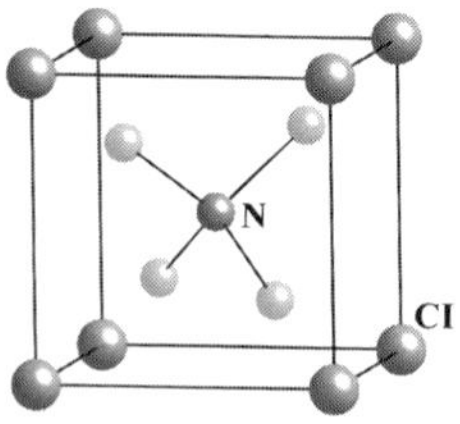

Figure 7.14. The **3·2PTOT** structure of NH_4Cl. The tetrahedral NH_4^+ ion is at the center of the cube.

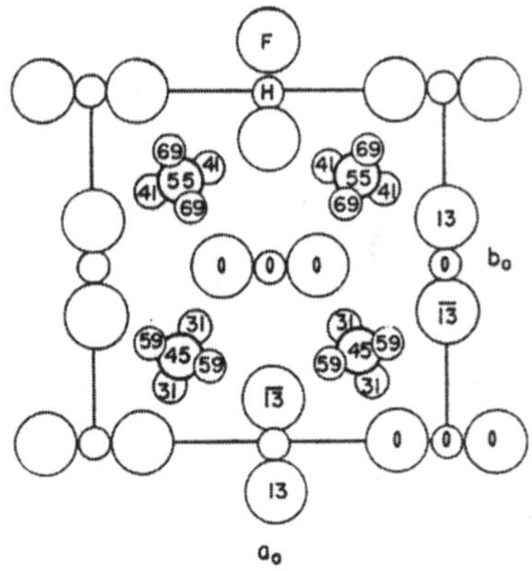

Figure 7.15. A projection of the orthorhombic cell of NH_4HF_2 along its c_o axis.
(*Source*: R.W.G. Wyckoff, *Crystal Structures*, Vol. 2, ed., Wiley, New York, 1964, p. 280.)

and **O** sites with alternate orientations of HF_2^- ions for hydrogen bonding to NH_4^+. This structure is found for NH_4N_3 containing the linear N_3^- ion that aligns for formation of strong hydrogen bonds to NH_4^+.

7.2.8. The 3·2PTOT(t) Crystal Structure of NH_4CN

The structure of NH_4CN is similar to that of NH_4Cl (**3·2PTOT**), but the CN^- ions are in planes parallel to the base of the tetragonal cell (Figure 7.16) ($\mathbf{D}_{4h}^{10}$, $P4/mcm$, $a_o = 4.16$, and $c_o = 7.64\text{Å}$, at 35°C). There are two molecules in the unit cell and the CN^- ions are aligned at small angles to one a_o axis. The positions of the hydrogens of NH_4^+ ions are not shown. The crystal of NH_4CN shrinks significantly only along the c_o

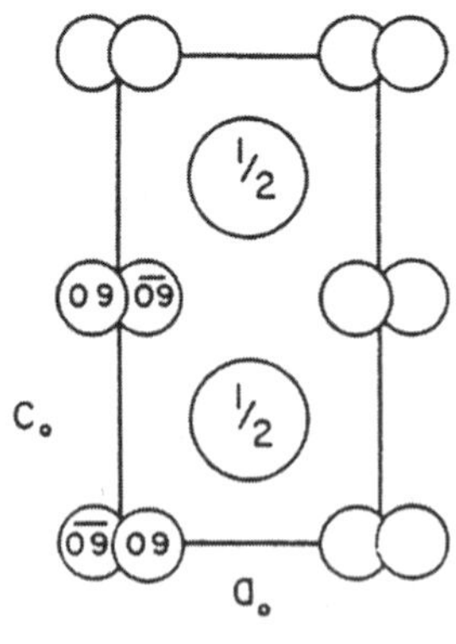

Figure 7.16. A projection along an a_o axis of the tetragonal structure of NH_4CN. The large circles represent NH_4^+ ions. The C and N atoms of cyanide ion are not distinguished.
(*Source*: R.W.G. Wyckoff, *Crystal Structures*, Vol. 1, 2nd ed., Wiley, New York, 1963, p. 106.)

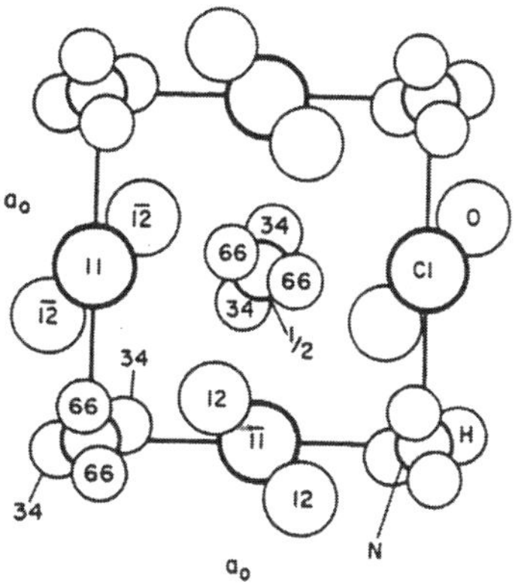

Figure 7.17. A projection along the c_o axis of the tetragonal structure of NH_4ClO_2. Heights are shown.

(*Source*: R.W.G. Wyckoff, *Crystal Structures*, Vol. 2, ed., Wiley, New York, 1964, p. 306.)

axis as the temperature is lowered. This is expected if the preferential vibration of the CN^- ion is normal to its axis.

7.2.9. The 3·2PTOT(t) Crystal Structure of NH_4ClO_2

The NH_4ClO_2 structure is the **3·2PTOT(t)** or CsCl-type, but it is bimolecular and tetragonal because the angular ClO_2^- ions (110.5°) have their C_2 axis aligned with the c_o axis (Figure 7.17) ($\mathbf{D}_{2d}^3$, $P\bar{4}2_1m$, $a_o = 6.362$, and $c_o = 3.823$Å, at 24°C). The orientations of ClO_2^- and NH_4^+ provide strong hydrogen bonding. The ClO_2^- ions have their "centers" at height zero (and 100) (O at 12 and −12 and Cl at 11 and −11) forming an elongated cube with NH_4^+ at the center (height 50) and the NH_4^+ ions form an elongated cube with ClO_2^- ions at the center.

7.2.10. The 3·2PTOT(t) Crystal Structure of $N(CH_3)_4ICl_2$

Tetramethylammonium dichloroiodate(1−), $N(CH_3)_4ICl_2$, and PCl_4ICl_2 have a **3·2PTOT(t)** CsCl-like structure with some distortion to accommodate the linear ICl_2^- ion (Figure 7.18). The bimolecular unit cell is tetragonal for $N(CH_3)_4ICl_2$, $\mathbf{D}_{2d}^3$, $P\bar{4}2_1m$, $a_o = 9.26$, and $c_o = 5.68$Å. The ICl_2^- ions are in planes parallel to the base of the cell with two orientations at 90°, and are slightly above or below the center of each octant of the **PTOT** in Figure 7.1.

7.2.11. The 3·2PTOT(h) Crystal Structure of $[Ni(H_2O)_6][SnCl_6]$

$[Ni(H_2O)_6][SnCl_6]$ has two large octahedral ions and, as expected, it has the **3·2PTOT(h)** or CsCl-like structure (Figure 7.19). The index (**3**) indicates that the sequence of layers is *A*, *B*, *C*. The cell is distorted to give a unimolecular rhombohedral unit cell, $\mathbf{C}_{3i}^2$, $R\bar{3}$, $a_o = 7.09$Å,

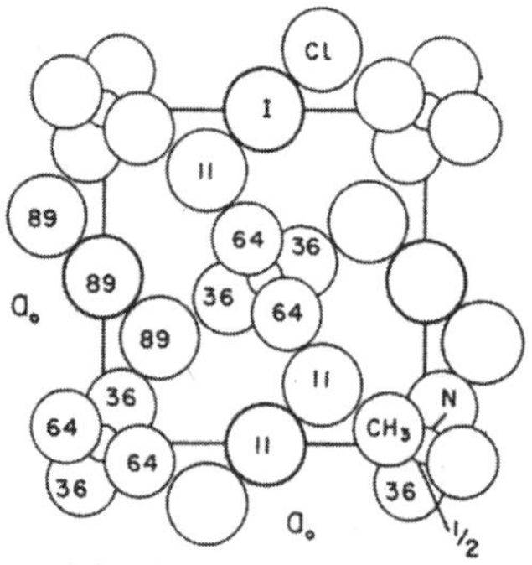

Figure 7.18. A projection along the c_o axis of the tetragonal structure of $N(CH_3)_4ICl_2$. Heights are shown.

(*Source*: R.W.G. Wyckoff, *Crystal Structures*, Vol. 2, 2nd ed., Wiley, New York, 1964, p. 298.)

and $\alpha = 96.75°$. This structure is found for many complexes of six-coordinate cations and anions (Table 7.3).

7.2.12. The 3•4PTOT(t) Crystal Structure of KN_3

The **3•4PTOT(t)** CsCl-like structure of KN_3 [potassium azide or potassium trinitride(1–)] is tetragonal with four molecules per unit cell, $\mathbf{D}_{4h}^{18}$, $I4/mcm$, $a_o = 6.094$, and $c_o = 7.056$Å. It is distorted because of the linear azide ion. The azide ions are in planes parallel to the base

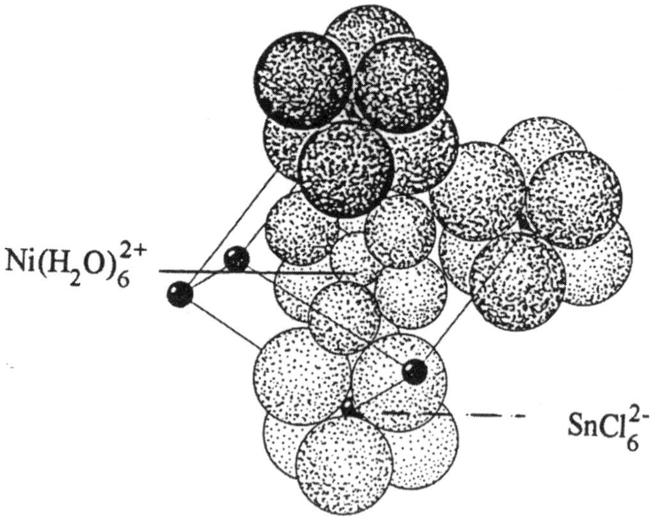

Figure 7.19. A packing drawing showing part of the unit rhombohedron of $[Ni(H_2O)_6][SnCl_6]$. Sn atoms are black and three of them are shown as $[SnCl_6]^{2-}$ octahedra. The $[Ni(H_2O)_6]^{2+}$ octahedron is at the center (H atoms are not shown).

(*Source*: R.W.G. Wyckoff, *Crystal Structures*, Vol. 3, 2nd ed., Wiley, New York, 1965, p. 796.)

TABLE 7.3. **Complexes with the 3·2PTOT structure of $[Ni(H_2O)_6]$ $[SnCl_6]$.**

$[Co(NH_3)_6][Co(CN)_6]$	$[Mg(H_2O)_6][SiF_6]$	$[Ni(H_2O)_6][SiF_6]$
$[Co(NH_3)_5H_2O][Co(CN)_6]$	$[Mg(H_2O)_6][SnF_6]$	$[Zn(H_2O)_6][SiF_6]$
$[Co(NH_3)_4(H_2O)_2][Co(CN)_6]$	$[Mg(H_2O)_6][TiF_6]$	$[Zn(H_2O)_6][SnF_6]$
$[Co(NH_3)_5H_2O][Fe(CN)_6]$	$[Mn(H_2O)_6][SiF_6]$	$[Zn(H_2O)_6][TiF_6]$
$[Co(H_2O)_6][SiF_6]$	$[Ni(H_2O)_6][SnCl_6]$	$[Zn(H_2O)_6][ZrF_6]$

with two orientations at 90° (see Figure 7.20). The figure is not drawn to show that it is a CsCl-type structure. A model of CsCl shows a cube of one set of ions with the other ion at the center. We can visualize this type of cell using the central K^+ (at one-quarter), those at the corners of the right edge (one-quarter) and one in the center of the cell to the right (one-quarter) as the base (and top at three-quarters) with N_3^- (at one-half) at the center. This structure is encountered in RbN_3, KHF_2, α-$RbHF_2$, α-$CsHF_2$, and the cyanates KCNO, RbCNO, CsCNO, and TlCNO.

7.2.13. The 3·4PTOT Crystal Structure of K_3C_{60}

The almost spherical C_{60} molecule gives the face-centered cubic (*ccp* or **3P**) structure (Figure 4.11). C_{60} reacts with K vapor to form K_3C_{60}. The K atoms fill the **O** and both **T** layers to give the **3·4PTOT** structure with C_{60} molecules in **P** layers. Figure 7.21*a* shows the NaCl structure showing the **P** (dark spheres) and **O** positions (light spheres). Figure 7.21*b* shows the **T** positions for both **T** layers. The K atoms fill all of the **O** and **T** sites. These are combined with the *ccp* structure of solid C_{60} in Figure 7.21*c*.

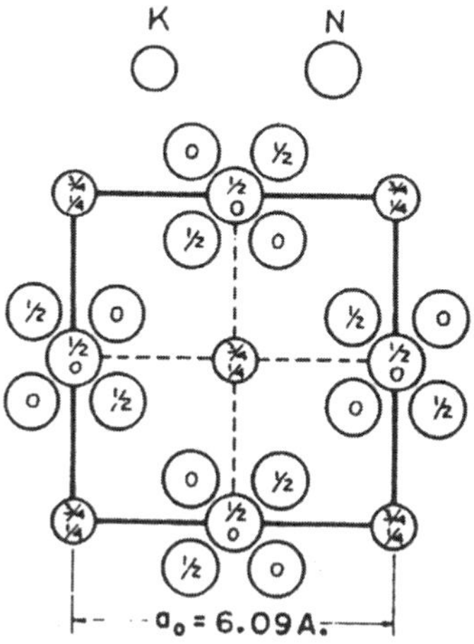

Figure 7.20. A projection on the base of the tetragonal structure of KN_3. (*Source*: R.W.G. Wyckoff, *Crystal Structures*, Vol. 2, 2nd ed., Wiley, New York, 1964, p. 278.)

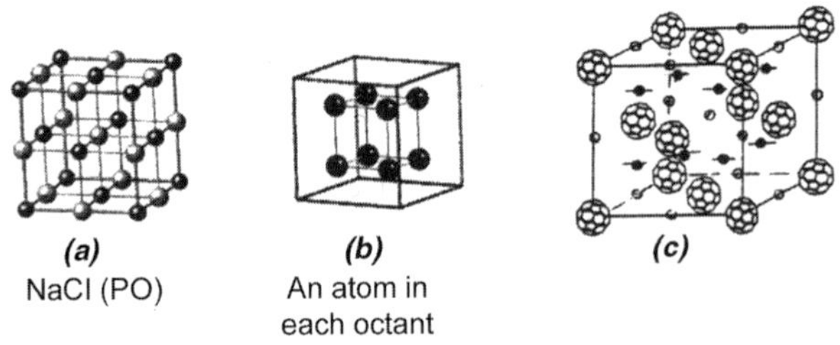

Figure 7.21. (*a*) The NaCl structure using **P** and **O** sites. (*b*) The sites of both **T** layers. (*c*) The **3·4PTOT** structure of K_3C_{60}. The C_{60} molecules are in **P** sites with K in **O** (open circles) and **T** (filled circles) sites.

(*Source*: *CrystalMaker*, by David Palmer, *CrystalMaker* Software Ltd., Begbroke Science Park, Bldg. 5, Sandy Lane, Yarnton, Oxfordshire, OX51PF, UK.)

7.2.14. The $3{\cdot}4P^{Tl}TO^{Tl}T(t)$ Structure of TlSe

The compound TlSe is unexpected for these elements because neither $Tl^{+}Se^{-}$ nor $Tl^{2+}Se^{2-}$ is likely. It is actually $Tl^{+}[Tl^{3+}Se_2]^{-}$, showing expected oxidation states for each element. The space group of the tetragonal cell is $\mathbf{D}_{4h}^{18}$, $I4/mcm$, $a = 8.02$, and $c = 6.791$ Å. The unit cell has Tl at corners and centers of faces (**P** positions) and in centers of edges and at the center (**O** positions). Se atoms fill both **T** layers. Tl(I) ions are at the centers of Archimedes antiprisms, *CN* 8 with Tl—Se distance 3.403 Å. Tl(III) ions are in $TlSe_4$ tetrahedra with the Tl—Se distance 2.641 Å. There are eight TlSe molecules in the unit cell with 12 packing layers to give the notation $\mathbf{3{\cdot}4P^{Tl}TO^{Tl}T(t)}$. In Figure 7.22*a*, the antiprisms are shown at the corners with another not outlined in the center. Tetrahedra are in the centers of cell edges linking the antiprisms. The antiprisms are stacked along the *c* axis, sharing square

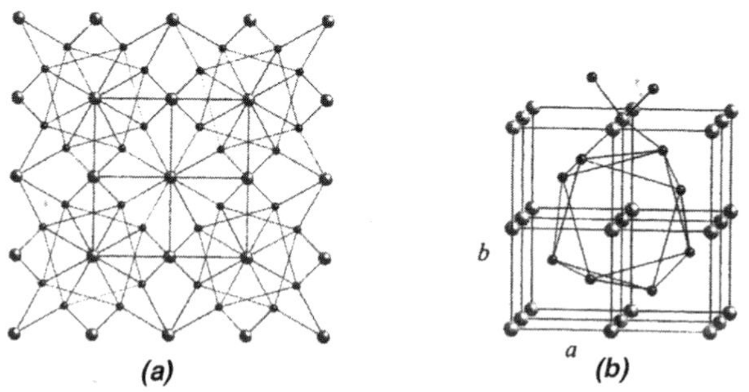

Figure 7.22. (*a*) A projection down the *c* axis of the TlSe cell showing $TlSe_8$ Archimedes antiprisms at the center and corners. $TlSe_4$ tetrahedra are at centers of edges. Tl atoms are larger atoms. (*b*) The cell showing antiprisms sharing square faces along *c*.

(*Source*: Y. Matsushita, *Chalcogenide* crystal structure data library, Version 5.5B, Institute for Solid State Physics, The University of Tokyo, Tokyo, Japan, 2004; and *CrystalMaker*, by David Palmer, *CrystalMaker* Software Ltd., Begbroke Science Park, Bldg. 5, Sandy Lanc, Yarnton, Oxfordshire, OX51PF, UK.)

faces. The Tl—Tl distance is 3.395 Å. The tetrahedra share edges to form chains along *c*. If the Se atoms were in the centers of the octants, the Tl atoms would be in $TlSe_8$ cubes as for CsCl. In Figure7.22*b*, an antiprism is outlined in the center and a tetrahedron is shown at the center of the top edge.

7.3. Complex PTOT Crystal Structures

7.3.1. The 2·2P^{Re}TOT (or 2P^{Re}IIIP) Crystal Structure of ReB_3

The crystal structure of ReB_3 is hexagonal with two molecules in the cell, $\mathbf{D}_{6h}^4$, $P6_3/mmc$, $a_o = 2.900$, and $c_o = 7.475$Å. The structure (Figure 7.23) is very similar to that of ReB_2 (Section 6.3.17, Figure 6.29). For both structures, the locations of Re are the same and the boron atoms within the cell are in the same *A* or *B* positions. The difference is that for ReB_3 boron atoms are added at the corners at heights 0 and 50. We follow the usual designations of positions for a hexagonal cell—*A* at the corners and *B* and *C* inside the cell. Each layer is filled by boron or Re. The heights and positions are below:

Height:	0	5	25	45	50	55	75	95	100	105
Atom:	B	B	Re	B	B	B	Re	B	B	B
Position:	$\mathbf{O}_A$	$\mathbf{T}_B$	$\mathbf{P}_C$	$\mathbf{T}_B$	$\mathbf{O}_A$	$\mathbf{T}_C$	$\mathbf{P}_B$	$\mathbf{T}_C$	$\mathbf{O}_A$	$\mathbf{T}_B$

The boron atoms at 0 and 50 have octahedral arrangements of six Re atoms. These boron layers are exactly halfway between **P** layers of Re atoms. Other boron atoms are in irregular **T** sites. A boron atom at

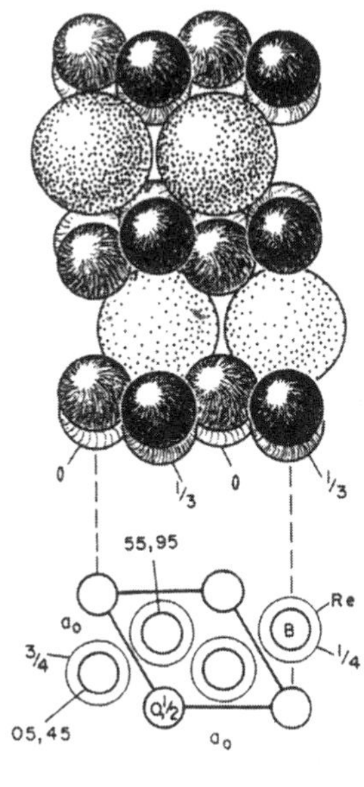

Figure 7.23. The **2·2P^{Re}TOT** structure of ReB_3. Two projections of the hexagonal cell are shown. Re atoms are larger.

(*Source*: R.W.G. Wyckoff, *Crystal Structures*, Vol. 2, ed., Wiley, New York, 1964, p. 116.)

45 has three Re at 25 in the $\mathbf{P}_C$ layer and one Re at 75 in the $\mathbf{P}_B$ layer. The ratio of the distances of the boron atom from the apex and from the base is 1.5 (30/20). **T** layers are normally one-quarter or three-quarters between **P** layers, giving a normal ratio of 3/1. It is likely that the distortion results from the close boron layers (45, 50, 55 or 95, 100, 105). Each of these boron layers is filled, and they are at different positions. Boron at 45 has a triangular arrangement of three boron atoms at height 55. A boron atom at 50 has three boron atoms at 45 and three boron atoms at 55 giving a trigonal antiprism—a very flattened octahedron. An Re atom at 25 has a trigonal prism of six boron atoms (three at 0 and three at 50). There is another trigonal prism of six boron atoms (three at 5 and three at 45) about this Re. The triangular faces of both prisms are capped by boron atoms at 55 and −5 (95). The network of trigonally arranged boron atoms suggests boron–boron bonding. The notation for the ReB_3 structure is $\mathbf{2{\cdot}2P^{Re}TOT}$. However, because of the close spacing of the **T–O–T** layers this can be described as a triple layer (**IIIP**), giving the notation $\mathbf{2P^{Re}}$**IIIP**. The **IIIP** designation loses the designation of regular and proper designations of the positions (*BC* or *CB*) of the **O** layer. Also, the *B* or *C* positions of **T** layers are those predicted from the **PTOT** system.

We have noted that a **PTOT** structure based on an *hcp* sequence of **P** layers is not expected if all layers are filled because of repulsion between close layers as in $\mathbf{T}_B\ \mathbf{P}_C\ \mathbf{T}_B$ or $\mathbf{T}_C\ \mathbf{P}_B\ \mathbf{T}_C$. ReB_3 is a notable exception because of the triple boron layers. Bonding among the three boron layers results in height separations of 50 between $\mathbf{T}_B$ and $\mathbf{T}_B$ or $\mathbf{T}_C$ and $\mathbf{T}_C$ layers.

7.3.2. The $3{\cdot}2P_{1/2}TO_{1/2}T(t)$ Crystal Structure of K_2PtCl_4

Potassium tetrachloroplatinate(II) (K_2PtCl_4) contains the square planar $PtCl_4^{2-}$ ion. These planar ions pack in parallel planes, resulting in an unimolecular tetragonal unit cell, $\mathbf{D}_{4h}^1$, $P4/mmm$, $a_o = 6.99$, and $c_o = 4.13$Å. The $PtCl_4^{2-}$ ions, treated as units, half-fill **P** and **O** layers with K^+ ions filling both **T** layers, giving the notation $\mathbf{3{\cdot}2P_{1/2}TO_{1/2}T(t)}$ (see Figure 7.24). The $PtCl_4^{2-}$ unit is in the center of

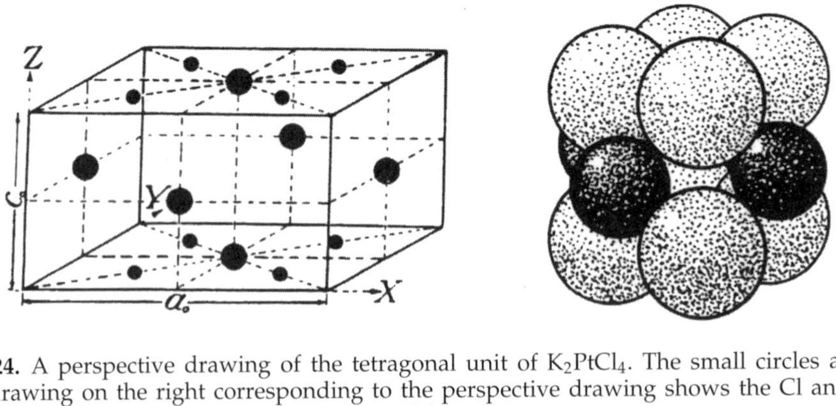

Figure 7.24. A perspective drawing of the tetragonal unit of K_2PtCl_4. The small circles are Cl. The packing drawing on the right corresponding to the perspective drawing shows the Cl and K (dark) atoms using their ionic radii. Pt atoms are not visible.

(*Source*: R.W.G. Wyckoff, *Crystal Structures*, Vol. 3, 2nd ed., Wiley, New York, 1965, p. 70.)

a compressed cube formed by eight K^+ ions. Each K^+ ion has eight chlorine atoms as closest neighbors. The K_2PtCl_6 complex contains octahedral $PtCl_6^{2-}$ ions, giving a more simple structure, **3·3PTT** (Figure 6.14).

7.3.3. The Crystal Structure of Spinel [$MgAl_2O_4$]

The mineral spinel ($MgAl_2O_4$) is the structural model for hundreds of minerals and synthesized compounds known as spinels. They share the formula $M^{II}M_2^{III}O_4$. In a normal spinel M^{II} occurs in **T** sites and M^{III} are in **O** sites. In an inverse spinel such as $NiFe_2O_4$ the Ni^{2+} ions are in **O** sites and 1/2 of the Fe^{3+} ions are in **T** sites and 1/2 are in **O** sites.

The spinel structure is complex. It helps to recognize that the metal ions (Mg and 2Al) are in the same positions of those of the Laves phase, $MgCu_2$ (Section 9.4.1, Figure 9.44). Neglecting the oxide ions of $MgAl_2O_4$, the Mg ions are in a face-centered cubic arrangement and in the centers of four octants of the cube. These fill the **P** layer and one **T** layer. Each of the other four octants contains Al_4 tetrahedra centered at the positions of the sites of the other **T** layer (Figure 7.25). The oxide ions form a *ccp* framework providing tetrahedral coordination of Mg atoms and octahedral coordination of Al. Considering this *ccp* framework of O^{2-} ions, the **P** layers are filled by O^{2-} ions with Mg^{2+} ions in the **T** sites and Al^{3+} ions in **O** sites. In Figure 7.26***a*** we see that **O** layers alternate in occupancy ($\mathbf{O}_{1/4}$ and $\mathbf{O}_{3/4}$). Both **T** layers adjacent to the $\mathbf{O}_{1/4}$ layer are one-quarter occupied by Mg^{2+}, giving $\mathbf{3{\cdot}6PO_{3/4}PT_{1/4}O_{1/4}T_{1/4}}$. Figure 7.26***b*** shows the stacking of polyhedra. The cubic unit cell contains eight molecules with 18 layers repeating, $\mathbf{O}_h^7$, $Fd3m$, $a_o = 8.08\text{Å}$, for $MgAl_2O_4$.

The spinels have been investigated intensively. They are important ferromagnetic materials. Some compounds with the spinel structure are shown in Table 7.4. Inverse spinels are designated by (I) in the table.

7.3.4. The Crystal Structure of Chrysoberyl [$BeAl_2O_4$]

Chrysoberyl, $BeAl_2O_4$, has an orthorhombic structure, $\mathbf{D}_{2h}^{16}$, $Pnma$, $a = 9.407$, $b = 5.4781$, and $c = 4.4285\text{Å}$, with four molecules in the unit cell. The structure is similar to that of olivine (Section 10.3.1). Al_I

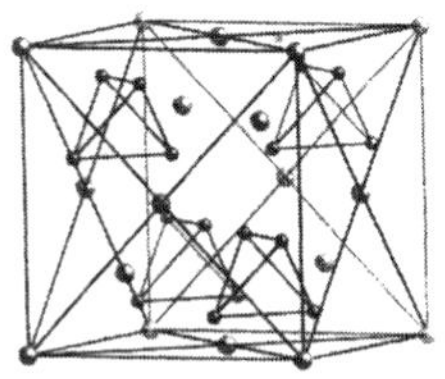

Figure 7.25. Positions of Mg atoms (larger, light color) are shown in a *fcc* cell and in one set of **T** positions for spinel, $MgAl_2O_4$. Al_4 tetrahedra are in the other **T** positions. Oxygen atoms are omitted.

(*Source*: *CrystalMaker*, by David Palmer, *CrystalMaker* Software Ltd., Begbroke Science Park, Bldg. 5, Sandy Lane, Yarnton, Oxfordshire, OX51PF, UK.)

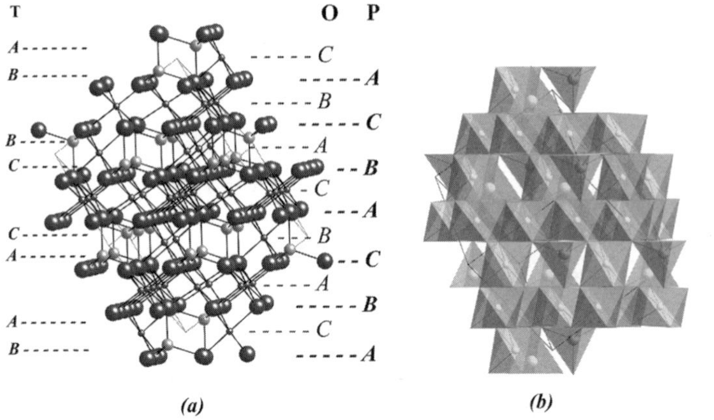

Figure 7.26. A model of spinel, $MgAl_2O_4$, with layers labeled and a similar view of the model showing the stacking of polyhedra.

(*Source*: *CrystalMaker*, by David Palmer, *CrystalMaker* Software Ltd., Begbroke Science Park, Bldg. 5, Sandy Lane, Yarnton, Oxfordshire, OX51PF, UK.)

TABLE 7.4. **Compounds with the cubic spinel structure.**

Ag_2MoO_4	Al_2CdO_4	Al_2CoO_4	Al_2CrS_4	Al_2CuO_4	$AlFeNiO_4$
Al_2FeO_4	Al_2MnO_4	Al_2NiO_4	Al_2SnO_4	Al_2ZnS_4	$BaIn_2O_4$
$CaIn_2O_4$	$CdCr_2O_4$	$CdFe_2O_4$	$CdGa_2O_4$	$CdIn_2O_4$	$CdMn_2O_4$
$CdRh_2O_4$	CdV_2O_4	Co_2CuO_4	Co_2CuS_4	Co_2GeO_4	Co_2MgO_4
Co_2MnO_4	Co_2NiO_4	Co_2SnO_4(I)	Co_2TiO_4	Co_2ZnO_4	Co_3O_4
$(Co,Ni)_3O_4$	Co_3S_4	$(Co,Ni)_3S_4$	Co_3Se_4	Cr_2CdO_4	Cr_2CdS_4
Cr_2CdSe_4	Cr_2CoO_4	Cr_2CoS_4	Cr_2CuS_4	Cr_2CuSe_4	Cr_2CuTe_4
Cr_2FeO_4	$Cr_2(Fe,Mg)O_4$	Cr_2FeS_4	Cr_2HgS_4	Cr_2MgO_4	Cr_2MnO_4
Cr_2MnS_4	Cr_2NiO_4	Cr_2ZnO_4	Cr_2ZnS_4	Cr_2ZnSe_4	Cu_2SnS_4
$FeAlMgO_4$	$FeCrMnO_4$	Fe_2CdO_4	Fe_2CoO_4(I)	Fe_2CuO_4(I)	$FeMn(Zn_{0.5}Ge_{0.5})O_4$
Fe_2GeO_4	Fe_2MgO_4(I)	Fe_2MnO_4	Fe_2MoO_4	Fe_2NiO_4(I)	$Fe_2(Mg,Mn,Fe)O_4$
Fe_2PbO_4	Fe_2TiO_4(I)	Fe_2ZnO_4	Fe_3O_4	Fe_3S_4	Ga_2CdO_4
Ga_2CoO_4	Ga_2CrS_4	Ga_2CuO_4	Ga_2MgO_4(I)	Ga_2MnO_4	Ga_2NiO_4(I)
Ga_2ZnO_4	In_2CaS_4	In_2CdO_4(I)	In_2CdS_4	In_2CoS_4(I)	In_2CrS_4(I)
In_2FeS_4(I)	In_2HgS_4	In_2MgO_4(I)	In_2MgS_4(I)	In_2NiS_4(I)	In_2S_3
$K_2Cd(CN)_4$	$K_2Hg(CN)_4$	$K_2Zn(CN)_4$	$LiAlMnO_4$	$LiAlTiO_4$	$LiCoMnO_4$
$LiCoSbO_4$	$LiCoVO_4$	$LiCrGeO_4$	$LiCrMnO_4$	$LiCrTiO_4$	$LiFeTiO_4$
$LiGaTiO_4$	$LiGeRhO_4$	$LiMnTiO_4$	$LiNiVO_4$	Li_2NiF_4	$LiRhGeO_4$
$LiRhMnO_4$	$LiTiRhO_4$	$LiZnSbO_4$	Mg_2GeO_4	Mg_2SnO_4	Mg_2TiO_4(I)
Mg_2VO_4	Mn_2CuO_4	Mn_2MgO_4	Mn_2LiO_4	Mn_2SnO_4	Mn_2NiO_4(I)
Mn_2TiO_4	Mn_3O_4	Na_2WO_4	Ni_2FeS_4	Ni_2GeO_4	Ni_3S_4
Ni_2SiO_4	Rh_2CdO_4	Rh_2CoO_4	Rh_2CuO_4	Rh_2MgO_4	Rh_2MnO_4
Rh_2NiO_4	Rh_2ZnO_4	Ti_2CuS_4	Ti_2MgO_4	Ti_2MnO_4	V_2CdO_4
V_2CuS_4	V_2FeO_4	V_2LiO_4	V_2MgO_4	V_2MnO_4	V_2ZnO_4
Zn_2SnO_4(I)	Zn_2TiO_4(I)				

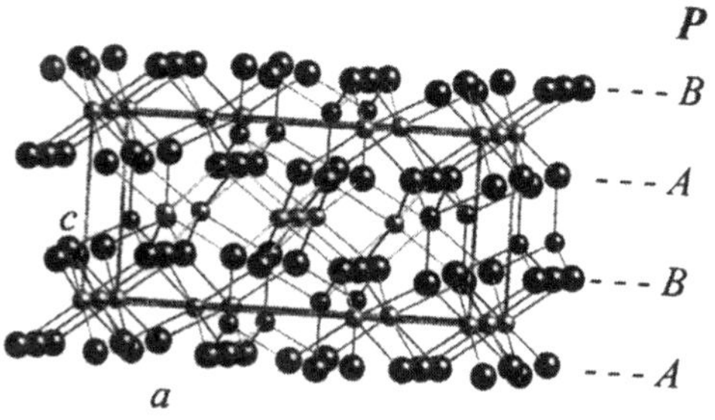

Figure 7.27. The structure of chrysoberyl, $BeAl_2O_4$.
(*Source*: *CrystalMaker*, by David Palmer, *CrystalMaker* Software Ltd., Begbroke Science Park, Bldg. 5, Sandy Lane, Yarnton, Oxfordshire, OX51PF, UK.)

sites form infinite chains of edge-linked octahedra along the *b* axis. These chains are crosslinked by $Al_{II}O_6$ octahedra and BeO_4 tetrahedra. There is less distortion for the Al_IO_6 octahedra along *b* edges and in the center of the cell (see Figure 7.27). Oxygen atoms are close packed with an *AB* sequence. Al atoms occupy half of each **O** layer. All **T** layers are one-eighth occupied by Be, giving $\mathbf{T}^+$ (pointing upward) and $\mathbf{T}^-$ (pointing downward) tetrahedra. The notation is $\mathbf{2{\cdot}4PT_{1/8}O_{1/2}T_{1/8}(o)}$.

7.3.5. The Crystal Structures of AgI

There are two modifications of AgI at ordinary temperatures, β-AgI has the wurtzite (**2·2PT**) structure and γ-AgI has the zinc blende (**3·2PT**) structure. For both of these structures Ag^+ and I^- ions have *CN* 4 (tetrahedral). Above 145.8°C, α-AgI is formed with a *bcc* (**3·2PTOT**) structure for I^- ions. For a *bcc* structure all **P**, **T**, and **O** layers are filled by I^- ions for α-AgI. There are secondary interstitial sites for a *bcc* structure—four distorted tetrahedral sites (**T′**) in each face of the cube and distorted octahedral sites (**O′**) in the centers of the edges (12) and in the centers of the faces (6) of the cube. The **T′** sites are shown as squares in each face of the *bcc* cube in Figure 7.28. The *bcc* cell

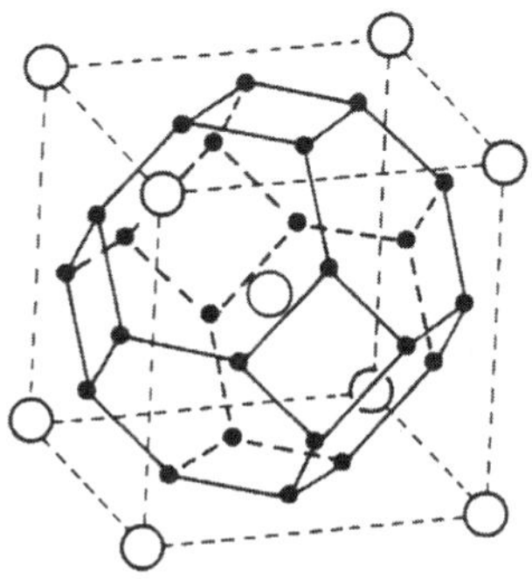

Figure 7.28. The 24 secondary **T′** sites available for Ag^+ (solid circles) in a *bcc* cell for α-AgI. In the *bcc* cell the positions of I^- ions are large open circles.

contains two I^- ions and the two Ag^+ ions are randomly distributed among the 12 **T**′ sites. To visualize the **T**′ sites in a **PTOT** structure requires a model showing several cells such as that of CsCl in Figure 7.10 (all Cs^+ and Cl^- positions are occupied by I^- in α-AgI).

The remarkable thing about AgI is that it "half melts" at 145.8°C and the electrical conductivity increases by a factor of 10^4! It is said to half melt because one ion (Ag^+) becomes mobile. The tetrahedra (**T** sites) share faces so Ag^+ can move through a trigonal face to a neighboring **T**′ site with a small barrier. The $2Ag^+$ ions can move among 12 **T**′ sites per cell. At high temperature AgI gives the NaCl (**3·2PO**) structure with both ions having *CN* 6.

7.3.6. The Crystal Structure of $K_2Pb_2Ge_2O_7$

The crystal structure of $K_2Pb_2Ge_2O_7$ is hexagonal with one molecule per cell, $\mathbf{C}_{3i}^1$, $P\bar{3}$, $a_o = 5.775$, and $c_o = 7.81$Å. Figure 7.29 shows a projection of the $K_2Pb_2Ge_2O_7$ cell. Packing layers are stacked along c_o. The sequence of layers is:

Height:	22	29	39	50	61	71	78	92	108
Atom:	3O	Ge	Pb	1O	Pb	Ge	3O	K	K
Position:	*A*	*B*	*C*	*B*	*A*	*B*	*C*	*C*	*A*
Layer:	$\mathbf{P}_{3/4}$	$\mathbf{T}_{1/4}$	$\mathbf{O}_{1/4}$	$\mathbf{P}_{1/4}$	$\mathbf{O}_{1/4}$	$\mathbf{T}_{1/4}$	$\mathbf{P}_{3/4}$	$\mathbf{O}_{1/4}$	$\mathbf{O}_{1/4}$

The $Ge_2O_7^{6-}$ ions are aligned along c_o as $O_3Ge{-}O{-}GeO_3^{6-}$, with both Ge atoms and the oxygen atom bridging the tetrahedra at *B* positions. Pb atoms are in **O** layers, although the three sites of oxygen atoms in layer $\mathbf{P}_B$ to complete the octahedron are vacant. This gives Pb three oxygen close neighbors at 2.13 Å and causes the **O** layer to be closer to one **P** layer rather than the normal situation where **O** layers are halfway between **P** layers. K atoms occur at the same positions as Pb in different layers. The *CN* of K is 9, with the O atoms at 2.89 or 3.10 Å. Each K atom is at the apex joining a trigonal pyramid and a hexagonal pyramid. The layers for K are adjacent without oxygens between them. The K atoms are in staggered positions (*C* and *A*). The K atom in

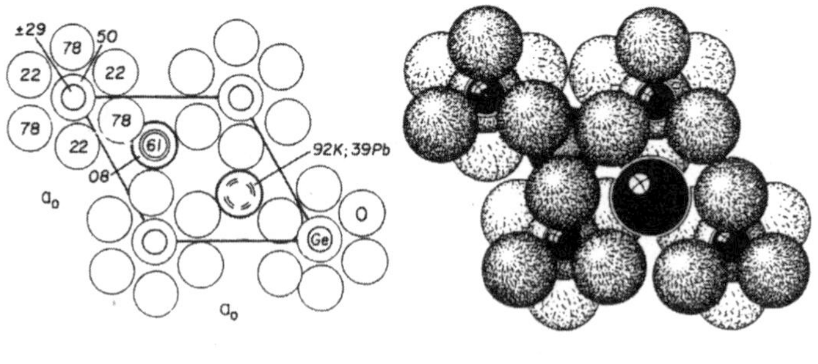

Figure 7.29. The structure of $K_2Pb_2Ge_2O_7$. On the left there is a projection of the hexagonal cell along the c_o axis. On the right there is a similar view of a packing drawing.

(*Source*: R.W.G. Wyckoff, *Crystal Structures*, Vol. 4, 2nd ed., Wiley, New York, 1965, p. 224.)

position C at height 92 is directly over the vacancy at the center of a hexagon of oxygen atoms in $\mathbf{P}_C$ at 78 and under three oxygen atoms the $\mathbf{P}_A$ layer at 22. We choose to identify the K layers as **O** layers. The positions and spacings of adjacent K layers suggest that a $\mathbf{P}_A$ layer of oxygen at about height 100 is missing. However, the structure is dictated by the chemistry and stoichiometry of $K_2Pb_2Ge_2O_7$! The notation is $\mathbf{3{\cdot}3P^{O}_{3/4}T^{Ge}_{1/4}O^{Pb}_{1/4}P^{O}_{1/4}O^{Pb}_{1/4}T^{Ge}_{1/4}P^{O}_{3/4}O^{K}_{1/4}O^{K}_{1/4}(h)}$.

7.4. POT Crystal Structures

7.4.1. The Crystal Structure of CaF_2 Expressed as 3•3POT

The structures of CaF_2 (fluorite) and Li_2O were discussed in Section 6.3.1 as **3•3PTT** structures. Normally we assign the larger anions to **P** positions because in a close-packed structure the **O** and **T** interstitial sites are smaller than the sites for packing atoms. For Li_2O, the larger oxide ions are assigned to **P** positions with the small Li^+ ions filling all **T** sites giving **3•3PTT**. For CaF_2, the notation **3•3PTT** requires Ca^{2+} ions in **P** positions and F^- ions in all **T** sites. Alternatively, we can assign F^- to **P** and **O** layers with Ca^{2+} in one **T** layer giving **3•3POT**. The structure is the same for both notations. Each Ca^{2+} has *CN* 8, a cube made up of a tetrahedral arrangement of four F^- in **P** positions and a tetrahedral arrangement of four F^- in **O** positions. In Figure 7.30 Ca—F bonds are shown in **P** positions (darker atoms). Each F^- has a tetrahedral arrangement of four Ca^{2+} in one of the **T** layers. Many structures are found for compounds varying considerably in anion/cation ratios. For CaF_2, the preference for sites is not very important because the ionic radius for Ca^{2+} (*CN* 8) is 1.26 Å and that of F^- is 1.19 Å.

7.4.2. The $3{\cdot}3PT_{1/3}O_{1/3}(o)$ Crystal Structure of $CuGeO_3$

$CuGeO_3$ has an orthorhombic unit cell (Figure 7.31) with two molecules per cell, C^2_{2v}, $Pb2_1m$, $a_o = 4.8$, $b_o = 8.5$, and $c_o = 2.94$Å. The oxygen atoms are in *ABC* positions in a deformed *ccp* structure. Cu atoms are in regular octahedral sites, and Ge atoms are in regular tetrahedral sites. The view is along a C_2 axis of the CuO_6 octahedron and along the direction of one edge of the GeO_4 tetrahedron, the other oxygen atoms are eclipsed at 0 and 100. The tetrahedra share corners to form chains. Each

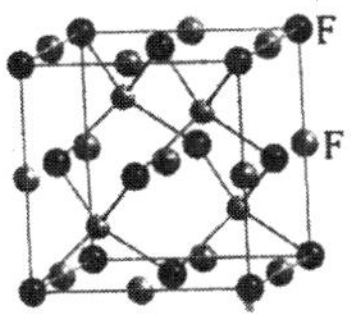

Figure 7.30. A model of CaF_2 with F atoms in **P** sites (dark atoms) and **O** sites (lighter). The very light atoms within the cell are Ca.

(*Source*: *CrystalMaker*, by David Palmer, *CrystalMaker* Software Ltd., Begbroke Science Park, Bldg. 5, Sandy Lane, Yarnton, Oxfordshire, OX51PF, UK.)

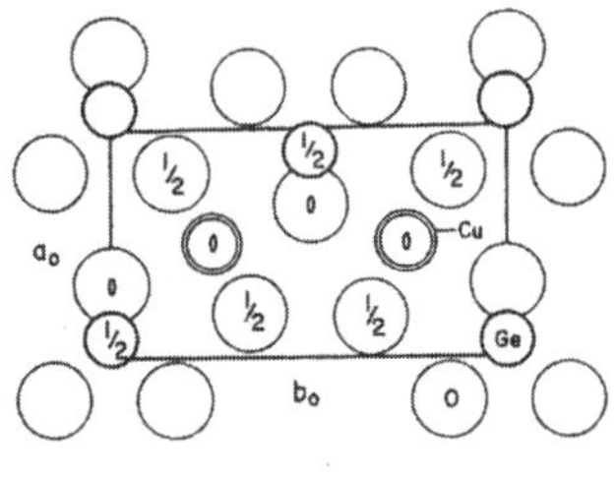

Figure 7.31. A projection of the orthorhombic cell of $CuGeO_3$ along c_o.

(*Source*: R.W.G. Wyckoff, *Crystal Structures*, Vol. 2, 2nd ed., Wiley, New York, 1964, p. 444.)

octahedron shares two oxygens with other octahedra, and all oxygens are shared with the tetrahedra. The sequence in close-packed layers is:

O	Ge	Cu	O	Ge	Cu	O	Ge	Cu
$\mathbf{P}^A$	$\mathbf{T}^B_{1/3}$	$\mathbf{O}^C_{1/3}$	$\mathbf{P}^B$	$\mathbf{T}^C_{1/3}$	$\mathbf{O}^A_{1/3}$	$\mathbf{P}^C$	$\mathbf{T}^A_{1/3}$	$\mathbf{O}^B_{1/3}$

The notation is $\mathbf{3{\cdot}3PT_{1/3}O_{1/3}(o)}$.

7.4.3. The $\mathbf{2{\cdot}4PP^{Cu}\ PO^{Ga}(h)}$ Hexagonal Crystal Structure of $CuGaO_2$

The compound $CuGaO_2$ has a hexagonal crystal structure, $\mathbf{D}^4_{6h}$, $P6_3/mmc$, $a = 3.223$, and $c = 11.413$Å with unusual features as seen in Figure 7.32. Oxygen packing layers occur in the sequence *AA BB* with Ga atoms in octahedral layers at *C* positions. Cu atoms are between oxygen layers at the *same* positions, giving the O–Cu–O linear arrangement. Because the Cu layers are not **T** or **O** layers, we designate them as **P** layers. The full sequence is $\mathbf{P}_A\mathbf{P}^{Cu}_A\ \mathbf{P}_A\mathbf{O}^{Ga}_C\ \mathbf{P}_B\mathbf{P}^{Cu}_B\ \mathbf{P}_B\mathbf{O}^{Ga}_C$ and the notation is $\mathbf{2{\cdot}4PP^{Cu}PO^{Ga}(h)}$. The unusual occurrence of

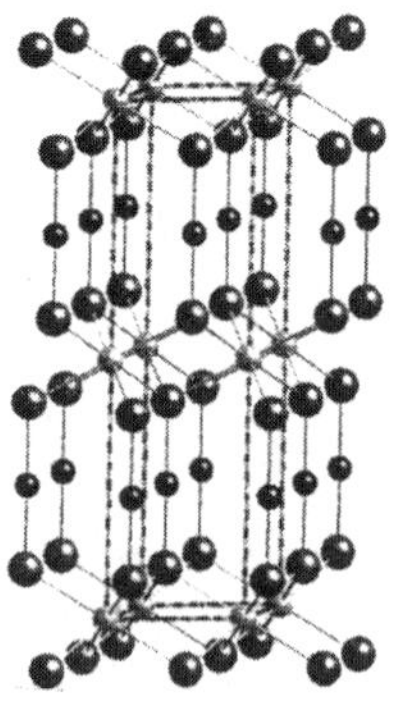

Figure 7.32. The $\mathbf{2{\cdot}4PP^{Cu}PO}^{Ga}$ structure of $CuGaO_2$. Cu atoms are black and the stacking of layers is along the *c* axis.

(*Source*: Y. Matsushita, *Chalcogenide* crystal structure data library, Version 5.5B, Institute for Solid State Physics, The University of Tokyo, Tokyo, Japan, 2004; and *CrystalMaker*, by David Palmer, *CrystalMaker* Software Ltd., Begbroke Science Park, Bldg. 5, Sandy Lanc, Yarnton, Oxfordshirc, OX51PF, UK.)

TABLE 7.5. **Compounds with the $CuGaO_2$ structure.**

$AgBiSe_2$	$AgCrO_2$	$AgFeO_2$	$CaCN_2$	$CuAlO_2$	$CuCoO_2$
$CuCrO_2$	$CuFeO_2$	$CuGaO_2$	$CuRhO_2$	$KCrS_2$	$KTlO_2$
$LiAlO_2$	$LiCoO_2$	$LiCrO_2$	$LiHF_2$	$LiNO_2$	$LiNiO_2$
$LiRhO_2$	$LiVO_2$	$NaCrO_2$	$NaCrS_2$	$NaCrSe_2$	$NaFeO_2$
$NaInO_2$	$NaInS_2$	$NaInSe_2$	$NaNiO_2$(H.T.)	$NaTiO_2$	$NaTlO_2$
$NaVO_2$	$RbCrS_2$	$RbCrSe_2$	$RbTlO_2$	$TlBiTe_2$	$TlSbTe_2$

P layers at the same position occurs because of the preference of linear bonding for Cu(I). In $CuGeO_3$ (previous section) the copper as Cu(II) occurs in octahedral sites. Each oxygen atom is surrounded tetrahedrally by three Ga and one Cu. This is not a **POT** structure because Cu is not in a **T** layer, but the structure is included here to show the great effect of replacing Cu(II) by Cu(I) in comparison with $CuGeO_3$. Table 7.5 lists many of the $MM'X_2$ type compounds with the $CuGaO_3$ structure.

Chapter 8

Structures with Multiple Layers

In earlier chapters we dealt with crystal structures of elements and molecules involving close packing of **P** layers, compounds involving **P** and **O** layers or **P** and **T** layers, and those with combinations of **P**, **O**, and **T** layers. The body-centered cubic (*bcc*) structure is common for metals, and molecular crystals, and the related CsCl structure is common for ionic compounds. It has been shown that the *bcc* structure can be included in the **PTOT** system with all **P**, **T**, and **O** sites occupied. In this chapter we examine cases that are variations of the regular **PTOT** pattern. Nature's preference for the regular **PTOT** pattern seems to be based on efficiency of packing, high symmetry, and high coordination. Deviations occur for chemical bonding not well adapted to the regular **PTOT** pattern and for unusual sizes or shapes of building units. In some cases large groups of atoms, molecules, or ions are the repeating units. Some structures included here contain double layers consisting of puckered layers or a layer consisting of two packing positions. A few structures included have triple layers.

8.1. Crystal Structures Involving Unusual Combinations of P Layers

8.1.1. The $3{\cdot}2\mathbf{P}^{Hg}\mathbf{P}^{S}(h)$ Crystal Structure of Cinnabar [HgS]

Cinnabar, HgS, is a common mineral of mercury. The crystal contains HgS helices. It is trigonal, $\mathbf{D}_3^6$, P3$_2$21, $a_o = 4.149$, and $c_o = 9.495$Å. The other enantiomorph belongs to the space group $\mathbf{D}_3^4$. The helices are aligned with the *c* axis, with two bonds to each atom. The S—Hg—S units are nearly linear (172.82°) as seen in Figure 8.1***a***. The bond angles at S are 104.74°. The positions of atoms correspond to *A*, *B*, and *C* positions in a close-packed arrangement. Each close-packed layer has a regular hexagonal pattern, but there are slight shifts of adjacent layers because of bonding within the helices. Alternate **P** layers are filled by Hg and S. The sequence is $\mathbf{P}_A^{Hg}\mathbf{P}_B^{S}\mathbf{P}_C^{Hg}$ or for Hg it is $\mathbf{P}_A\mathbf{P}_C\mathbf{P}_B$ and for S it is $\mathbf{P}_B\mathbf{P}_A\mathbf{P}_C$. Figure 8.1***b*** is a projection in the *ab* plane showing the triangular arrangement of the helices. Both Hg and S atoms occur at

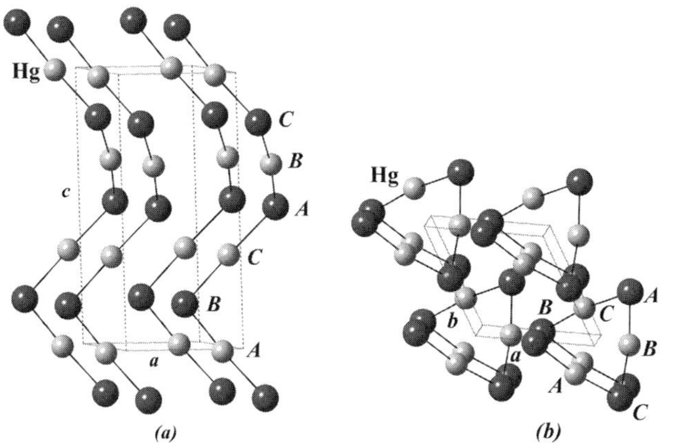

Figure 8.1. (*a*) The spirals of HgS. (*b*) A perspective view of the helices showing the *ABC* positions of atoms.

(*Source*: *CrystalMaker*, by David Palmer, *CrystalMaker* Software Ltd., Begbroke Science Park, Bldg. 5, Sandy Lane, Yarnton, Oxfordshire, OX51PF, UK.)

A, *B*, and *C* positions. A perspective view of the helices is very similar to those of Te in Figure 4.19*c*. However, the triangles contain six atoms for HgS. The Hg—S distance within the helices is 2.36 Å and between helices it is 3.2 Å. The notation is $\mathbf{3{\cdot}2P^{Hg}P^{S}(h)}$.

8.1.2. The 2P Crystal Structure of Tungsten Carbide [WC]

Tungsten carbide, WC, has a hexagonal structure with one molecule per cell, $\mathbf{D}^1_{3h}$, $P\bar{6}m2$, $a_o = 2.9065$, and $c_o = 2.8366$Å. Figure 8.2 shows that the hexagonal cell has W in *A* layers with carbon atoms in *B* layers giving a simple **2P** structure. The usual *c*/*a* ratio for a hexagonal structure is about 1.63. For WC the ratio is less than 1 because of the very small carbon atom compared to W. Each atom is at the center of a trigonal prism formed by six of the other atoms. WC is extremely hard, and has a melting point of 2,720° C. RuC, MoP, NbS, and some MN-type nitrides have the WC structure.

8.1.3. The $3P_{1/2\ 1/2}$(t) Crystal Structure of $N(CH_3)_4Cl$

The unit cell of $N(CH_3)_4Cl$ is tetragonal with $N(CH_3)_4^+$ ions at the corners and centers of the top and bottom faces of the cell. Opposite

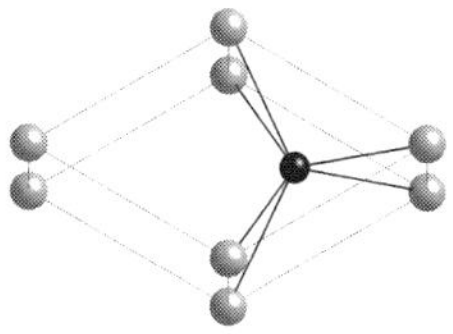

Figure 8.2. The **2P** structure of tungsten carbide, WC.

(*Source*: Y. Matsushita, *Chalcogenide* crystal structure data library, Version 5.5B, Institute for Solid State Physics, The University of Tokyo, Tokyo, Japan, 2004: and *CrystalMaker*, by David Palmer, *CrystalMaker* Software Ltd., Begbroke Science Park, Bldg. 5, Sandy Lanc, Yarnton, Oxfordshire, OX51PF, UK.)

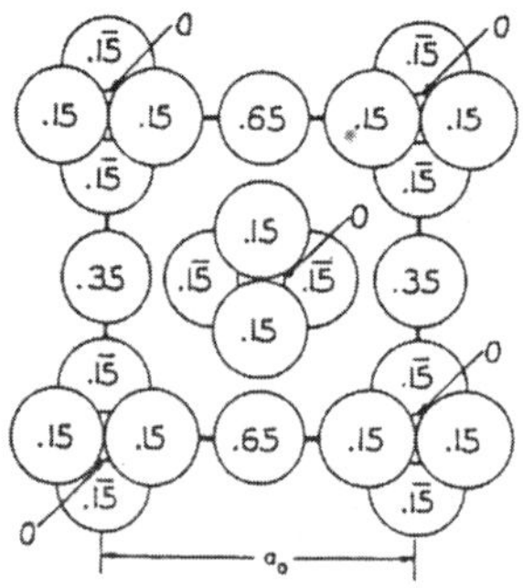

Figure 8.3. A projection of the tetragonal unit cell of $N(CH_3)_4Cl$ along the c_o axis. The tetrahedrally arranged CH_3 groups are at the corners and at the center. N atoms are at the centers of the tetrahedra and Cl atoms are in side faces.

(*Source*: R.W.G. Wyckoff, *Crystal Structures*, Vol. 1, 2nd ed., Wiley, New York, 1963, p. 109.)

edges of the $N(CH_3)_4^+$ tetrahedra are aligned with the a_o axes (see Figure 8.3). Cl^- ions are in the side faces, but not quite at the centers. A face-centered tetragonal cell can be described as $\mathbf{3P_{1/2\,1/2}(t)}$ with half of the **P** layers occupied by each ion. Two Cl^- ions are at 0.35 (height along c_o). For these Cl^- ions the closest groups are eight CH_3, four at −0.15 and four at 0.85. The Cl^- ions are located halfway between these CH_3 groups at 0.35 (−0.15 + 0.50) . The other two Cl^- ions are at 0.65. For these Cl^- ions the closest eight CH_3 groups are at 0.15 and 1.15. The Cl^- ions are centered relative to the closest CH_3 groups. The displacements of Cl^- by 0.15 are relative to the N atoms. The notation $\mathbf{3P_{1/2\,1/2}(t)}$ describes the roles of each ion with the expected displacement of Cl^- ions. This structure is found for $CsNH_2$, NH_4SH, NH_4Br, NH_4I, PH_4Br, PH_4I, $N(CH_3)_4Br$, $N(CH_3)_4I$, $N(CH_3)_4ClO_4$, and $N(CH_3)_4MnO_4$. LiN_3 has the same $\mathbf{P_{1/2\,1/2}}$ arrangement in a monoclinic cell.

8.1.4. The Special $2(3P^B_{1/4}P^A_{1/4\,3/4}P^C_{1/4})[H_2O;Fe,Cl/H_2O;H_2O]$ Crystal Structure of $[Fe(CH_3NC)_6]Cl_2{\cdot}3H_2O$

The isocyanomethyl complex of Fe(II), $[Fe(CH_3NC)_6]Cl_2{\cdot}3H_2O$, gives a hexagonal unit cell containing one molecule, D^3_{3d}, $P\bar{3}m1$, $a_o = 10.47$, and $c_o = 5.315$Å (see Figure 8.4). Four cells are shown to clarify the location of groups around the Fe complex. The octahedral complex ion contains six isocyanomethyl groups coordinated to Fe(II) through carbon. The Fe atoms are at the corners of the cell so they are stacked in layers along the c_o axis at the *same* positions ($\mathbf{P}_A$). The two Cl atoms and one H_2O molecule occupy randomly (2:1) the positions in the cell edges (heights 0 and 100) and the center of the top and bottom faces. These are also *A* positions in the same layers occupied by Fe. Each Fe complex is surrounded by a hexagonal arrangement of six Cl/H_2O in the same plane. There is a larger puckered hexagon of six H_2O molecules outside the inner one. The heights of H_2O molecules alternate at ±10. Each Cl/H_2O position is at the center of a hexagon formed by four

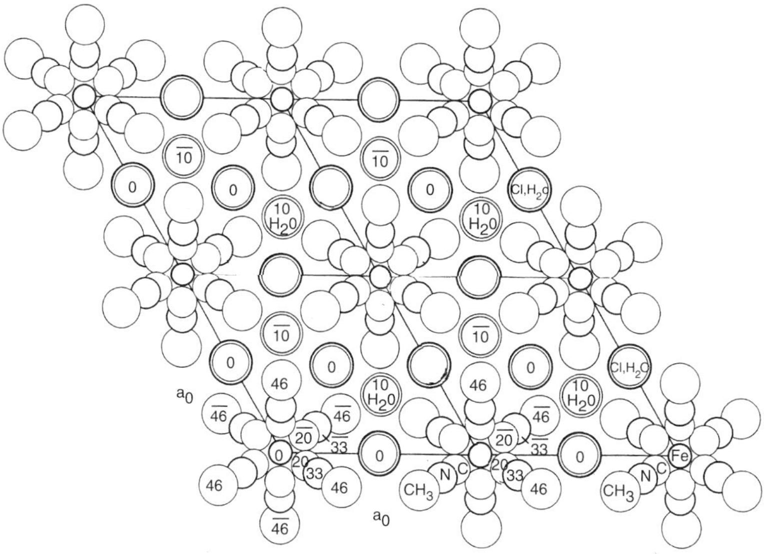

Figure 8.4. A projection along the c_o axis for the structure of $[Fe(CH_3NC)_6]Cl_2{\cdot}3H_2O$. Carbon atoms are bonded to Fe.

Cl/H_2O positions and two Fe atoms (of the complexes). In the figure, looking down the c_o axis, we are looking down the C_3 axis of each octahedral complex. The $\mathbf{P}_A$ layer is filled by Fe atoms (one Fe per cell) and a random distribution of Cl atoms and H_2O molecules (two Cl + one H_2O) in the edges and two faces. There are three of these Cl/H_2O sites per cell ($2 \times 1/2 = 1$ in faces and $8 \times 1/4 = 2$ in edges). The notation for this layer is $\mathbf{P}_{1/4\ 3/4}$, one-quarter for the Fe atoms and three-quarter for the Cl/H_2O positions.

Two H_2O molecules are in unique positions in layers just above (height 10) and below (height -10 or 90) the $\mathbf{P}_A$ layer. The H_2O sites at height 10 are at B positions and those at height -10 (90) are at C positions. Each of these H_2O molecules forms a trigonal pyramid with two Cl atoms and another H_2O molecule in the base. These layers (at ± 10) contain neither octahedral nor tetrahedral sites so they are considered as **P** layers, giving the notation $\mathbf{P}^{B}_{1/4}\mathbf{P}^{A}_{1/4\ 3/4}\mathbf{P}^{C}_{1/4}$. The closely spaced layers are in an ABC sequence with a large gap between the triple layers. This situation is unusual so the six layers in the repeating unit are designated as $\mathbf{2(3P^{B}_{1/4}P^{A}_{1/4\ 3/4}P^{C}_{1/4})[H_2O;Fe,Cl/H_2O;H_2O]}$ rather than using the usual **2·3** index.

Each $\mathbf{P}_A$ layer is a well-ordered close-packed layer. It is very unusual to have a $\mathbf{P}_A$ layer stacked with other $\mathbf{P}_A$ layers with the large complexes directly aligned along the packing direction. The CNC group is linear, and the $CNCH_3$ ligands project out from Fe along octahedral apices. The three ligands projecting above one layer are staggered relative to those projecting below the next layer above. It is CH_3 groups that come closest, and they are staggered. H_2O molecules in the H_2O-only positions might be involved in H-bonding with H_2O molecules in Cl/H_2O positions and N of $CNCH_3$ ligands.

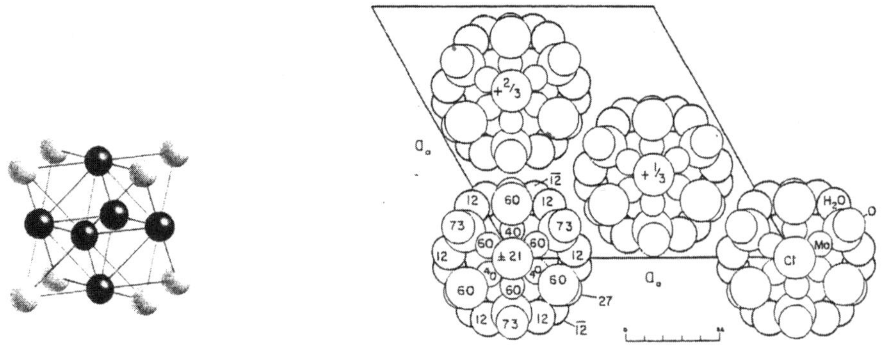

(a) (b)

Figure 8.5. (*a*) The structure of the $[Mo_6Cl_8]^{4+}$ ion. (*b*) The hexagonal unit cell of $[Mo_6Cl_8(OH)_4(H_2O)_2]{\cdot}12H_2O$ with only two complex units at corners shown.

(*Source*: R.W.G. Wyckoff, *Crystal Structures*, Vol. 3, 2nd ed., Wiley, New York, 1965, p. 880.)

8.1.5. The Complex 3P(h) Crystal Structure of $[Mo_6Cl_8][(OH)_4(H_2O)_2]{\cdot}12H_2O$

The crystal structure of $[Mo_6Cl_8][(OH)_4(H_2O)_2]{\cdot}12H_2O$ is much simpler than might be expected from the complicated formula. The $[Mo_6Cl_8]^{4+}$ ion (Figure 8.5*a*) is encountered in several compounds. In this compound the $[Mo_6Cl_8]^{4+}$ ions are surrounded by 14 H_2O molecules and four OH^- ions (to balance the charge of the cation) to complete the coordination of Mo atoms, giving large, almost spherical, clusters. These clusters are in *ABC* positions in the hexagonal cell with three molecules per cell, $\mathbf{D}_{3h}^5$, $R\bar{3}m$, $a_o = 15.15$, and $c_o = 11.02$Å. Figure 8.5*b* shows the hexagonal cell showing the simple packing of the large clusters. The notation is **3P(h)**.

8.2. Crystal Structures with Double Layers

8.2.1. The $2\mathrm{P}\mathbb{P}_{BC}$ Crystal Structure of AlB_2

Aluminum boride, AlB_2, has a hexagonal unit cell with one molecule per cell, $\mathbf{D}_{6h}^1$, $P6/mmm$, $a_o = 3.006$, and $c_o = 3.252$Å. The structure is similar to that of WC but with two boron atoms within the cell giving a graphite-like layer between Al layers (Figure 8.6*a*). The Al atoms occur only at *A* positions. Each boron atom is at the center of a trigonal prism and each Al atom is surrounded by 12 boron atoms forming a hexagonal prism. The figure shows the hexagons of boron atoms. The structure of AlB_2 is also similar to that of CdI_2 (Section 5.5.4, Figure 5.58) with both boron atoms (replacing I^-) in the same plane so that AlB_2 does not have a layered structure with adjacent layers occupied by the same atoms. More than 75 compounds have the AlB_2 structure, including at least 30 borides, and many intermetallic compounds (see Table 8.1).

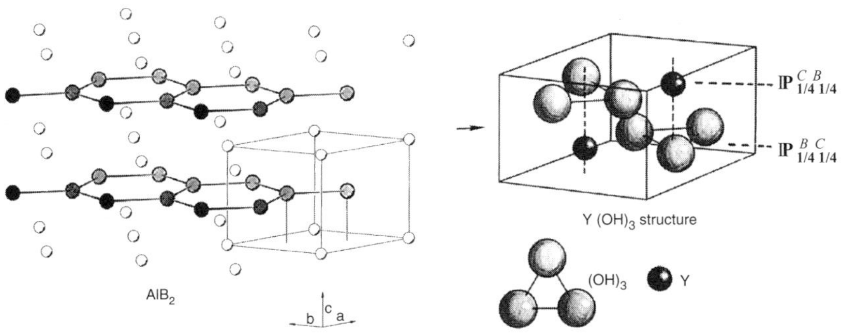

Figure 8.6. (*a*) A pictorial view of the AlB_2 structure. Smaller open circles are Al atoms. (*b*) The hexagonal unit cell is shown in two projections. Al atoms are the larger circles.

(*Source*: W.P. Pearson, *The Crystal Chemistry and Physics of Metals and Alloys*, Wiley-Interscience, New York, 1972, p. 494.)

(*Source:* R.W.G. Wyckoff, *Crystal Structures*, Vol. 1, 2nd ed., Wiley, New York, 1963, p. 365.)

Because the boron layer is similar to graphite it can be considered as a double layer, giving the notation $\mathbf{2P}_A\mathbf{I\!P}_{BC}$. The Al $\mathbf{P}_A$ layer is filled with the boron double layer involving filled *B* and *C* positions. The unit cell shown in Figure 8.6***b*** shows that the Al layer looks like a normal filled layer, as for the W layers in WC. It seems reasonable to consider the boron layer to be a double layer, because the double layer of boron atoms is favored by the small size and tendency of boron to form three bonds in a plane.

8.2.2. The $2{\cdot}2\mathbf{I\!P}_{3/3\ 1/3}\mathbf{O}_{1/3}$ Crystal Structure of $YbBO_3$

Crystals of ytterbium borate, $YbBO_3$, are trigonal, $\mathbf{D}^6_{3d}$, $R\bar{3}c$, $a = 4.921$, and $c = 16.304$ Å, with six molecules in the unit cell. The oxygen layers in Figure 8.7 are close-packed in an *AB* sequence. Yb atoms are in *C* positions, occupying one-third of **O** sites in each layer. Boron atoms are in *C* positions in the oxygen layers giving double layers, $\mathbf{I\!P}^{A\ \ C}_{3/3\ 1/3}$ or $\mathbf{I\!P}^{B\ \ C}_{3/3\ 1/3}$. The projection of two oxygen layers in the figure shows Yb atoms at corner positions (one-third of sites for this **O** layer) and boron atoms at two *C* sites within the cell. Each of these represents one-third occupancy of *C* sites for the $\mathbf{P}_A$ or $\mathbf{P}_B$ layers. The positions of

Table 8.1. Compounds with the AlB_2 structure.

AgB_2	θ-CrB_2	LuB_2	$PrGa_2$	TaB_2	UB_2
AuB_2	$EuGa_2$	MgB_2	PuB_2	$ThAl_2$	USi_2
$BaGa_2$	$EuHg_2$	MnB_2	RuB_2	$ThCu_2$	VB_2
$CaGa_2$	HfB_2	MoB_2(H.T.)	ScB_2	$ThNi_2$	VHg_2
$CaHg_2$	$LaGa_2$	NbB_2	$SrGa_2$	TiB_2	ZrB_2
$CeGa_2$	$LaHg_2$	OsB_2	$SrHg_2$	TiV_2	$ZrBe_2$

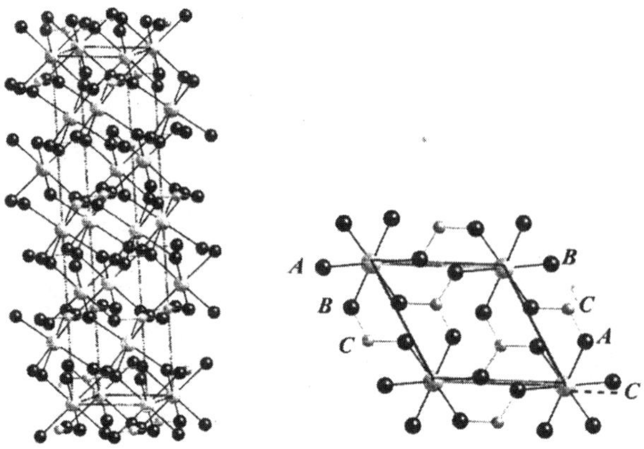

Figure 8.7. The unit cell for $YbBO_3$. Oxygen atoms are in *A* and *B* layers with Yb and boron in *C* positions. The projection of three layers shows the positions. (*Source*: Y. Matsushita, *Chalcogenide* crystal structure data library, Version 5.5B, Institute for Solid State Physics. The University of Tokyo, Tokyo, Japan, 2004; and *CrystalMaker*, by David Palmer, *CrystalMaker* Software Ltd., Begbroke Science Park, Bldg. 5, Sandy Lane, Yarnton, Oxfordshire, OX51PF, UK.)

Yb atoms are staggered so each has vacant *C* sites in adjacent **P** or **O** layers. Because the oxygen atoms are closer together in planar BO_3^{3-} ions than those in the trigonal faces of YbO_6 octahedra, BO_3^{3-} ions share only apices with the YbO_6 octahedra. This is a beautiful structure, showing the importance of avoiding having cations adjacent at *C* sites in an *hcp* arrangement. The notation is $\mathbf{2{\cdot}2 I\!P_{3/3}\ {}_{1/3}O_{1/3}}$.

8.2.3. The 2·3 $\mathbf{I\!P^{Na\,As}\ T^{Na}\ T^{Na}}$ Crystal Structure of Na_3As

Sodium arsenide, Na_3As, has a hexagonal crystal structure with two molecules per cell, C_{6v}^3, $P6_3cm$, $a_o = 8.7838$, and $c_o = 8.999$Å. There are double **IP** layers occupied by As and Na, followed by two adjacent **T** layers, each occupied by Na:

Height:	25	42	58	75	92	08
Position:	*C* *B*	*A*	*B*	*C* *A*	*B*	*A*
Atom:	Na_I **As**	Na_{II}	Na_{II}	Na_I **As**	Na_{II}	Na_{II}
Layer:	**IP**	**T**	**T**	**IP**	**T**	**T**

Figure 8.8 shows the structure. The As layers are in an *hcp* sequence (*BA*) with Na_I atoms in the same layers (**IP**) at *C* positions. The Na_{II} atoms fill all **T** layers at *A* or *B* positions. Each Na_I atom has a triangular array of three As atoms in the same layer and a trigonal prism formed by three Na of each adjacent layer. The As atoms cap the rectangular faces of the prism. As atoms have six Na in a trigonal prism at 3.205 or 3.387 Å. There are three Na atoms in the plane of As at 3.922 or 3.938 Å. These atoms cap faces of the trigonal prism. There are also Na atoms directly above and below at same packing positions at 2.947 and 3.030 Å. Na_{II} atoms are in **T** sites. An $NaAs_4$ unit in a $\mathbf{T^+}$ site is shown on the right of the cell. $\mathbf{T^+}$ and $\mathbf{T^-}$ sites can be seen within

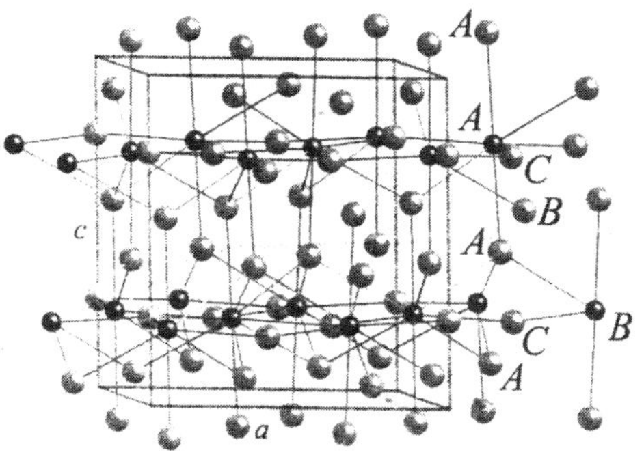

Figure 8.8. A pictorial view of the Na_3As structure, small dark circles are As. (*Source*: *CrystalMaker*, by David Palmer, *CrystalMaker* Software Ltd., Begbroke Science Park, Bldg. 5, Sandy Lane, Yarnton, Oxfordshire, OX51PF, UK.)

the cell. The Na–As distances are 3.205, 3.205, and 3.430 Å for the base of the tetrahedron and 3.030 Å for the axial position. There are three Na_I atoms at 3.2 Å. These Na_I atoms and As atoms form a hexagon around the Na_{II} atom. There is an Na_I atom at the same packing position and in the opposite direction of the axial As atom. There are three Na_{II} atoms of the other **T** layer in a triangular arrangement at 3.27 and 3.29 Å. These Na_{II} atoms form the base of a tetrahedron with the axial position occupied by the Na at the same packing position as the central Na_{II} and axial As atoms. This axial Na_{II} atom is in a **T** layer on the opposite side of the double Na_I–As layer. There are six repeating packing layers, giving the notation $\mathbf{2{\cdot}3P}^{Na\ As}\mathbf{T}^{Na}\mathbf{T}^{Na}$. This notation has the double layer fully occupied, Na_I fills *C* positions and *B* or *A* positions are filled by As. Overfilling could be avoided by considering all positions of **P** and **T** to be one-quarter filled, $\mathbf{2{\cdot}3P}^{Na\ As}_{1/4\ 1/4}\mathbf{T}^{Na}_{1/4}\mathbf{T}^{Na}_{1/4}$. This structure is found for M_3P compounds of Li, Na, and K; for M_3As and M_3Sb, compounds of Li, Na, K, and Rb; and for M_3 Bi compounds, of Na, K, and Rb.

8.2.4. The $\mathbf{2P}^{Y\ (OH)_3}_{1/4\ 1/4}$ Crystal Structure of $Y(OH)_3$

The crystal structure of yttrium hydroxide, $Y(OH)_3$, is hexagonal with two molecules per cell, $\mathbf{C}^2_{6h}$, $P6_3/m$, $a_o = 6.24$, and $c_o = 3.53$Å. The structure can be visualized in terms of $(OH)_3$ groups as units. Figure 8.9***a*** shows that the Y^{3+} ions and triangular $(OH)_3$ groups occur at two positions (*B* and *C*) in the hexagonal cell. The *A* positions in the corners of the cell are vacant. Figure 8.9***b*** shows that the Y^{3+} ions are halfway between eclipsed triangles of $(OH)_3$ groups, giving a trigonal prism of OH^- ions. The ***CN*** of Y^{3+} is 6, with OH^- ions at 2.42 Å. There are three more OH^- ions of other $(OH)_3$ groups in the plane of Y^{3+} (heights one-quarter and three-quarter) at 2.54 Å. These three OH^- ions cap the tetragonal faces of the trigonal prism. The double **P** layers are very open because $(OH)_3$ groups are assigned one site. The **P** layers are one-quarter occupied by Y^{3+} and one-quarter by $(OH)_3$

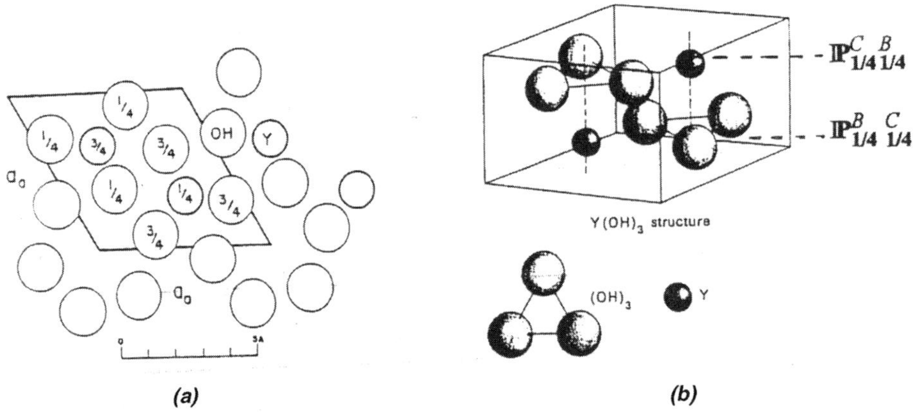

Figure 8.9. (*a*) A projection along the c_o axis of the hexagonal unit cell of $Y(OH)_3$. (*b*) A pictorial view of the cell.

(*Source*: R.W.G. Wyckoff, *Crystal Structures*, Vol. 2, 2nd ed., Wiley, New York, 1964, p. 77.)
(*Source*: F.S. Galasso, *Structure and Properties of Inorganic Solids*, Pergamon, Oxford, 1970, p. 125.)

groups in the sequence $\mathbb{P}_{BC}\mathbb{P}_{CB}$ giving the notation $2\mathbb{P}_{1/4\ 1/4}[Y, (OH)_3]$ or $2\mathbb{P}^{Y\ (OH)_3}_{1/4\ 1/4}$. The OH^- ions are not close packed because they are grouped in close triangular groups. More accurately, these are positions of oxygen atoms because the hydrogens were not located. The $Y(OH)_3$ crystal structure is common for hydroxides of Gd, La, Nd, Pr, Sm, and Yb; chlorides and bromides of Ac, Ce, La, Np, Pr, and U; and chlorides of Am, Eu, Gd, Nd, Pu, and Sm.

8.2.5. The $2\mathbb{P}_{1/4\ 1/4}(h)[NO_2^+, NO_3^-]$ Crystal Structure of N_2O_5

The crystal structure of N_2O_5 is hexagonal with two molecules in the unit cell, D^4_{6h}, $C6/mmc$, $a_o = 5.45$, and $c_o = 6.66$Å at ca. 20°C. In the solid, N_2O_5 exists as linear NO_2^+ and planar NO_3^- ions. Figure 8.10 shows the NO_3^- ions at heights one-quarter and three-quarter at the corners of the cell parallel to c_o with NO_2^+ ions aligned with the c_o axis within the cell centered at heights one-quarter and three-quarter. The NO_3^- ions are in only *A* positions in each layer. Each packing layer must expand for O atoms of NO_3^- and for N atoms of NO_2^+ ions. In the first layer the NO_2^+ ions are at *B* positions. In the next layer the NO_2^+ ions are at *C* positions. Because the linear NO_2^+ ions are aligned perpendicular to the packing layers their positions must be staggered. The notation is $2\mathbb{P}_{1/4\ 1/4}(h)$ for the *AB* and *AC* double layers.

8.2.6. The Crystal Structure of $Ba(ClO_4)_2 \cdot 3H_2O$

The crystal structure of $Ba(ClO_4)_2 \cdot 3H_2O$ is hexagonal with two molecules per cell, C_6^6, $P6_3$, $a_o = 7.294$, and $c_o = 9.675$ Å. Figure 8.11 shows a projection of the cell. Each Ba^{2+} ion is octahedrally surrounded by six H_2O (H_2O molecules are shared along c_o) with six ClO_4^- ions completing an icosahedron of oxygen atoms around Ba^{2+}. Ba^{2+} ions occur only at *A* positions, with ClO_4^- ions at *B* and *C*

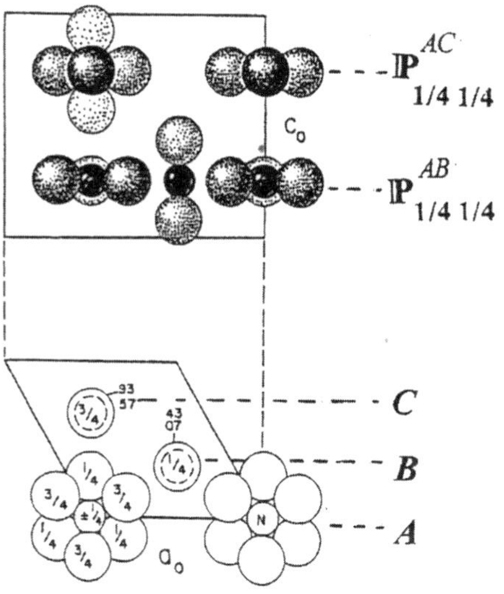

Figure 8.10. Two views of the hexagonal unit cell of the N_2O_5 structure are shown.

(*Source*: R.W.G. Wyckoff, *Crystal Structures*, Vol. 2, 2nd ed., Wiley, New York, 1964, p. 181.)

positions. The positions of Ba^{2+} and Cl atoms of ClO_4^- ions with heights along c_o are:

Height:	0	0	42	50	50	92
Atom:	Ba	Cl	Cl	Ba	Cl	Cl
Position:	*A*	B	B	A	C	C

There are two double layers of Ba and Cl of ClO_4^- at *AB* and *AC* positions with ClO_4^- ions in two layers at *B* and *C*. The sequence is $\mathbb{P}_{AB}\mathbf{P}_B\mathbb{P}_{AC}\mathbf{P}_C$ giving the notation $\mathbf{2{\cdot}2}\mathbb{P}_{1/4\ 1/4}\mathbf{P}_{1/4}\mathbf{(h)}$. Each double layers has a near layer of ClO_4^- ions at heights separation 8, with the next layer separated by 42 units. H_2O molecules occur at heights 25 and 75 completing the octahedra around Ba^{2+} ions. They are counted with Ba as $Ba(H_2O)_6^{2+}$. The heights of the oxygen atoms are shown in the figure.

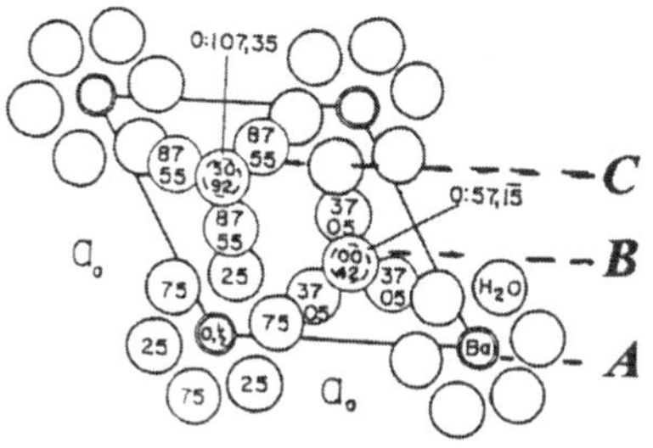

Figure 8.11. The hexagonal unit cell of $Ba(ClO_4)_2{\cdot}3H_2O$.

(*Source*: R.W.G. Wyckoff, *Crystal Structures*, Vol. 3, 2nd ed., Wiley, New York, 1965, p. 699.)

8.2.7. The $8\mathbf{P}^{S_2}_{1/4}\mathbf{T}^{Cu}_{1/4}\mathbb{P}^{S}_{1/4}{}^{Cu}_{1/4}\mathbf{T}^{Cu}_{1/4}$ Crystal Structure of Covellite [CuS]

The mineral covellite, CuS, is an important ore for copper. Its crystal structure is much more complex than the simple formula suggests. S occurs as S^{2-} and S_2^{2-}, and there are two roles for Cu, presumably Cu^{+} and Cu^{2+}. The crystal structure is an elongated hexagonal cell with six molecules per cell, $\mathbf{D}^4_{6h}$, $P6_3/mmc$, $a_o = 3.78813$, and $c_o = 16.33307$ Å. A projection of the hexagonal cell along c reveals that S_2^{2-} ions are at corners of the cell (position A) with Cu and S^{2-} in B and C positions. The structure in Figure 8.12 shows that S_2^{2-} ions only occur at A positions, while Cu and single S atoms occur at B and C positions in the same double layer. The S_2^{2-} ions are aligned along c_o, so we can consider their centers as positions in A sites in $\mathbf{P}_A$ layers. The double layers consist of single S atoms in B or C positions and Cu atoms in C or B positions, $\mathbb{P}_{BC}$ or $\mathbb{P}_{CB}$. The double layer contains planar hexagons formed by three Cu and three S atoms. The vacant centers of the hexagons are at A positions. Cu atoms also occur in tetrahedra, $\mathbf{T}_B$ or $\mathbf{T}_C$, with three S atoms of S_2^{2-} ions in the base and with S of a double layer at the apex. The sequence is:

Height:	**0**	**11**	**25**	**39**	**50**	**61**	**75**	**89**
Atom:	S_2	**Cu**	**S,Cu**	**Cu**	S_2	**Cu**	**S,Cu**	**Cu**
Layer:	$\mathbf{P}_A$	$\mathbf{T}_B$	$\mathbb{P}_{BC}$	$\mathbf{T}_B$	$\mathbf{P}_A$	$\mathbf{T}_C$	$\mathbb{P}_{CB}$	$\mathbf{T}_C$

The eight layers give the notation $8\mathbf{P}^{S_2}_{1/4}\mathbf{T}^{Cu}_{1/4}\mathbb{P}^{S}_{1/4}{}^{Cu}_{1/4}\mathbf{T}^{Cu}_{1/4}$. This can be described as a structure with all layers filled and with the double layer having filled A and B positions. The notation would be $8\mathbf{P}^{S_2}\ \mathbf{T}^{Cu}\mathbb{P}^{S,\ Cu}\ \mathbf{T}^{Cu}$. Cu_{II} atoms are in $\mathbf{T}$ sites

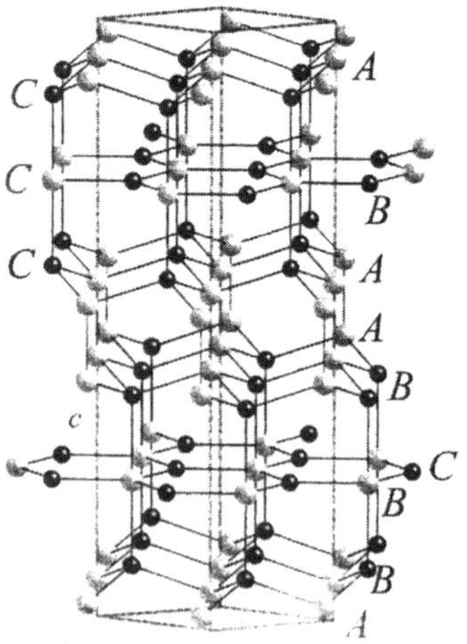

Figure 8.12. The hexagonal cell of CuS showing two roles of Cu and S.

(*Source*: Y. Matsushita, *Chalcogenide* crystal structure data library, Version 5.5B, Institute for Solid State Physics, The University of Tokyo, Tokyo, Japan, 2004; and *CrystalMaker*, by David Palmer, *CrystalMaker* Software Ltd., Begbroke Science Park, Bldg. 5, Sandy Lane, Yarnton, Oxfordshire, OX51PF, UK.)

and Cu_I atoms are in the double layer with three close S neighbors as planar CuS_3. S_I atoms are in the double layer at the centers of the trigonal bipyramids (*CN* 5). Each S_{II} atom of S_2^{2-} ions has one S atom (of S_2^{2-}) at 2.067 Å and three Cu_{II} atoms at 2.303 Å, giving a tetrahedron. Two tetrahedra are bonded by the S_2^{2-} ions. CuSe has the covellite structure.

8.2.8. The Crystal Structure of KAs_4O_6I

The KAs_4O_6I structure has one **P** layer one-quarter occupied by I ($\mathbf{P}_{1/4}$), two **P** layers three-quarter occupied by oxygen ($\mathbf{P}_{3/4}$), and one **P** layer one-quarter occupied by K ($\mathbf{P}_{1/4}$). These **P** layers are all at *A* positions. The K and I are in positions over or under vacancies in oxygen layers. The As atoms are in double layers ($\mathbb{P}_{1/4\ 1/4}$) at *B* and *C* positions between layers of I and oxygen (see Figure 8.13). The sequence of layers is:

Height:	0	21	32	50	68	79	100
Atom:	I	As	O	K	O	As	I
Layer:	$\mathbf{P}_{1/4}$	$\mathbb{P}_{1/4\ 1/4}$	$\mathbf{P}_{3/4}$	$\mathbf{P}_{1/4}$	$\mathbf{P}_{3/4}$	$\mathbb{P}_{1/4\ 1/4}$	$\mathbf{P}_{1/4}$
Position:	*A*	*BC*	*A*	*A*	*A*	*BC*	*A*

The As atoms are close to an oxygen layer, giving pyramidal AsO_3 units with each oxygen atom shared by two As atoms. The As layers are one-half filled, but using *B* and *C* positions. These can be consi-

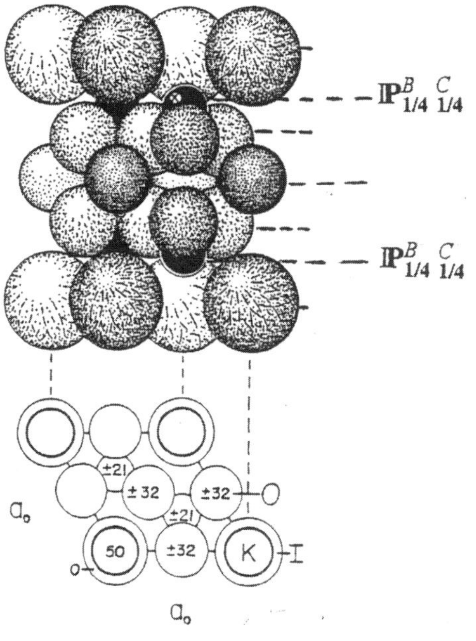

Figure 8.13. Two projections of the hexagonal cell of KAs_4O_6I are shown; As atoms are smallest circles. Oxygen and K atoms are of about the same size; are distinguished by heavily outlined circles for K.

(*Source*: R.W.G. Wyckoff, *Crystal Structures*, Vol. 4, 2nd ed., Wiley, New York, 1965, p. 41.)

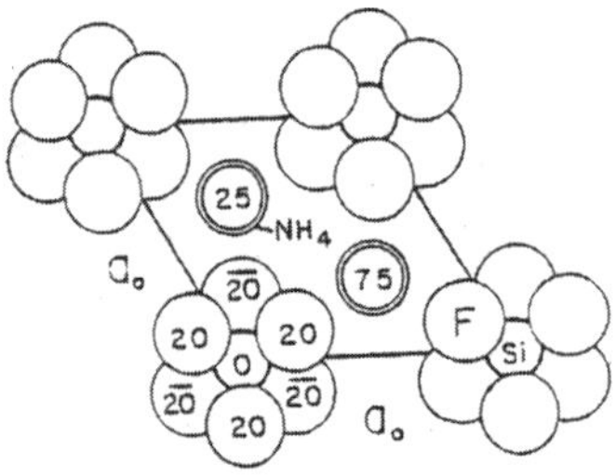

Figure 8.14. The hexagonal cell of β-$(NH_4)_2SiF_6$.

(*Source*: R.W.G. Wyckoff, *Crystal Structures*, Vol. 3, 2nd ed., Wiley, New York, 1965, p. 351.)

dered as **P** layers giving the $\mathbf{6P}^{\mathrm{I}}_{1/4}\mathbb{P}^{B}_{1/4}{}^{\mathrm{C}}_{1/4}{}^{\mathrm{As}}\mathbf{P}^{\mathrm{O}}_{3/4}\mathbf{P}^{\mathrm{K}}_{1/4}\ \mathbf{P}^{\mathrm{O}}_{3/4}\mathbb{P}^{B}_{1/4}{}^{\mathrm{C}}_{1/4}{}^{\mathrm{As}}$. The I layers are between As layers, and K layers are between oxygen layers. K and I atoms are in hexagonal prisms of 12 oxygen atoms. The hexagonal cell contains one molecule, $\mathbf{D}^1_{6h}$, $P6/mmm$ with $a_o = 5.277$, and $c_o = 9.157$ Å.

8.2.9. The $2{\cdot}3/2\mathbf{P}^{\mathrm{SiF_6}}_{A}\mathbf{P}^{\mathrm{NH_4}}_{B}\mathbf{P}^{\mathrm{NH_4}}_{C}$(h) Crystal Structure of β-$(NH_4)_2SiF_6$

In Section 5.5.6 β-ammonium hexafluorosilicate, β-$(NH_4)_2SiF_6$, was treated as an M_2X-type compound with a layer structure such as $Mg(OH)_2$, treating SiF_6^{2-} as a unit in sandwiches between NH_4^+ layers. There is an unusual wide separation (height 50) between NH_4^+ layers at *B* and *C* positions. This is because of the packing of large SiF_6^{2-} ions at the same *A* positions in each layer and hydrogen bonding is expected between the NH_4^+ and F^- ions. The heights of the NH_4^+ and SiF_6^{2-} ions and the atoms N (hydrogens are not located), Si and F are:

Height:	−20	0	20	25	75	80	100	120
Ion:			SiF_6	NH_4	NH_4	SiF_6		
Atom:	F	Si	F	N	N	F	Si	F
Position:	*B*	*A*	*C*	*C*	*B*	*B*	*A*	*C*

The usual approach for β-$(NH_4)_2SiF_6$, considering the F^- ions as close packed, works well. The F^- ions fill three-quarter of two **P** layers with the other one-quarter sites filled by NH_4^+ ions forming puckered double layers $(\mathbb{P}^{\mathrm{F,\ N}}_{3/4\ 1/4})_C$ and $(\mathbb{P}^{\mathrm{N,\ F}}_{1/4\ 3/4})_B$ with Si in one-quarter of $\mathbf{O}_A$ layers (Figure 8.14). The heights along c_o of the N atoms and F atoms in the $\mathbb{P}$ layers differ by five units. The notation is $\mathbf{2{\cdot}3/2}\mathbb{P}^{\mathrm{F}}_{3/4}{}^{\mathrm{N}}_{1/4}\mathbf{O}^{\mathrm{Si}}_{1/4}\mathbb{P}^{\mathrm{N}}_{1/4}{}^{\mathrm{F}}_{3/4}$. Si atoms are only in $\mathbf{O}_A$ sites surrounded by six F^- ions. The close-packed F layers are staggered in *CBCB* sequence.

8.2.10. The $4\mathbb{P}^{\mathrm{Li}}_{A}{}^{\mathrm{O_2}}_{C}\mathbf{O}^{\mathrm{Li}}_{B}\mathbb{P}^{\mathrm{Li}}_{C}{}^{\mathrm{O_2}}_{A}\mathbf{O}^{\mathrm{Li}}_{\mathrm{B}}$ Crystal Structure of Li_2O_2

Lithium peroxide, Li_2O_2, has a hexagonal unit cell with two molecules per cell, $\mathbf{C}^1_{3h}$, $P\bar{6}$, $a_o = 3.142$, and $c_o = 7.650$ Å. Two projections are shown in Figure 8.15. The heights along c_o are:

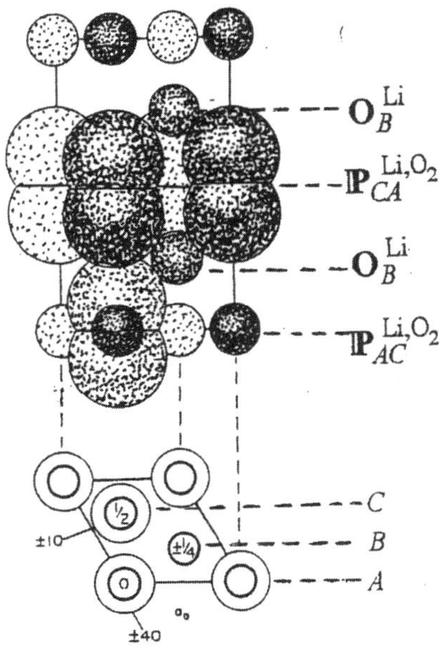

Figure 8.15. Two views of the hexagonal cell of lithium peroxide, Li_2O_2. (*Source*: R.W.G. Wyckoff, *Crystal Structures*, Vol. 1, 2nd ed., Wiley, New York, 1964, p. 168.)

Height:		0		25		50		75
Ion:		Li, O_2		Li		Li, O_2		Li
Position:		AC		B		CA		B
Atom:	O	Li	O	Li	O	Li	O	Li
Height:	−10	0	10	25	40	50	60	75
Position:	*C*	*A*	*C*	*B*	*A*	*C*	*A*	*B*

There are two different sites for Li. Li_I, at heights 25 and 75, are in octahedral sites. Li_I at 25 in position *B* has three oxygen atoms at height 40 (from O_2^{2-} at 50) in positions *A* and three oxygen atoms at height 10 (from O_2^{2-} at zero) in position *C*. Li_{II} atoms occur at heights 0 and 50. Both *A* and *C* positions are filled in the same layer by Li and O_2^{2-}. The centers of the O_2^{2-} bonds are in this layer with an oxygen above and one below. Each Li_{II} atom at zero is in the same layer as three O_2^{2-} ions and heights of oxygen atoms at ± 10 give a small trigonal prism. There are two more distant oxygen atoms from different O_2^{2-} ions at the same position, capping the prism, at ± 40. The relative spacings are the same for Li_{II} atoms at height 50. There are four repeating layers including two double layers, $\mathbf{4}\mathbb{P}_{AC}^{Li,O_2}\mathbf{O}_B^{Li}\mathbb{P}_{CA}^{Li,O_2}\mathbf{O}_B^{Li}$. If the oxygen atoms are considered individually, these double layers would be triply layered structure giving the notation $\mathbf{2{\cdot}2}\mathbb{P}_{CAC}^{O,Li,O}\mathbf{O}_B^{Li}\mathbb{P}_{ACA}^{O,Li,O}\mathbf{O}_B^{Li}$.

8.2.11. The Crystal Structure of Na_2O_2

The unit cell of sodium peroxide, Na_2O_2, is hexagonal with three molecules per cell, $\mathbf{D}_{3h}^3$, $C\bar{6}2m$, $a_o = 6.208$, and $c_o = 4.460$ Å. Two projections of the unit cell are shown in Figure 8.16. Oxygen atoms are at heights along $c_o \pm 67$ and ± 17. These correspond to heights 50 (1/2) and zero for O_2^{2-} ions aligned along c_o. Thus, all Na^+ and O_2^{2-} ions occur at zero and one-half. The lower projection of the cell in the figure corresponds to a close-packed layer with all *A* positions filled, but the ions are not in one layer. There is slight distortion of the layer caused by difference in sizes of the ions. It also corresponds to a projection with O_2^{2-} in *A* positions with Na^+ ions in *B* (one-half) and *C* (0) positions. The layer at zero is filled by Na^+ at *B* positions with O_2^{2-} in two-thirds of *A* positions. The layer at one-half is filled by Na^+ in *C* positions with O_2^{2-} in one-third of *A* positions. Each O_2^{2-} ion at heights zero or one-half has a triangular arrangement of Na^+ ions above and below with three Na^+ ions in the same plane as the center of O_2^{2-} giving a tricapped trigonal prism. The Na^+ ions have trigonal prismatic arrangements of six Na^+ ions as neighbors and each Na^+ ion has a distorted trigonal prismatic arrangement of six oxygen atoms. For example, the central Na^+ at height zero has two O_2^{2-} ions at zero (± 17 for the oxygen atoms) and one O_2^{2-} at 1/2 (33 for one of the oxygen atoms of the O_2^{2-} at one-half and -33 from one of the oxygen atoms from O_2^{2-} at $-1/2$. In the cell there are three Na^+ ions at zero and three Na^+ at one-half. Assigning O_2^{2-} ions to *A* positions with Na^+ ions in *B* and *C* positions, the notation is $\mathbf{1{\cdot}2\mathbb{P}}^{Na\ O_2}_{3/3\ 2/3}\mathbb{P}^{Na\ O_2}_{3/3\ 1/3}(\mathbf{h})$. This overfills the layers, but the oxygen atoms are above and below these layers. Alternatively we can use fractional filling by Na^+ and O_2^{2-} ions to give two double layers, giving the notation, $\mathbf{1{\cdot}2\mathbb{P}}^{Na\ O_2}_{3/9\ 2/9}\mathbb{P}^{Na\ O_2}_{3/9\ 1/9}(\mathbf{h})$. The sum of the fractions is 9/9 corresponding to the projection appearing as a full layer.

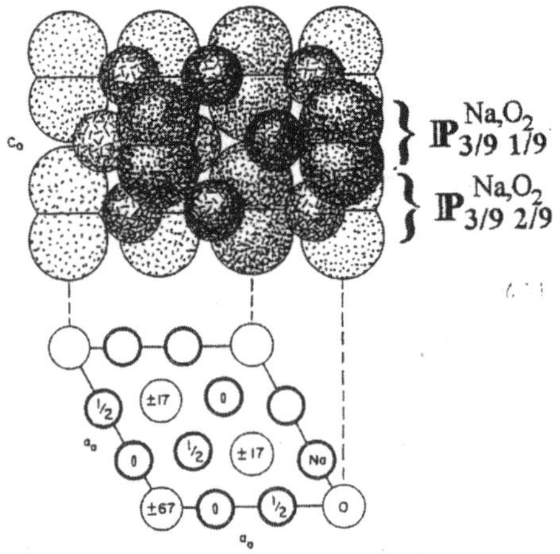

Figure 8.16. Two views of the hexagonal cell of Na_2O_2.

(*Source*: R.W.G. Wyckoff, *Crystal Structures*, Vol. 1, 2nd ed., Wiley, New York, 1964, p. 169.)

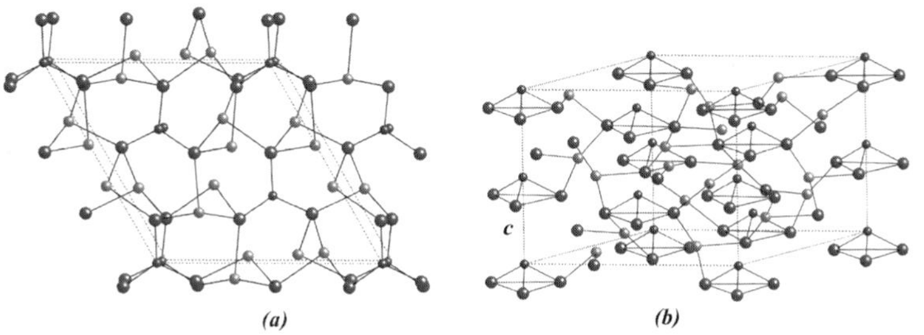

Figure 8.17. (*a*) A projection of the upper half of the Ag_3AsS_3 cell along the c_o axis. As are small, dark atoms, Ag atoms are lighter, and S atoms are largest. (*b*) A view of the cell showing the AsS_3 pyramids.

(*Source*: Y. Matsushita, *Chalcogenide* crystal structure data library, Version 5.5B, Institute for Solid State Physics, The University of Tokyo, Tokyo, Japan, 2004; and *CrystalMaker*, by David Palmer, *CrystalMaker* Software Ltd., Begbroke Science Park, Bldg. 5, Sandy Lane, Yarnton, Oxfordshire, OX51PF, UK.)

8.2.12. The Crystal Structure of Proustite [Ag_3AsS_3] and Pyrargyrite [Ag_3SbS_3]

Proustite (Ag_3AsS_3) forms blood-red crystals, and pyrargyrite (Ag_3SbS_3) occurs as deep ruby-red to black crystals. They occur together with native Ag in Ag ores. They have the same trigonal structure, $\mathbf{C}_{3v}^6$, R3*c*. For proustite the hexagonal cell has $a_o = 10.767$ and $c_o = 8.713$ Å., with are six molecules in the cell. Figure 8.17 shows two views of the cell. Figure 8.17***a*** is a projection of the top half of the cell and the heights along c_o for the atoms are:

Height:	**50**	**55**	**56**		**66**	**71**	**73**		**83**	**88**	**90**		**100**
Atom:	**As**	**S**	**Ag**		**As**	**S**	**Ag**		**As**	**S**	**Ag**		**As**
Position:	$\mathbf{P}_B$	$\mathbb{P}_{AC}$			$\mathbf{P}_B$	$\mathbb{P}_{AC}$			$\mathbf{P}_B$	$\mathbb{P}_{AC}$			$\mathbf{P}_B$

S atoms occur only at *A* positions, and As atoms occur only at *B* positions. These occur as AsS_3 flattened trigonal pyramids with As at the apex (Figure 8.17***b***). The As heights are 0, one-half, 100, one-sixth (17), one-third, two-thirds, and five-sixths (83). The top and lower halves of the cell are identical. The As positions are identified as $\mathbf{P}_B$ sites. The As atoms are in pyramids required by the chemistry of As and S. A missing S $\mathbf{P}_B$ layer between $\mathbf{P}_A$ layers could provide $\mathbf{T}_B$ sites for As, but only three bonds are required for As. The S and Ag atoms are in puckered double layers. The positions of Ag atoms are assigned as *C* sites. The positions of S and As are quite regular, as required for AsS_3 units. The Ag positions are displaced for bonding to S atoms. Each S is bonded to one As (at 2.25 Å) and two Ag atoms at 2.40 Å. The projection of the top of the cell corresponds to a filled $\mathbf{P}_A$ layer for S, but it includes S layers at 55, 71, and 88. Each $\mathbb{P}$ layer is one-third filled by S and one-third by Ag. The $\mathbf{P}_B$ layers are one-nineth filled by As giving the notation $\mathbf{1{\cdot}3}\mathbb{P}_{1/3\ 1/3}^{A(S)\ C(Ag)}\mathbf{P}_{1/9}^{B(As)}(\mathbf{h})$. The positions of the S and Ag atoms are shifted for the three layers in the top and lower halves of the cell, corresponding to the one-third of the sites occupied, and repeated in the other half. The *A*, *B*, and *C* positions are unchanged, so there are three layers repeating giving the index **1·3**. This is an open structure with no layer more than one-third filled because the S atoms are in MS_3 pyramids.

8.2.13. The $4\mathbf{IP}_{3/9\ 4/9}$ Crystal Structure of α-Si_3N_4

The α-Si_3N_4 crystal structure is hexagonal, with four molecules in the unit cell, $\mathbf{C}_{3v}^4$, $P31c$, $a_o = 7.753$, and $c_o = 5.618$ Å. Si and N atoms occur in layers at heights along c_o as follows:

Height:	0	25	50	75	100
Si atoms:	$\mathbf{P}_{3/9}^{C}$	$\mathbf{P}_{3/9}^{B}$	$\mathbf{P}_{3/9}^{B}$	$\mathbf{P}_{3/9}^{C}$	$\mathbf{P}_{3/9}^{C}$
N atoms:	$\mathbf{P}_{4/9}^{A}$	$\mathbf{P}_{4/9}^{A}$	$\mathbf{P}_{4/9}^{A}$	$\mathbf{P}_{4/9}^{A}$	$\mathbf{P}_{4/9}^{A}$
Double layers:	$\mathbf{IP}_{3/9\ 4/9}^{C\ \ A}$	$\mathbf{IP}_{3/9\ 4/9}^{B\ \ A}$	$\mathbf{IP}_{3/9\ 4/9}^{B\ \ A}$	$\mathbf{IP}_{3/9\ 4/9}^{C\ \ A}$	$\mathbf{IP}_{3/9\ 4/9}^{C\ \ A}$

Nitrogen atoms occur only at *A* positions and the projection of all N atoms in the unit cell in Figure 8.18***a*** corresponds to a full close-packed layer. The Si atoms are near the centers of distorted tetrahedral arrangements of four N atoms, for example, Si at one-half has two N at one-half, one at 0 and one at three-quarters. For each double layer the N atoms fill 4/9 of the *A* sites and Si atoms fill 3/9 of the *B* or *C* sites. N atoms are in planar triangular NSi_3 units. Other sites are more irregular. The notation for the double layer structure is $\mathbf{4IP}_{3/9\ 4/9}$.

8.2.14. The $2\mathbf{IP}_{3/9\ 4/9}$ Crystal Structure of β-Si_3N_4

The β-Si_3N_4 crystal structure is similar to that of α-Si_3N_4, but simpler. The unit cell for β-Si_3N_4 is rhombohedral. The corresponding hexagonal unit cell contains two molecules, $\mathbf{C}_{3i}^2$, $R\bar{3}$, $a_o = 7.606$, and $c_o = 2.909$ Å. The structure corresponds to that of phenakite (Be_2SiO_4, Section 10.3.8). The cell is larger for Be_2SiO_4, with three different atoms. The projection of the unit cell of β-Si_3N_4 (Figure 8.18***b***) shows the N atoms at *A* positions with vacancies at the corners of the cell. Each $\mathbf{P}^A$ layer (at heights one-quarters and three-quarters) is four-

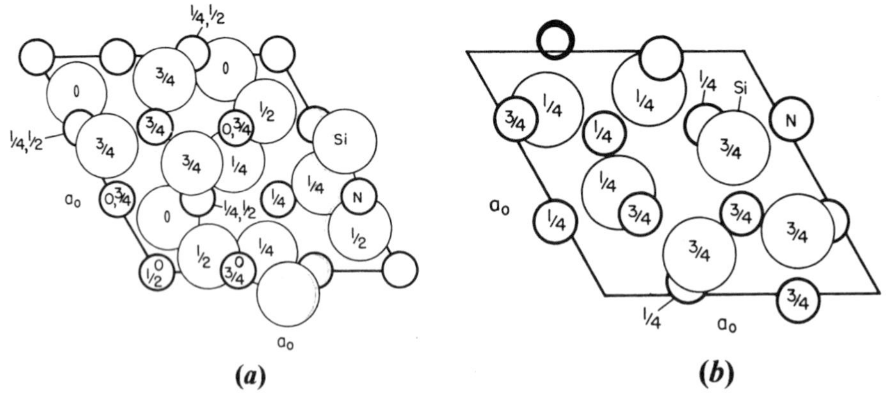

Figure 8.18. (*a*) A projection of the hexagonal cell of α-Si_3N_4 along c_o. (*b*) A projection of the hexagonal cell of β-Si_3N_4 along c_o.

(*Source*: R.W.G. Wyckoff, *Crystal Structures*, Vol. 2, 2nd ed., Wiley, New York, 1964, p. 158.)
(*Source*: R.W.G. Wyckoff, *Crystal Structures*, Vol. 2, 2nd ed., Wiley, New York, 1964, p. 159.)

nineths filled by N atoms with one-nineth of all A sites are vacant (positions in corners of the cell). Si atoms have a tetrahedral arrangement of four N atoms. Each layer is three-nineth filled by Si atoms. These positions and heights along c_o are:

Height:	**25**	**75**
Si atoms:	$\mathbf{P}^{B}_{3/9}$	$\mathbf{P}^{C}_{3/9}$
N atoms:	$\mathbf{P}^{A}_{4/9}$	$\mathbf{P}^{A}_{4/9}$
Double layers:	$\mathbb{P}^{B\ \ A}_{3/9\ 4/9}$	$\mathbb{P}^{C\ \ A}_{3/9\ 4/9}$

Some N atoms are at the centers of triangular arrangements of Si atoms in the same plane. Other N atoms are in triangles in planes perpendicular to the plane of the projection. The Si atoms are in tetrahedral arrangements of four N atoms, two N in the same plane with the opposite N—N edge vertical. For example, the Si at three-quarters in the lower right of the cell has two N atoms in the same plane with one N at one-quarter and the fourth at 125 (1-1/4). The notation for β-Si_3N_4 is $\mathbf{2}\mathbb{P}_{3/9\ 4/9}$.

8.2.15. The $\mathbf{2}\mathbb{P}^{K\ \ Re}_{3/3\ 1/3}\mathbb{P}^{K\ \ Re}_{3/3\ 1/3}$ Crystal Structure of K_2ReH_9

K_2ReH_9 is an unusual metal hydride, and the crystal structure is unusual. The ReH_9^{2-} ion is a trigonal prism of 6 H^- ions around Re with three H^- ions capping the rectangular faces ($\mathbf{D}_{3h}$). The ReH_9^{2-} ions also are in tricapped trigonal prisms formed by nine K^+ ions. The K^+ ions are in filled $\mathbf{P}_B$ layers at zero and $\mathbf{P}_C$ at 50, in an *hcp* arrangement with slight displacement of one K^+ within the cell. ReH_9^{2-} ions are in the same layers (0 and 50) giving double layers ($\mathbb{P}$) (see Figure 8.19). The A positions for ReH_9^{2-} ions are one-third filled at 0 and A positions are two-thirds filled at 50, giving $\mathbf{2}\mathbb{P}^{K\ \ Re}_{3/3\ 1/3}\mathbb{P}^{K\ \ Re}_{3/3\ 1/3}$. The index **2** refers to two double layers. There are three molecules per hexagonal cell, $\mathbf{D}^3_{3h}$, $P\bar{6}2m$, $a_o = 9.607$, and $c_o = 5.508$ Å. The compound K_2TaH_9 has the same structure.

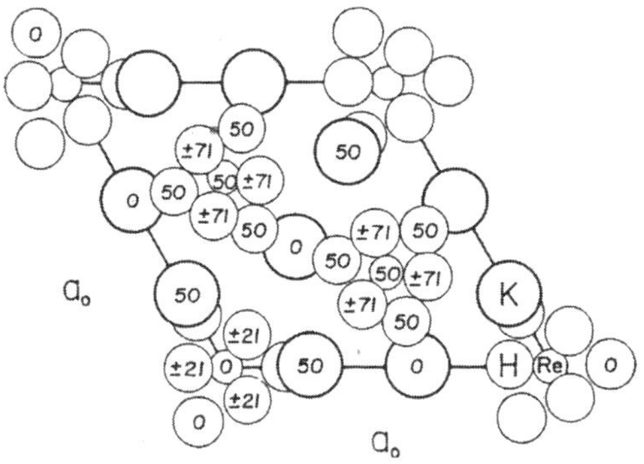

Figure 8.19. A projection of the hexagonal cell of K_2 ReH_9.
(*Source*: R.W.G. Wyckoff, *Crystal Structures*, Vol. 4, 2nd ed., Wiley, New York, 1965, p. 52.)

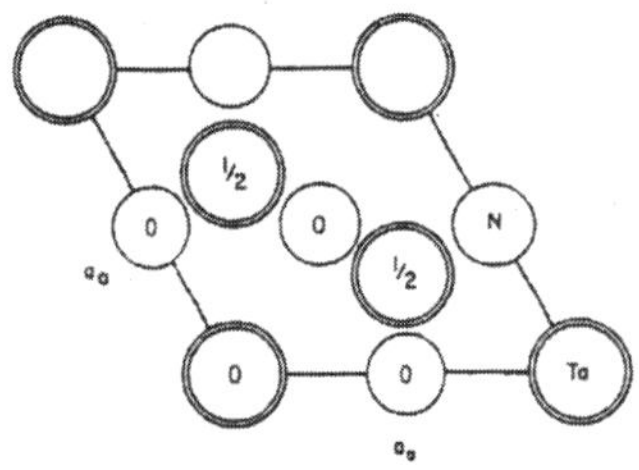

Figure 8.20. A projection on the base of the hexagonal cell of ε-TaN.
(*Source*: R.W.G. Wyckoff, *Crystal Structures*, Vol. 1, 2nd ed., Wiley, New York, 1963, p. 150.)

8.2.16. The $2\mathbf{P}^{\mathrm{Ta}\ \ \mathrm{N}}_{1/4\ 3/4}\mathbb{P}^{B\ \ C}_{1/4\ 1/4}$ Crystal Structure of ε-TaN

The structure of ε-TaN is fairly simple as seen in Figure 8.20. The hexagonal cell contains three molecules, $\mathbf{D}^1_{6h}$, $P6/m\ mm$, $a_o = 5.191$, and $c_o = 2.908$ Å. There are two different environments for Ta. In the first layer (the base of the cell at 0) there are four $\mathrm{Ta_I}$ atoms at the corners, one N at the center and four N in the edges, all at *A* positions, giving $\mathbf{P}^{\mathrm{Ta}\ \ \mathrm{N}}_{1/4\ 3/4}$. The next layer contains two $\mathrm{Ta_{II}}$ atoms at 50, a half-filled **P** layer with Ta in *B* (one-quarters filled) and *C* (one-quarters filled) positions, giving the notation $2\mathbf{P}^{\mathrm{Ta}\ \ \mathrm{N}}_{1/4\ 3/4}\mathbb{P}^{B\ \ C}_{1/4\ 1/4}$. $\mathrm{Ta_I}$ (one-third of the total Ta) is surrounded by six N atoms in the same plane.

$\mathrm{Ta_{II}}$ atoms (two-thirds of the total Ta) alone half fill the layer at 50, but divided into *B* and *C* sites. The $\mathrm{Ta_{II}}$ atoms are at the centers of trigonal prisms formed by three N atoms in $\mathbf{P}_A$ planes above and below. We can see from the figure that the central N has as neighbors two $\mathrm{Ta_{II}}$ in layers above and below (at 2.08 Å) and two $\mathrm{Ta_I}$ atoms in the same plane, giving a distorted octahedron.

8.2.17. The $4{\cdot}2\mathbf{P}^{\mathrm{Be}}\mathbb{P}^{\mathrm{Be,\ N}}\mathbf{P}^{\mathrm{Be}}\mathbf{O}^{\mathrm{N}}$ Crystal Structure of β-Be_3N_2

The hexagonal modification of β-Be_3N_2, beryllium nitride, is prepared by heating the cubic form above 1,400°C. There are two molecules per cell, $\mathbf{D}^4_{6h}$, $P6_3/mmc$, $a_o = 2.8413$, and $c_o = 9.693$ Å. Essentially it is a close-packed arrangement of Be layers with N atoms in interstices. There are three types of layers: those containing only N, only Be, and both Be and N. The **O** layers are filled by N atoms, the double $\mathbb{P}$ layers have *A* positions filled by Be atoms and *C* or *B* positions filled by N atoms ($\mathbb{P}^{\mathrm{Be,\ N}}$), and there are **P** layers filled by Be (see Figure 8.21). The Be atoms in layers shared with N have a triangular arrangement of three N atoms in the plane of the layer at 1.64 Å. There is a Be atom in each adjacent layer completing a trigonal bipyramid. The Be atoms in layers alone have a tetrahedral arrangement of four N atoms, three from the **O** layer, and one from the $\mathbb{P}^{\mathrm{Be,\ N}}$ layer. The sequence of layers is:

Height:	0	7	25	43	50	57	75	93	100
Atom:	N	Be	Be, N	Be	N	Be	Be, N	Be	N
Layer:	**O**	**P**	$\mathbb{P}$	**P**	**O**	**P**	$\mathbb{P}$	**P**	**O**
Position:	*A*	*C*	*A*(*C*)	*C*	*A*	*B*	*A*(*B*)	*B*	*A*

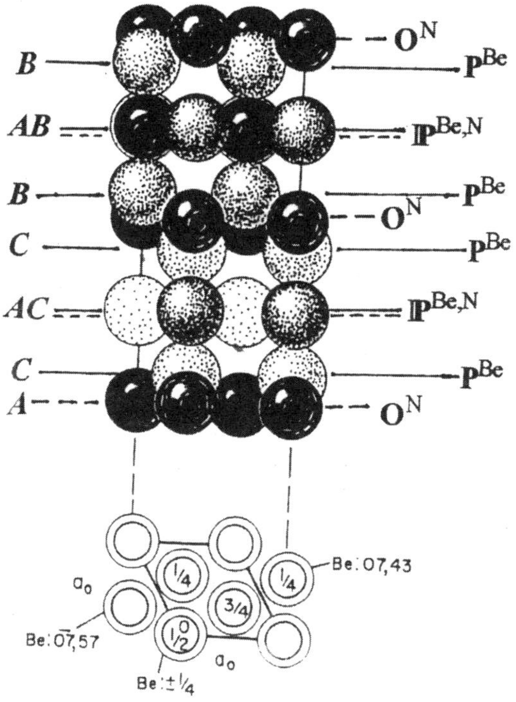

Figure 8.21. A projection on the base of the hexagonal cell of β-Be_3N_2 and a packing drawing along the c axis are shown. N atoms are black and smaller circles.

(*Source*: R.W.G. Wyckoff, *Crystal Structures*, Vol. 2, 2nd ed., Wiley, New York, 1964, p. 34.)

The N atoms in **O** layers are at the centers of flattened octahedra (trigonal antiprisms) formed by six Be atoms. The notation is $\mathbf{4{\cdot}2P^{Be}\mathbb{P}^{Be,N}P^{Be}O^{N}}$. The layers are filled. The double layer has filled A and B positions for the small Be and N atoms. Alternatively, to avoid the overfilling the double layer, all layers can be considered as one-third filled with two double layers, giving $\mathbf{4{\cdot}2P_{1/3}\mathbb{P}_{1/3\,1/3}P_{1/3}O_{1/3}}$.

8.2.18. The $BaFe_{12}O_{19}$ Crystal Structure

The structure of $BaFe_{12}O_{19}$ (Figure 8.22) is described as a complex spinel structure. The structure is hexagonal, D^{3}_{6h}, $P6_3/m\,mc$, $a_o = 5.888$, and $c_o = 23.22$ Å. It combines some features of spinel and perovskite. Layers 6 and 16 are double layers with Fe atoms in A positions, different from those for O and Ba. The sequence of **P** layers combines *hcp* and *ccp* patterns (*BABABCACAC*) with the *ABC* sequence, changing *hcp* patterns from *AB* to *AC*. The Fe atoms occur in **O** and **P** sites and some **T** sites. The sequence beginning with layer 2 and ending with 21 (or 1) is $\mathbf{P}^{B}|\mathbf{O}^{C}_{3/4}\mathbf{P}^{A}\mathbf{O}^{C}_{1/4}|\mathbb{P}^{B}_{3/4}{}_{1/4}{}^{A}_{1/4}|\mathbf{O}^{C}_{1/4}\mathbf{P}^{A}\mathbf{O}^{C}_{3/4}|$ $\mathbf{P}^{B}|\mathbf{O}^{A}_{1/4}\mathbf{T}^{B}_{1/4}{}^{C}_{1/4}|\mathbf{P}^{C}|\mathbf{O}^{B}_{3/4}\mathbf{P}^{A}\mathbf{O}^{B}_{1/4}|\mathbb{P}^{C}_{3/4}{}_{1/4}{}^{A}_{1/4}|\mathbf{O}^{B}_{1/4}\ \mathbf{P}^{A}\mathbf{O}^{B}_{3/4}|\ \mathbf{P}^{C}|\mathbf{O}^{A}_{1/4}\mathbf{T}^{B}_{1/4}{}^{C}_{1/4}|$. The pattern for layers 2–11 repeats with most positions changing, giving the notation $\mathbf{20PO_{3/4}PO_{1/4}}\mathbb{P}^{O}_{3/4}{}^{Ba}_{1/4}{}^{Fe}_{1/4}\mathbf{PO_{3/4}PO_{1/4}T_{1/4}T_{1/4}}$ with

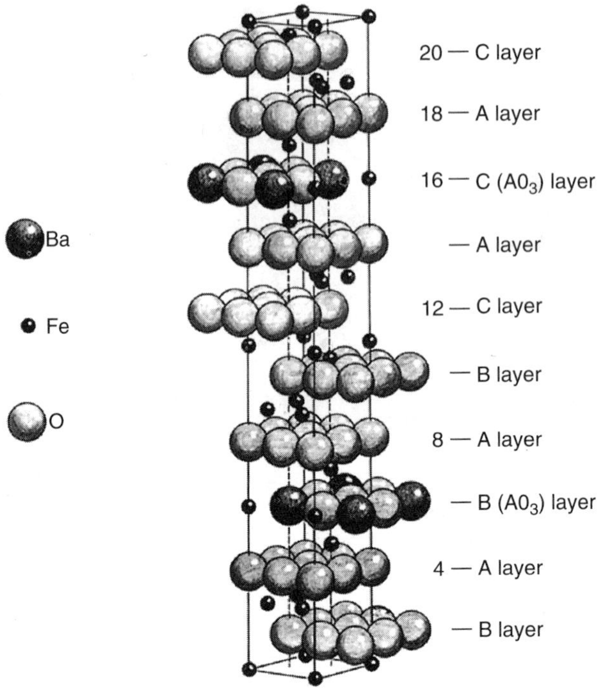

Figure 8.22. The structure of $BaFe_{12}O_{19}$ with layers labeled.
(*Source*: F.S. Galasso, *Structure and Properties of Inorganic Solids*, Pergamon, Oxford, 1970, p. 227.)

Fe in **O** and **T** sites plus one-quarter of *A* sites in the double layer (**IIP**). Layers 11 and 21 (or 1 and 11) have Fe atoms in $\mathbf{O}_A$, $\mathbf{T}_B$, and $\mathbf{T}_C$ sites; these could be considered as triple layers. The positions of layers 10 and 12 are *B* and *C*, respectively, and those of 20 and 22 (or 2) are *C* and *B*, respectively. In each case the **O** sites between **P** layers are at *A* positions and **T** positions are *B* and *C*. The oxygen and Ba atoms in double layers are in *B* positions for layer 6 and *C* positions for layer 16. In both cases Fe atoms are at *A* positions forming FeO_3 units. These also could be considered as triple layers. The oxygen atoms in layers 4 and 8 and 14 and 18 are at *A* positions completing trigonal bipyramids about Fe in layers 6 and 16, respectively. Changing the positions of two adjacent **P** layers from *B* and *C* to *C* and *B* changes the positions of intermediate **T** and **O** positions from $\mathbf{T}^C\mathbf{O}^A\mathbf{T}^B$ to $\mathbf{T}^B\mathbf{O}^A\mathbf{T}^C$. Thus, the positions are shifted after layer 11.

8.3. Crystal Structures with Triple Layers (IIIP)

8.3.1. The 2·3/2PO$_{2/3}$P or IIIP$_{3/3\ 2/3\ 3/3}(h)$ Layered Crystal Structure of BiI$_3$

The bismuth triiodide crystals are hexagonal (or rhombohedral), with six molecules in the hexagonal cell, C_{3i}^2, $R\bar{3}$, $a_o' = 7.498$, and

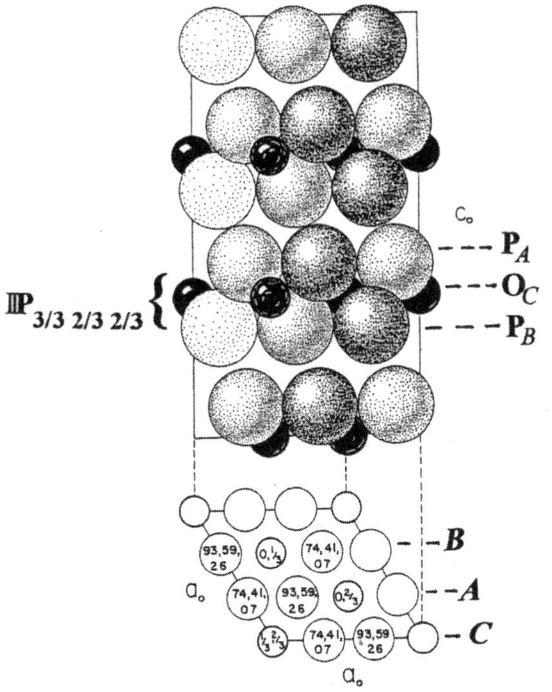

Figure 8.23. Two projections of the hexagonal cell of BiI_3, smaller circles are Bi atoms.

(*Source*: R.W.G. Wyckoff, *Crystal Structures*, Vol. 2, 2nd ed., Wiley, New York, 1964, p. 46.)

$c'_o = 20.68$ Å. BiI_3 has a layered structure similar to that of CdI_2 except that **O** layers are only two-thirds occupied by Bi (see Figure 8.23). The **P** layers are in an *AB* sequence, but three layers repeat. The notation is $\mathbf{2{\cdot}3/2PO_{2/3}P(h)}$. The BiI_3 crystal structure can also be described as a triple layer structure, $\mathbf{IIIP}_{3/3\ 2/3\ 3/3}$, but this does not designate the Bi layer as an **O** layer. There is larger spacing between I layers of adjacent I–Bi–I sandwiches. The layers of the I–Bi–I sandwiches are very close. The layers and heights along c_o are:

Height:	**–07**	**0**	**07**	**26**	**33**	**41**	**59**	**66**	**74**
Atom:	**I**	**Bi**	**I**	**I**	**Bi**	**I**	**I**	**Bi**	**I**
Position:	P_B	O_C	P_A	P_B	O_C	P_A	P_B	O_C	P_A

The I atoms are *hcp* with *CN* 12. Bi atoms are in **O** sites only at *C* positions. The BiI_3 structure is found for MCl_3 compounds of Sc, Ti, V, and Fe; MBr_3 compounds of Ti, Cr, and Fe; and MI_3 compounds of As, Sb, and Bi.

8.3.2. The β-MoS_2 Crystal Structure Described as 2IIIP and α-MoS_2 Described as 3IIIP

β-MoS_2, molybdenite, was described in Section 6.3.16 as a **2·3TPT** layered structure with Mo in **P** layers. The triply layered structure is

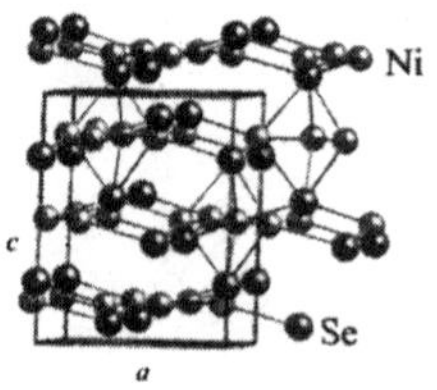

Figure 8.24. The hexagonal cell of Ni_3Se_2.

(*Source*: *CrystalMaker*, by David Palmer, *CrystalMaker* Software Ltd., Begbroke Science Park, Bldg. 5, Sandy Lane, Yarnton, Oxfordshire, OX51PF, UK.)

consistent with the chemistry of MoS_2. The wide separation between S—Mo—S layers results from a "missing" layer of Mo atoms between the S—Mo—S layers, as required by the formula MoS_2. There is only van der Waals interaction between adjacent S layers. However, β-MoS_2 can be described as a triple layer structure, giving **2IIIP** with the sequence $S_A Mo_B S_A$ $S_B Mo_A S_B$.

The crystal structure of α-MoS_2 is similar to that of β-MoS_2 except that α-MoS_2 is based on an *ABC* pattern for Mo layers. The heights of layers are:

Height:	−08	0	8	25	33	41	59	67	75
Atom:	S	Mo	S	S	Mo	S	S	Mo	S
Position:	*B*	A	B	C	B	C	A	C	A

As for β-MoS_2, the Mo atoms are at the centers of trigonal prisms formed by six S atoms. This is a layered structure, with three triple layers giving the notation **3IIIP**for α-MoS_2.

8.3.3. The $1{\cdot}3IIIP_{1/3\ 3/3\ 1/3}(h)$ Layer Crystal Structure of Ni_3Se_2

Trinickel diselenide, Ni_3Se_2, crystals are trigonal, D_3^7, R32, $a = 6.033$, and $c = 7.257$ Å, with three molecules in the unit cell. The structure is shown in Figure 8.24. In a projection along *c* the Se atoms are at the corners and, for each Se layer, Se is in one *A* position within the cell. In each Ni layer Ni atoms fill *B* positions. This is a triple layer structure and there are three triple layers in the cell. Each triple layer has a filled Ni layer between one-third filled Se layers or a Ni layer centered within a puckered Se layer. The separation of these Se layers is about 7% of the cell height along *c*. The notation is $1{\cdot}3IIIP_{1/3\ 3/3\ 1/3}(h)$. The index is written as **1·3** because the same triple layers are repeated, but there are three triple layers in the cell. Ni atoms have *CN* 8, four Se atoms at 2.359 and 2.382 Å, and four Ni atoms at 2.571 and 2.612 Å. It is not close to a regular polyhedron. Se has three neighbor Ni atoms in the closer layer and three Ni atoms in the farther layer.

Chapter 9

Crystal Structures of Some Intermetallic Compounds

Most of the compounds we have considered are sometimes known as "normal valence" compounds, even though the valence or oxidation numbers in many cases are not those usually encountered. For ionic compounds the structural limitations in crystals are severe. Ions of one charge must be well screened by those of opposite charge, and the required coordination number increases with the size of an ion. Ions of opposite charge attract one another, giving the minimum distance determined by repulsive forces. Because anions are usually larger than cations, for a structure based on close packing, the anions are usually in **P** layers with cations in **O** and/or **T** layers. For nonpolar compounds we encounter molecular structures such as those of H_2, N_2, O_2, CO_2, halogens (Chapter 4), and organic compounds (Chapter 11). Some elements or compounds form giant molecules with three-dimensional bonding determined by valence orbitals and electrons available. These include diamond, graphite, Si, and SiO_2.

Some intermetallic compounds have the same structures as those of simple polar compounds. Quite a few AM type intermetallic compounds have the NaCl (**3·2PO**, Section 5.1.1) structure, but usually for those differing significantly in electronegativities. Table 5.1 includes many compounds of the types MP, MAs, MSb, and MBi. The NiAs structure (**2·2PO**, Section 5.2.1) is found for a few MAs and MSb compounds (Table 5.5), the MSn compounds of Fe, Ni, Cu, Pd, Pt, Rh, and Au, and MnBi, NiBi, PtBi, and RhBi. The ZnS structures (*CN* 4) are not usually encountered for intermetallic compounds. The compounds of Al, Ga, and In with P, As, and Sb have the zinc blende (ZnS, **3·2PT**, Section 6.1.1) structure. These are semiconductors or insulators. Because the *bcc* structure is common for metals, it is not surprising that many 1:1 intermetallic compounds have the CsCl structure (**3·2PTOT**, Section 7.2.1). A few of these intermetallic compounds are included in Table 7.1; a more extensive list is given in Table 9.1.

Some AM_2 type structures have the fluorite structure, but for metals differing significantly in electronegativities. Table 6.4 contains many intermetallic compounds with the fluorite structure. Table 9.2 contains

TABLE 9.1. **Intermetallic compounds with the CsCl structure (3·2PTOT).**

β-AgCd	β-AuCd	CdCe	ErTl	InLa	β-MnPd
AgCe	AuMg	CdEu	EuZn	InPd	β-MnRh
AgLa	β-AuMn	CdLa	β-GaNi	InPr	OsTi
β-AgLi	AuYb	CdPr	GaRh	InTm	PrZn
β-AgMg	BaCd	CrSr	HgLi	InYb	RhY
AgNd	BaHg	CeHg	HgMg	IrLu	α-RuSi
AgY	BeCo	CeMg	HgMn	LaMg	RuTi
AgYb	BeCu	CeZn	HgNd	LaZn	SbTi
β-AgZn	BeNi	CoFe	HgPr	β-LiPb(H.T.)	ScRh
β-AlCo	BePd	CoSc	HgSr	LiTi	SrTi
AlFe	CaTi	CuEu	HoIn	LuRh	TeTh
AlNd		β-CuPd	HoTl	MgPr	TlTm
β-AlNi		CuY		MgSc	TmRh
AlPd(H.T.)		β-CuZn		MgSr	
AlSc				MgTi	

several intermetallic compounds with the pyrite structure (FeS_2, **3·2PO**). The structure of CaC_2 (Section 5.1.2) is similar to that of pyrite, but with elongation in one direction because of alignment of C_2^{2-} ions. Intermetallic compounds with the CaC_2 structure are listed in Table 9.3. Thus intermetallic compounds can follow the rules for "normal valence" compounds or those of the metallic state. The "normal valence" compounds have lower ***CN*** than found in *ccp* or *hcp* structures and are expected to be less "metallic" in terms of electrical conductance, etc.

Metals have been considered in Chapter 4 on the elements. Some alloys are simple solid solutions, and others have the composition of intermetallic compounds. If two metals are melted together we almost always obtain a liquid solution. If the liquid is cooled it can form a solid solution, give two or more phases of the metal(s) and/or intermetallic compound(s), or a single intermetallic compound might be formed if the composition corresponds to that of the compound. Chemically similar metals of similar size have the greatest tendency to form solid solutions. The following pairs, of similar size and the same periodic group, form solid solutions for any proportions: K–Rb, Ag–Au, Cu–Au, As–Sb, Mo–W, and Ni–Pd. Hume-Rothery noted that usually a metal of lower valence is likely to dissolve more of one of

TABLE 9.2. **Intermetallic compounds with the pyrite [FeS_2] structure (3·2PO).**

$AuSb_2$	$PdBi_2$	$PtAs_2$	$PtSb_2$
$PdAs_2$	$PdSb_2$	$PtBi_2$	$RuSn_2$

TABLE 9.3. **Intermetallic compounds with the CaC_2 structure (3·2PO).**

Ag_2Er	$AlCr_2$	Au_2Yb	$MoSi_2$
Ag_2Ho	$AuEr_2$	β-Ge_2Mo	$ReSi_2$
Ag_2Yb	$AuHo_2$	Hg_2MgSi_2W	

higher valence than *vice versa,* for example, the limit of solubility of Zn in Cu is 38.4% atomic Zn, but for Cu in Zn it is 2.3% atomic Cu. Solubilities decrease with increase in separation of periodic groups of metals, for example, maximum 38.4% Zn in Cu, 19.9% Ga in Cu, 11.8% Ge in Cu, and 6.9% As in Cu.

The alloys just considered are substitutional solid solutions. Interstitial solid solutions are alloys with small atoms, for example, H, C, N, and O, in the interstitial sites, usually **O** and **T** sites. Some alloys have random distribution (disordered) if the melt is quenched but become ordered if heated and annealed or if cooled slowly. An example is the 1:1 alloy CuAu. The disordered structure is *ccp,* and the ordered structure is also *ccp,* except alternate layers parallel to a cell face contain Cu or Au.

The Cu–Zn system (brass) is complex as shown in Figure 9.1. The α phase is a *ccp* solid solution of Zn in Cu. The β-brass is body-centered cubic, the composition corresponding to CuZn. Each phase exists over a range of Cu/Zn ratios corresponding to a solid solution with Zn or Cu added to the compound. The γ-brass, Cu_5Zn_8, has a complex cubic structure and ε-brass, $CuZn_3$, has an *hcp* structure. Hume-Rothery found that many intermetallic compounds have structures similar to β-, γ-, and ε-brass at the same electron-to-atom ratio as the corresponding brass compounds. Some examples of these so-called *electron*

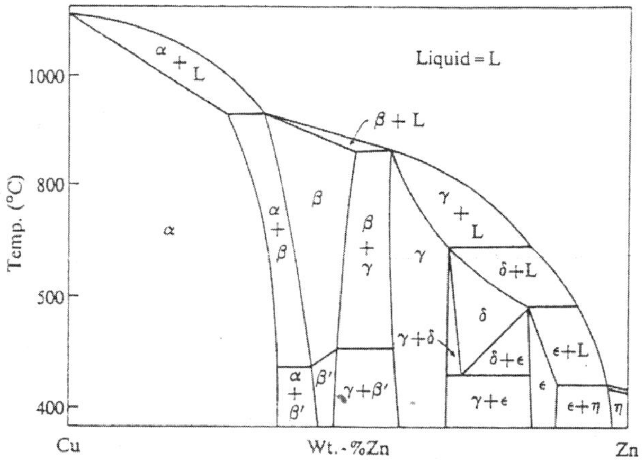

Figure 9.1. Phase diagram for the Cu–Zn system.
(*Source*: J.C. Slater, *Introduction to Chemical Physics* McGraw-Hill, New York, 1939, p. 287.)

TABLE 9.4. **Electron compounds (Hume-Rothery phases).**

Electron/atom	β-Brass (*bcc*) 3/2 or 21/14	γ-Brass (*complex cubic*) 21/13	ε-Brass (*hcp*) 7/4 or 21/12
	CuZn	Cu_5Zn_8	$CuZn_3$
	AgZn	Ag_5Zn_8	$AgZn_3$
	AuZn	Cu_9Al_4	Ag_5Al_3
	AgCd	$Cu_{31}Sn_8$	Cu_3Sn
	Cu_3Al	$Na_{31}Pb_8$	Cu_3Si
	Cu_5Sn	Rh_5Zn_{21}	
	CoAl	Pt_5Zn_{21}	
	FeAl		
	NiAl		

compounds (Hume-Rothery phases) are shown in Table 9.4. Some of the compounds with the same electron/atom ratio have much different formulas. The electrons are counted by assigning each metal atom the number of valence electrons equal to the periodic group number (old periodic groups I–VIII) with zero valence electrons for Group VIII. Thus, NiAl ($0 + 3$ valence electrons) has an electron-to-atom ratio of 3:2, the same as that for β-brass, CuZn. A few intermetallic compounds with 3:2 electron/atom ratios, Ag_3Al, Au_3Al, and $CoZn_3$, have the β-Mn structure. Although the range of stability of β-Mn is 727 to 1,095°C, it can be retained at room temperature by quenching. The structure is cubic with a large cell, 20 atoms per unit cell.

The metallic state is usually treated in terms of band theory, considering a metal as a framework of metal ions in a sea of electrons. The filling of electronic bands is important in determining the crystal structure. Figure 9.2 shows a distribution of energy states vs. the population of the states for an electron/atom ratio for *ccp* and *bcc* lattices. Here, the *bcc* structure is favored. The extent of filling and separation between energy bands determines whether a substance is an insulator or an electrical conductor. Mg_2Si and Mg_2Pb have the

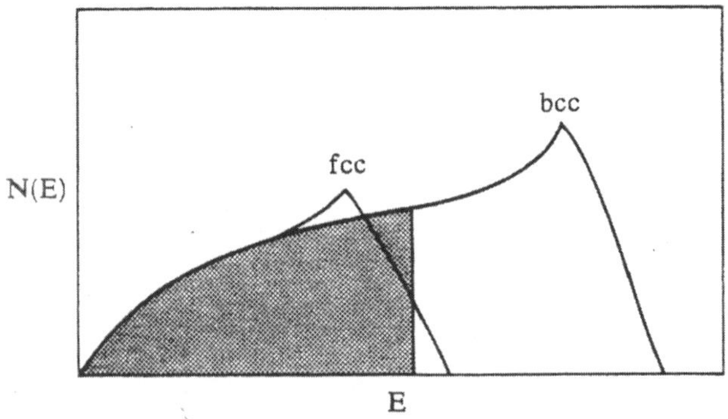

Figure 9.2. Distribution of energy states for an electron/atom ratio of 3/2 favoring the *bcc* structure.

CaF_2 structure, and while Mg_2Si is almost an insulator, Mg_2Pb is a good metallic conductor.

In addition to structures corresponding to simple ionic compounds and those for electron compounds, there are structures for intermetallic compounds depending on atomic sizes. Laves phases (Section 9.4) with the same structure have relatively constant radius ratios. For example, $MgCu_2$, ZrW_2, KBi_2, and $BiAu_2$ have radius ratios of about 1.25. Zintl phases are intermetallic compounds of an alkali metal or an alkaline metal and a metal with an underlying d^{10} configuration, for example, NaTl, CaCd, and LiZn. $CaZn_2$ has Zn atoms in graphite-like layers with Ca^{2+} ions between Zn layers. There are also more complex Zintl phases such as AM_{13}, A = Na, K, Ca, Sr, or Ba with M = Zn, or K, Rb; or Cs with Cd.

9.1. Compounds Involving Only P Layers

9.1.1. The $2P_{1/4\ 3/4}$ Crystal Structure of ε-Brass, $CuZn_3$

ε-Brass, $CuZn_3$, is hexagonal, *hcp*. It is an electron compound with 21 valence electrons for 12 atoms. A compound is formed rather than part of a continuous range of solid solutions of Cu in Zn because the limit of solubility of Cu in Zn is 2.3% atomic Cu. Zn is *hcp* (**2P**) and Cu is *ccp* (**3P**).

9.1.2. The $2P_{1/4\ 3/4}$ Crystal Structure of $MgCd_3$

$MgCd_3$ forms a hexagonal superstructure with Mg and Cd in a 1:3 ratio in each layer. The Cd and Mg atoms are hexagonally arranged as shown in Figure 9.3. The notation is $\mathbf{2P_{1/4\ 3/4}}$. The structure of $CdMg_3$ is very similar with the ratio reversed.

9.1.3. The $4P_{1/4\ 3/4}$ Crystal Structure of $TiNi_3$

$TiNi_3$ is hexagonal with four molecules per unit cell, $a = 5.101$ and $c = 8.3067$ Å. The close-packed layers are in the double hexagonal

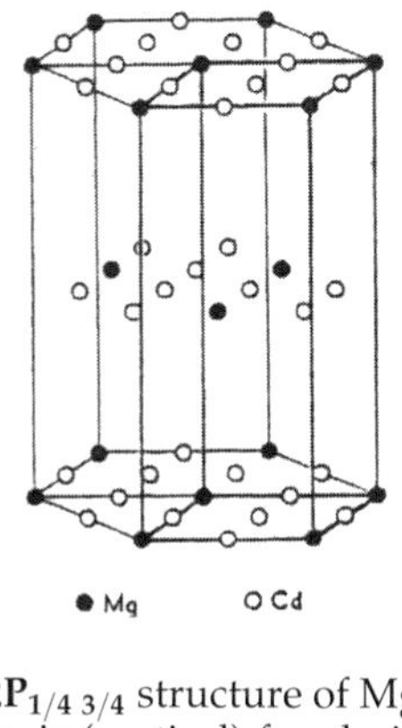

Figure 9.3. The hexagonal $\mathbf{2P_{1/4\ 3/4}}$ structure of $MgCd_3$. The hexagonal cell has been elongated along the c axis (vertical) for clarity.

(*Source*: W. Hume-Rothery, R.E. Smallman, and C.W. Haworth, *The Structure of Metals and Alloys*, Institute of Metals and the Institution of Metallurgists, London, 1969, 175.)

sequence *ABAC* giving the notation $\mathbf{4P_{1/4\ 3/4}}$. Each Ti has 12 Ni neighbors at 2.55 Å and Ni has four Ti and eight Ni neighbors.

9.1.4. The 3·11P(m)(3:8) Crystal Structure of Mo_3Al_8

The Mo_3Al_8 crystal structure is monoclinic with two molecules per unit cell, $a = 9.208$, $b = 3.638$, $c = 10.065$ Å, and $\beta = 100.78°$. It is a superstructure of distorted *fcc* cells.

Figure 9.4.***a*** shows a projection along *c*. The structure is similar to that of Pb_3Li_8 (Section 9.2.9) except it is based on a *fcc* subcell and there are no Mo–Mo pairs. The Mo and Al atoms are distributed among all sites with no close Mo–Mo neighbors. A close-packed layer normal to {111} of the pseudocell is shown in Figure 9.4***b***. Because of the 3:8 filling of layers, there are 33 layers repeating. Figure 9.4***b*** shows positions of other layers by solid dots. The notation is **3·11P(m)(3:8)**.

9.1.5. The $2P_{1/2\ 1/2}$(o) Crystal Structure of HgNa

The crystal structure of HgNa is orthorhombic with eight molecules per unit cell, $a = 7.19$, $b = 10.79$, and $c = 5.21$ Å. Figure 9.5 shows two distorted close-packed layers with Hg and Na distributed equally in each layer. It is unusual to have close packing along an axis of the cell of an orthorhombic cell, but this is a good description of HgNa. The simple notation is $\mathbf{2P_{1/2\ 1/2}(o)}$, the fractions indicating equal population by Hg and Na in each layer.

9.1.6. The $3P_{1/2\ 1/2}$(t) Crystal Structure of BiIn

BiIn is tetragonal with two molecules per unit cell, $a = 5.015$ and $c = 4.781$ Å. The face-centered tetragonal cell shown in Figure 9.6 has In at corners and in the centers of the top and bottom faces. The Bi atoms are displaced from the center of side faces. Two Bi in opposite

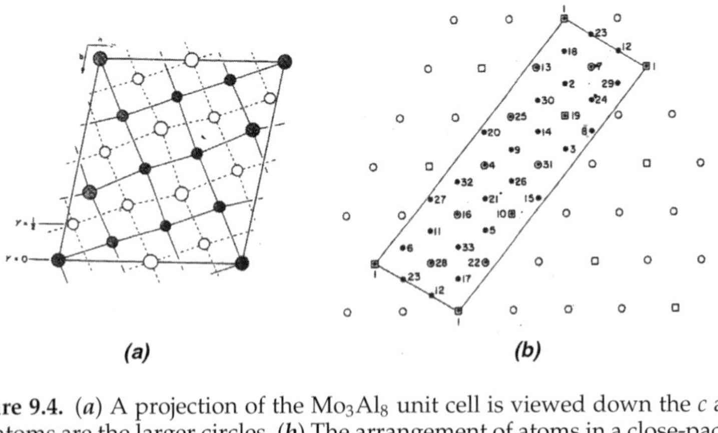

Figure 9.4. (***a***) A projection of the Mo_3Al_8 unit cell is viewed down the *c* axis. Mo atoms are the larger circles. (***b***) The arrangement of atoms in a close-packed layer of Mo_3Al_8. There are open circles for Al and squares for Mo in this layer. The repeating positions are solid dots in the cell for 33 layers.

(*Source*: W.P. Pearson, *The Crystal Chemistry and Physics of Metal and Alloys*, Wiley-Interscience, New York, 1972, p. 350.)

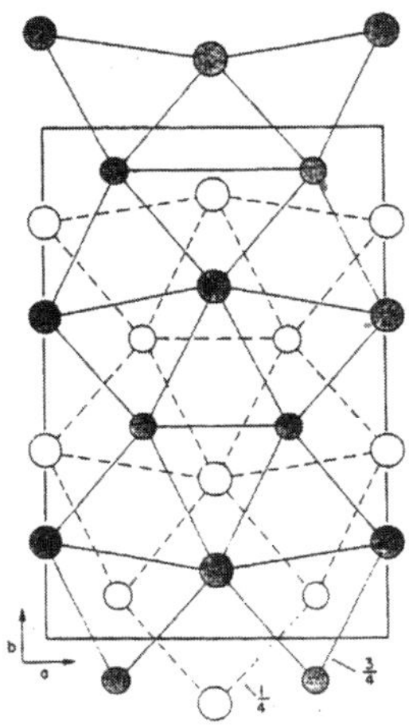

Figure 9.5. A projection of the cell of HgNa for layers at one-quarter and three-fourth along the *c* axis is shown. Na atoms are the larger circles.

(*Source*: W.P. Pearson, *The Crystal Chemistry and Physics of Metals and Alloys*, Wiley-Interscience, New York, 1972, p. 316.)

faces are at $z = 0.38$, and the other two Bi in opposite faces are at $z = 0.62$. The close-packed layers are equally occupied by Bi and In giving the notation $\mathbf{3P_{1/2\ 1/2}(t)}$.

9.1.7. The $\mathbf{9P_{1/2\ 1/2}(m)}$ Crystal Structure of NbRh

NbRh is monoclinic with nine molecules per unit cell, $a = 2.806$, $b = 4.772$, $c = 20.250$ Å, and $\beta = 90.53°$. Close-packed layers are stacked along *c*. Figure 9.7 shows a projection of *A*, *B*, and *C* positions. The *A* layer is shown by solid points and solid lines, *B* positions are joined by dotted lines, and *C* positions are joined by dashed lines. Each close-packed layer has equal numbers of Nb and Rh atoms, with rows of Nb and Rh atoms giving rectangular patterns for Nb or Rh. This is the pattern for $\mathbf{L_{1/2}}$ shown in Figure 3.11***a***. The unusual close-packed sequence is *ABA BCB CAC*, giving nine repeating layers and the notation $\mathbf{9P_{1/2\ 1/2}(m)}$. Odd sequences such at this

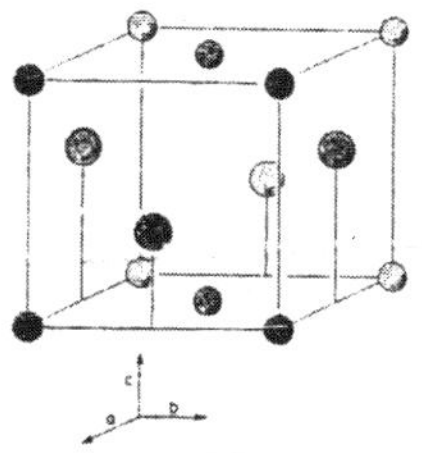

Figure 9.6. The distorted face-centered tetragonal cell of BiIn is shown. Bi atoms are the larger circles.

(*Source*: W.P. Pearson, *The Crystal Chemistry and Physics of Metals and Alloys*, Wiley-Interscience, New York, 1972, p. 317.)

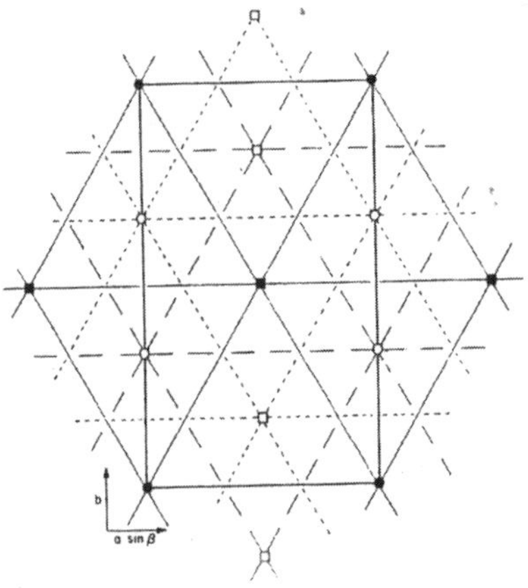

Figure 9.7. A projection of the NbRh structure along the c axis is shown, Rh atoms are squares.

(*Source*: W.P. Pearson, *The Crystal Chemistry and Physics of Metals and Alloys*, Wiley-Interscience, New York, 1972, p. 315.)

interest pattern are more common for intermetallic compounds than for ordinary valence compounds. Long-range interactions are more important for intermetallic compounds.

9.1.8. The $2P_{1/4\ 3/4}$ Crystal Structure of $SnNi_3$

The crystal structure of $SnNi_3$ is hexagonal with two molecules per unit cell, $a = 5.286$ and $c = 4.243$ Å. The hexagonal cell is shown in Figure 9.8. There are close-packed layers of Sn and Ni at A (height three-quarters along c) and B (height one-quarter along c) positions (*hcp* sequence). Each atom has ***CN*** 12. For Sn the 12 neighbors are Ni. For Ni there are four Ni and two Sn atoms in the same plane with two Ni and one Sn in planes above and below. The notation is $\mathbf{2P_{1/4\ 3/4}}$.

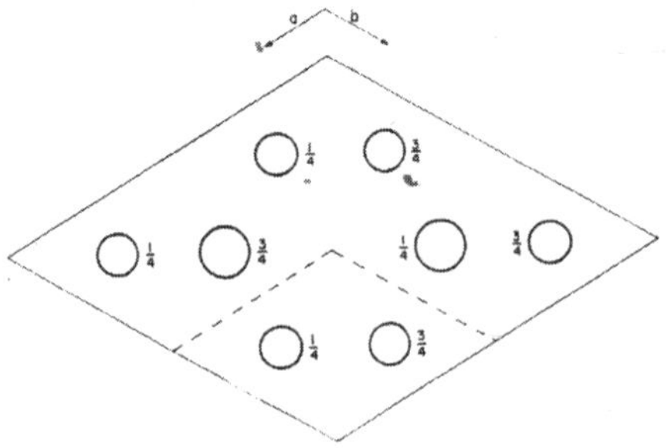

Figure 9.8. A projection of the $SnNi_3$ cell down the c axis is Shown, Sn atoms are the larger circles

(*Source*: W.P. Pearson, *The Crystal Chemistry and Physics of Metals and Alloys*, Wiley-Interscience, New York, 1972, p. 324.)

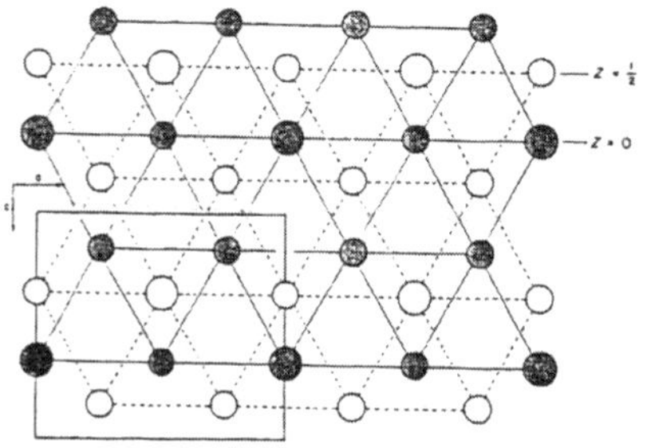

Figure 9.9. A projection of the β-$TiCu_3$ structure down the *b* axis is shown. Ti atoms are the larger circles.

(*Source*: W.P. Pearson, *The Crystal Chemistry and Physics of Metals and Alloys*, Wiley-Interscience, New York, 1972, p. 331.)

9.1.9. The $2P_{1/4\ 3/4}$(o) Crystal Structure of β-$TiCu_3$

The unit cell for β-$TiCu_3$ is orthorhombic with two molecules per cell, $a = 5.162$, $b = 4.347$, and $c = 4.531$ Å. Figure 9.9 shows a projection of the cell down the *b* axis, the packing direction. The layers are at *A* and *B* positions similar to those of $SnNi_3$. Each Ti has 12 Cu neighbors at 2.58, 2.61, and 2.67 Å. Each Cu has four Cu and two Ti atoms in the same plane plus one Ti and two Cu atoms in planes above and below. There are no Ti—Ti close neighbors, giving Ti 12 Cu neighbors at 2.58, 2.61, and 2.67 Å. The notation is $\mathbf{2P_{1/4\ 3/4}(o)}$.

9.1.10. The $3P_{1/4\ 3/4}$ Crystal Structure of $AuCu_3$

The crystal structure of $AuCu_3$ is cubic, $\mathbf{O}_h^1$, $Pm3m$, $a_o = 3.748$ Å. The unit cell (Figure 9.10) is face-centered cubic with Cu atoms in the centers of the faces. The **P** layers are *ccp* with Cu in three-quarters of the **P** sites giving $\mathbf{3P_{1/4\ 3/4}}$. Each Au atom has 12 Cu neighbors at 2.65 Å arranged in a cuboctahedron and Cu has four Au and eight Cu at the same distance and in the same arrangement. This is a common structure of many AM_3 intermetallic compounds or alloys, as shown in Table 9.5.

9.1.11. The $3{\cdot}3P_{1/4\ 3/4}$(t) Crystal Structure of $TiAl_3$

$TiAl_3$ is tetragonal with two molecules per unit cell, $a = 3.848$ and $c = 8.596$ Å. There are two distorted $AuCu_3$ (previous section) type

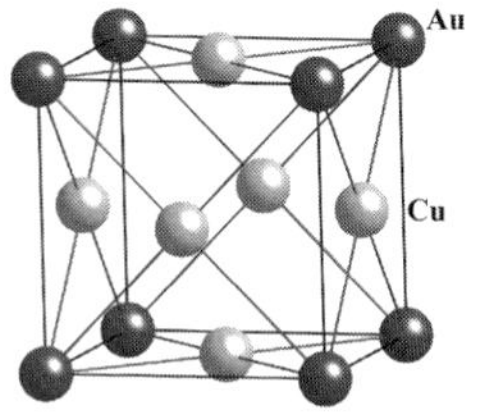

Figure 9.10. The face-centered lattice of $AuCu_3$ with Cu in the face centers is shown.

(*Source*: *CrystalMaker*, by David Palmer, *CrystalMaker* Software Ltd., Begbroke Science Park, Bldg. 5, Sandy Lanc, Yarnton, Oxfordshirc, OX51PF, UK.)

TABLE 9.5. **Intermetallic compounds with the $AuCu_3$ structure.**

α''-Ag_3Pt	Cd_3Nb	$FeNi_3$	In_3Pr	γ'-Mn_3Rh	Pt_3Sm
γ-$AgPt_3$	α'-$CdPt_3$	$FePd_3$	$InPu_3$	β-$NaPb_3$	Pt_3Sn
α-$AlCo_3$	$CeIn_3$	$GaNi_3$	In_3Pu	Nb_3Si	Pt_3Tb
Al_3Er	Ce_3In	Ga_3U	In_3Sc	$NdPt_3$	Pt_3Ti
Al_3Ho	$CePb_3$	$GeNi_3$	In_3Tb	$NdSn_3$	Pt_3Tm
α'-$AlNi_3$	$CePt_3$	Ge_3Pu	In_3Tm	β-Ni_3Si	Pt_3Y
Al_3Np	$CeSn_3$	Ge_3U	In_3U	$PbPd_3$	Pt_3Yb
Al_3U	$CoPt_3$	δ-$HgTi_3$	In_3Yb	Pb_3Pr	Pt_3Zn
Al_3Yb	α'-$CrIr_3$	Hg_3Zr	γ'-$IrMn_3$	$PbPu_3$	$PuSn_3$
$AlZr_3$	Cr_3Pt	$HoIn_3$	$LaPb_3$	Pd_3U	Rh_3Sc
$AuCu_3$	α'-Cu_3Pd	$HoPt_3$	$LaPd_3$	Pd_3Sn	Ru_3U
α'-Au_3Pt	Cu_3Pt	$HoTl_3$	$LaPt_3$	Pd_3Y	Si_3U
$CaPb_3$	$ErIn_3$	$InLa_3$	$LaSn_3$	$PrPt_3$	Sn_3U
$CaSn_3$	$ErPt_3$	$InNd_3$	$MnPt_3$	$PrSn_3$	$TbTl_3$
$CaTl_3$	$ErTl_3$	$InPr_3$	γ'-Mn_3Pt	Pt_3Sc	$TiZn$

subcells stacked along the *c* axis (Figure 9.11). The subcells are tetragonally distorted face-centered cells with no Ti—Ti nearest neighbors. Layers are in an *ABC* sequence, but spacings are unequal giving nine layers repeating. The notation is $\mathbf{3{\cdot}3P_{1/4\ 3/4}(t)}$. VNi_3, VPd_3, VPt_3, $NbPd_3$, and $TaPd_3$ have the $TiAl_3$ structure.

9.1.12. The $3{\cdot}4P_{1/4\ 3/4}(t)$ Crystal Structure of $ZrAl_3$

The $ZrAl_3$ crystal structure is tetragonal with four molecules in the unit cell, $a = 4.014$ and $c = 17.32$ Å. Figure 9.12***a*** shows that there are four *fcc* pseudocells stacked along *c*. The lowest pseudocell has Zr in two side faces and the next one has Zr in the top and bottom faces in the projection in Figure 9.12***a***. The upper half of the cell along *c* is the mirror image

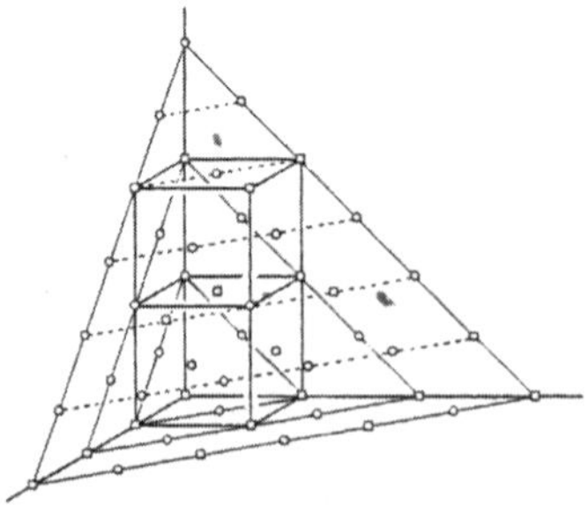

Figure 9.11. The structure of $TiAl_3$. Two $AuCu_3$-type cells stacked along the *c* axis are shown. The close-packed layers are along the body diagonal.

(*Source*: W.P. Pearson, *The Crystal Chemistry and Physics of Metals and Alloys*, Wiley-Interscience, New York, 1972, p. 329.)

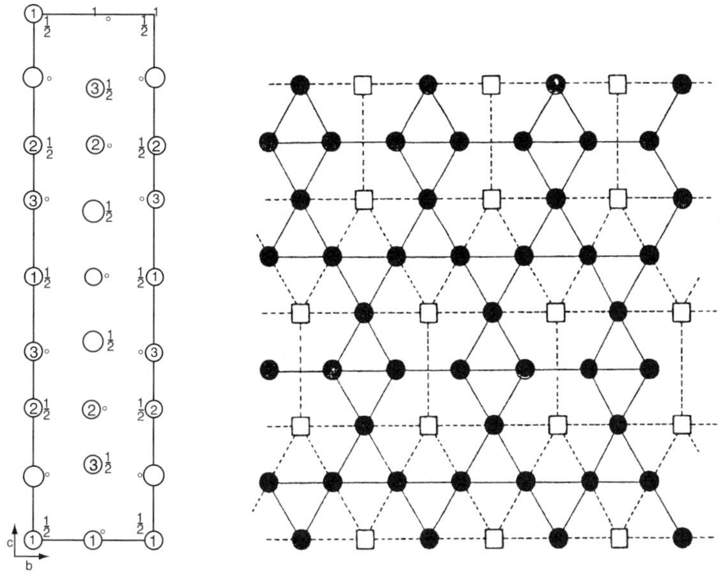

(*a*) (*b*)

Figure 9.12. The structure of $ZrAl_3$. (*a*) The view along the *a* axis is shown. Pseudo *fcc* cells are stacked along the *c* axis. Zr atoms are the larger circles. (*b*) Arrangements of Zr (squares) and A1 atoms (circles) in a close-packed layer of $ZrAl_3$.

(*Source*: W.P. Pearson, *The Crystal Chemistry and Physics of Metals and Alloys*, Wiley-Interscience, New York, 1972, p. 332.)

of the lower half. Figure 9.12*b* shows a close-packed layer with Zr in one-fourth of sites and Al in three-quarter of sites. The Zr atoms are in the rectangular arrangement for $\mathbf{L}_{1/4}$ shown in Figure 3.10***b***. Because this is based on *fcc* pseudocells, the packing sequence *ABC* gives the notation $\mathbf{3{\cdot}4P}_{1/4\ 3/4}(\mathbf{t})$. $ZnAu_3$ has this structure. The compounds $CdAu_3$, $MgAu_3$, and $Mg_{24}Au_{76}$ also have related structures.

9.1.13. The $3{\cdot}5P_{1/5\ 4/5}(t)$ Crystal Structure of $MoNi_4$

$MoNi_4$ is tetragonal with two molecules per unit cell, $a = 5.727$ and $c = 3.566$ Å. Figure 9.13 shows the structure. A face-centered subcell is shown by dotted lines. In the center of the large square there is a body-centered subcell with Mo at the center. Adjacent body-centered subcells have Ni at the center. We have seen (Figure 4.5) that a face-centered cubic structure contains body-centered tetragonal subcells. Close-packing is along the body diagonal of the unit cell with a 15-layer close-packed repeat sequence. The notation is $\mathbf{3{\cdot}5P}_{1/5\ 4/5}(\mathbf{t})$.

9.1.14. The $2P_{1/5\ 4/5}(o)$ or $2{\cdot}5P_{1/5\ 4/5}(o)$ Crystal Structure of $ZrAu_4$

The unit cell of $ZrAu_4$ is orthorhombic with four molecules per cell, $a = 5.006$, $b_o = 4.845$, and $c_o = 14.294$ Å. Figure 9.14 shows a projection

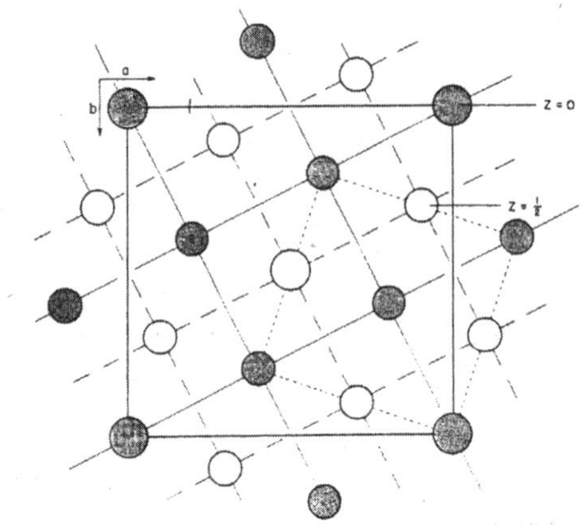

Figure 9.13. A projection of the of $MoNi_4$ cell along the *c* axis is shown. The *fcc* subcell is outlined by dotted lines.

(*Source*: W.P. Pearson, *The Crystal Chemistry and Physics of Metals and Alloys*, Wiley-Interscience, New York, 1972, p. 341.)

of the close-packed layers along the *b* axis. Each layer contains Zr and Au in the 1:4 ratio. Zr atoms are staggered to avoid Zr–Zr near neighbors. Each Zr has 12 Au close neighbors at 2.86 to 2.94 Å. Au atoms have Au and Zr neighbors giving ***CN*** 12. For two close-packed layers at *A* and *B* positions the notation is $\mathbf{2P}_{1/5\ 4/5}(\mathbf{o})$. The figure shows that in the direction of the *c* axis there are 10 repeating layers because of partial filling by Zr and Au. Along the packing direction the sequence is *ABAB*···, but partial filling can cause this to be more complex

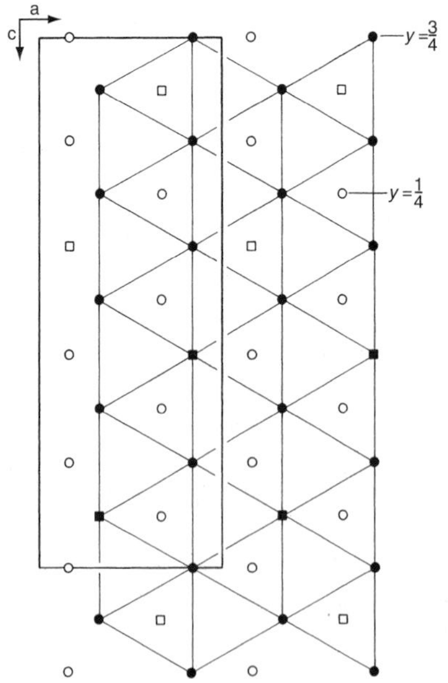

Figure 9.14. Two close-packed layers of the $ZrAu_4$ structure are viewed along the *b* axis, squares are Zr atoms.

(*Source*: W.P. Pearson, *The Crystal Chemistry and Physics of Metals and Alloys*, Wiley-Interscience, New York, 1972, p. 342.)

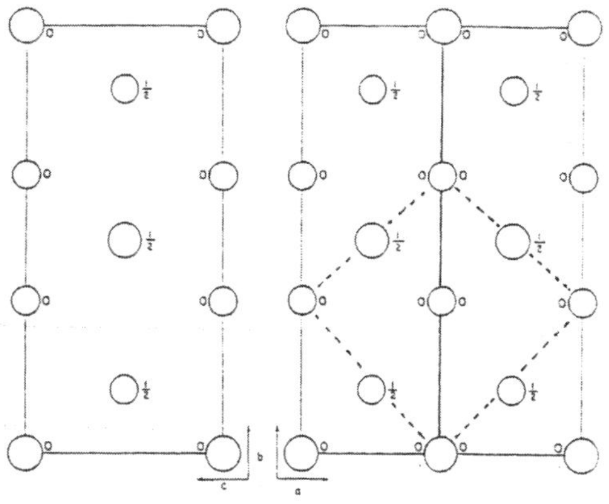

Figure 9.15. The $\mathbf{3{\cdot}3P_{1/3\ 2/3}}$ **(o)** structure of $MoPt_2$. Views of the structure are viewed along the *a* axis on the left and along the *c* axis on the right. Heights are shown. Larger circles are Mo atoms.

(*Source*: W.P. Pearson, *The Crystal Chemistry and Physics of Metals and Alloys*, Wiley-Interscience, New York, 1972, p. 320.)

because locations of Zr and Au atoms differ from one layer to the next. It is likely that $\mathbf{2{\cdot}5P_{1/5\ 4/5}(o)}$ is a better description.

9.1.15. The $3{\cdot}3P_{1/3\ 2/3}(o)$ Crystal Structure of $MoPt_2$

$MoPt_2$ is orthorhombic with two molecules per unit cell, $a = 2.748$, $b = 8.238$, and $c = 3.915$ Å. It is a close-packed superstructure based on *fcc* subcells. Figure 9.15 shows two projections of the cell. Body-centered subunits are apparent in each projection. Displacements of body-centered atoms show the distortion from *bcc*. Two unit cells are shown for the projection on the right. The face-centered subcell is outlined by dotted lines. We have noted (Section 4.2.3, Figure 4.5) that a *fcc* structure contains body-centered tetragonal subcells. Along the body diagonal there are close-packed layers in an *ABC* sequence. Each close-packed layer is one-third filled by Mo, and each Mo is at the center of a orthorhombohedron formed by eight Pt atoms. This is the regular pattern for $\mathbf{L_{1/3\ 2/3}}$ layers (Figure 3.9). The notation is $\mathbf{3{\cdot}3P_{1/3\ 2/3}(o)}$.

9.1.16. The $3{\cdot}3P_{1/3\ 2/3}(o)$ Crystal Structure of $ZrGa_2$

$ZrGa_2$ is orthorhombic with four molecules per unit cell, $a = 12.894$, $b = 3.994$, and $c = 4.123$ Å. It is a slightly distorted superstructure of close-packed layers. In the *a* direction there are three *fcc* pseudocells. A close-packed layer normal to {111} of the pseudocell is shown in Figure 9.16. The sequence of close-packed layers is *ABC*, but actually *ABC A′B′C′ A″B″C″* because of changing environments causing by partial filling by Zr and Ga. Positions *B*, *C*, *B′*, *C′*, *A″*, *B″*, and *C″* are shown by solid dots. For the close-packed layer shown, one horizontal row is fully occupied by Ga and one-half occupied by Ga in the next two rows. In rows parallel to the other cell faces one row is filled

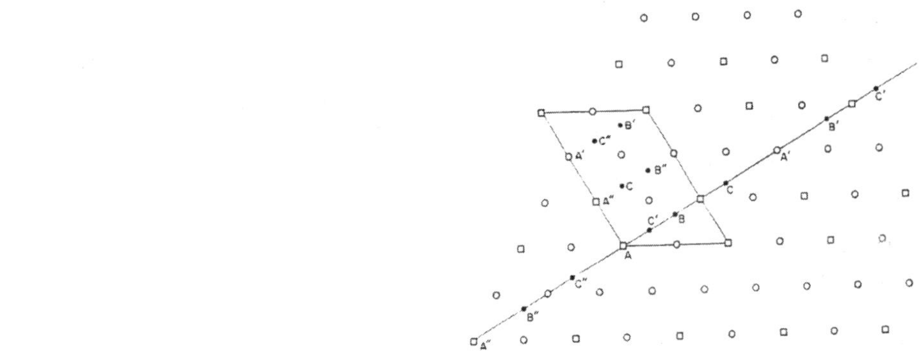

Figure 9.16. The view of a close-packed layer normal to {111} of the pseudocell of $ZrGa_2$ is shown. Repeating positions of the other eight layers are shown by solid dots (B, C, B', C', B'', C'').

(*Source*: W.P. Pearson, *The Crystal Chemistry and Physics of Metals and Alloys*, Wiley-Interscience, New York, 1972, p. 338.)

by Ga and the next is one-third filled by Ga. There are nine close-packed layers repeating, giving the notation **3·3P$_{1/3\ 2/3}$(o)**.

9.1.17. The 3·7P(m)(2:5) Crystal Structure of Mn_2Au_5

Mn_2Au_5 is monoclinic with two molecules per unit cell, $a = 9.188$, $c = 6.479$ Å, and $\beta = 97.56°$. It is a superstructure of face-centered subcells. In Figure 9.17***a*** we can see the face-centered subunit (dotted lines) and the smaller body-centered subunits (solid lines). There are close-packed layers parallel to {111} planes of the pseudocell. The close-packed layers are in an *ABC* sequence with 21 layers repeating (Figure 9.17***b***). The notation is **3·7P(m)(2:5)**.

9.1.18. The 3·4P(o)(3:5) Crystal Structure of Ga_3Pt_5

Ga_3Pt_5 is orthorhombic with two molecules per unit cell, $a = 8.031$, $b = 7.440$, and $c = 3.948$ Å. The unit cell corresponds to a block of four distorted *fcc* pseudocells. Figure 9.18 shows the close-packed sequence is *ABC*. The figure shows locations of one point in each layer *B* and *C*. Each layer is occupied in the 3:5 ratio. In the horizontal rows one row is filled by Pt and the next three are one-half occupied by Pt. In rows parallel to the other cell faces one row is filled by Pt and the next is one-quarter filled by Pt. The 3:5 ratio is maintained for both sets of layers. There are 12 layers repeating, giving the notation **3·4P(3:5)(o)**. Each atom has *CN* 12. The closest Ga—Pt distance is 2.59 Å.

9.1.19. The 1·2$P^{Li}_{1/2}$ $P^{Sn}_{1/2}$(m) Crystal Structure of LiSn

Crystals of LiSn are monoclinic, C^1_{2h}, $P2/m$, $a = 5.17$, $b = 3.18$, $c = 7.74$ Å, and $\beta = 104.5°$ with three molecules in the unit cell. A projection along b gives a good approximation of a close-packed layer allowing

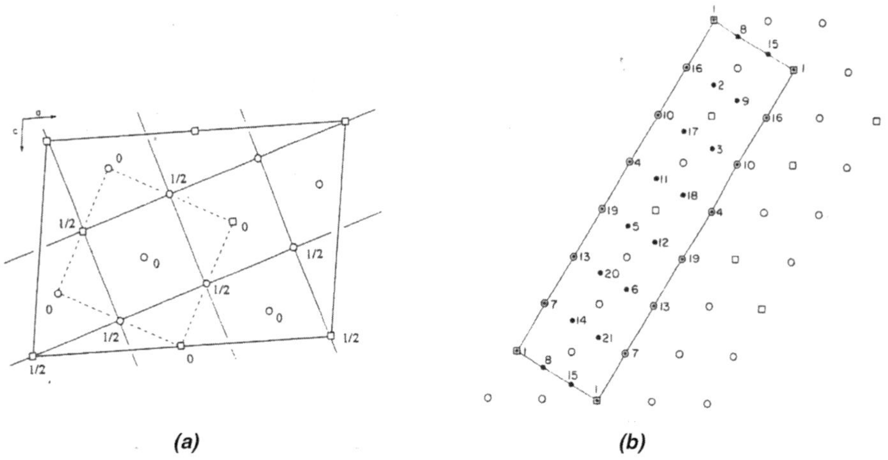

Figure 9.17. (*a*) A projection of the Mn_2Au_5 structure along the *b* axis is shown, circles are for Au and squares are for Mn. A face-centered subcell is outlined by dashed lines. (*b*) A close-packed layer of Mn_2Au_5 (open circles and squares) and positions for the other 20 layers are shown by solid dots.

(*Source*: W.P. Pearson, *The Crystal Chemistry and Physics of Metals and Alloys*, Wiley-Interscience, New York, 1972, p. 347.)

for the differences in size of Li and Sn and for vacancies in each layer. Sn and Li separately form one-half filled packing layers. The vacancies in each layer coincide for Sn layers and for Li layers. Vacancies are aligned with positions of atoms in adjacent layers. All atoms occupy *A* positions. Sn atoms occur at corners and two positions within the cell in a projection along *b*. Li atoms are in two edges parallel to *a* and two positions within the cell (see Figure 9.19). Half of the Sn atoms occur in pairs at the distance 2.998 Å compared to 3.160 Å for other neighbors in the layer and 3.180 Å to nearest Sn atoms in an adjacent layer. Sn has an elongated cubic arrangement of eight Li atoms at 3.035 and 3.061 Å. Opposite faces are capped along *b* by Sn atoms at 3.180 Å. Li has a similar arrangement of eight Sn and two Li atoms. The notation

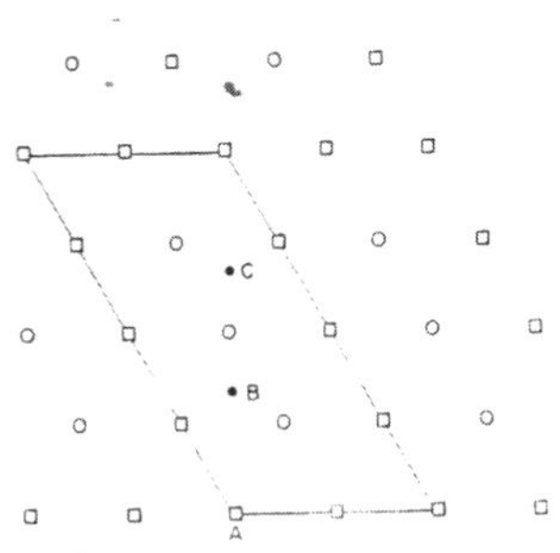

Figure 9.18. A close-packed layer of the structure of Ga_3Pt_5 is shown (squares are for Pt and circles for Ga). Packing positions in the other two close-packed layers are shown as solid dots labeled B and C.

(*Source*: W.P. Pearson, *The Crystal Chemistry and Physics of Metals and Alloys*, Wiley-Interscience, New York, 1972, p. 348.)

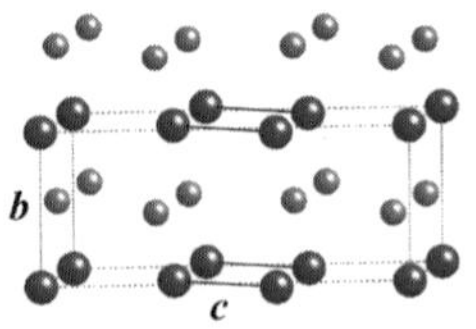

Figure 9.19. The structure of LiSn. Larger atoms are Sn. Li atoms and Sn atoms are aligned separately along the *b* direction.

(*Source*: Y. Matsushita, *Chalcogenide* crystal structure data library, Version 5.5B, Institute for Solid State Physics. The University of Tokyo, Tokyo, Japan, 2004; and *CrystalMaker*, by David Palmer, *CrystalMaker* Software Ltd., Begbroke Science Park, Bldg. 5, Sandy Lanc, Yarnton, Oxfordshire, OX51PF, UK.)

is $\mathbf{1{\cdot}2P^{Li}_{1/2}P^{Sn}_{1/2}(m)}$, indicating only one position (*A*) occupied and alternating half-filled packing layers.

9.1.20. The $4P^{Al}_{2/3}{}^{W}_{1/3}P^{Al}$ Crystal Structure of WAl_5

The WAl_5 intermetallic compound has one of several structures for AM_5 type compounds. There are layers of Al and W atoms at *ABAC* positions, a double *hcp* pattern. In one layer ($\mathbf{P}_A$) two-thirds of the sites are occupied by Al_I and one-third by W, giving $\mathbf{P}_{2/3\ 1/3}$. The next layer following $\mathbf{P}_A$ is filled by Al_{II} at *B* positions. The next layers are at *A* and *B* positions. The positions and heights along *c* are:

Height:	0	25	50	75
Atom:	Al_I, W	Al_{II}	Al_I, W	Al_{II}
Position:	*A*	*B*	*A*	*C*
	$\mathbf{P}_{2/3\ 1/3}$	$\mathbf{P}_{1/3}$	$\mathbf{P}_{2/3\ 1/3}$	$\mathbf{P}_{1/3}$

The notation is $\mathbf{4P^{Al}_{2/3}{}^{W}_{1/3}P^{Al}}$. Figure 9.20 shows two layers of four cells of the hexagonal structure of WAl_5 with two molecules per cell, $a_o = 4.902$ and $c_o = 8.851$ Å. Each atom has *CN* 12. There are no W—W close neighbors. $MoAl_5$ has a rhombohedral structure related to WAl_5.

9.1.21. The $2\mathbb{I}\mathbb{P}^{A\,Ca}_{1/4}{}^{B\,Cu}_{1/4}{}^{C\,Cu}_{1/4}\mathbf{P}^{A\,Cu}_{3/4}$ Crystal Structure of $CaCu_5$

The structure of $CaCu_5$ is different from that for WAl_5, as another example of an AM_5 intermetallic compound. It is hexagonal with one

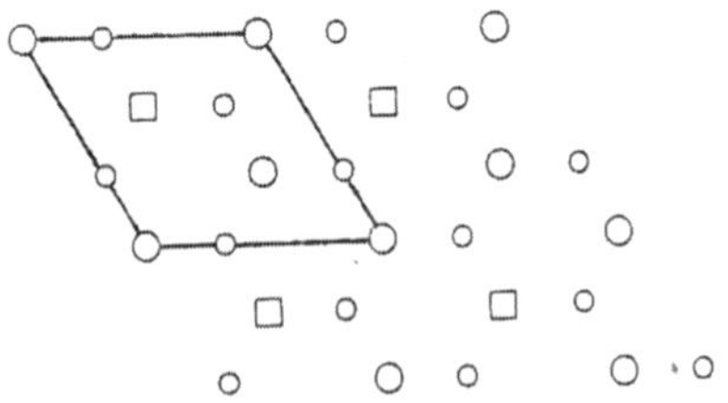

Figure 9.20. Two close-packed layers of the WAl_5 structure are shown. The larger symbols are for one layer and smaller symbols for the other layer occupied only by Al. Squares are for W.

(*Source*: W.P. Pearson, *The Crystal Chemistry and Physics of Metals and Alloys*, Wiley-Interscience, New York, 1972, p. 343.)

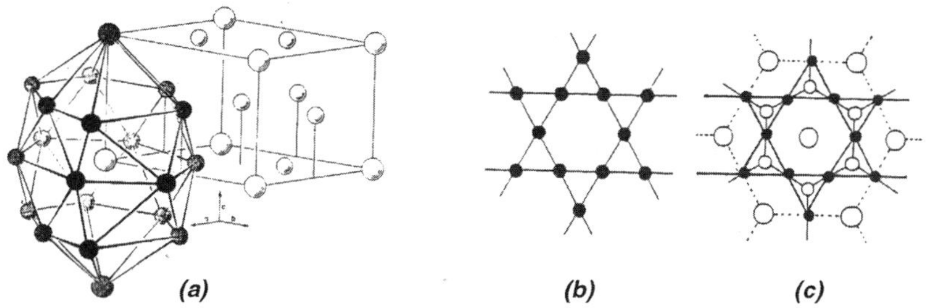

Figure 9.21. The structure of $CaCu_5$. (*a*) The unit cell with the polyhedron around Ca is shown. The larger circles are for Ca. (*b*) A close-packed layer for Cu only. (*c*) A close-packed layer for Ca (larger open circles) and Cu (smaller open circles) superimposed on the Cu-only layer (solid dots).

(*Source*: W.P. Pearson, *The Crystal Chemistry and Physics of Metals and Alloys*, Wiley-Interscience, New York, 1972, p. 644.)

molecule per cell, $a_o = 5.092$ and $c_o = 4.086$ Å. As seen in Figure 9.21*a* each Ca is at the center of a planar hexagon of Cu atoms with hexagons of Cu atoms in layers above and below. These two hexagons are eclipsed forming a hexagonal prism. Ca has *CN* 18. The Cu-only layers contain Cu_6 hexagons joined in a $\mathbf{P}_{3/4}$ layer (assigned as *A* positions) as shown in Figure 9.21*b*. The arrangements of Cu and Ca in the other layer are shown by open circles in Figure 9.21*c*. The Ca positions correspond to the vacancies in the hexagons in Figure 9.21*b*, so this layer is one-quarter filled by Ca in *A* positions. The filled circles in Figure 9.21*c*, the same positions as in Figure 9.21*b*, show the *A* positions for Cu. The Cu atoms (open circles) in the same plane with Ca are in triangles formed by the filled circles. The Cu atoms in triangles (in plane) pointed upward are in *B* positions and those pointed downward are in *C* positions. One of the rows of *B* or *C* positions is half filled and the next row is empty. This indicates that for the **IIIP** layers *B* and *C* sites are one-quarter filled by Cu and *A* sites are one-quarter filled by Ca. The notation is $\mathbf{2IIIP}P^{ACa}_{1/4}{}^{Cu}_{1/4}{}^{Cu}_{1/4}\,\mathbf{P}^{A}{}^{Cu}_{3/4}$. Note that Cu_6 hexagon in the layer with Ca is slightly larger than that in the $\mathbf{P}_{3/4}$ layer (filled circles). The $CaCu_5$ structure is found for $CaZn_5$, $CaNi_5$, $LaZn_5$, $CeCo_5$, $ThCo_5$, and $ThZn_5$.

9.1.22. The $2{\cdot}3\mathbf{P}^{Sn}_{1/4}{}^{Co}_{3/4}\mathbf{IIP}^{Sn}_{1/4}{}_{1/4}$ Crystal Structure of CoSn

The intermetallic compound CoSn has an intricate crystal structure involving fascinating patterns. The structure is hexagonal with three molecules per cell, $a_o = 5.279$ and $c_o = 4.258$ Å. The *c/a* ratio is only 0.807. Alternate layers are half filled by Sn_{II} atoms in a graphite-like array in *B* and *C* positions, a double layer. The other layers contain Sn_I atoms and Co atoms in a 1:3 ratio, $\mathbf{P}^{Sn}_{1/4}{}^{Co}_{3/4}$, at *A* positions. Sn_I atoms are at the center of a Co_6 hexagon and between $(Sn_{II})_6$ hexagons as shown in the upper right of Figure 9.22. Co is at the center of a polygon, shown on the lower left, consisting of a hexagon formed by four Co and two Sn_I atoms at 2.64 Å with two Sn_{II} atoms above and below.

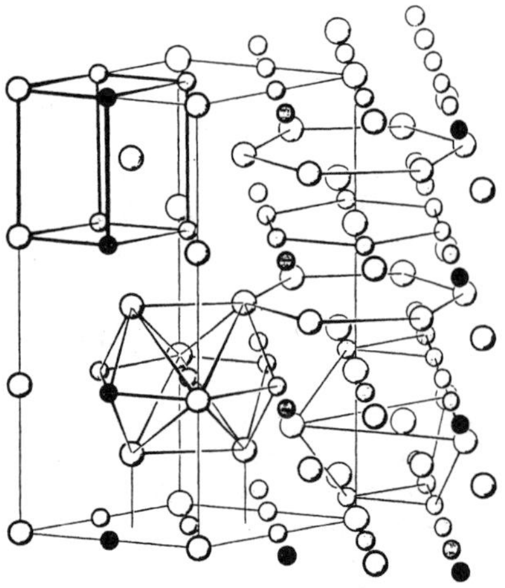

Figure 9.22. A pictorial view of the CoSn structure is shown with Sn as larger circles. The hexagonal unit cell is outlined at the left. An Sn_I atom in a hexagon of six Co with hexagons of six Sn above and below are shown on the upper right. Sn_{II} shown on the lower right is at the center of a triangle of three Sn in the same plane and a trigonal prismatic arrangement of six Co. The polyhedron of six Sn and four Co about Co is shown on the lower left.

(*Source*: W.P. Pearson, *The Crystal Chemistry and Physics of Metals and Alloys*, Wiley-Interscience, New York, 1972, p. 537.)

On the lower right we see a Sn_{II} atom surrounded by three Sn_{II} atoms in the same plane and triangular groups of three Co atoms above and below, all at 3.05 Å. The notation for the complex structure is simply $\mathbf{2{\cdot}3P}\,{}^{\mathrm{Sn}}_{1/4}\,{}^{\mathrm{Co}}_{3/4}\,\mathbf{I\!P}^{\mathrm{Sn}}_{1/4}\,{}_{1/4}$.

9.1.23. The $3{\cdot}3P_{1/9\ 8/9}$ Crystal Structure of $NbNi_8$

$NbNi_8$ is cubic with four molecules per unit cell, $\mathbf{O}_h^5$, $Fm\bar{3}m$, $a = 2.12$ Å. It is a superstructure of face-centered pseudocells as shown in Figure 9.23. Each close-packed layer contains Nb and Ni (1:8) with a nine

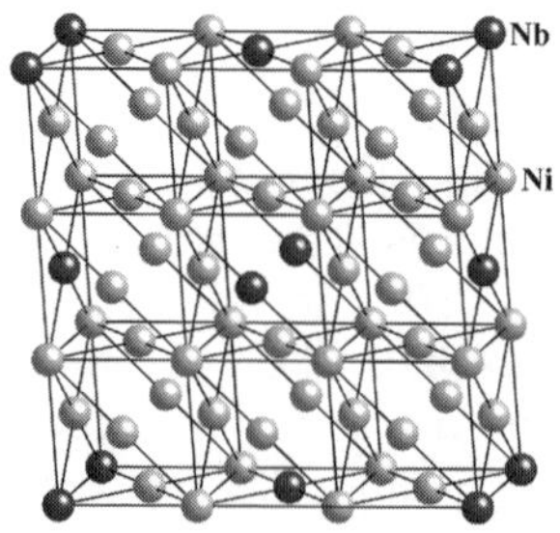

Figure 9.23. The superstructure of $NbNi_8$ of face-centered subcells is shown.

(*Source*: *CrystalMaker*, by David Palmer, *CrystalMaker* Software Ltd., Begbroke Science Park, Bldg. 5, Sandy Lane, Yarnton, Oxfordshire, OX51PF, UK.)

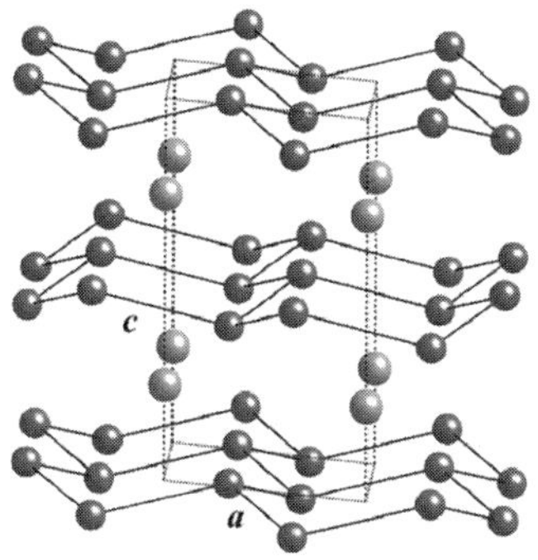

Figure 9.24. The structure of $CaIn_2$. Ca atoms are centered between hexagons in puckered layers Larger circles are Ca atoms.

(*Source*: Y. Matsushita, *Chalcogenide* crystal structure data library, Version 5.5B, Institute for Solid State Physics, The University of Tokyo, Tokyo, Japan, 2004; and *CrystalMaker*, by David Palmer, *CrystalMaker* Software Ltd., Begbroke Science Park, Bldg. 5, Sandy Lane, Yarnton, Oxfordshire, OX51PF, UK.)

layer stacking repeat sequence. In close-packed layers at least two Ni atoms separate Nb atoms. The notation is $\mathbf{3{\cdot}3P}_{1/9\ 8/9}$.

9.1.24. The 22ℙP^{Ca} Crystal Structure of $CaIn_2$

The unit cell of $CaIn_2$ is hexagonal with two molecules per cell, $\mathbf{D}^4_{6h}$, $P6_3/mmc$, $a = 4.895$, and $c = 7.750$ Å. Figure 9.24 shows two projections of the unit cell. The heights along c_o and positions are:

Height:	05	25	45	55	75	95	105
Atom:	In	Ca	In	In	Ca	In	In
Position:	P_B	P_A	P_B	P_C	P_A	P_C	P_B

The projection of the cell along c is the same at that of AlB_2 (Section 8.2.1, Figure 8.6) described as **2ℙ**$_{BC}$**P**^{C}. The double layer of $CaIn_2$ has In in B and C positions. This is a puckered layer of hexagons sharing edges with each In atom bonded to three neighbors. Ca atoms are at A positions at the corners of the cell. In has six Ca neighbors, forming a trigonal prism. The In atom is slightly below or above the center of the prism. There are three In neighbors capping faces of the prism. They are on the opposite of the center of the prism compared to the off-center In. Ca is centered between two hexagons of six In. The Ca-In distances are 3.242 and 3.636 Å.

9.2. Compounds Involving PTOT Layers

9.2.1. The 3·2PTOT Body-Centered Crystal Structures of β-Brass, CuZn

β-Brass, CuZn, has the CsCl cubic structure. It is an electron compound with 21 valence electrons for 14 atoms. The disordered structure (stable above ~460°C) has each site equally occupied by Cu and Zn. The ordered structure has Zn (or Cu) at the center of the cube. For both cases the notation is **3·2PTOT**. For the ordered

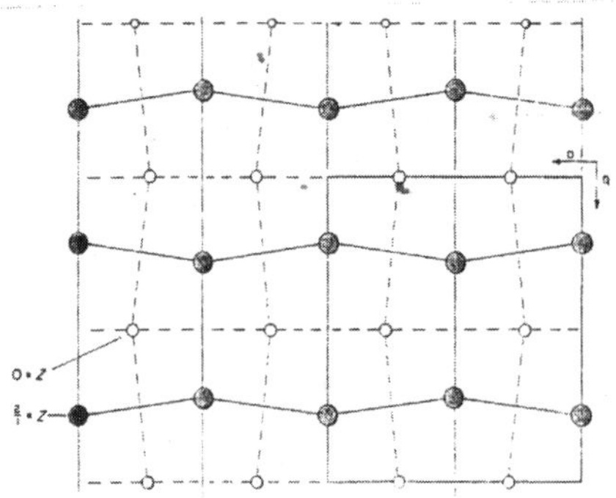

Figure 9.25. The CsCl-type structure of α-IrV. A view of the structure looking down the *c* axis is shown. The larger shaded circles are Ir atoms.

(*Source*: W.P. Pearson, *The Crystal Chemistry and Physics of Metals and Alloys*, Wiley-Interscience, New York, 1972, p. 573.)

structure Cu is in **P** and **O** sites with Zn in all **T** sites. CaIn has the same structure.

9.2.2. The 3·2P^{Ir}TO^{Ir}T(o) CsCl-Type Crystal Structure of α-IrV

α-IrV is orthorhombic with four molecules per unit cell, $a = 5.791$, $b = 6.756$, and $c = 2.976$ Å. It is a superstructure of the CsCl type containing eight distorted CsCl type cells. Figure 9.25 shows a projection of the cell along *c*. Ir atoms occupy **P** and **O** layers with V in both **T** layers. Because of the distortion, the positions of Ir and V are not equivalent. Ir has four V at 2.61 and four V at 2.675 Å and five Ir at 2.80–2.97 Å. V has eight Ir neighbors, one V at 2.55, and two V at 2.80 Å. The notation is **3·2P^{Ir}TO^{Ir}T(o)** CsCl-type.

9.2.3. The 3·2PTOT Crystal Structure of $MoAl_{12}$

$MoAl_{12}$ is an intermetallic compound with an icosahedral arrangement of 12 Al atoms around the central Mo atom. This large "molecule" has high symmetry, and can pack almost as spheres. The crystal of $MoAl_{12}$ (Figure 9.26) has a regular *bcc* or **3·2PTOT** structure. The figure shows only two of the icosahedra to avoid cluttering and allow visualization of the simplicity of the structure.

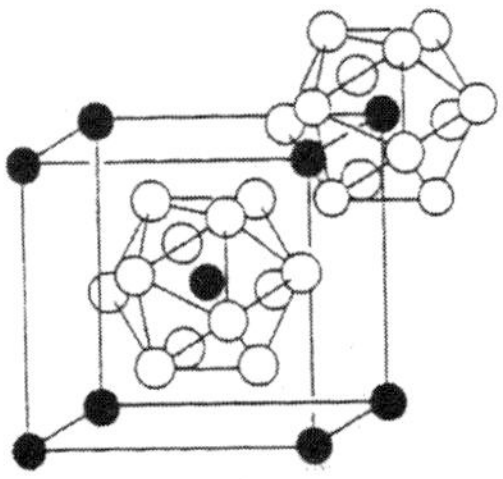

Figure 9.26. The body-centered cubic structure of $MoAl_{12}$ shows Mo atoms at the corners and center. Only two $MoAl_{12}$ molecules are shown.

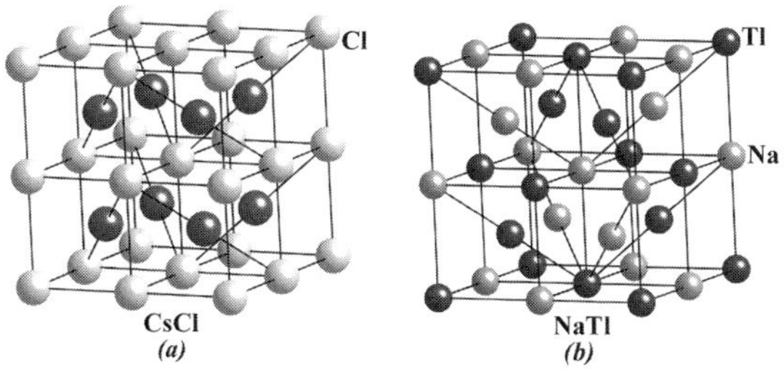

Figure 9.27. (*a*) The CsCl structure. (*b*) The NaTl structure.

(*Source*: *CrystalMaker*, by David Palmer, *CrystalMaker* Software Ltd., Begbroke Science Park, Bldg. 5, Sandy Lane, Yarnton, Oxfordshire, OX51PF, UK.)

9.2.4. The 3·4$P^{M}T^{M}O^{M'}T^{M'}$ Crystal Structure of NaTl

In the previous section we saw that the complex formula of $MoAl_{12}$ leads to a simple basis for the structure. The structure of NaTl is more complex than the simple formula suggests. The distinction is in the chemical nature of the elements. Ionic compounds demand that each ion be surrounded by ions of opposite charges in crystals. For intermetallic compounds the neighbors can be atoms of the same element. Figure 9.27*a* shows the CsCl structure where open circles (Cl^-) represent **P** and **O** layers and the black circles (Cs^+) represent both **T** layers (or we can reverse the positions of the ions). If all circles are the same, the structure is *bcc*. In NaTl (Figure 9.27*b*) the Na and Tl atoms alternate along each edge. This corresponds to having Tl atoms in **P** and one **T** layer with Na atoms in **O** and one **T** layer (or reverse the pairs). The Na and Tl atoms alternate in the center of the octants shown. Each Na or Tl atom has four neighbors of each element (two tetrahedral arrangements forming a cube). The unit cell has 12 packing layers with eight molecules giving **3·4$P^{Na}T^{Na}O^{Tl}T^{Tl}$**. The packing sequences with heights along the body diagonal are given below.

Height:	0	8	17	23	33	42	50	58	67	75	83	92
Atom:	Na	Tl	Tl	Na	Na	Tl	Tl	Na	Na	Tl	Tl	Na
Position:	P_A	T_B	O_C	T_A	P_B	T_C	O_A	T_B	P_C	T_A	O_B	T_C

Again, the complex structure is greatly simplified in terms of the **PTOT** system.

9.2.5. The Crystal Structures of 3·4PTOT $BiLi_3$ and $AlMnCu_2$

The structures of $BiLi_3$ and $AlMnCu_2$ are similar to NaTl (previous section) with different roles for the elements. Figure 9.28 shows these together for comparison. The structure for $BiLi_3$ differs from that for NaTl in that the centers of all octants and the centers of each edge are filled by the same atom (Li). This corresponds to a **3·4P^{Bi}TOT** structure with Bi in all **P** positions (face-centered cubic) with Li filling **O** and both **T** layers. This structure is compatible with the rules for ionic compounds, and the same structure is found for BiF_3 where each

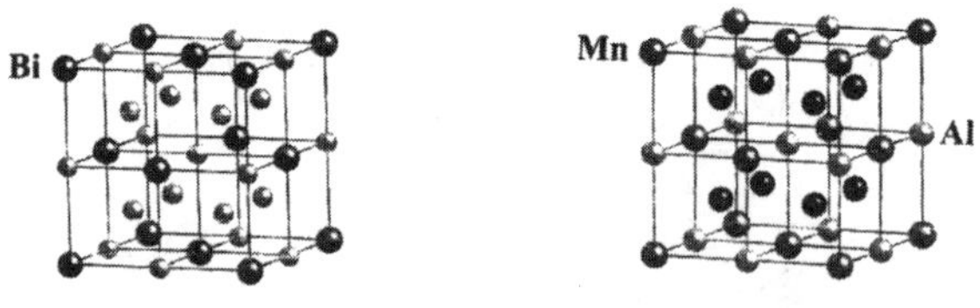

(a) (b)

Figure 9.28. (*a*) The $BiLi_3$ structure. (*b*) The $AlMnCu_2$ structure.

(*Source*: *CrystalMaker*, by David Palmer, *CrystalMaker* Software Ltd, Begbroke Science Park, Bldg. 5, Sandy Lane, Yarnton, Oxfordshire, OX51PF, UK.)

Bi^{3+} ion has a cubic arrangement of eight F^- ions (in **T** sites) and an octahedral arrangement of six F^- ions (in **O** sites). $TlLi_3$ has the same structure as $BiLi_3$.

The $AlMnCu_2$ structure is also a **3·4PTOT** structure with Cu in both **T** layers with Al in the **P** layer and Mn in the **O** layer, $\mathbf{3{\cdot}4P^{Al}T^{Cu}\,O^{Mn}T^{Cu}}$. It is equivalent to assign Cu to both **P** and **O** layers with Al and Mn in **T** layers.

9.2.6. The 3·2PTOT(t)(4:5) Crystal Structure of V_4Zn_5

V_4Zn_5 is tetragonal with two molecules per unit cell, $a = 8.910$ and $c = 3.224$ Å. Figure 9.29 shows that the structure contains distorted *bcc* pseudocells. In the center of the figure there is a Zn at the center of a distorted *bcc* unit with Zn at the corners, while others have four Zn and four V at corners with Zn or V at the center. Each atom has 14 close neighbors. The closest V—V distance is 2.50 Å and for Zn—Zn 2.70 Å. The closest V—Zn distance is 2.72 Å. Because this structure is based on distorted pseudocells the notation is **3·2PTOT(t)(4:5)**.

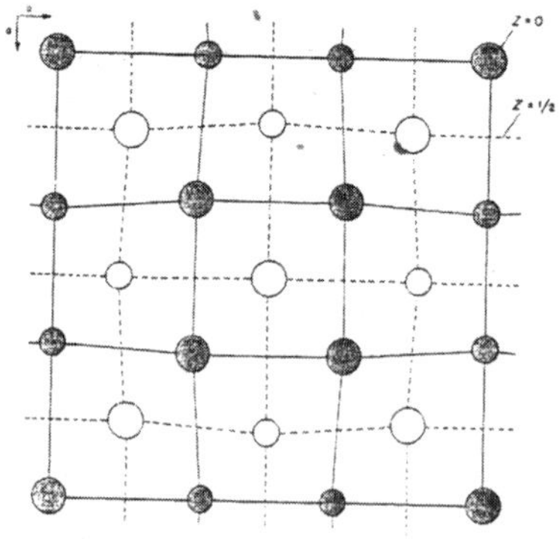

Figure 9.29. The structure of V_4Zn_5. A projection of the structure is viewed down the *c* axis. One layer is represented by open circles and the larger circles are Zn.

(*Source*: W.P. Pearson, *The Crystal Chemistry and Physics of Metals and Alloys*, Wiley-Interscience, New York, 1972, p. 581.)

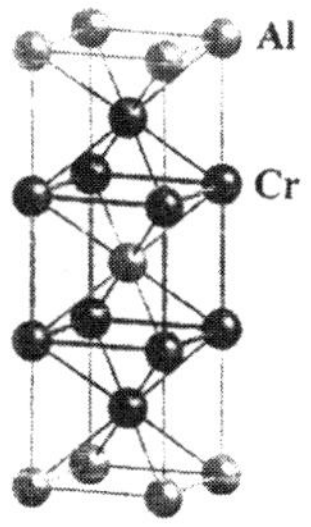

Figure 9.30. The superstructure of $AlCr_2$.

(*Source*: *CrystalMaker*, by David Palmer, *CrystalMaker* Software Ltd., Begbroke Science Park, Bldg. 5, Sandy Lane, Yarnton, Oxfordshire, OX51PF, UK.)

9.2.7. The Body-Centered Type Crystal Structure of $AlCr_2$

$AlCr_2$ is cubic, a superstructure of body-centered subcells, $\mathbf{O}_h^9$, $Im\bar{3}m$, $a = 1.73$ Å. Figure 9.30 shows that the top and bottom subcells have Cr at the center with four Cr and four Al neighbors. The central subunit has Al at the center with eight Cr neighbors. The notation is **3·4PTOT(1:2)**.

9.2.8. The $3{\cdot}3P^{Hg}T_{1/4}O^{Hg}T_{1/4}$ Crystal Structure of $PtHg_4$

The crystal structure of $PtHg_4$ is cubic, $a_o = 6.186$ Å, with two molecules in the unit cell. Figure 9.31 shows two views of the unit cell. In the center of Figure 9.31*b*, Pt is in a cube formed by eight Hg at 2.68 Å. Hg is in an octahedron formed by six Hg at 3.09 Å with two Pt neighbors arranged linearly. Figure 9.31*b* corresponds to a structure formed by Pt in one-quarter of **P** and **O** sites with Hg in all **T** sites. Because **P** and **O** layer are interchangeable with both **T** layers, we can

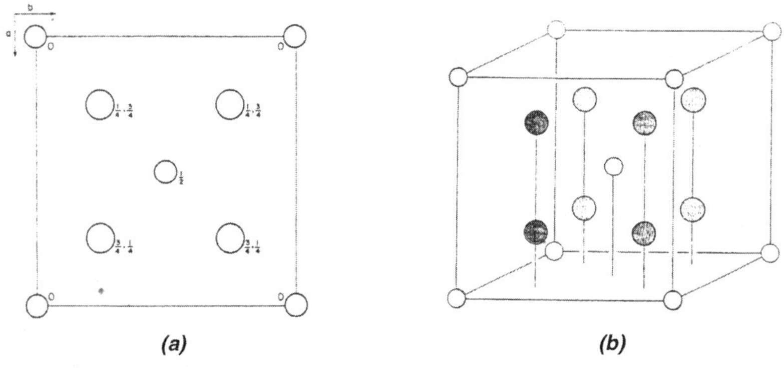

Figure 9.31. (*a*) A projection of the $PtHg_4$ structure along the *c* axis is shown. (*b*) A view of the whole cell. Larger circles are Hg atoms.

(*Source*: W.P. Pearson, *The Crystal Chemistry and Physics of Metals and Alloys*, Wiley-Interscience, New York, 1972, p. 567.)

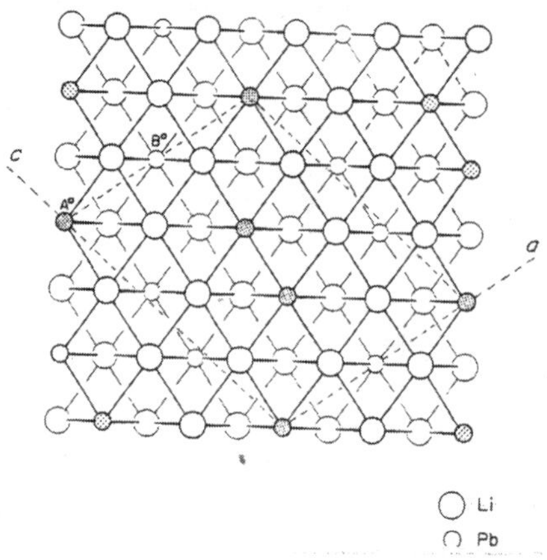

Figure 9.32. A projection of the Pb_3Li_8 structure along the *b* axis is shown. (*Source*: W.P. Pearson, *The Crystal Chemistry and Physics of Metals and Alloys*, Wiley-Interscience, New York, 1972, p. 580.)

fill **P** and **O** layers with Hg with Pt in one-quarter of all **T** layers, giving the notation $\mathbf{3{\cdot}3P^{Hg}T_{1/4}O^{Hg}T_{1/4}}$.

9.2.9. The 3·4PTOT(m)(3:8) Crystal Structure of Pb_3Li_8

The Pb_3Li_8 crystal structure is monoclinic with two molecules per unit cell, $a = 8.240$, $b = 4.757$, $c = 11.03$ Å, and $\beta = 104.42°$. It is a superstructure based on a *bcc*-type structure. Figure 9.32 shows a view of the structure along *b*. Pb and Li are distributed through all **PTOT** sites of a body-centered structure with some ordering of Pb atoms to provide about half of the Pb atoms adjacent in pairs at 2.91 Å. This does not cause distortions because Pb–Li and even Li–Li distances are also 2.91 Å. Each atom has ***CN*** 14 corresponding to the neighbors at the corners of the cube and the six through the faces. The notation is **3·4PTOT(m)(3:8)**, with Pb and Li in the 3:8 ratio.

9.2.10. The Body-Centered Crystal Structures of Ti_2Cu_3 and Ti_3Cu_4

Ti_2Cu_3 is tetragonal with two molecules per unit cell, $a = 3.13$ and $c = 13.95$ Å. Ti_3Cu_4 is also tetragonal with two molecules per unit cell, $a = 3.13$ and $c = 19.94$ Å. Both structures contain cubic pseudocells stacked along *c*. There are five squashed body-centered subunits for Ti_2Cu_3 and seven subunits for Ti_3Cu_4. These ordered structures are shown in Figure 9.33, with one body-centered subunit outlined for each. The notation for Ti_2Cu_3 is **3·5(PTOT)(t)(2:3)** and for Ti_3Cu_4 it is **3·7(PTOT)(t)(3:4)**.

9.2.11. The $\mathbf{4{\cdot}2IIP^{B}_{1/4}\,{}^{C}_{1/4}P^{Al}_{1/4}P^{Cu}_{1/2}P^{Al}_{1/4}(t)}$ Crystal Structure of $CuAl_2$

The $CuAl_2$ tetragonal crystal structure, $\mathbf{D}^{18}_{4h}$, $I4/m\,cm$ is found for many AM_2 type intermetallic and boride compounds (see Table 9.6). The tetragonal cell has four molecules per unit cell, $a = 6.067$ and

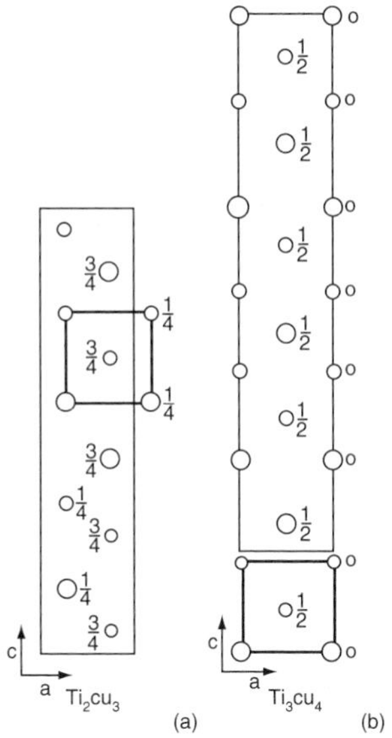

Figure 9.33. (*a*) A projection of the Ti_2Cu_3 structure along the *b* axis, a body-centered subcell is outlined. Cu atoms are smaller circles. (*b*) A projection of the Ti_3Cu_4 structure along the *b* axis. A body-centered subcell is outlined.

(*Source*: W.P. Pearson, *The Crystal Chemistry and Physics of Metals and Alloys*, Wiley-Interscience, New York, 1972, p. 582.)

$c = 4.877$ Å. Figure 9.34***a*** shows a projection down the *c* axis. There are alternate sheets of Cu and Al atoms stacked along *c*. The Cu atoms are at the centers of square antiprisms stacked along *c* and sharing square faces. Each antiprism in the same layer shares four side edges. The Cu–Al distance in the antiprism is 2.590 Å. The Cu–Cu distance in the chain along *c* is 2.438 Å. There are packing layers along the vertical direction of the paper and between Cu layers there are three Al layers. In a projection along this vertical direction the layers labeled *BC* have a hexagonal pattern like a close-packed layer without atoms at the centers of hexagons. The atoms of both adjacent layers labeled *A* are centered above and below the centers of the hexagons for a *BC* layer. It is tempting to assign all of the Al atoms to *A* positions. However, if the atoms in the lower *BC* layer are assigned *A* positions, the higher *BC*

TABLE 9.6. Intermetallic and boride compounds with the $CuAl_2$ structure.

ϕ-$AgIn_2$	$BeTa_2$	$FeSn_2$	$SiTa_2$	Co_2B	Mo_2B
$AgTh_2$	$CoSc_2$	$MnSn_2$	$SiZr_2$	Cr_2B	Ni_2B
$AlTh_2$	$CoSn_2$	$PdPb_2$	$TiSb_2$	Fe_2B	β-Ta_2B
$AuNa_2$	θ-$CuAl_2$	$PdTh_2$	VSb_2	Mn_2B	W_2B
$AuPb_2$	$CuTh_2$	$RhPb_2$	$ZnTh_2$		
$AuTh_2$	$FeGe_2$				

layer does not correspond to *A*, *B*, or *C* positions. The hexagonal pattern of the higher *BC* layer is shifted in one direction relative to the lower one. Also, the positions of Cu do not correspond to *A*, *B*, or *C* positions relative to the lower *BC* layer if it is assigned as an *A* layer. The *CrystalMaker* computer program allows us to examine layers in any orientation individually and in any combination. A projection along the vertical direction in Figure 9.34***a*** of the Cu layer and the adjacent layers of Al gives a hexagonal pattern corresponding to a filled close-packed layer (Figure 9.34***b***). Each of these individual layers is partially filled; the Cu layer is one-half filled, and each of these Al_I layers is one-quarter filled. The power of *CrystalMaker* and the wonder of nature reveal that each Al layer halfway between Cu layers is not a simple hexagonal close-packed layer because the positions of Cu atoms are not accommodated. The positions of Al (Al_{II}) atoms in each layer halfway between Cu layers correspond to *B* and *C* positions in the projection of an *A* layer made up of the Cu layer and both adjacent Al layers (Al_I). These positions are labeled in one hexagon of Figure 9.34***b***. The Al_{II} layer is a double layer ($\mathbf{IP}^{B}_{1/4}\,{}^{C}_{1/4}$). The notation is $\mathbf{4{\cdot}2IP}^{B}_{1/4}\,{}^{C}_{1/4}\mathbf{P}^{Al}_{1/4}\,\mathbf{P}^{Cu}_{1/2}\,\mathbf{P}^{Al}_{1/4}(\mathbf{t})$.

The Al_I atoms nearest the Cu layer are centered over a hexagon of six Al_{II} atoms in the double layer. Four Al–Al distances are 3.103 and two are 3.232 Å. On the other side of Al_I there are four Cu atoms (at 2.590 Å), two Al atoms of the next Al_I layer at 2.904 Å, and two Al of the next Al_{II} layer at 3.232 Å. Al_{II} atoms have three Al in the same double layer at 2.904 Å and a trigonal prism of six Al_I of two adjacent layers, four at 3.103, and two at 3.232 Å. The three Al_{II} atoms of the same layer of the central Al_{II} atom cap faces of the trigonal prism.

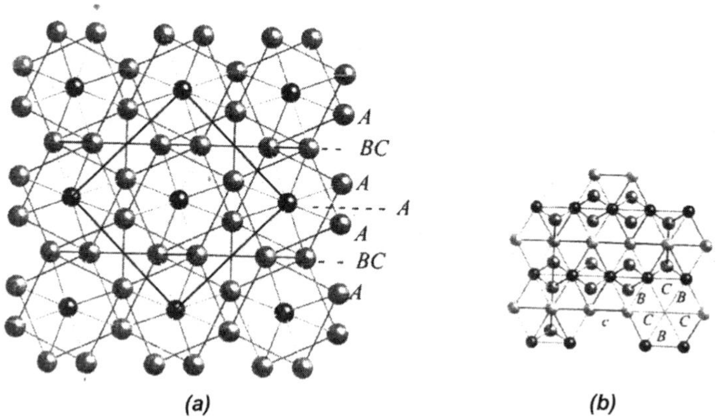

Figure 9.34 (***a***) A projection of the $CuAl_2$ cell along *c*. (***b***) A projection of packing layers with Al_I (light balls) and Cu (black balls) atoms filling *A* positions. Al_{II} atoms occupy *B* and *C* positions with labels shown in one hexagon.

(*Source*: Y. Matsushita, *Chalcogenide* crystal structure data library, Version 5.5B, Institute for Solid State Physics, The University of Tokyo, Tokyo, Japan, 2004; and *CrystalMaker*, by David Palmer, *CrystalMaker* Software Ltd, Begbroke Science Park, Bldg. 5, Sandy Lane, Yarnton, Oxfordshire, OX51PF, UK.)

There are two Cu atoms above and below, and there is one Al_{II} atom above and below of the next Al_{II} layers. In projection along the packing direction (Figure 9.34***b***) these atoms are aligned and form a triangle with the aligned pairs of Cu atoms. In this projection the three Al_{II} and three eclipsed pairs of Al_{I} atoms form a hexagon.

Most compounds with the $CuAl_2$ structure have a radius ratio of about 0.8. For $CuAl_2$, the radius ratio is 0.895. The borides BFe_2 and BCo_2, with the $CuAl_2$ structure, are ferromagnetic.

9.2.12. The $3{\cdot}4\mathbf{P}^{Ti}\mathbf{T}^{Cu}\mathbf{O}^{Cu}\mathbf{T}^{Ti}$(t) Crystal Structure of γ-CuTi

γ-CuTi is tetragonal with two molecules per unit cell, $a = 3.118$ and $c = 5.921$ Å. Figure 9.35 shows a plane parallel to {110}. This is a distorted CsCl-type structure involving two cells stacked in the {001} direction. The figure corresponds to a plane through the diagonal of a body-centered cube with one atom at the center (**T** position) of a rectangle formed by two **P** sites and two **O** sites. The positions are ordered with Ti in **P** and one **T** layer and Cu in **O** and one **T** layer (these site designations for Ti and Cu can be reversed). Spacings between Cu layers are less than spacings between Cu and Ti layers and those between Ti layers are more. Cu has five Ti at~2.66 and four Cu at 2.50 Å. Ti has five Cu and four Ti at 2.93 Å. The notation is $3{\cdot}4\mathbf{P}^{Ti}\mathbf{T}^{Cu}\mathbf{O}^{Cu}\mathbf{T}^{Ti}$(t).

9.2.13. The $3{\cdot}4\mathbf{P}_{3/4}\mathbf{T}^{Na}_{1/4}\mathbf{O}_{1/4}\mathbf{T}^{Na}_{1/4}$ Crystal Structure of $NaPt_2$

Crystals of $NaPt_2$ are cubic, $\mathbf{O}_h^7$, $Fd\bar{3}m$, and $a = 7.482$ Å. Pt atoms occupy three-quarters of **P** layers and one-quarter of **O** layers and Na atoms occupy one-quarter of all **T** layers. The **P** layers have an *ABC* sequence. The positions of **T** and **O** sites are those required of geometry in $\mathbf{P}_A\mathbf{T}_B\mathbf{O}_C\mathbf{T}_A\mathbf{P}_B\mathbf{T}_C$. In Figure 9.36***a*** **P** planes are labeled $\mathbf{P}_A\mathbf{P}_B\mathbf{P}_C\mathbf{P}_A$. Halfway between the **P** layers are the **O** layers. The six bonds to Pt in **P** atoms are shown for three octahedra. Between the **P** layers the lower layer of Na atoms is for $\mathbf{T}^+$ sites (apex up) and the upper layer is for $\mathbf{T}^-$ sites (apex down). The fractional occupancies

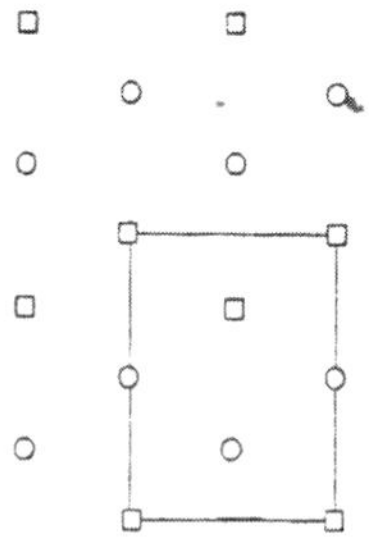

Figure 9.35. The structure of γ-CuTi. The arrangement of Cu and Ti atoms in a plane parallel to {110} planes is shown.

(*Source*: W.P. Pearson, *The Crystal Chemistry and Physics of Metals and Alloys*, Wiley-Interscience, New York, 1972, p. 574.)

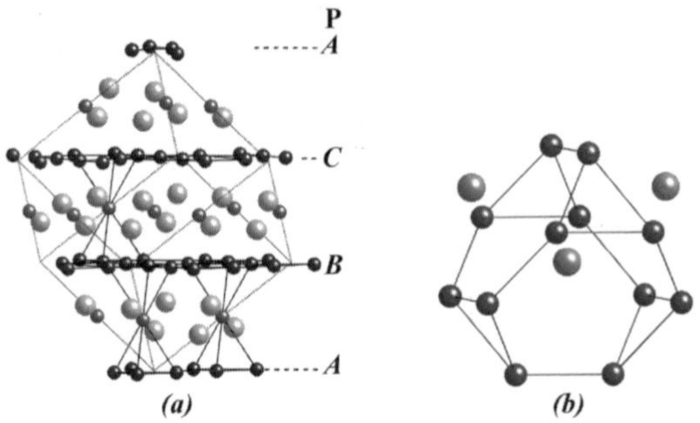

Figure 9.36. (*a*) The cell of $NaPt_2$ showing **P**, **T**, and **O** layers. (*b*) The arrangement of 12 Pt atoms around Na.

(*Source*: Y. Matsushita, *Chalcogenide* crystal structure data library, Version 5.5B, Institute for Solid State Physics, The University of Tokyo, Tokyo, Japan, 2004; and *CrystalMaker*, by David Palmer, *CrystalMaker* Software Ltd, Begbroke Science Park, Bldg. 5, Sandy Lane, Yarnton, Oxfordshire, OX51PF, UK.)

were determined using *CrystalMaker*. The cell was expanded greatly, and projections of one or two **P** layers with **T** and **O** layers were printed. The patterns indicated $\mathbf{P}_{3/4}$, $\mathbf{T}_{1/4}$, and $\mathbf{O}_{1/4}$ occupancies. The Na atoms are in **T** sites, but they are in open areas of the cell with 12 Pt neighbors at 3.10 Å. Three of these are in one **P** layer ($\mathbf{P}_1$), corresponding to the base of a tetrahedron. The Na is above (or below) a vacancy of the other **P** ($\mathbf{P}_2$) layer; the missing Pt atom would be at the apex of the tetrahedron. The hexagonal arrangement of Pt atoms in $\mathbf{P}_2$ layer are at 3.10 Å from the Na atom. The other three Pt atoms are in **O** sites. The arrangement of Pt atoms around Na is shown in Figure 9.36***b***. There are tetrahedral arrangements of Pt_6 hexagons, Pt_3 triangles, and Na atoms. The Na atoms are centered with the hexagons and the two lower Na neighbors are not shown because in this view those would appear to be inside the polyhedron. The $PtPt_6$ octahedra are not distorted. The notation is $\mathbf{3{\cdot}4P_{3/4}T^{Na}_{1/4}O_{1/4}T^{Na}_{1/4}}$.

9.2.14. The 3·3PTT Crystal Structure of $PbMg_2$

Crystals of $PbMg_2$ are cubic, $\mathbf{O}_h^5$, $Fm\bar{3}m$, and $a = 6.815$ Å, with four molecules in the unit cell. It is the fluorite structure (Section 6.3.1) with Pb atoms in a face-centered cubic arrangement with Mg atoms in all **T** sites (Figure 9.37). The notation is **3·3PTT**.

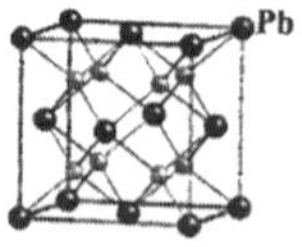

Figure 9.37. The **3·3PTT** fluorite structure of $PbMg_2$.

(*Source*: Y. Matsushita, *Chalcogenide* crystal structure data library, Version 5.5B, Institute for Solid State Physics, The University of Tokyo, Tokyo, Japan, 2004; and *CrystalMaker*, by David Palmer, *CrystalMaker* Software Ltd, Begbroke Science Park, Bldg. 5, Sandy Lane, Yarnton, Oxfordshire, OX51PF, UK.)

9.3. Compounds Involving Only P and O Layers

9.3.1. The 2·2ℙO Structure of $InNi_2$

The structure of $InNi_2$ is closely related to NiAs (**2·2PO**), with In replacing As and with one Ni added above one In and another Ni added below the other In within the cell. Ni_I and In atoms are in the same double-packing layers at heights 25 and 75. Double-layer (**ℙ**) positions B and C are filled by In and Ni_I. The $\mathbf{O}_A$ layers at heights 0 and 50 are filled by Ni_{II} atoms. There are four layers in the repeating unit, giving the notation **2·2ℙO**. The Ni_{II} atoms have octahedral arrangements of six In atoms and six Ni_I atoms. The Ni_I atoms in **ℙ** layers have three neighboring In atoms in the same plane. The dodecahedron with six square and six triangular faces formed by Ni and In is shown in Figure 9.38*a*. The polyhedron around Ni_I is shown in Figure 9.38*b*. The hexagonal unit contains two molecules, $\mathbf{D}_{6h}^3$, $P6_3/m\,cm$, $a_o = 4.179$, and $c_o = 5.131$ Å. Heights of atoms along c are:

Height:	0	25	50	75
Atom:	Ni_{II}	**Ni In**	Ni_{II}	**Ni In**
Position:	A	C B	A	B C

Intermetallic phases with the $InNi_2$ structure are given in Table 9.7.

9.3.2. The $3{\cdot}2P_{1/4}O_{3/4}$ Crystal Structure of $SnNb_3$

A body-centered cubic structure is **3·2PTOT** with all sites filled. For $SnNb_3$ the addition of Nb to the *bcc* cell of Sn atoms requires expansion of the cell with partial filling of layers. Expansion to give eight subcells places Sn at the corners and in the center. Pairs of Nb atoms occur along lines parallel to the axes in the faces. Lines joining Nb atoms are parallel in opposite faces (see Figure 9.39). The cell for $SnNb_3$ is cubic, $\mathbf{T}_h^1$, $Pm3$, $a = 5.289$ Å, containing two molecules. The sequence of close-packed layers along the body-diagonal in the expanded cell is $\mathbf{P}_A\mathbf{O}_C\mathbf{P}_B\mathbf{O}_A\mathbf{P}_C\mathbf{O}_B$. This is a **PO** structure, with Sn occupying one-quarter of **P** sites and Nb occupying three-quarters of **O** sites, $\mathbf{P}_{1/4}\mathbf{O}_{3/4}$. The

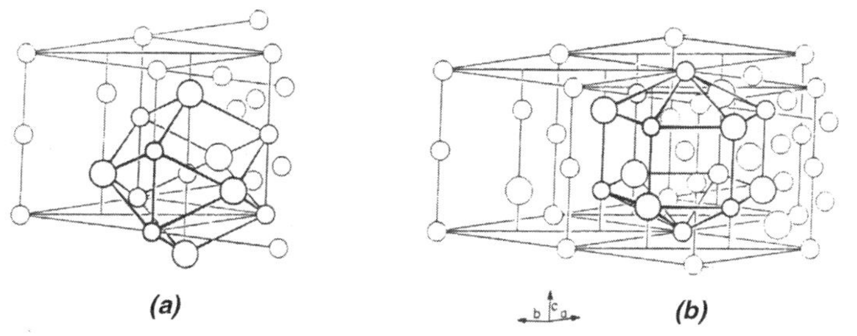

Figure 9.38. (*a*) A projection of the unit cell of $InNi_2$ along c is shown. (*b*) An expanded cell to show the polyhedron for Ni_I.

(*Source*: F.S. Galasso, *Structure and Properties of Inorganic Solids*, Pergamon, Oxford, 1970, p. 143.) (*Source*: W.P. Pearson, *The Crystal Chemistry and Physics of Metals and Alloys*, Wiley-Interscience, New York, 1972, p. 531.)

TABLE 9.7. **Intermetallic phases with the $InNi_2$ structure.**

BeSiZr	CoNiSb	FeGeNi	$InCu_2$	Pd_3Sn_2
CoFeGe	CoNiSn	Fe_3Sb_2	Ni_3Sn_2	$SnMn_2$
CoGeMn	FeGeMn	GeMnNi	Pb_2Pd_3	$SnTi_2$

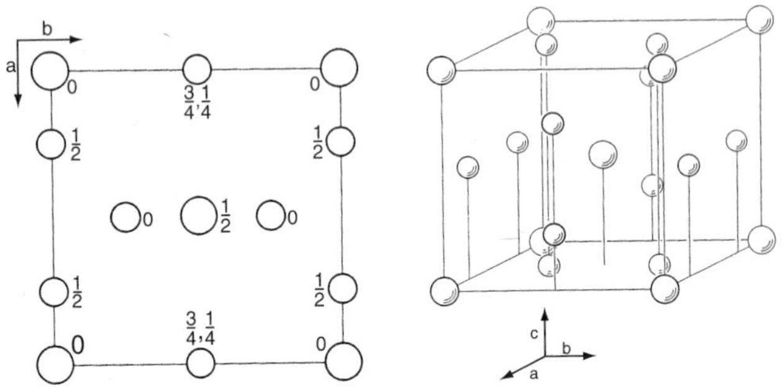

Figure 9.39. (*a*) A projection of the unit cell of $SnNb_3$ along the *c* axis. Larger circles are Sn atoms. (*b*) A view of the unit cell.

(*Source*: W.P. Pearson, *The Crystal Chemistry and Physics of Metals and Alloys*, Wiley-Interscience, New York, 1972, p. 671.)

notation is $\mathbf{3{\cdot}2P_{1/4}O_{3/4}}$. $SnNb_3$ is an important superconductor. Many intermetallic compounds have this structure (Table 9.8). The structure is sometimes known as the Cr_3O, Cr_3Si, or β-tungsten structure. It was reported as the structure of β-W, but the sample was found to be W_3O.

9.3.3. The Layer Crystal Structure of Pb_2Li_7

The crystal structure of Pb_2Li_7 is hexagonal with one molecule per unit cell, $a = 4.751$ and $c = 8.589$ Å. Figure 9.40 shows a projection of the unit cell. The heights along *c* are:

Height:	−8	0	8	25	33	42	58	67	75	92
Atom:	Li	Li	Li	Pb	Li	Li	Li	Li	Pb	Li
Position:	P_A	O_C	P_B	P_A	O_C	P_B	P_A	O_C	P_B	P_A

TABLE 9.8. **Intermetallic phases with the $SnNb_3$ (or β-tungsten) structure.**

$AlMo_3$	$BiNb_3$	$GeNb_3$	$IrTi_3$	PdV_3	$RuCr_3$
$AlNb_3$	CoV_3	GeV_3	IrV_3	$PtCr_3$	$SbNb_3$
AsV_3	$GaCr_3$	γ-$HgTi_3$ (540–760° C)	NiV_3	$PtNb_3$	$SbTi_3$
$AuNb_3$	$GaMo_3$	$HgZr_3$	$OsCr_3$	$PtTi_3$	SbV_3
$AuTa_3$	$GaNb_3$	$InNb_3$ (H.P.)	$OsMo_3$	PtV_3	$SnMo_3$
$AuTi_3$	GaV_3	$IrCr_3$	$OsNb_3$	$RhCr_3$	$SnNb_3$
AuV_3	$GeCr_3$	$IrMo_3$	$PbNb_3$	$RhNb_3$	SnV_3
$AuZr_3$	$GeMo_3$	$IrNb_3$	PbV_3	RhV_3	

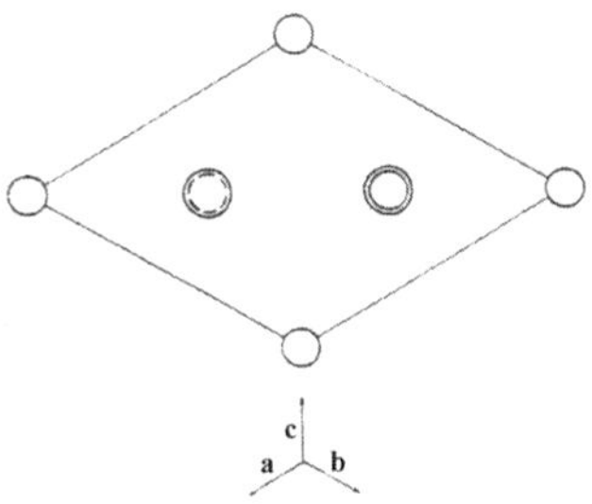

Figure 9.40. The hexagonal unit cell of the layer structure of Pb_2Li_7. Larger circles are Pb atoms.

(*Source*: W.P. Pearson, *The Crystal Chemistry and Physics of Metals and Alloys*, Wiley-Interscience, New York, 1972, p. 398.)

The **P** layers, occupied by Pb or Li, are in an *AB* sequence with uniform spacing. The Li atoms at the corners of the cell are in **O** layers, and, as required in the **PTOT** pattern, they are at *C* positions. There are triple-layered **POP** sandwiches. The Li and Pb atoms in **P** layers are in the sequence $(\mathbf{P}^{Li}\mathbf{O}\mathbf{P}^{Li})(\mathbf{P}^{Pb}\mathbf{O}\mathbf{P}^{Li})(\mathbf{P}^{Li}\mathbf{O}\mathbf{P}^{Pb})$, with Li in all of these **O** layers. The notation is $\mathbf{2{\cdot}9/2}(\mathbf{P}^{Li}\mathbf{O}\mathbf{P}^{Li})(\mathbf{P}^{Pb}\mathbf{O}\mathbf{P}^{Li})(\mathbf{P}^{Li}\mathbf{O}\mathbf{P}^{Pb})$. Each Li at heights 0, 33, and 67 are in octahedra formed by six Li or three Pb and three Li atoms. Li atoms at 8 and 92 have a trigonal prismatic arrangement of three Li and three Pb atoms with the "central" Li closer to the three Li. Li atoms at 42 and 58 have a trigonal prismatic arrangement of six Li with the "central" Li closer to one triangular face. Pb atoms are also off-center of a trigonal prism of six Li atoms with several other Li atoms more distant.

9.3.4. The 3·5/3 POPPO(h) Crystal Structure of Ni_2Al_3 and α-Bi_2Mg_3

The crystal structure of Ni_2Al_3 is hexagonal, with one molecule per unit cell, $\mathbf{D}_{3d}^3$, $C\bar{3}m$, $a_o = 4.036$, and $c_o = 4.901$ Å. The structure is the same as La_2O_3 (Section 5.5.10) with different spacings. A projection of the cell is shown in Figure 9.41***a***. Aluminum atoms alone are in *A* positions, with Ni and Al in *B* and *C* positions. These layers are filled. The heights along c_o are:

Height:	0	15	35	65	85	100	115
Atom:	Al	Ni	Al	Al	Ni	Al	Ni
Position:	$\mathbf{P}_A$	$\mathbf{O}_C$	$\mathbf{P}_B$	$\mathbf{P}_C$	$\mathbf{O}_B$	$\mathbf{P}_A$	$\mathbf{O}_C$

Ni atoms are displaced toward one **P** layer from the center of an octahedron of six Al atoms. There is a seventh Al atom capping the more distant face of the octahedron. Al_I atoms at zero and 100 are at the centers of octahedra formed by six Ni atoms at 2.44 Å with six Al atoms at 2.90 Å in the same plane as the central Al. Al_{II} atoms in positions *B* and *C* have three Al atoms at 2.90 Å and three at 2.745 Å, an octahedron with the "central" atom displaced toward one face. Al_{II} has five Ni neighbors in the form of a very distorted

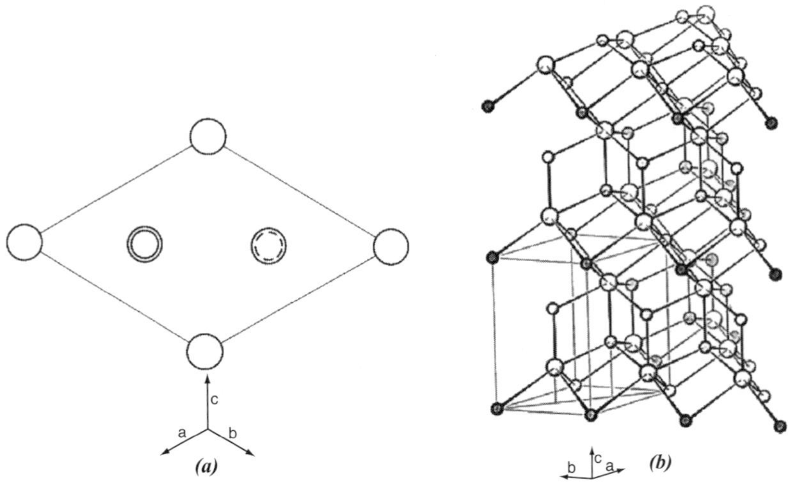

Figure 9.41. (*a*) The hexagonal unit cell of Ni_2Al_3, smaller circles are Ni atoms.(*b*) The crystal structure of α-Bi_2Mg_3, larger circles are Bi atoms. A unit cell is outlined.

(*Source*: W.P. Pearson, *The Crystal Chemistry and Physics of Metals and Alloys*, Wiley-Interscience, New York, 1972, p. 400.)
(*Source*: W.P. Pearson, *The Crystal Chemistry and Physics of Metals and Alloys*, Wiley-Interscience, New York, 1972, p. 401.)

trigonal bipyramid. α-Bi_2Mg_3 has this structure, as shown in Figure 9.41*b*. We see that Bi atoms have ***CN*** 7, with a Mg atom capping one of the faces of the $BiMg_6$ octahedron. The notation is **3·5/3 POPPO(h)** for both structures.

9.4. The Laves Phases

There are three crystal structures of compounds of the AM_2 type known as Laves phases, $MgCu_2$, $MgZn_2$, and $MgNi_2$. Laves and co-workers did much of the early work on these phases. All of these structures have frameworks of tetrahedra sharing apices, and for $MgZn_2$ and $MgNi_2$, apices and tetrahedral faces are shared. These patterns are similar to those for silica (see Figure 10.1). The A metal of AM_2 is larger than metal M, Mg is the larger metal of the parent structures of Laves phases. The frameworks of tetrahedra are formed by Cu, Zn, or Ni.

The Mg atom is much too large for **T** sites; they fit into large cages formed by the tetrahedra. There are three patterns encountered for these phases, as shown in Figure 9.42. The packing layers are partially filled, $\mathbf{P}_{3/4}$ and $\mathbf{P}_{1/4}$, alternating. $MgCu_2$ is cubic and the sequence of layers along the packing direction (the body diagonal of the cube) is *ACB ACB*. $MgZn_2$ and $MgNi_2$ are hexagonal. The sequence of packing layers is *AB CB AB* for $MgZn_2$ and *AB CB AC BC* for $MgNi_2$.

The large Mg atom has ***CN*** 16. Twelve M atoms (Cu, Zn, or Ni) form a truncated tetrahedron seen in Figure 9.43. The ***CN*** 16 is completed by four tetrahedrally arranged Mg atoms through the triangular faces.

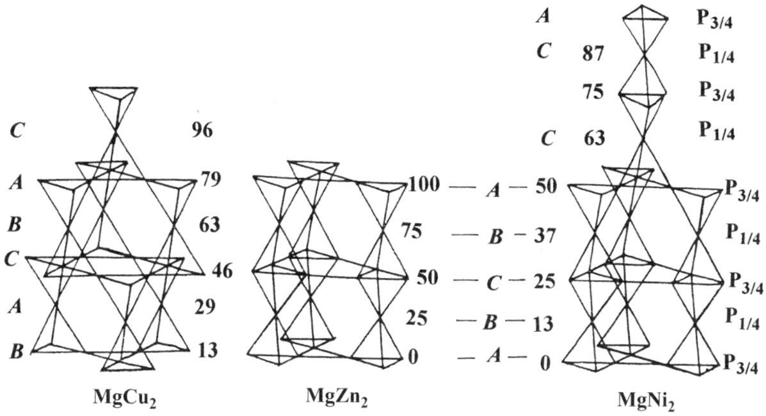

Figure 9.42. The frameworks of tetrahedra for Laves phases, $MgCu_2$, $MgZn_2$, and $MgNi_2$.

The smaller M atoms have *CN* 12 with an icosahedral arrangement of equal numbers of each metal as neighbors. Each of these structures will be discussed more fully in sections to follow.

The structures of the Laves phases, AM_2, are determined by size to a great extent. The suitable radius ratio (r_A/r_M) is given as 1.2, but it varies in the range 1.06–1.67. Using radii for *CN* 12 the ratios are 1.25 for $MgCu_2$, 1.17 for $MgZn_2$, and 1.28 for $MgNi_2$. The size considerations are much different from those for many structures we have covered where the relative sizes of **P**, **O**, and **T** sites are important. For Laves phases, the smaller M metals form the framework of shared tetrahedra with larger A metal atoms in the large truncated tetrahedra formed by shared M_4 tetrahedra. As the **T** and **O** sites are vacant, we are dealing only with **P** layers.

It is very interesting that size is not the only important factor. For $MgZn_2$ and $MgNi_2$ the first five **P** layers of Zn or Ni are in the same *AB CB A* sequence, giving two cages for Mg stacked along the packing direction. The *AB CB* sequence continues for $MgZn_2$, but changes to *AC BC* for $MgNi_2$. There must be long-distance interactions. For mixed Laves phases such as A(M′M″), where M′ and M″ are Cu, Ag, Zn, Al, or Si, an increase in electron concentration favors the Laves phases in the order $MgCu_2$, $MgNi_2$, and $MgZn_2$. The $MgCu_2$ structure is favored over a range from 1.33 (for $MgCu_2$) to about 1.8 electrons per atom. $MgZn_2$ is favored above about 1.9 electrons per atom. The $MgNi_2$ structure occurs for intermediate electron/atom ratios.

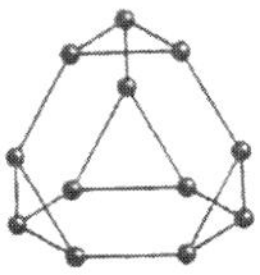

Figure 9.43. The truncated tetrahedron, the cage for Mg, in Laves phases.

9.4.1. The $3{\cdot}2\mathbf{P}^{\mathrm{Mg}}\mathbf{T}^{\mathrm{Mg}}\mathbf{T}^{\mathrm{Cu_4}}$ Crystal Structure of $MgCu_2$

The unit cell of $MgCu_2$ is cubic with eight molecules (24 atoms) per cell, $\mathbf{O}_h^7$, $Fd3m$, $a_o = 7.027$ Å. At least 200 intermetallic compounds have the $MgCu_2$ structure, and some of these are listed in Table 9.9. The rather complex structure of $MgCu_2$ can be visualized in terms of a simplification by treating a tetrahedral group of four Cu atoms as one unit. The Mg atoms form a face-centered framework. This cell is divided into octants. Mg atoms also occupy two of the octants arranged diagonally across the top of the cell and two octants arranged along the opposite diagonal of the bottom of the cell. These four Mg atoms form a tetrahedron. The framework of Mg atoms and those in four octants correspond to the diamond structure (**3·2PT**). Those in the octants correspond to the sites of one of the **T** layers of the **PTOT** sequence. The other four octants of the cube are occupied by Cu_4 tetrahedra. The centers of these tetrahedra correspond to the sites of the second **T** layer (see Figure 9.44). The face-centered Mg framework corresponds to **3P**. Mg atoms occupy $\mathbf{T}^+$ positions and Cu_4 units occupy $\mathbf{T}^-$ positions giving $3{\cdot}3\mathbf{P}^{\mathrm{Mg}}\mathbf{T}^{\mathrm{Mg}}\mathbf{T}^{\mathrm{Cu_4}}$. This is similar to the structure of fluorite involving a *ccp* (**3P**) framework of Ca^{2+} ions with F^- ions filling both **T** layers.

TABLE 9.9. Compounds with the $MgCu_2$ structure.

δ-$AgBe_2$	$BaPd_2$	$BaPt_2$	$BaRh_2$	$BiAu_2$	$CaAl_2$	$CaPd_2$
$CaPt_2$	$CeAl_2$	$CeCo_2$	$CeFe_2$	$CeIr_2$	$CeMg_2$	$CeNi_2$
$CeOs_2$	$CePt_2$	$CeRh_2$	$CeRu_2$	$CsBi_2$	$CuBe_2$	$DyCo_2$
$DyFe_2$	$DyIr_2$	$DyMn_2$	$DyNi_2$	$DyPt_2$	$DyRh_2$	$ErCo_2$
$ErFe_2$	$ErIr_2$	$ErMn_2$	$ErNi_2$	$ErRh_2$	$EuIr_2$	ε-FeBe
$GdCo_2$	$GdFe_2$	$GdIr_2$	$GdMn_2$	$GdNi_2$	$GdPt_2$	$GdRh_2$
$HfCo_2$	$HfCr_2$	$HfFe_2$	$HfMo_2$	HfV_2	HfW_2	$HoCo_2$
$HoFe_2$	$HoIr_2$	$HoMn_2$	$HoNi_2$	$HoRh_2$	KBe_2	KBi_2
$LaAl_2$	$LaIr_2$	$LaMg_2$	$LaNi_2$	$LaOs_2$	$LaPt_2$	$LaRh_2$
$LaRu_2$	$LuCo_2$	$LuFe_2$	$LuNi_2$	$LuRh_2$	$MgCu_2$	MgNi
$MgSnCu_4$	$NaAu_2$	$NaPt_2$	$NdCo_2$	$NdCr_2$	$NdIr_2$	$NdNi_2$
$NdPt_2$	$NdRh_2$	$NdRu_2$	$NpAl_2$	$PbAu_2$	$PrCo_2$	$PrIr_2$
$PrMg_2$	$PrNi_2$	$PrOs_2$	$PrPt_2$	$PrRh_2$	$PrRu_2$	$PtMg_2$
$PuAl_2$	$PuCo_2$	$PuFe_2$	$PuMn_2$	$PuNi_2$	$PuRu_2$	$PuZn_2$
$RhBi_2$	$ScCo_2$	$ScIr_2$	$ScNi_2$	$SmCo_2$	$SmFe_2$	$SmIr_2$
$SmRu_2$	$SrIr_2$	$SrPd_2$	$SrPt_2$	$SrRh_2$	β-$TaCo_2$	$TaCr_2$
TaCuV	TaFeNi	$TbCo_2$	$TbIr_2$	$TbFe_2$	$TbIr_2$	TbMn
$TbNi_2$	$ThIr_2$	$ThMg_2$	$ThOs_2$	$ThRu_2$	$TiBe_2$	$TiCo_2$
$TiCr_2$	$TmCo_2$	$TmFe_2$	$TmIr_2$	$TmNi_2$	UAl_2	UCo_2
UFe_2	UMn_2	UNiFe	UOs_2	VIr_2	YAl_2	YCo_2
YFe_2	YIr_2	YMn_2	YNi_2	YPt_2	YRh_2	$YbCo_2$
$YbIr_2$	$YbNi_2$	$ZrCo_2$	$ZrCr_2$	$ZrFe_2$	$ZrIr_2$	$ZrMo_2$
ZrV_2	ZrW_2	$ZrZn_2$				

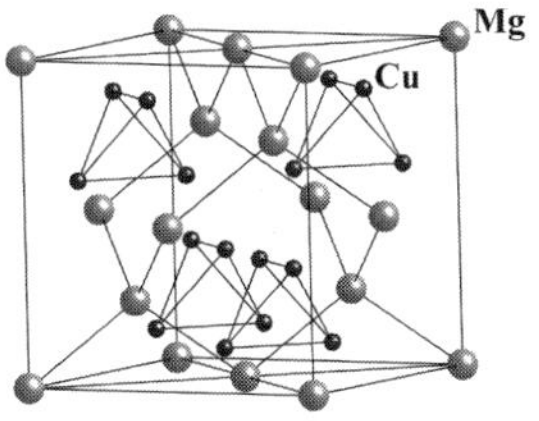

Figure 9.44. The cubic cell for $MgCu_2$ with the Cu_4 tetrahedra as units.
(*Source*: Y. Matsushita, *Chalcogenide* crystal structure data library, Version 5.5B, Institute for Solid State Physics, The University of Tokyo, Tokyo, Japan, 2004; and *CrystalMaker*, by David Palmer, *CrystalMaker* Software Ltd, Begbroke Science Park, Bldg. 5, Sandy Lane, Yarnton, Oxfordshire, OX51PF, UK.)

For a full description of the structure of $MgCu_2$ and its relationship to $MgZn_2$ and $MgNi_2$ we need to consider the framework of M atoms (Cu, Zn, or Ni). The full framework for $MgCu_2$ consists of Cu tetrahedra stacked along the packing direction, as shown in Figure 9.42. For $MgCu_2$, only apices of tetrahedra are shared. The partially filled **P** layers occupied by Cu. alternate as $\mathbf{P}_{1/4}$ and $\mathbf{P}_{3/4}$ in an *BAC* sequence (same as *ABC*, labels are chosen to agree with Figure 9.42). The $\mathbf{P}_{3/4}$ layers have hexagons with the center positions vacant (see Figure 3.10***a*** for $\mathbf{L}_{3/4}$). The heights along the packing direction and layer sequence for $MgCu_2$ are:

Height:	0	13	25	29	33	46	58	63	67	79	92	96	100
Atom:	Mg	Cu	Mg	Cu	Mg	Cu	Mg	Cu	Mg	Cu	Mg	Cu	Mg
Position:	*B*	*B*	*B*	*A*	*C*	*C*	*C*	*B*	*A*	*A*	*A*	*C*	*B*
Layers:		$\mathbf{P}_{3/4}$		$\mathbf{IIIP}_{1/4\ 1/4\ 1/4}$		$\mathbf{P}_{3/4}$		$\mathbf{IIIP}_{1/4\ 1/4\ 1/4}$		$\mathbf{P}_{3/4}$		$\mathbf{IIP}_{1/4\ 1/4\ 1/4}$	

The height separations between Cu layers ($\mathbf{P}_{1/4}$ and $\mathbf{P}_{3/4}$) are 16 or 17. The very close Mg–Cu–Mg layers are treated as triple layers (**IIIP**). The Mg atoms are above or below the vacant centers of Cu_6 hexagons of the closest $\mathbf{P}_{3/4}$ Cu layer. Mg atoms at 25 are above the vacant centers of Cu_6 hexagons of the $\mathbf{P}_{3/4}$ Cu layer at 13. Mg atoms at 33 are below the vacant centers of Cu_6 hexagons of the closest $\mathbf{P}_{3/4}$ Cu layer at 46. The cage around each Mg is a truncated tetrahedron, as shown in Figure 9.43. One of these cages has six Cu at height 13 ($\mathbf{P}^{\mathbf{B}}_{3/4}$), three Cu at 29 ($\mathbf{P}^{\mathbf{A}}_{1/4}$) and three Cu at 46 ($\mathbf{P}^{\mathbf{C}}_{3/4}$). In addition, Mg has four Mg neighbors arranged tetrahedrally through the triangular faces. The full cell in Figure 9.45 shows the cell and the icosahedra. There are 12 close-packed layers repeating or six layers repeating, counting a triple layer as one, giving the notation $\mathbf{3{\cdot}2P}^{\mathbf{Cu}}_{\mathbf{3/4}}\ \mathbf{IIIP}^{\mathbf{Mg}\ \mathbf{Cu}\ \mathbf{Mg}}_{\mathbf{1/4}\ \mathbf{1/4}\ \mathbf{1/4}}$. The packing direction is along the body diagonal of the cube.

9.4.2. The $\mathbf{3{\cdot}2P}^{\mathbf{Be}}_{\mathbf{3/4}}\mathbf{IIIP}^{\mathbf{Be}\ \mathbf{Be}\ \mathbf{Au}}_{\mathbf{1/4}\ \mathbf{1/4}\ \mathbf{1/4}}$ Crystal Structure of $AuBe_5$

The crystal structure of $AuBe_5$ is derived from the $MgCu_2$ structure by forming two arrays of Mg sites. One array is occupied by Au and the other by Be_I. Be_{II} atoms occupy the Cu sites. Au has ***CN*** 16, 12 Be at 2.78, and four Be at 2.90 Å. Be_I has 12 Be_{II} at 2.78 and four Au at 2.90 Å. Be_{II} is icosahedrally surrounded by three Au, three Be_I, and six Be_{II}. The notation is $\mathbf{3{\cdot}2P}^{\mathbf{Be}}_{\mathbf{3/4}}\mathbf{IIIP}^{\mathbf{Be}\ \mathbf{Be}\ \mathbf{Au}}_{\mathbf{1/4}\ \mathbf{1/4}\ \mathbf{1/4}}$.

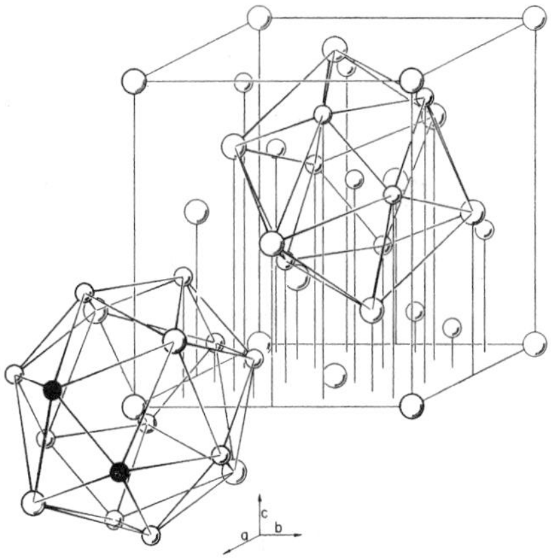

Figure 9.45. A pictoral view of the $MgCu_2$ structure shows the *CN* 16 for Mg at the corners of the cell and the icosahedron (*CN* 12) about Cu within the cell.

(*Source*: W.P. Pearson, *The Crystal Chemistry and Physics of Metals and Alloys*, Wiley-Interscience, New York, 1972, p. 656.)

9.4.3. The $3\cdot2\mathbf{P}^{Cu}_{3/4}\mathbf{IIP}^{Mg\ Cu\ Sn}_{1/4\ 1/4\ 1/4}$ Crystal Structure of $MgSnCu_4$

The crystal structure of $MgSnCu_4$ is derived from the $MgCu_2$ structure by dividing Mg sites into two sublattices. These sites are occupied by Mg at 0, 0, 0 and Sn at 1/4, 1/4, 1/4. Cu is surrounded icosahedrally by three Mg and three Sn at 2.92 and six Cu at 2.49 Å. Mg has *CN* 16 with 12 Cu at 2.92 and four Sn at 3.05 Å. Sn also has *CN* 16 with 12 Cu and four Mg at 3.05 Å. The notation is $\mathbf{3\cdot2P}^{Cu}_{3/4}\mathbf{IIP}^{Mg\ Cu\ Sn}_{1/4\ 1/4\ 1/4}$.

9.4.4. The $4\mathbf{P}^{Zn}_{3/4}\mathbf{IIP}^{Mg\ Zn\ Mg}_{1/4\ 1/4\ 1/4}$ Crystal Structure of $MgZn_2$

The unit cell of $MgZn_2$ is hexagonal with four molecules, $P6_3/mmc$, $\mathbf{D}^4_{6h}$, $a = 5.223$, and $c = 8.566$Å. As for $MgCu_2$, there are Zn_4 tetrahedra, but they are joined along the packing directions, alternatively sharing faces and apices (Figure 9.42). The larger Mg atoms occupy the cages formed by six tetrahedra. These truncated tetrahedra have Mg surrounded by four Mg at 3.17 and 3.20 Å and 12 Zn at about 3.04 Å. The Zn atoms have *CN* 12 with six Mg and six Zn atoms forming an icosahedron. Figure 9.42 shows the stacking of $\mathbf{P}_{3/4}$ layers of Zn and the triple layers of Mg, Zn, and Mg. The heights along c are:

Height:	0	6	25	44	50	56	75	94	100	106
Atom:	Zn	Mg	Zn	Mg	Zn	Mg	Zn	Mg	Zn	Mg
Position:	*A*	*B*	*B*	*B*	*A*	*C*	*C*	*C*	*A*	*B*
Layer:			$\mathbf{P}_{3/4}$		$\mathbf{IIP}_{1/4\ 1/4\ 1/4}$		$\mathbf{P}_{3/4}$		$\mathbf{IIP}_{1/4\ 1/4\ 1/4}$	

The Zn atoms are in the sequence *AB AC*, with eight **P** layers repeating or four layers repeating, counting a triple layer as one, gives the notation $\mathbf{4P}^{Zn}_{3/4}\mathbf{IIP}^{Mg\ Zn\ Mg}_{1/4\ 1/4\ 1/4}$. The close-packed layers for $MgZn_2$ are shown in Figure 9.46***a***. The positions of the atoms in the hexagonal cell are shown in Figure 9.46***b***. At least 170 intermetallic compounds have the $MgZn_2$ structure, and some of these are included in Table 9.10.

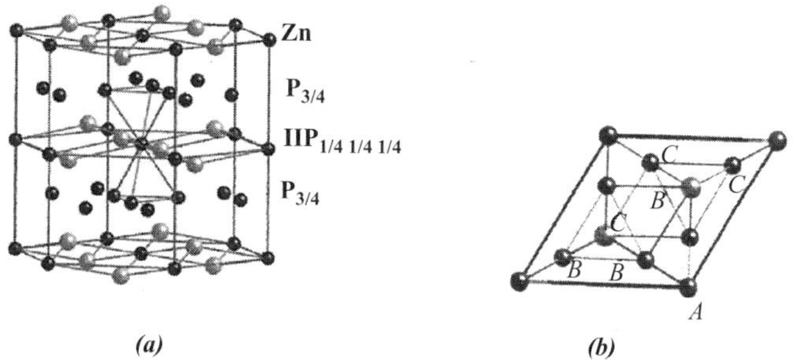

Figure 9.46. (*a*) The alternating layers for $MgZn_2$. Two Zn_4 tetrahedra sharing an apex are outlined. (*b*) A projection of the hexagonal cell along *c*.

(*Source*: Y. Matsushita, *Chalcogenide* crystal structure data library, Version 5.5B, Institute for Solid State Physics, The University of Tokyo, Tokyo, Japan, 2004; and *CrystalMaker*, by David Palmer, *CrystalMaker* Software Ltd, Begbroke Science Park, Bldg. 5, Sandy Lane, Yarnton, Oxfordshire, OX51PF, UK.)

9.4.5. The $4P^{Mn}_{3/4}IIIP^{Lu}_{1/4}\,{}^{Mn}_{1/4}\,{}^{Mn}_{1/4}$ Crystal Structure of $LuMn_5$

The unit cell of $LuMn_5$ is hexagonal, $a = 5.186$ and $c = 8.566$Å with two molecules. The $LuMn_5$ structure is related to that of $MgZn_2$ as $AuBe_5$ is related to $MgCu_2$. The Mg sites of $MgZn_2$ are occupied equally by Lu and Mn. Zn sites are occupied by Mn. The notation is $4P^{Mn}_{3/4}IIIP^{Lu}_{1/4}\,{}^{Mn}_{1/4}\,{}^{Mn}_{1/4}$.

9.4.6. The $8P^{Ni}_{3/4}IIIP^{Mg}_{1/4}\,{}^{Ni}_{1/4}\,{}^{Mg}_{1/4}$ Crystal Structure of $MgNi_2$

The unit cell of $MgNi_2$ is hexagonal with eight molecules, $a = 4.815$ and $c = 15.80$Å. Similar to $MgZn_2$, $MgNi_2$ involves Ni tetrahedra sharing faces and apices (Figure 9.42). The larger Mg atoms occupy the truncated tetrahedra formed by the framework of Ni_4 tetrahedra. Mg has *CN* 16, a truncated tetrahedron formed by four Mg and 12 Ni. Ni has *CN* 12 with six Mg and six Ni forming an icosahedron. The heights along *c* are:

Height:	0	9	13	16	25	34	37	41	50	59	63	66	75	84	87	91
Atom:	Ni	Mg	Ni	Mg	Ni	Mg	Ni	Mg	Ni	Mg	Ni	Mg	Ni	Mg	Ni	Mg
Position:	*A*	*A*	*B*	*C*	*C*	*C*	*B*	*A*	*A*	*A*	*C*	*B*	*B*	*B*	*C*	*A*
Layers:	$P_{3/4}$	$IIIP_{1/4\,1/4\,1/4}$			$P_{3/4}$	$IIIP_{1/4\,1/4\,1/4}$			$P_{3/4}$	$IIIP_{1/4\,1/4\,1/4}$			$P_{3/4}$	$IIIP_{1/4\,1/4\,1/4}$		

Table 9.10. Compounds with the $MgZn_2$ structure.

$BaMg_2$	$CaAg_2$	$CaCd_2$	$CaLi_2$	$CaMg_2$	$CrBe_2$	$CrMg_2$
β-$FeBe_2$	$HfFe_2$	$HfOs_2$	$HfRe_2$	KNa_2	KPb_2	$LiOs_2$
$LuRu_2$	$MgZn_2$	$MnBe_2$	$MoBe_2$	$MoFe_2$	$NbFe_2$	$NbMn_2$
$ReBe_2$	$ScMn_2$	$ScOs_2$	$SrMg_2$	$TaCo_2$	TaCoCr	TaCoTi
TaCoV	$TaCr_2$	$TaFe_2$	$TaMn_2$	$ThMn_2$	$TiCr_2$	$TiFe_2$
$TiMn_2$	UNi_2	VBe_2	VNi_2	WBe_2	WFe_2	$ZrAl_2$
$ZrCr_2$	$ZrIr_2$	$ZrMn_2$	$ZrRe_2$	$ZrRu_2$	$ZrOs_2$	ZrV

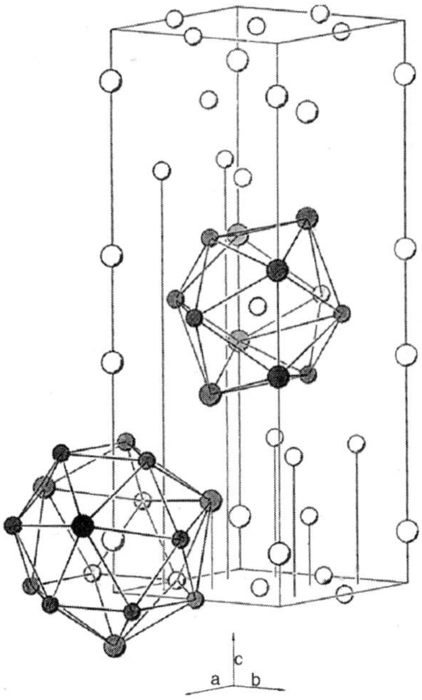

Figure 9.47. A pictorial view of the $MgNi_2$ structure shows the ***CN*** 16 for Mg and ***CN*** 12 for Ni.

(*Source*: W.P. Pearson, *The Crystal Chemistry and Physics of Metals and Alloys*, Wiley-Interscience, New York, 1972, p. 659.)

The sequence of Ni atoms is *AB CB AC BC*, giving eight layers repeating, counting a triple layer as one layer, and has the notation $\mathbf{8P^{Ni}_{3/4}\,\mathbb{IP}^{Mg}_{1/4}\,{}^{Ni}_{1/4}\,{}^{Mg}_{1/4}}$. Figure 9.47 shows the arrangements of both polyhedra in the cell.

Chapter 10

Crystal Structures of Silica and Metal Silicates

10.1. Structures of Silica [SiO_2]

The formula for silica is very simple—SiO_2. The silica structures (and those of metal silicates) have only SiO_4 tetrahedra, and the 1:2 ratio of SiO_2 requires that each oxygen atom is shared by two tetrahedra in silica. Even though the basis of the structures is very simple, there are three polymorphic structures of silica, and each of these has a low temperature form (α) and a high temperature form (β). The three polymorphic structures and their temperature ranges are shown below.

$$\textbf{Quartz} \xleftrightarrow{870^\circ C} \textbf{Tridymite} \xleftrightarrow{1470^\circ C} \textbf{Cristobalite} \xleftrightarrow{1710^\circ C} \textbf{melting point}$$

The three forms are not readily interconvertible, and all three are found as minerals. Only quartz is common. Liquid silica does not readily crystallize but solidifies to a glass. Glassy silica softens on heating rather than melting at a sharp melting point as crystals do. It can be softened (~ 1,500°C) to form crucibles, boats, etc., that are useful as containers because of the small coefficient of thermal expansion, and they are usable over a wide temperature range. Silica glass is found in nature as obsidian. The transitions from low to high temperature modifications are: α-β-quartz, 573°; α-β-tridymite, 110–180°; and α-β-cristobalite, 218 ± 2°C. These transformations can be achieved even though the transition temperatures are outside the range of stabilities of tridymite and cristobalite. The differences of the structures of cristobalite and tridymite are similar to the differences between the polymorphic forms of ZnS, zinc blende (**3·2PT**), and wurtzite (**2·2PT**), and there are minor variations such as slight rotations for α to β transformations. The quartz to tridymite transformation requires breaking Si—O—Si bonds to change the linking of the SiO_4 tetrahedra.

Interesting models of many close-packed structures can be made by stacking octahedra (see ReO_3, Section 5.3.7) or tetrahedra (see Figures B.4–B.6). Because the forms of silica and silicates are formed by packing tetrahedra it is helpful to examine patterns involving joined tetrahedra. A.F. Wells examined various different connections of tetrahedra

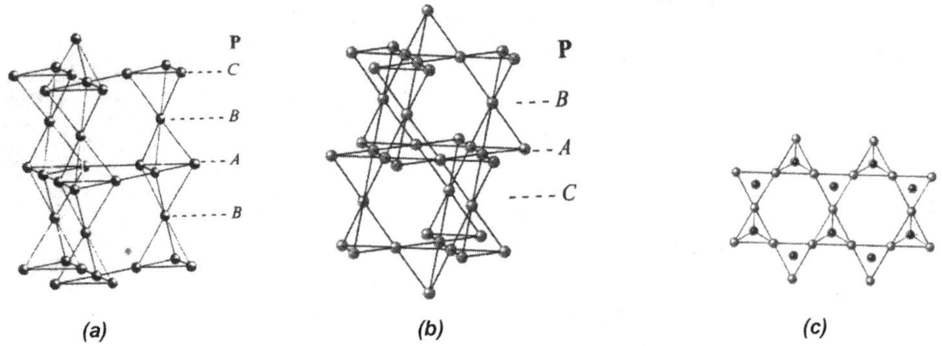

Figure 10.1. Arrangements of tetrahedra between $\mathbf{P}_{3/4}$ and $\mathbf{P}_{1/4}$ layers. (*a*) Tetrahedra sharing apices and bases. (*b*) Tetrahedra sharing apices. (*c*) Hexagons of tetrahedra in $\mathbf{P}_{3/4}$ layers for (*b*).

in close-packed arrangements involving vacancies in packing sites for Laves phases (intermetallic phases, see Section 9.4). This is a general treatment, not just the case involving the sharing of tetrahedral apices encountered for forms of silica.

Figure 10.1*a* shows tetrahedra sharing alternatively in one direction apices and bases as in $MgZn_2$ (Section 9.4.4). The layers of the bases ($\mathbf{P}_A$ and $\mathbf{P}_C$ positions) are three-quarters occupied, and the layers of apices ($\mathbf{P}_B$ positions) are one-quarter occupied. The tetrahedra do not share apices in the $\mathbf{P}_{3/4}$ layers. The hexagonal arrangement of six atoms in the $\mathbf{P}_{3/4}$ layers are from edges of three tetrahedra. In Figure 10.1*b* the **P** layers are in the *BAC* sequence as in β-cristobalite (Section 10.1.2). The atoms in the one-quarter occupied layers join tetrahedra with bases in layers in layers three-quarters occupied, but the bases are not shared by two tetrahedra. The tetrahedra share apices in the $\mathbf{P}_{3/4}$ layers forming hexagons as shown in Figure 10.1*c*, with tetrahedra alternatively pointing up and down. Now let us consider the structures of the modifications of silica where tetrahedra share only apices. All of these structures result from joining SiO_4 tetrahedra sharing all apices.

10.1.1. The $4{\cdot}2\mathbf{P}_{3/4}\mathbf{T}^{+}_{1/4}\mathbf{P}_{1/4}\mathbf{T}^{-}_{1/4}$ Crystal Structure of β-Tridymite [SiO_2]

The hexagonal structure of β-tridymite shown in Figure 10.2 was studied by R.E. Gibbs and J.E. Fleming over many years (1927–1960). The space group is $\mathbf{D}^4_{6h}$, $P6_3/mmc$, $a_o = 5.03$, and $c_o = 8.22$ Å above 200°C. The oxygen atoms occupy three-quarters of sites in $\mathbf{P}_A$ layers, separated alternatively by $\mathbf{P}_B$ and $\mathbf{P}_C$ layers (one-quarter occupied), in the sequence *ABAC*. The Si atoms occupy one-quarter of sites in **T** layers such that each oxygen atom of the SiO_4 tetrahedra is shared by two Si atoms. The relative heights and positions are:

Height:	0	1/16	1/4	7/16	2/4	9/16	3/4	15/16	100
Position:	$\mathbf{P}_{3/4}$	$\mathbf{T}^{+}_{1/4}$	$\mathbf{P}_{1/4}$	$\mathbf{T}^{-}_{1/4}$	$\mathbf{P}_{3/4}$	$\mathbf{T}^{+}_{1/4}$	$\mathbf{P}_{1/4}$	$\mathbf{T}^{-}_{1/4}$	$\mathbf{P}_{3/4}$
	A	*B*	*B*	*B*	*A*	*C*	*C*	*C*	*A*

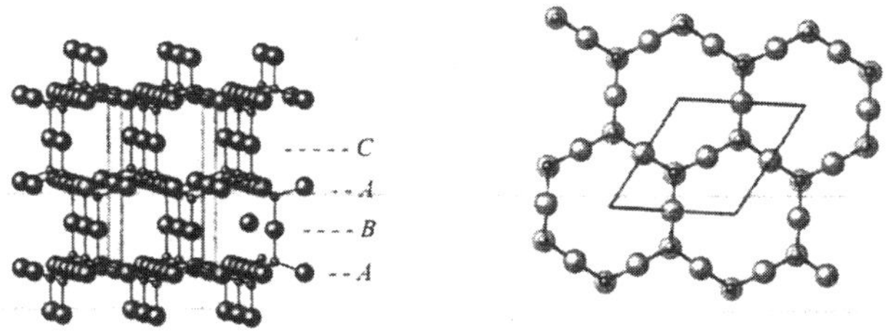

(a) (b)

Figure 10.2. (*a*) Packing layers in β-tridymite. Small Si atoms are seen on the right in $\mathbf{T}^+$ sites between $\mathbf{P}_A$ and $\mathbf{P}_B$ and in $\mathbf{T}^-$ sites between $\mathbf{P}_B$ and the next $\mathbf{P}_A$. (*b*) A projection along the *c* axis of the unit cell for β-tridymite.

(*Source*: *CrystalMaker*, by David Palmer, *CrystalMaker* Software Ltd., Begbroke Science Park, Bldg. 5, Sandy Lane, Yarnton, Oxfordshire, OX51PF, UK.)

Note how close together the $\mathbf{T}^-$ and $\mathbf{T}^+$ sites are on each side of a **P** layer. However, the sites occupied are staggered. The notation is $\mathbf{4{\cdot}2P_{3/4}T^+_{1/4}P_{1/4}T^-_{1/4}}$. The index 4 shows that the **P** layers are in the double hexagonal sequence, a more complicated sequence than for *hcp* (**2**) or *ccp* (**3**), and the product gives the total of eight layers in the repeating unit.

10.1.2. The $\mathbf{3{\cdot}4P_{3/4}T_{1/4}P_{1/4}T_{1/4}}$ Crystal Structure of β-Cristobalite [SiO_2]

The packing of oxygen atoms in the structure of β-cristobalite is similar to that of the cubic **3·2PT** structure of zinc blende (ZnS) or diamond. The space group is O_h^7, $Fd\bar{3}m$, $a_o = 7.16$Å, at 290°C. The structure as shown in Figure 10.3 was studied very early (1925) by R.W.G. Wyckoff. In 1958, T. Tokula used synthetic crystals to find that the **P** layers are distorted slightly from the positions found earlier. The tetrahedra point in one direction along the packing direction, the body diagonal. The **P** layers containing the bases of tetrahedra are three-quarters filled and those containing apices are one-quarter filled. Each oxygen atom is shared by two Si atoms. The notation is $\mathbf{3{\cdot}4P_{3/4}T_{1/4}P_{1/4}T_{1/4}}$.

β-Cristobalite can be represented by a face-centered cubic (*ccp*) arrangement of Si atoms in the Zn positions of the **3·2PT** structure of ZnS with SiO_4 tetrahedra in four octants of the cube. This representation emphasizes the high symmetry of β-cristobalite, the highest symmetry of the polymorphs of silica.

10.1.3. The $\mathbf{3{\cdot}2P_{2/4}T_{1/4}}(C_2)$ Crystal Structure of β-Quartz [SiO_2]

The structure of β-quartz is unique and much different from those of tridymite and cristobalite. It does not follow the rules for the **PTOT** system, but the notation can be applied with some qualification. In general, for close-packed structures between packing layers there are two **T** layers, and half way between the **P** layers there is an **O** layer. The

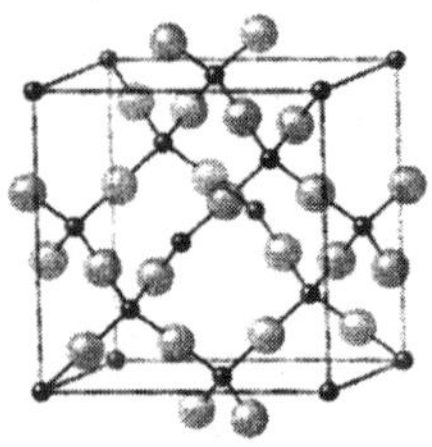

Figure 10.3. The cubic unit cell of β-cristobalite. Oxygen atoms outside of the cell are shown on top, bottom and right and left side faces.

(*Source: CrystalMaker*, by David Palmer, *CrystalMaker* Software Ltd., Begbroke Science Park, Bldg. 5, Sandy Lane, Yarnton, Oxfordshire, OX51PF, UK.)

tetrahedra point up ($\mathbf{T}^+$) or down ($\mathbf{T}^-$) with one atom of one **P** layer and three atoms of the other **P** layer forming the base of the tetrahedron. A C_3 axis of the tetrahedron is in the packing direction. For β-quartz (see Figure 10.4***a***) two atoms of each **P** layer form the opposite edges of a tetrahedron. The center of the tetrahedron is halfway between the **P** layers. For this structure there is only one **T** layer between adjacent **P** layers and because the **P** layers are only one-half filled, there is no **O** layer! Along the packing direction the axis of rotation of each tetrahedron is C_2; hence, we describe these sites as $\mathbf{T}(C_2)$. The usual **T** site is $\mathbf{T}(C_3)$. The half-filled **P** layers are in a $\mathbf{P}_A\mathbf{P}_B\mathbf{P}_C$ sequence. The $\mathbf{T}(C_2)$ sites are not aligned at *A*, *B*, or *C* positions, and these layers are one-quarter filled. The notation is $\mathbf{3{\cdot}2P_{2/4}T_{1/4}}(C_2)$. The heights and positions are:

Height:	0	17	33	50	67	83	100
Atom:	Si	O	Si	O	Si	O	Si
Position:	$\mathbf{T}^{B}_{1/4}$	$\mathbf{P}^{A}_{2/4}$	$\mathbf{T}^{C}_{1/4}$	$\mathbf{P}^{B}_{2/4}$	$\mathbf{T}^{A}_{1/4}$	$\mathbf{P}^{C}_{2/4}$	$\mathbf{T}^{B}_{1/4}$

The structures of tridymite and cristobalite are more open than that of quartz as shown by their densities: quartz, 2.655; tridymite, 2.26, and cristobalite, 2.32 g/cm^3. Quartz crystals are optically active, that is,

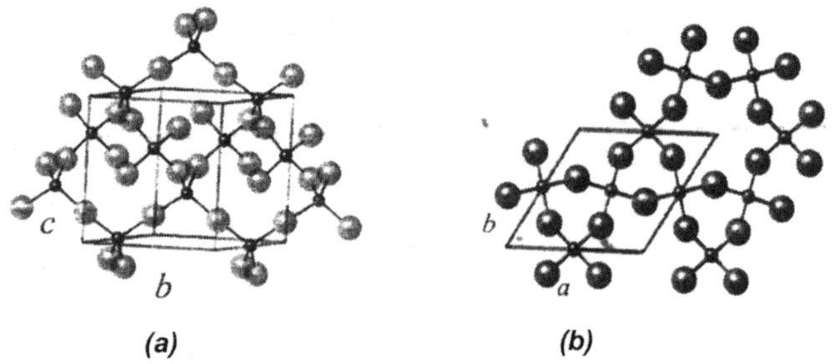

Figure 10.4. (***a***) The unit cell of β-quartz showing Si in tetrahedra with a C_2 axis aligned with the *c* axis. (***b***) A projection along *c* of the β-quartz unit cell.

(*Source*: *CrystalMaker*, by David Palmer, *CrystalMaker* Software Ltd., Begbroke Science Park, Bldg. 5, Sandy Lane, Yarnton, Oxfordshire, OX51PF, UK.)

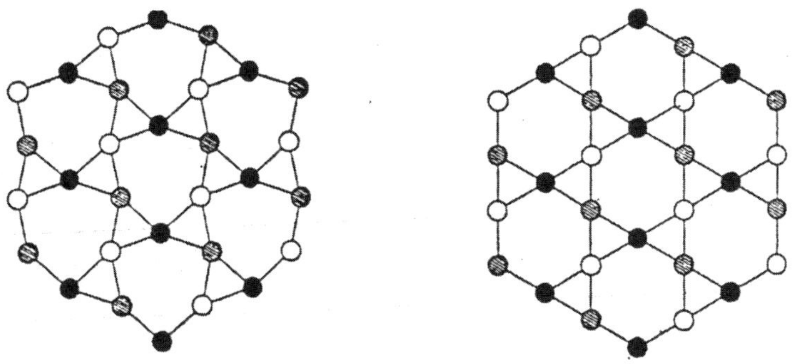

(a) (b)

Figure 10.5. Projections of Si positions only for (*a*) α-quartz and (*b*) β-quartz.

each crystal is either right handed or left handed (individual crystals rotate the plane of linearly polarized light right or left). The optical activity results from a spiral arrangement of linked SiO_4 units through the crystal. The hexagonal cell contains three molecules, $\mathbf{D}_6^4$ ($P6_222$) and $\mathbf{D}_6^5$ ($P6_422$) for the enantiomorphic pair, $a_o = 5.01$ and $c_o = 5.47\text{Å}$, at 600°C. A projection of the unit cell is shown in Figure 10.4*b*. α- and β-quartz are optically active, and tridymite and cristobalite are not active. If a single crystal of α-quartz is heated cautiously above 575°C, gradually it is transformed into a single crystal of β-quartz, which is still optically active with the *same handedness*. The symmetry of the crystal changes from threefold to sixfold. We can see this change in symmetry of the silicon positions in Figure 10.5.

Quartz is one of the most common minerals. Large crystals of α-quartz are used for lenses and optical windows because it is transparent to infrared and ultraviolet light. Quartz crystals exhibit piezoelectricity, the ability to develop an electric charge when subjected to external pressure. This is useful in pressure gauges. Quartz plates have long been used to control radio frequencies. The quartz plate in a "quartz" watch regulates time by vibrating at a set frequency when charged by a battery. White sand, usually quartz sand of high purity, is used for producing glass. Even though quartz is a common mineral, over 700 tons of optical-quality quartz crystals are grown each year. Chips of quartz dissolve in H_2O and form single crystals by a hydrothermal process when heated in a "bomb" at pressure over 20,000 atm.

Colored quartz crystals are common gems. Amethyst is a semiprecious gem containing a trace of Fe^{3+}. Citrine is less common and also contains Fe^{3+}. Amethyst can be converted to citrine by heating at 550°C. This supplements the limited supply of citrine using low-quality amethyst. Rose quartz, pink to deep rose-red, contains traces of Ti and sometimes contains tiny crystals of rutile.

10.1.4. The $2{\cdot}2PO_{1/2}(t)$ Crystal Structure of Stishovite [SiO_2]

The mineral stishovite, SiO_2, was found in the Canyon Diablo meteorite. It is the form of silica formed at very high pressure. The crystal structure

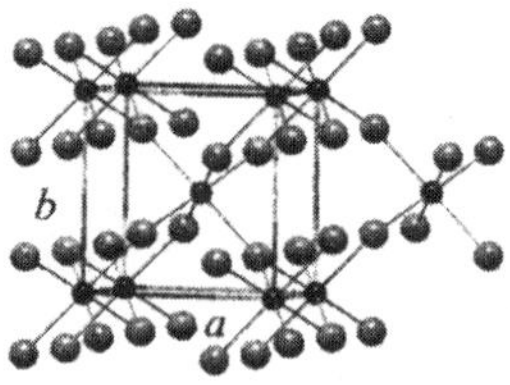

Figure 10.6. The tetragonal cell of stishovite, SiO_2, isolated SiO_6 octahedra are in the center and right of the cell.

(*Source*: *CrystalMaker*, by David Palmer, *CrystalMaker* Software Ltd., Begbroke Science Park, Bldg. 5, Sandy Lane, Yarnton, Oxfordshire, OX51PF, UK.)

has been studied at high pressure and atmospheric pressure. Crystals are tetragonal, $\mathbf{D}_{4h}^{14}$, $P4_2/mnm$, $a = 4.1787$, and $c = 2.6669$Å. It is the rutile structure with Si in octahedral sites (Figure 10.6). Octahedra share edges along the c axis. Along the packing direction (b axis) the oxygen atoms are in A and B positions with Si in C positions. The notation is $\mathbf{2{\cdot}2PO_{1/2}(t)}$. The density of stishovite is much greater (4.28 g/cm^3) than the polymorphic structure containing SiO_4 tetrahedra.

10.2. General Patterns of Structures of Silicates

Metal silicates provide an interesting array of structural types. The silicates are also of great technical importance. The crust of the earth is made up primarily of metal silicates. Silicon and oxygen account for almost 75% by weight of the earth's crust. Moreover, the eight most abundant elements in the earth's crust usually occur in silicates. Although most of these abundant elements are found in silicates in the igneous rocks, these are not their primary ores. Sedimentary rocks are more important as ores. Thus, we take advantage of the concentration of the elements by the weathering process in forming sedimentary rocks.

All silicates contain tetrahedral SiO_4 units, which may be linked together by sharing corners, but never by sharing edges or faces. When other cations (alkali or alkaline earth metal ions, Fe^{2+}, etc.) are present in the structure, they usually share oxygens of the SiO_4 groups, to give an octahedral configuration around the cation. Large cations such as K and Ca usually have higher coordination numbers. Aluminum commonly occupies octahedral sites, but it can also replace Si in the SiO_4 tetrahedra, requiring the addition of another cation or the replacement of one by another of higher charge to maintain charge balance.

10.2.1. Small Anions

Some silicates contain small anions containing one to six Si atoms. The simplest silicate anion is the orthosilicate ion, $SiO_4{}^{4-}$. In the orthosilicates forsterite, $Mg_2[SiO_4]$, and olivine, $(Mg, Fe)_2[SiO_4]$, stacking of the

SiO_4 tetrahedra around the divalent cation produces an octahedral configuration (see Section 10.3.1). Forsterite has oxide ions *hcp* (**P** layers), one-eighth of the **T** sites occupied by Si, and half of the **O** sites occupied by Mg. Topaz, $[Al(F, OH)]_2SiO_4$ (Section 10.3.2) contains SiO_4 tetrahedra and $Al(O, F)_6$ octahedra. Orthosilicates (SiO_4^{4-}) are the most compact of the silicate structures.

The uncommon mineral phenakite (phenacite), $Be_2[SiO_4]$, contains the very small Be^{2+} ions in tetrahedral sites (Section 10.3.8). In zircon, $Zr[SiO_4]$ (Section 10.3.3), the large Zr^{4+} ion has ***CN*** 8. The garnets, $M_3^{II}M_2^{III}[SiO_4]_3$, where M^{II} is Ca, Mg, or Fe and M^{III} is Al, Cr, or Fe, have a more complex structure, in which the M^{II} ions have ***CN*** 8 and the M^{III} ions have ***CN*** 6. Sillimanite (Section 10.3.4), kyanite (Section 10.3.5), and andalusite (Section 10.3.6) have the same composition, Al_2SiO_5 or $Al_2O_3 \cdot SiO_2$, and all contain SiO_4 and AlO_6 groups, but in sillimanite Al also has ***CN*** 4 and in andalusite Al also has ***CN*** 5.

There are few examples of silicates containing short chains of SiO_4 tetrahedra. The $Si_2O_7^{6-}$ anion is encountered in thortveitite, $Sc_2[Si_2O_7]$ (Section 10.3.10), akermanite, $Ca_2MgSi_2O_7$ (Section 10.3.11), and in a few other minerals of greater complexity. In condensing two SiO_4^{4-} tetrahedra one O^{2-} is excised and the two tetrahedra share the O at one vertex, giving $Si_2O_7^{6-}$. Gehlenite, $Ca_2Al(Si,Al)_2O_7$, contains X_2O_7 anions with some Si replaced by Al. Ca and other Al ions form octahedra. There are very few examples of short chains containing more than two SiO_4 groups. Anions consisting of rings of SiO_4 groups are encountered more commonly. Rings of three tetrahedra (to give six-membered rings) containing the anion $[Si_3O_9]^{6-}$ (see Figure 10.7) are encountered in wollastonite, $Ca_3[Si_3O_9]$, and benitoite, $BaTi[Si_3O_9]$. The anion $[Si_6O_{18}]^{12-}$, a ring of six tetrahedra, is found in the mineral beryl, $Al_2Be_3[Si_6O_{18}]$ (Section 10.3.12). In beryl, an oxygen is shared by one Si, one Al (***CN*** 6), and one Be (***CN*** 4). Beryl is the only important Be ore. Gem quality crystals of beryl contain Cr^{3+} in emerald, Fe^{2+} in blue aquamarine, and Mn in morganite (pink).

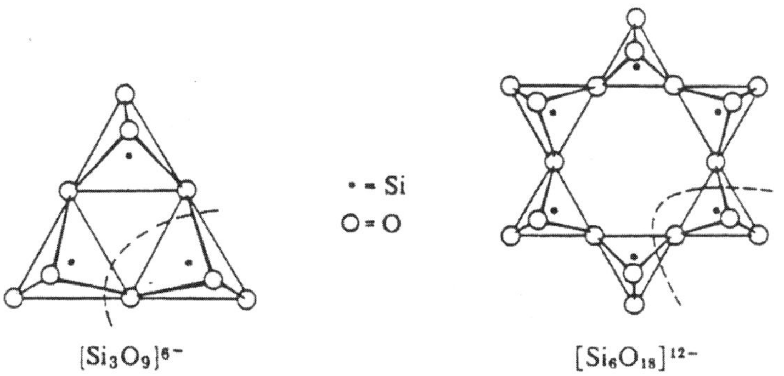

Figure 10.7. Cyclic silicate anions. The repeated SiO_3^{2-} unit is marked by a dashed line.

(*Source*: B.E. Douglas, D.H. McDaniel, and J.J. Alexander, *Concepts and Models of Inorganic Chemistry*, 3rd ed., Wiley, New York, 1994, p. 241.)

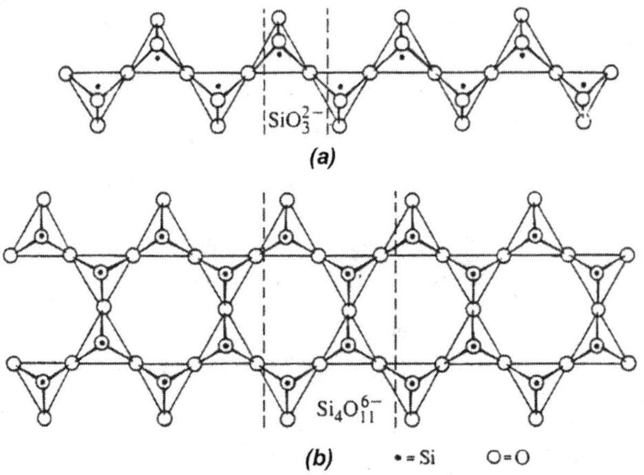

Figure 10.8. Chains of SiO_4 tetrahedra in (*a*) pyroxenes (repeating unit SiO_3^{2-}) and (*b*) amphiboles (repeating unit $Si_4O_{11}^{6-}$). The tetrahedra for ***b*** are viewed down one O—Si bond, where Si (•) is below the O. The repeating units are marked by dashed lines.

(*Source*: B.E. Douglas, D.H. McDaniel, and J.J. Alexander, *Concepts and Models for Inorganic Chemistry*, 3rd ed., Wiley, New York, 1994, p. 242.)

10.2.2. Chains

Each SiO_4 tetrahedron can share two oxygens to form long single chains (Figure 10.8). The formula of the anion can be represented by the repeating unit, SiO_3^{2-}. Minerals of this type are called the *pyroxenes,* which include enstatite, $Mg[SiO_3]$, diopside, $CaMg[SiO_3]_2$ (Section 10.3.17), and spodumene, $LiAl[Si_2O_6]$ (Section 10.3.16). The nonbridging oxygens are shared with Mg^{2+} (***CN*** 6) in enstatite. In diopside, the Mg has ***CN*** 6 and the Ca has ***CN*** 8. Both Li and Al are six-coordinate in spodumene.

A class of minerals known as the *amphiboles* contain double chains of SiO_4 tetrahedra (Figure 10.7) joined to form rings of six tetrahedra. The repeating unit is $Si_4O_{11}^{6-}$, with half of the silicon atoms sharing three oxygens with other Si atoms and half sharing only two oxygens with other Si atoms. The amphiboles always contain some OH^- groups associated with the metal ion. Tremolite, $Ca_2Mg_5[(OH)_2|(Si_4O_{11})_2]$ (Section 10.3.18), is a typical amphibole. The vertical bar in the formula indicates that the OH^- is not a part of the silicate framework.

The term **asbestos** was originally reserved for fibrous amphiboles such as tremolite, but now the term includes fibrous varieties of layer structures such as chrysotile, $Mg_3[(OH)_4|Si_2O_5]$. Asbestos was widely used for insulators at high temperature, for partitions and ceilings, roof shingles, and as a water-suspension to form mats for filtering in the laboratory. Now the uses are limited because of the hazard of lung cancer caused by inhaling the fibers.

10.2.3. Sheets

Each silicon atom can share three oxygen atoms of the SiO_4 units with other Si atoms, to give large sheets. Crosslinking similar to that in the

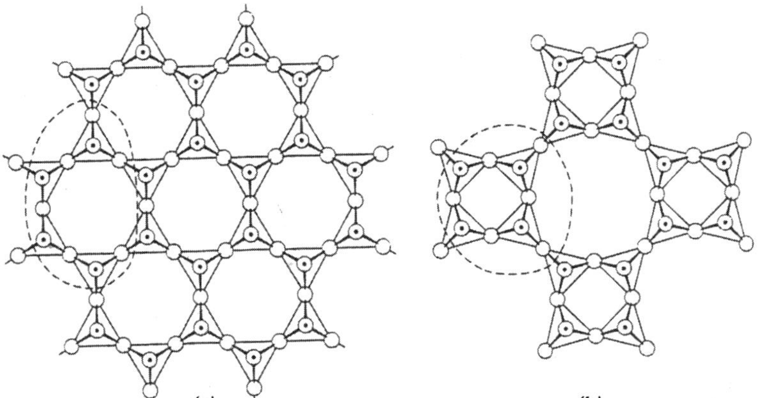

Figure 10.9. Sheets of SiO_4 tetrahedra. (***a***) Rings of six SiO_4^{2-} units in talc, $Mg_3[(OH)_2|Si_4O_{10}]$, and biotite, $K(Mg,Fe)_3[(OH)_2|AlSi_3O_{10}]$. (***b***) Alternating rings of four and eight SiO_4 units in apophyllite. The repeating $Si_4O_{10}^{4-}$ is outlined.

(*Source*: B.E. Douglas, D.H. McDaniel, and J.J. Alexander, *Concepts and Models for Inorganic Chemistry*, 3rd ed., Wiley, New York, 1994, p. 243.)

amphiboles, but extending indefinitely in two dimensions, produces rings of six Si in talc, $Mg_3[(OH)_2|Si_4O_{10}]$ (Section 10.3.24), and biotite, $K(Mg, Fe)_3[(OH)_2|AlSi_3O_{10}]$ (Figure 10.9***a*** and see Figure 10.42 and Section 10.3.29). Biotite is one of the mica minerals. The SiO_4 (and AlO_4) tetrahedra form sheets of interlocking rings with the unshared oxygens pointed in the same direction. Two of these sheets are parallel with the unshared oxygens pointed inward. These oxygens and the OH^- ions are bonded to Mg^{2+} and Fe^{2+} between the sheets. Double sheets weakly bonded together by the K^+ ions account for the characteristic cleavage into thin sheets.

Apophyllite, $Ca_4K[F|(Si_4O_{10})_2]\cdot 8H_2O$, contains sheets made up of SiO_4 tetrahedra linked to form alternating four- and eight-membered rings (Figure 10.9***b***). In this case the sheets are not doubled because the oxygens not shared between Si atoms do not all point in the same direction. The K^+ and Ca^{2+} ions lie between the puckered sheets associated with the oxygens uninvolved in the interlocking network of the sheets. Apophyllite is tetragonal, $\mathbf{D}_{4h}^6$, $P4/mnc$, $a_o = 9.00$, and $c_o = 15.8$ Å.

The sheet silicates containing only Si and Al or Mg, with oxide and hydroxide ions, contain layers of SiO_4 tetrahedra and Al or Mg in MO_6 octahedra. There are three common combinations of tetrahedral and octahedral layers (Figure 10.10). Type I sheet silicate is called a 1:1 layer silicate containing sheets of one layer of SiO_4 sharing oxygen atoms with a layer of MO_6 octahedra. The 1:1 layer (or **OT** for **POPTP**) is charge balanced so there is only weak bonding between the stacked layers. Type II sheet silicate is called a 2:1 layer (or **TOT** for **PTPOPTP**) silicate containing sheets of one layer of MO_6 octahedra between apices of two layers of SiO_4 tetrahedra. Type III sheet silicate is called a 2:1:1 (or **TOT O TOT** for **PTPOPTP POP PTPOPTP)** layer silicate

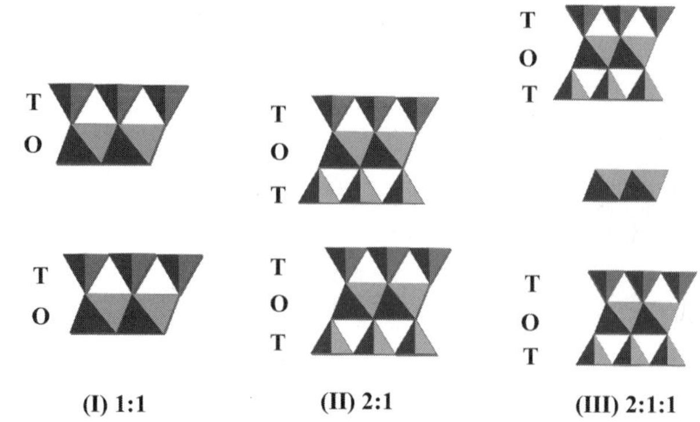

Figure 10.10. Three combinations of tetrahedral (**T**) and octahedral (**O**) layers in sheet silicates: ***(I)*** **TO**, ***(II)*** **TOT** and ***(III)*** **TOT O TOT**.

(*Source*: *CrystalMaker*, by David Palmer, *CrystalMaker* Software Ltd., Begbroke Science Park, Bldg. 5, Sandy Lane, Yarnton, Oxfordshire, OX51PF, UK.)

containing two 2:1 layers separated by a layer of MO_6 octahedra. For M = Mg, this is a layer of brucite, $Mg(OH)_2$ (Section 5.6.4). Other minerals are obtained by adding layers of K^+, Na^+, or other cations plus H_2O between silicate layers.

Type I (**OT**) silicates include kaolinite, $Al_2Si_2O_5(OH)_4$ (Section 10.3.21) and dickite with the same composition. These are true clay minerals. Clay minerals are layer silicates with a grain size $< 2\,\mu m$. The serpentines also are Type I minerals, include antigorite, $Mg_3Si_2O_5(OH)_4$ (Section 10.3.20) and chrysotiles with the same composition. Chrysotiles include clinochrysotile and orthochrysotile.

Type II (**TOT**) silicates include pyrophyllite, $AlSi_2O_5(OH)$ (Section 10.3.25) and talc, $Mg_3Si_4O_{10}(OH)_2$ (Section 10.3.24). Muscovite, $KAl_2(AlSi_3)O_{10}(OH)_2$ (Section 10.3.27), and phlogopite, $KMg_3(Si_3Al)O_{10}(F, OH)_2$ (Section 10.3.28) are Type II minerals with K^+ ions in interlayer sites.

Type III (**TOT O TOT**) silicates include the chlorite group. Chlorite, $Mg_5Al(Si_3Al)O_{10}(OH)_8$ (Section 10.3.30) is one of a group of minerals often containing Fe, Li, Mn, and Ni in octahedral sites.

10.2.4. Framework Silicates

The partial replacement of Si by Al requires another cation in the framework silicates. The SiO_4 tetrahedra, along with randomly distributed AlO_4, are linked by the sharing of all four oxygens, as in SiO_2. We encounter three large groups of framework silicates: the feldspars, the ultramarines with basketlike frameworks, and the zeolites with open structures.

The **feldspars**, comprising about two-thirds of the igneous rocks, are the most important of the rock-forming minerals. Granite consists of quartz, feldspars, and micas. The general formula for the feldspars is given as $M(Al, Si)_4O_8$. When the ratio of Si:Al is 3:1, M is Na or K; and when Si:Al is 2:2, M is Ca or Ba. The two classes of feldspars are based not on the charge on the cation, but on the crystal symmetry. The **monoclinic** feldspars, including orthoclase, $K[AlSi_3O_8]$, and celsian, $Ba[Al_2Si_2O_8]$, contain the larger cations. **Triclinic** feldspars are referred to as the plagioclase feldspars. There is a series of minerals varying continuously in composition from albite, $Na[AlSi_3O_8]$, to anorthite, $Ca[Al_2Si_2O_8]$, corresponding to the isomorphous substitution of Ca^{2+} for Na^+ and the required substitution of Al^{3+} for Si^{4+}.

The **ultramarines** are commonly colored and are used as pigments. The mineral lapis lazuli is of this type. The untramarines contain negative ions, such as Cl^-, SO_4^{2-}, or S^{2-}, that are not part of the framework. The typical basketlike framework of ultramarines is a truncated octahedron—the polyhedron obtained by cutting off the apices of an octahedron. Each of the 24 apices of the truncated octahedron is occupied by Si or Al with bridging O^{2-} along each edge. Sodalite, $Na_8Al_6Si_6O_{24}Cl_2$, is an ultramarine with a structure formed by sharing all square faces of the truncated octahedron (Figure 10.11).

The **zeolites** have very open structures formed by joining silicate chains or cages such as the truncated octahedron. The open structures have made zeolites useful as ion exchangers—softening water by replacing Mg^{2+} or Ca^{2+} by Na^+. The ion exchanger is regenerated by treating it with concentrated NaCl solution. Tetrahedral frameworks are shown in Figure 10.12 for mordenite, $NaAlSi_4O_{12}{\cdot}3H_2O$ (orthorhombic, $\mathbf{D}_{2h}^{17}$, *Cmcm*); heulandite, $CaNa_2Al_2Si_7O_{18}{\cdot}6H_2O$ (monoclinic, $\mathbf{C}_s^3$, *Cm*); laumonite, $CaAl_2Si_4O_{12}{\cdot}4H_2O$ (monoclinic, $\mathbf{C}_{2h}^3$, *C*2/*m*); and chabazite, $CaAl_2Si_4O_{12}{\cdot}6H_2O$ (rhombohedral, $\mathbf{D}_{3d}^5$, $R\bar{3}m$). Synthetic zeolites can provide channels of optimum size. The size of the channels are of molecular size. They are selective with respect to size, to function as *molecular sieves* or catalysts. Molecular sieves can separate gases based on molecular size, as in gas chromatography. Zeolites are now used in industry in great amounts.

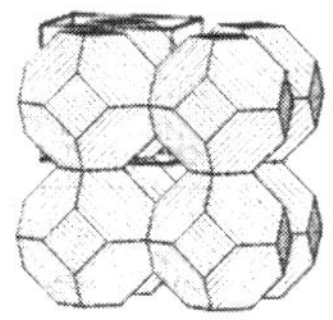

Figure 10.11. The crystal structure of sodalite, an ultramarine.

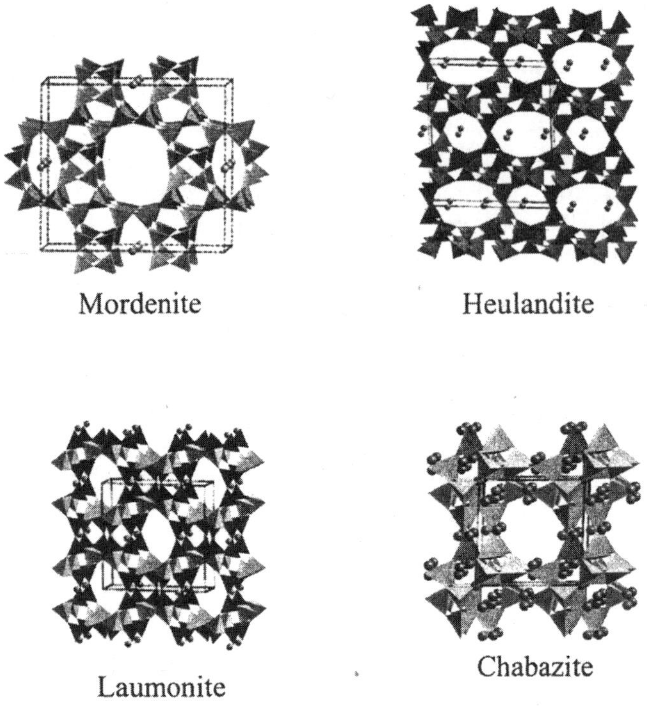

Figure 10.12. The crystal structures of four zeolites: mordenite, heulandite, laumonite, and chabazite.

10.3. Crystal Structures of Some Metal Silicates

Orthosilicates

10.3.1. The $2{\cdot}4PT_{1/8}O_{1/2}T_{1/8}(o)$ Crystal Structure of Olivine [$(Mg,Fe)_2SiO_4$] and Forsterite [Mg_2SiO_4]

The mineral olivine covers a wide range of ratios of Mg/Fe. The Mg-rich mineral (Mg_2SiO_4) is forsterite. The Fe-rich mineral [$(Fe,Mg)_2SiO_4$] is fayalite. Olivine occurs as green, white, brown, or black crystals. The bright-green gemstone is peridot, an olivine. The mantle of the earth is made up largely of olivine and its high-pressure polymorphs. Other minerals with the same structure of olivine (or forsterite) are tephroite (Mn_2SiO_4) and monticellite ($CaMgSiO_4$). The structure of forsterite is shown in Figure 10.13. The oxygen atoms are in the *hcp* (*AB*) sequence with some distortion. Mg atoms occupy half of each **O** layer at *C* positions. Silicon atoms occupy only one-eighth of each **T** layer, giving $\mathbf{T}^+$ (pointing upward) and $\mathbf{T}^-$ (pointing downward) tetrahedra. The octahedra are crosslinked by sharing edges to form ribbons parallel to the *c* axis and with other octahedra and SiO_4 tetrahedra. The octahedra in ribbons are less distorted than the other octahedra. The orthorhombic unit cell has four molecules with eight packing layers repeating, $\mathbf{D}_{2h}^{16}$, *Pnma*, $a_o = 4.861$, $b_o = 10.312$, and $c_o = 5.954$ Å, for forsterite. The structure of olivine and foresterite is very similar to that of chrysoberyl, Al_2BeO_4, (Section 7.34).

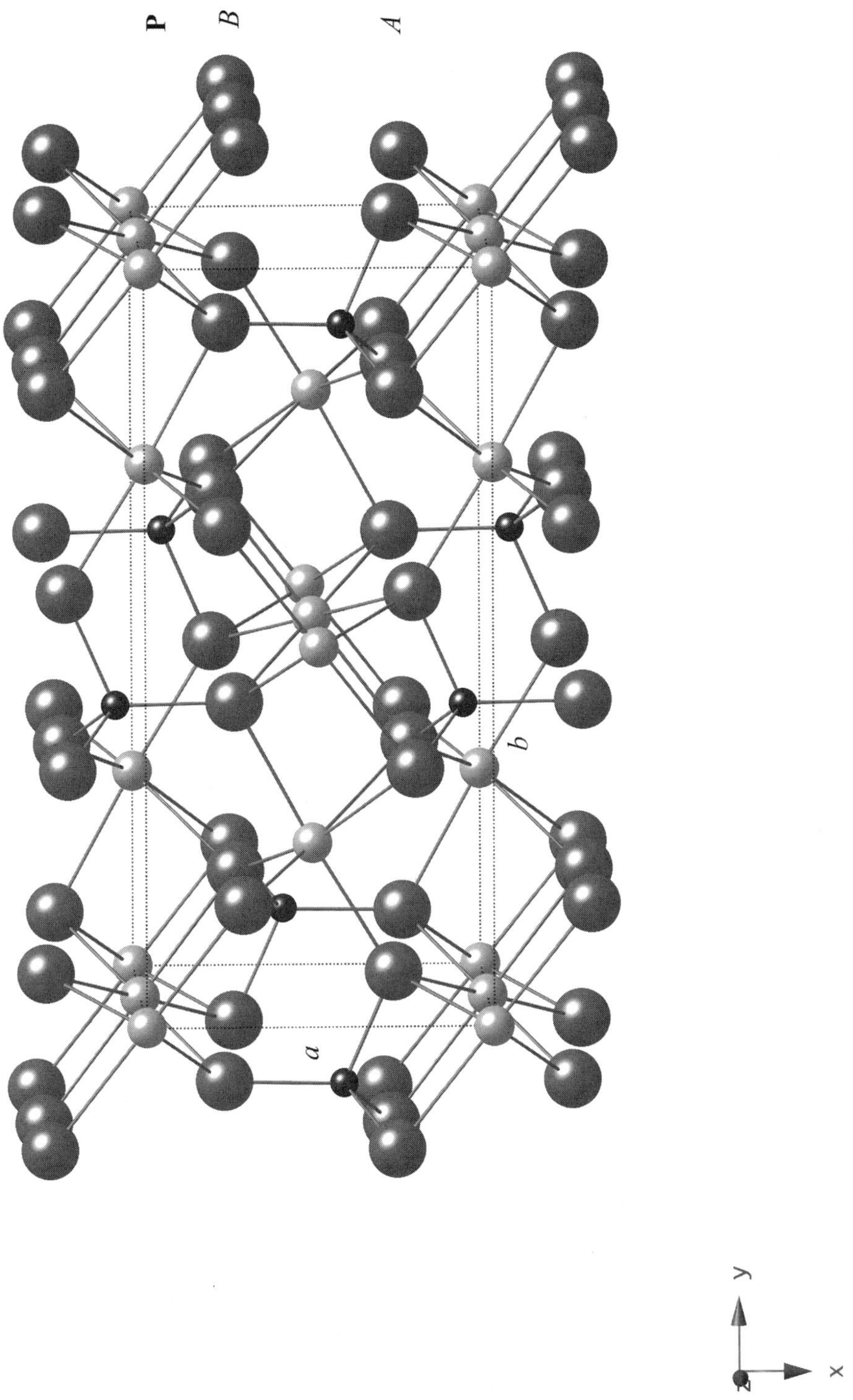

Figure 10.13. The unit cell for forsterite, Mg_2SiO_4. Small black circles are Si in $\mathbf{T}^+$ and $\mathbf{T}^-$ and light larger circles are Mg in octahedral sites. (*Source: CrystalMaker*, by David Palmer, *CrystalMaker* Software Ltd., Begbroke Science Park, Bldg. 5, Sandy Lane, Yarnton, Oxfordshire, OX51PF, UK.)

10.3.2. The 4·3P$T_{1/6}O_{2/6}PO_{2/6}T_{1/6}$(o) Crystal Structure of Topaz {$[Al(F, OH)]_2SiO_4$}

Topaz, $[Al(F, OH)]_2SiO_4$, is orthorhombic, $\mathbf{D}_{2h}^{16}$, *Pbnm*, $a = 4.6499$, $b = 8.7968$, and $c = 8.3909$ Å. It is colorless to aquamarine blue. The gem is golden-yellow or blue. Silicon occupies one-sixth of one **T** layer as SiO_4 and Al occupies two-sixths of the **O** layer between packing layers. Each octahedron is formed from four oxygens of the discrete SiO_4 tetrahedra and $2F^-$ or $2OH^-$. Octahedra share edges and apices are shared with SiO_4 tetrahedra. Along *c*, the tetrahedra are separated by pairs of octahedra. Tetrahedral layers alternate along the packing direction (*b*) as $\mathbf{T}^+$ and $\mathbf{T}^-$, causing every third packing layer to have **T** sites occupied close below and above the packing layer, but staggered to avoid close occupied **T** sites. Figure 10.14 shows that packing layers are in the double hexagonal sequence, *ABAC*, giving the nota-

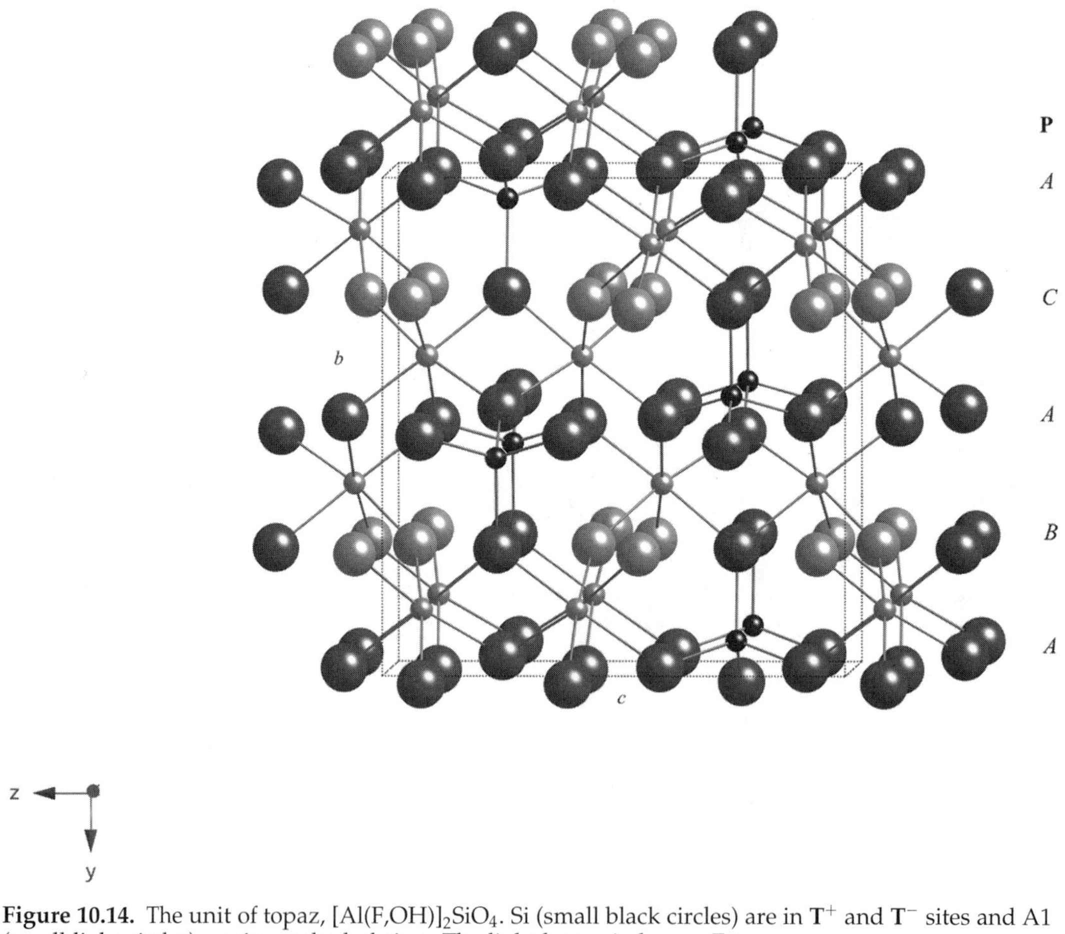

Figure 10.14. The unit of topaz, $[Al(F,OH)]_2SiO_4$. Si (small black circles) are in $\mathbf{T}^+$ and $\mathbf{T}^-$ sites and A1 (small light circles) are in octahedral sites. The light large circles are F^-.

(*Source*: *CrystalMaker*, by David Palmer, *CrystalMaker* Software Ltd., Begbroke Science Park, Bldg. 5, Sandy Lane, Yarnton, Oxfordshire, OX51PF, UK.)

tion **4·3PT$_{1/6}$ O$_{2/6}$ PO$_{2/6}$ T$_{1/6}$(o)**. The F^- and OH^- ions are light larger circles.

10.3.3. The 2·4PT(C_2)$_{1/4}$(Zr)$_{1/4}$(t) Crystal Structure of Zircon

Zircon, $ZrSiO_4$, is the primary ore for Zr, and because there are no independent hafnium minerals, the mineral zircon is the important ore for Hf as well. Zircon gems are important because the refractive index and dispersion approach those of diamond. Zircon is tetragonal with four molecules per cell, $\mathbf{D}_{4h}^{19}$, $I4_1/amd$, $a_o = 6.61$, and $c_o = 6.001$ Å. Si and Zr are in the same planes between the oxygen layers. The Si and Zr atoms are aligned in separate rows parallel to the *a* and *b* axes. They are aligned and alternate in the *c* direction. Si atoms are in tetrahedral sites, but with a C_2 axis aligned with the packing direction. Zr atoms are in ZrO_8 dodecahedra. Each oxygen layer is split into two levels because of the much shorter Si—O distance (1.627 Å) than the Zr—O distances (2.124 and 2.287 Å). The cell in Figure 10.15***a*** shows the split oxygen layers and the SiO_4 tetrahedra and ZrO_8 dodecahedra. Figure 10.15***b*** shows the two "ligands" of the ZrO_8 dodecahedra. Each oxygen of the unit is bonded to two Zr atoms and one Si atom.

The oxygen layers are distorted from the regular hexagonal arrangement for close packing because of the unusual bonding features of the ZrO_8 and the unusual C_2 orientation of SiO_4. Patterns of the oxygen layers repeat in alternate layers so they are designated as an *AB* sequence. The layers of Si and Zr are halfway between oxygen layers, as required for a tetrahedron with the C_2 orientation and the orientation of the ZrO_8 dodecahedra. This layer of Si and Zr is not a **T** or **O** layer in a close-packed structure. The **PTOT** notation is adapted to the situation to indicate that Si occupies one-quarter of **T**(C_2) sites in the layer and one-quarter of Zr sites are occupied, without describing it as a dodecahedral site. The notation is **2·4PT**(C_2)$_{1/4}$(**Zr**)$_{1/4}$(**t**) indicating an *AB* sequence and eight layers (four oxygen layers and four Si–Zr layers) repeating. The zircon structure is found for $MAsO_4$ compounds of Dy, Er, Eu, Gd, Ho, Lu, Sc, Sm, Tb, Tl, Tm, and Yb; for MPO_4 compounds of Dy, Er, Ho, Lu, Sc, Tb, Tm, Y, and Yb; for MVO_4 compounds of Ce, Dy, Er, Eu, Gd, Ho, Lu, Nd, Pr, Sc, Sm, Tb, Tm, Y, and Yb; for $CaCrO_4$, $TaBO_4$, $ThGeO_4$, and $ThSiO_4$.

10.3.4. The Crystal Structure of Sillimanite [Al_2O_3·SiO_2]

The three polymorphs of Al_2O_3·SiO_2, sillimanite, kyanite, and andalusite are found as ores in metamorphic rocks. Sillimanite is formed at high temperatures, above the stability range of andalusite or kyanite. At lower temperature andalusite forms at lower pressures than kyanite. All three forms are used in high-temperature porcelain products such as spark plugs, electrical insulators, refractory, tile, and acid-resistant containers such as crucibles.

In all three minerals straight chains of edge-sharing AlO_6 octahedra extend along the *c* axis. The differences of the three minerals with the same composition are extraordinary. Sillimanite has Al in **T** and **O**

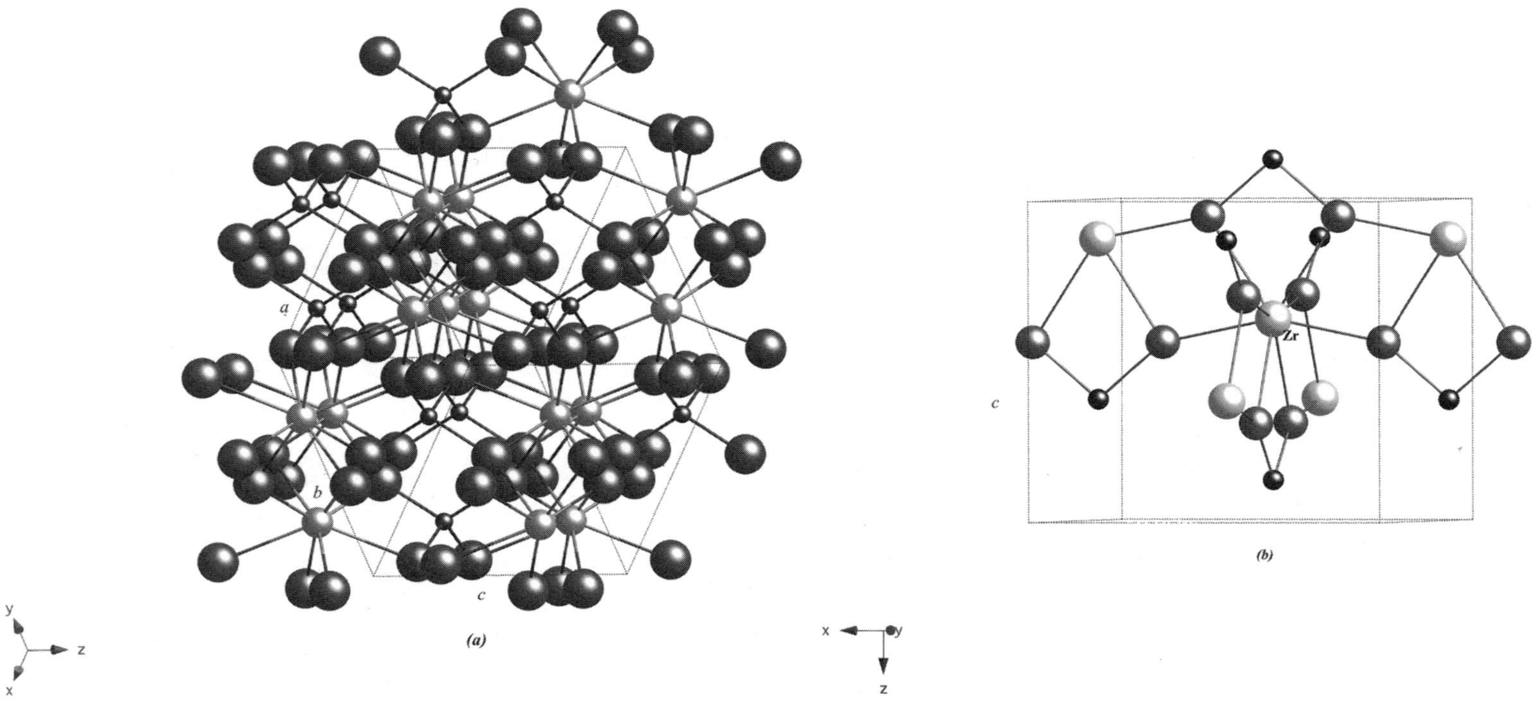

Figure 10.15. (*a*) Layers in the zircon crystal structure showing the orientation of SiO_4 tetrahedra with a C_2 axis perpendicular to the packing layers. (*b*) Orientation of two "ligands" in ZrO_8.

(*Source*: *CrystalMaker*, by David Palmer, *CrystalMaker* Software Ltd., Begbroke Science Park, Bldg. 5, Sandy Lane, Yarnton, Oxfordshire, OX51PF, UK.)

sites, kyanite (Section 10.3.5) has Al only in **O** sites and andalusite (Section 10.3.6) has Al in tetrahedral and trigonal bipyramidal sites. Sillimanite is orthorhombic, with four molecules per cell. The cell dimensions depend on the locality of the specimen. The case considered here ($\mathbf{D}_{2h}^{16}$, *Pbnm*) has $a_o = 7.4883$, $b_o = 7.6808$, and $c_o = 5.774$ Å. The layers of oxide ions have an *AB* sequence with Si and half of the Al atoms in both **T** layers with half of the Al atoms in **O** layers giving $\mathbf{2{\cdot}2PT_{1/10\ 1/10}O_{2/10}T_{1/10\ 1/10}(o)}$. The unusual small fractions result from Si and Al equally sharing **T** layers and the atom ratio in the formula $SiAl_2O_5$.

In Figure 10.16, the AlO_6 octahedra are at the corners and along edges in the c_o direction and through the center and faces in the same direction. The AlO_6 octahedra share edges to form chains. They are linked to the AlO_4 and SiO_4 tetrahedra by shared apices. The AlO_4 and SiO_4 tetrahedra are within the cell as seen in the figure. The AlO_6 octahedra at the corners form the orthorhombic cell. The four AlO_4 and four SiO_4 tetrahedra form an inner orthorhombic unit with AlO_6 at the center.

10.3.5. The Crystal Structure of Kyanite [$Al_2O_3{\cdot}SiO_2$]

Kyanite is triclinic, C_i^1, $\mathbf{P\bar{1}}$, $a = 7.1262$, $b = 7.852$, $c = 5.724$ Å, $\alpha = 89.99°$, $\beta = 101.11°$, and $\gamma = 106.03°$, with four molecules per cell. The packing sequence is *ABC*. Between the first two **P** layers each **T** layer is one-fifth occupied by Si only and the octahedral layer is one-fifth occupied by Al. Between the next two **P** layers, only octahedral sites are three-fifths occupied by Al. This alternating pattern continues, (see Figure 10.17). The notation is $\mathbf{3{\cdot}6PT_{1/5}O_{1/5}T_{1/5}PO_{3/5}(tri)}$. Kyanite is 14% more dense than andalusite and 11.5% more dense than sillimanite.

10.3.6. The Crystal Structure of Andalusite [$Al_2O_3{\cdot}SiO_2$]

The minerals sillimanite (Section 10.3.4), kyanite (Section 10.3.5), and andalusite are three polymorphs of Al_2SiO_5. They all contain SiO_4 tetrahedra and AlO_6 octahedra. In sillimanite, Al also occurs in tetrahedral sites, and in andalusite, Al occurs as AlO_6 octahedra and AlO_5 trigonal bipyramids. The trigonal bipyramids are in the oxygen packing layers giving double layers. Figure 10.18 shows the *AB* sequence of the oxygen packing layers. Between **P** layers there are alternate layers of Al in octahedral sites and Si in tetrahedral sites. However, the Si atoms are in tetrahedral sites with a C_2 axis aligned with the packing direction. In an *A* oxygen layer the Al sites are at *B* sites giving the AlO_3 equatorial grouping of the trigonal bipyramid. The oxygen atoms in axial positions are in adjacent *B* layers. Likewise, in *B* oxygen layers Al sites are at *A* positions aligned with axial oxygen atoms of adjacent *A* layers. This situation requires an *AB* sequence of oxygen layers.

There are chains of edge-linked AlO_6 octahedra parallel to the *c* axis. The octahedra are crosslinked by SiO_4 tetrahedra. The AlO_5 trigonal bipyramids are also aligned along the *c* axis. Two equatorial oxygen atoms of the trigonal bipyramid are shared with two octahedra. The

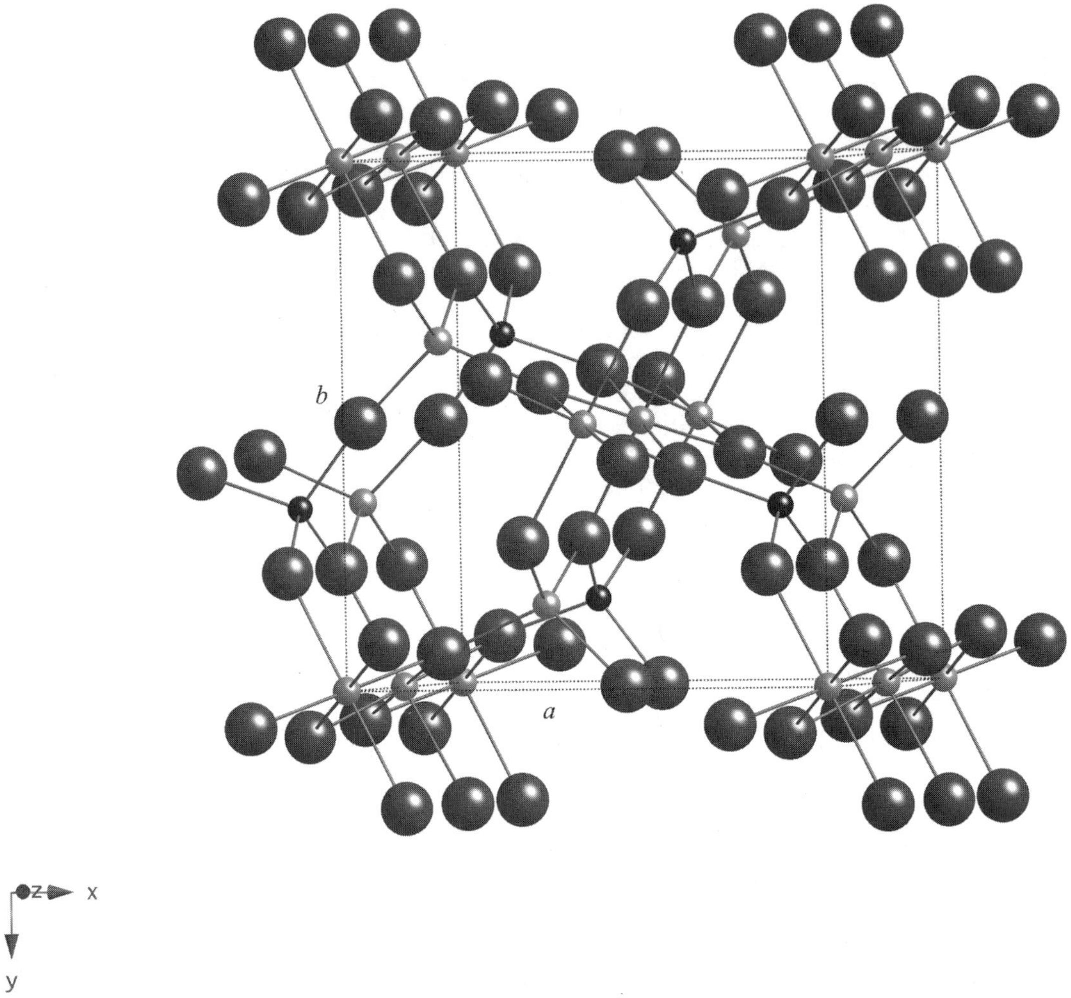

Figure 10.16. The unit cell of sillimanite, Al_2O_3•SiO_2. Si (small black circles) and Al (small light circles) are in **T** sites and Al is also in octahedral sites.

(*Source*: *CrystalMaker*, by David Palmer, *CrystalMaker* Software Ltd., Begbroke Science Park, Bldg. 5, Sandy Lane, Yarnton, Oxfordshire, OX51PF, UK.)

third equatorial oxygen atom is an axial oxygen atom of another AlO_5 group.

The Si—O distances in SiO_4 tetrahedra range from 1.618 to 1.646 Å. The Al—O distances in AlO_6 octahedra range from 1.827 to 2.086 Å. The equatorial Al—O distances in the trigonal bipyramids are 1.814 and 1.840 Å. The axial Al—O distances are 1.816 and 1.899 Å. Andalusite is orthorhombic, $\mathbf{D}_{2h}^{12}$, P*nnm*, $a = 7.48$, $b = 7.903$, and $c = 5.5566$, Å with four molecules in the unit cell. Double packing layers are five-sixths filled by oxygen atoms at *A* (or *B*) positions with Al filling one-sixth of *B* (or *A*) sites. All Al atoms in octahedral sites are in *C* positions, occupying one-third of sites. The Si layers are one-third filled. Like the Al layers, the Si layers are halfway between oxygen **P** layers. Two oxygen atoms form opposite edges of the tetrahedron.

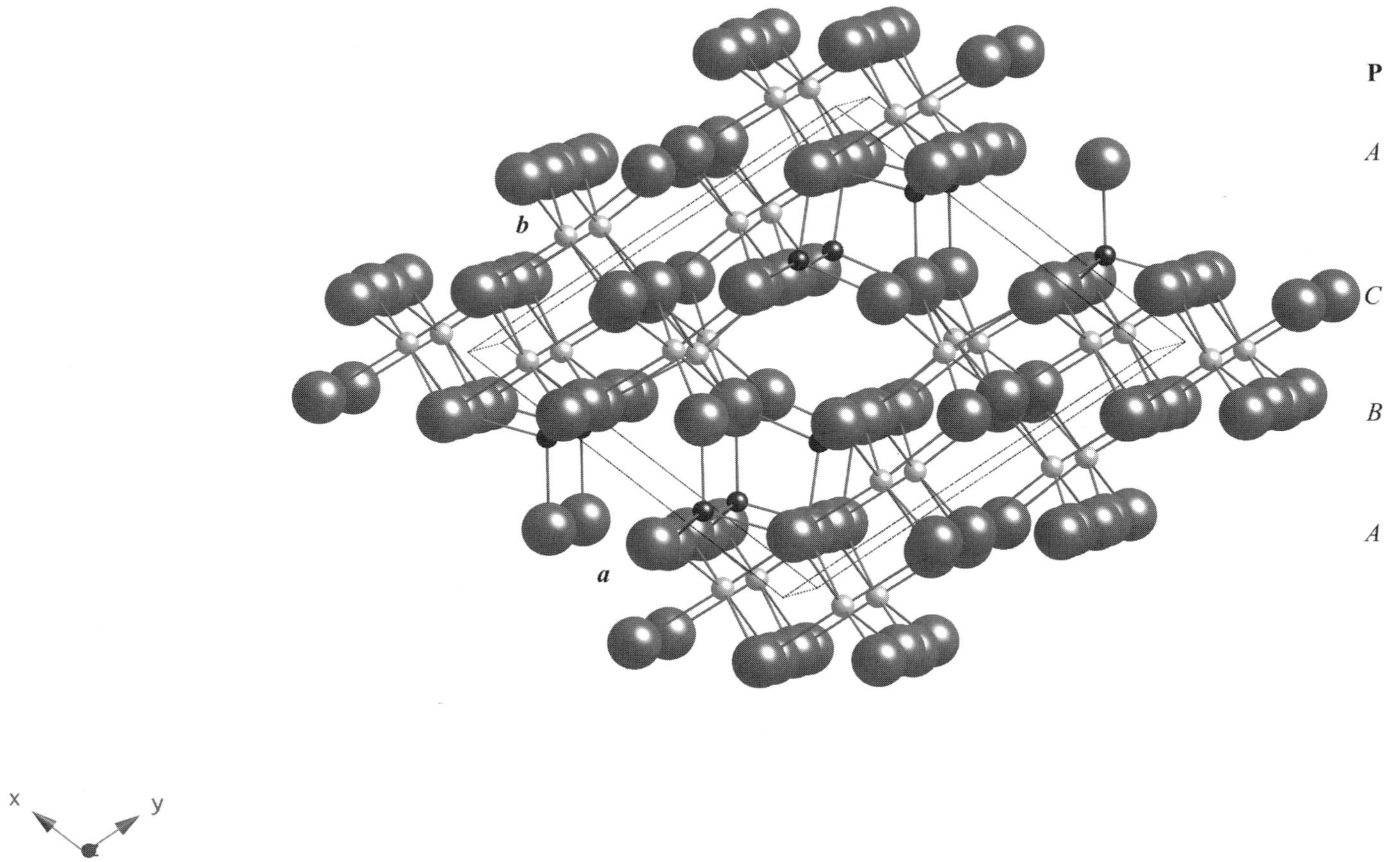

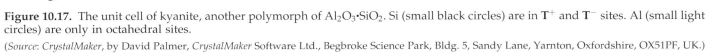

Figure 10.17. The unit cell of kyanite, another polymorph of $Al_2O_3{\cdot}SiO_2$. Si (small black circles) are in $\mathbf{T}^+$ and $\mathbf{T}^-$ sites. Al (small light circles) are only in octahedral sites.

(*Source*: *CrystalMaker*, by David Palmer, *CrystalMaker* Software Ltd., Begbroke Science Park, Bldg. 5, Sandy Lane, Yarnton, Oxfordshire, OX51PF, UK.)

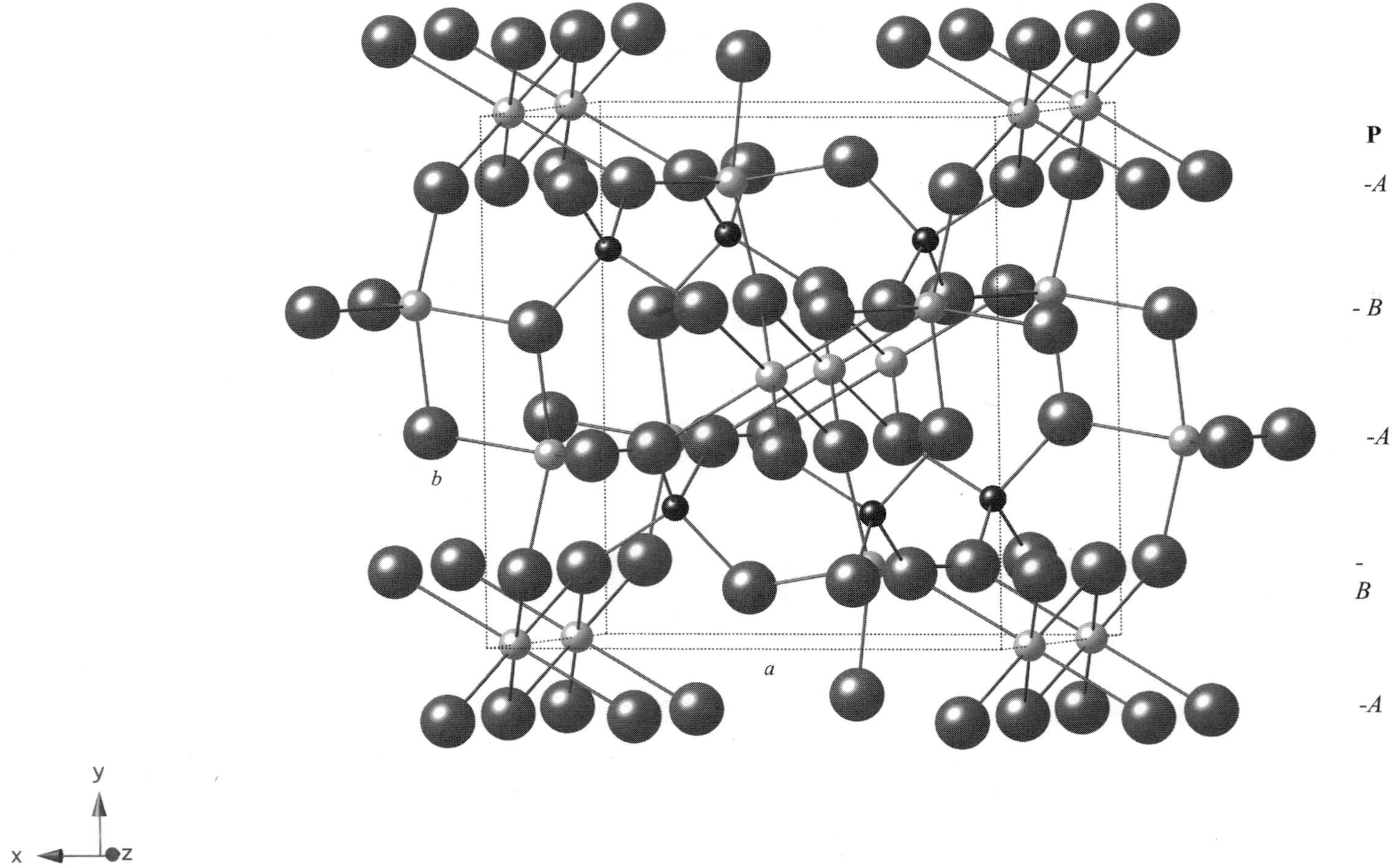

Figure 10.18. Layers in the unit cell of andalusite showing AlO_6 octahedra in the top and bottom faces and in the center of the cell. AlO_5 trigonal bipyramids are shown on the right and left of the unit cell.

(*Source*: *CrystalMaker*, by David Palmer, *CrystalMaker* Software Ltd., Begbroke Science Park, Bldg. 5, Sandy Lane, Yarnton, Oxfordshire, OX51PF, UK.)

The packing layers are distorted from the usual hexagonal pattern; they are partially filled, and oxygen atoms are grouped in AlO_6 octahedral faces, AlO_3 equatorial groups of the trigonal bipyramids, and in pairs for the edges of SiO_4 tetrahedra. The notation is $\mathbf{2{\cdot}4\mathbb{P}}^{\mathrm{O}}_{5/6}{}^{\mathrm{Al}}_{1/6}\mathbf{T}(C_2)_{1/3}\mathbb{P}^{\mathrm{O}}_{5/6}{}^{\mathrm{Al}}_{1/6}\mathrm{O}_{1/3}(\mathbf{o})$, indicating an *AB* sequence for five-sixths filled oxygen layers and one-sixth of *B* or A sites occupied by Al atoms in the same double layer. Si atoms occupy one-third of $\mathbf{T}(C_2)$ sites and Al atoms occupy one-third of octahedral sites in the **O** layers. Eight layers repeat.

10.3.7. The Crystal Structure of $LiAlSiO_4$

The structure of $LiAlSiO_4$ has a complex pattern of layers. **P** layers are one-half filled with one-quarter of the *B* sites occupied and one-quarter of *C* sites occupied by oxygen atoms. The Li, Al, and Si atoms occupy tetrahedral sites, but in an unusual way. There are three consecutive **T** layers one-quarter occupied in the sequence Si, Al, and Si beginning at height 33 and then three consecutive **T** layers are each one-quarter occupied by each of pairs, Li, Al, and Si in the sequence Al, Si; Li, Si; and Li, Al. Even more unusual is that the tetrahedra do not point up or down aligned with **P** positions. Instead, they occur with edges in **P** layers and this displaces the centers of the tetrahedra to positions between **P** positions of either layer, not at the same position as one of that of one **P** layer. The **P** positions are *B* and *C* so the sites of Li, Al, and Si are at *A* positions. These layers are also halfway between the **P** layers. These are the requirements for **O** layers. The situation for $LiAlSiO_4$ does not involve distortion of an **O** site. A metal, or Si, atom is bonded to two oxygen atoms of a $\mathbf{P}_A$ layer forming the edge of a tetrahedron and two oxygen atoms in a $\mathbf{P}_B$ layer forming the opposite tetrahedral edge. The center of the tetrahedron is half of the distance between the **P** layers. We recognize that this is a layer of tetrahedral sites. The tetrahedra have C_2 axes along the packing direction instead of the usual alignment of $\mathbf{T}(C_3)$ tetrahedra. In the notation we have labeled these as $\mathbf{T}(C_2)$ layers. This situation is similar to that of quartz (Section 10.1.3) and zircon (Section 10.3.3).

Figure 10.19 provides a wealth of information in heights of positions:

Height:	8	17	25	33	42	50	58	67	75	83	92	100
O Pos.:	$\mathbf{P}^{B}_{1/4}{}^{C}_{1/4}$		$\mathbf{P}^{B}_{1/4}{}^{C}_{1/4}$		$\mathbf{P}^{B}_{1/4}{}^{C}_{1/4}$		$\mathbf{P}^{B}_{1/4}{}^{C}_{1/4}$		$\mathbf{P}^{B}_{1/4}{}^{C}_{1/4}$		$\mathbf{P}^{B}_{1/4}{}^{C}_{1/4}$	
Atom Pos.:		Li,Al		Si		Al		Si		Al,Li		Li,Si
$\mathrm{T}(C_2)$ **Layers:**		$\mathbf{T}_{1/4\ 1/4}$		$\mathbf{T}_{1/4}$		$\mathbf{T}_{1/4}$		$\mathbf{T}_{1/4}$		$\mathbf{T}_{1/4\ 1/4}$		$\mathbf{T}_{1/4\ 1/4}$

The oxygen layers are equally spaced, and another unusual feature of this arrangement is that the layers are closely spaced. The center of a tetrahedron is closer to the edges than to an apex. For this situation there is only one orientation of the tetrahedra and only one **T** layer results. The sites of Li, Al, and Si form a regular hexagonal pattern. The sites of oxygen atoms are distorted from a hexagonal pattern because of vacancies and the fact that pairs of oxygen atoms of a layer span tetrahedral edges.

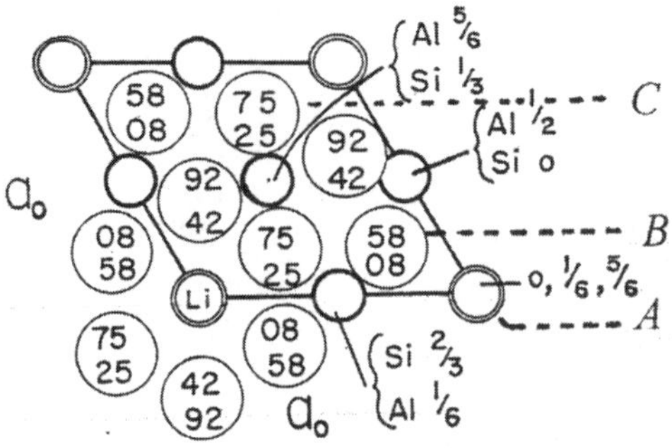

Figure 10.19. A projection of the $LiAlSiO_4$ structure viewed along the *c*. Atom heights are shown.

(*Source*: R.W.G. Wyckoff, *Crystal Structures*, Vol. 4, 2nd ed., Wiley, New York, 1965, p. 171.)

The full sequence for the 12 layers from the diagram for heights gives the notation $\mathbf{12P}^{B}_{1/4}\,{}^{C}_{1/4}\mathbf{T}^{\mathrm{Li}}_{1/4}\,{}^{\mathrm{Al}}_{1/4}\mathbf{P}^{B}_{1/4}\,{}^{C}_{1/4}\mathbf{T}^{\mathrm{Si}}_{1/4}\,\mathbf{P}^{B}_{1/4}\,{}^{C}_{1/4}\mathbf{T}^{\mathrm{Al}}_{1/4}\mathbf{P}^{B}_{1/4}\,{}^{C}_{1/4}\mathbf{T}^{\mathrm{Si}}_{1/4}\mathbf{P}^{B}_{1/4}\,{}^{C}_{1/4}\mathbf{T}^{\mathrm{Al}}_{1/4}\,{}^{\mathrm{Li}}_{1/4}\mathbf{P}^{B}_{1/4}\,{}^{C}_{1/4}\mathbf{T}^{\mathrm{Li}}_{1/4}\,{}^{\mathrm{Si}}_{1/4}[\mathbf{all\,T}(C_2)]$ or $\mathbf{2[3[P}^{B}_{1/4}\,{}^{C}_{1/4}\mathbf{T}_{1/4})\ \mathbf{3(P}^{B}_{1/4}\,{}^{C}_{1/4}\mathbf{T}_{1/4\ 1/4})][\mathbf{T}(\mathbf{C}_2)]$. The hexagonal cell of $LiAlSiO_4$ contains three molecules, $\mathbf{D}^{4}_{6}$, $P6_222$, or its enantiomorph $\mathbf{D}^{5}_{6}$, $P6_422$, $a_o = 5.27$, and $c_o = 11.25$ Å.

10.3.8. The Crystal Structure of Phenakite, [Be_2SiO_4]

Be_2SiO_4 (phenakite or phenacite) has a hexagonal cell containing 18 molecules, $\mathbf{C}^{2}_{3i}$, $R\bar{3}$, $a = 12.42$, and 8.24 Å. The oxygen **P** layers are in an $(A + C)(B + C)$ sequence. One description of the structure, involving overfilled layers, has the first layer *A* sites filled by oxygen and also with oxygen in one-third of *C* sites, Si in one-third of *B* sites, and Be in two-thirds of *B* sites. The next layer has filled *B* sites and one-third of *C* sites occupied by oxygen atoms with *A* sites one-third occupied by Si, and two-thirds occupied by Be. Phenakite is known as a "stuffed" silica, but we can avoid overfilling the layers by having a layer one-quarter occupied by oxygen in *A* (or *B*) sites, one-twelth occupied in *C* sites (for each layer), one-twelth of *B* (or *A*) sites occupied by Si and one-sixth of *B* (or *A*) sites occupied by Be. These are triple layers, each involving three positions for three elements. The full notation is $\mathbf{2{\cdot}3\,III\,P}^{A}_{1/4}\,{}^{C}_{1/12}\,{}^{B(\mathrm{Si})}_{1/12}\,{}^{B(\mathrm{Be})}_{1/6}\ \mathbf{III\,P}^{B}_{1/4}\,{}^{C}_{1/12}\,{}^{A(\mathrm{Si})}_{1/12}\,{}^{A(\mathrm{Be})}_{1/6}$. The cell with layers labeled is shown in Figure 10.20***a***. The XO_4 tetrahedra have two oxygen atoms in the same layer as that for Si or Be and with one oxygen atom from the layer above and one from the layer below. This centers Si or Be in a **P** layer and the Si and Be atoms are not quite aligned with the oxygen atoms bonded in adjacent layers. Tetrahedra are linked in chains along *c*. This is the only structure included in this book with tetrahedra centered in packing layers. The *C* position vacancies are aligned along *c* for the corners and two sites within the cell for each layer, as seen in Figure 10.20***b***. Willemite, Zn_2SiO_4, has the same structure.

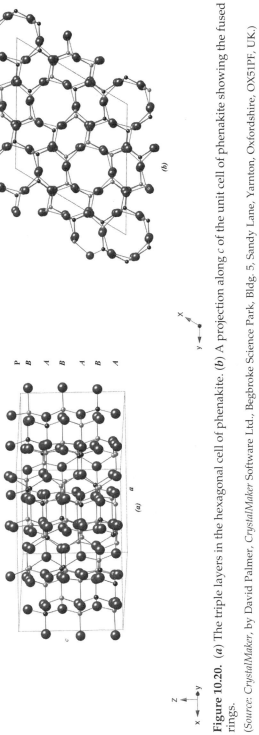

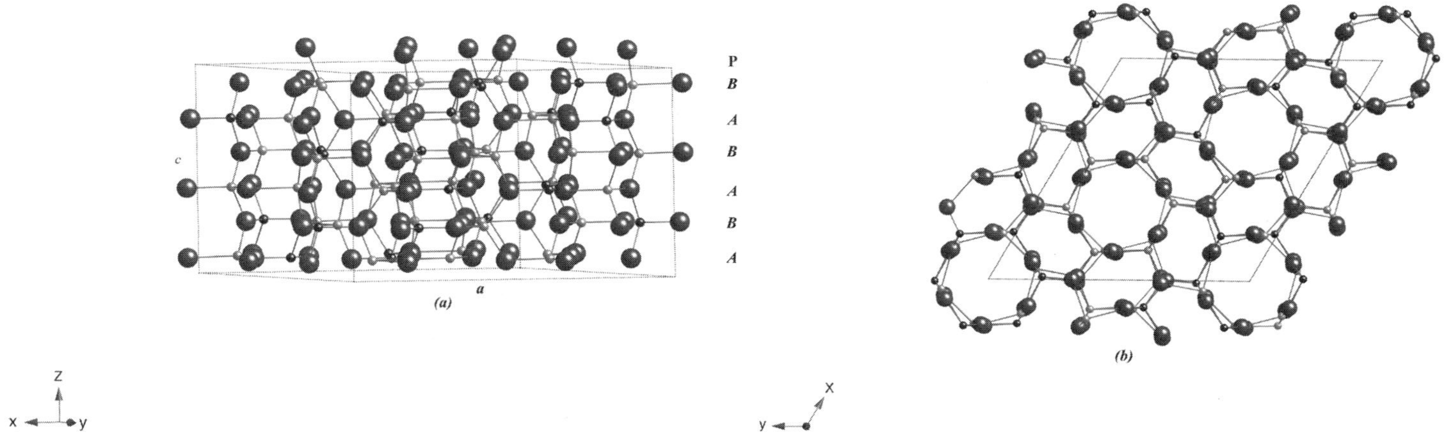

Figure 10.20. (*a*) The triple layers in the hexagonal cell of phenakite. (*b*) A projection along *c* of the unit cell of phenakite showing the fused rings.

(*Source*: *CrystalMaker*, by David Palmer, *CrystalMaker* Software Ltd., Begbroke Science Park, Bldg. 5, Sandy Lane, Yarnton, Oxfordshire, OX51PF, UK.)

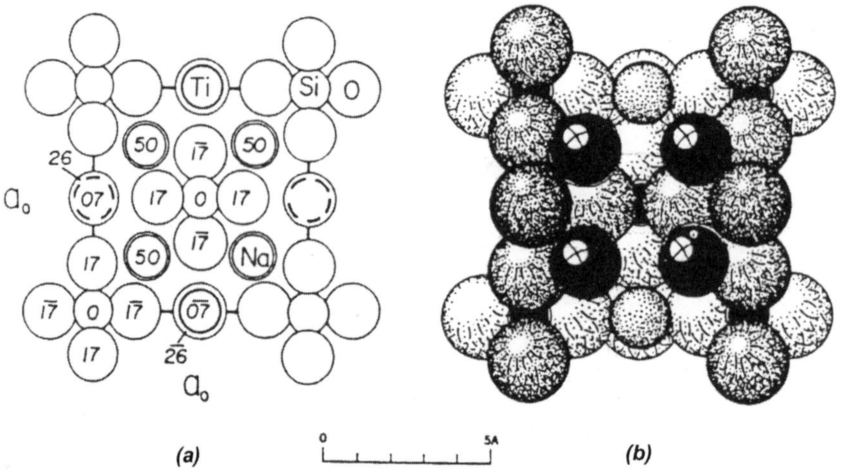

Figure 10.21. A projection on the base of the tetragonal structure of $Na_2(TiO)SiO_4$ **[3·4PTOT(t)]**. Heights are shown. The packing drawing on the right corresponds to the projection.
(*Source*: R.W.G. Wyckoff, *Crystal Structures*, Vol. 4, 2nd ed., Wiley, New York, 1965, p. 213.)

10.3.9. The 3·4PTOT(t) Crystal Structure of $Na_2(TiO)SiO_4$

The structure (Figure 10.21) of $Na_2(TiO)SiO_4$ is similar to that of NH_4HF_2 (Figure 7.15) but more complex. The tetragonal cell contains four molecules, $\mathbf{D}_{4h}^{7}$, $P4/nmm$, $a_o = 6.47$, and $c_o = 5.08\text{Å}$. Here, the **3·4PTOT(t)** structure contains three types of ions: TiO^{2+} ions in **P** layers, SiO_4^{4-} ions in **O** layers, and Na^+ ions filling both **T** layers: $\mathbf{3{\cdot}4P^{TiO}T^{Na}O^{SiO_4}T^{Na}(t)}$, treating TiO^{2+} and SiO_4^{4-} as units. In CsCl, the Cl^- ions fill **P** and **O** layers. Each TiO^{2+} or SiO_4^{4-} is at the center of a cube formed by eight Na^+ ions. Each Na^+ ion is at the center of a cube formed by four TiO^{2+} and four SiO_4^{4-} ions. The great advantage of the **PTOT** system is seen in reducing a complicated structure to its simple basis, **3·4PTOT(t)**.

Pyrosilicates

10.3.10. The Extraordinary Crystal Structure of Thortveitite [$Sc_2Si_2O_7$]

There are several uncommon minerals of pyrosilicates. One is these is thortveitite, $Sc_2Si_2O_7$. It is monoclinic, $\mathbf{C}_{2h}^{3}$, $C2/m$, $a = 6.542$, $b = 8.519$, $c = 4.669\,\text{Å}$, and $\beta = 102.55°$, with two molecules in the unit cell. The unit cell (Figure 10.22) is simple, but with unusual features. The $Si_2O_7^{6-}$ ions are seen clearly with the bridging oxygen atoms in four corners of the figure. The Si—O—Si bonds are nearly parallel to the *a* axis, and the bond angle at oxygen is 180°. In some pyrosilicates the bond angle is as small as 173°. The oxygen-packing layers, formed by the six terminal oxygen atoms of $Si_2O_7^{6-}$, are slightly distorted from regular hexagonal packing but in an *AB* sequence. In the center of the unit cell along *c* there is a layer of Sc^{3+} ions occupying two-thirds of the octahedral sites. The octahedra have three edges shared. The tetrahedral sites are approximately normally spaced

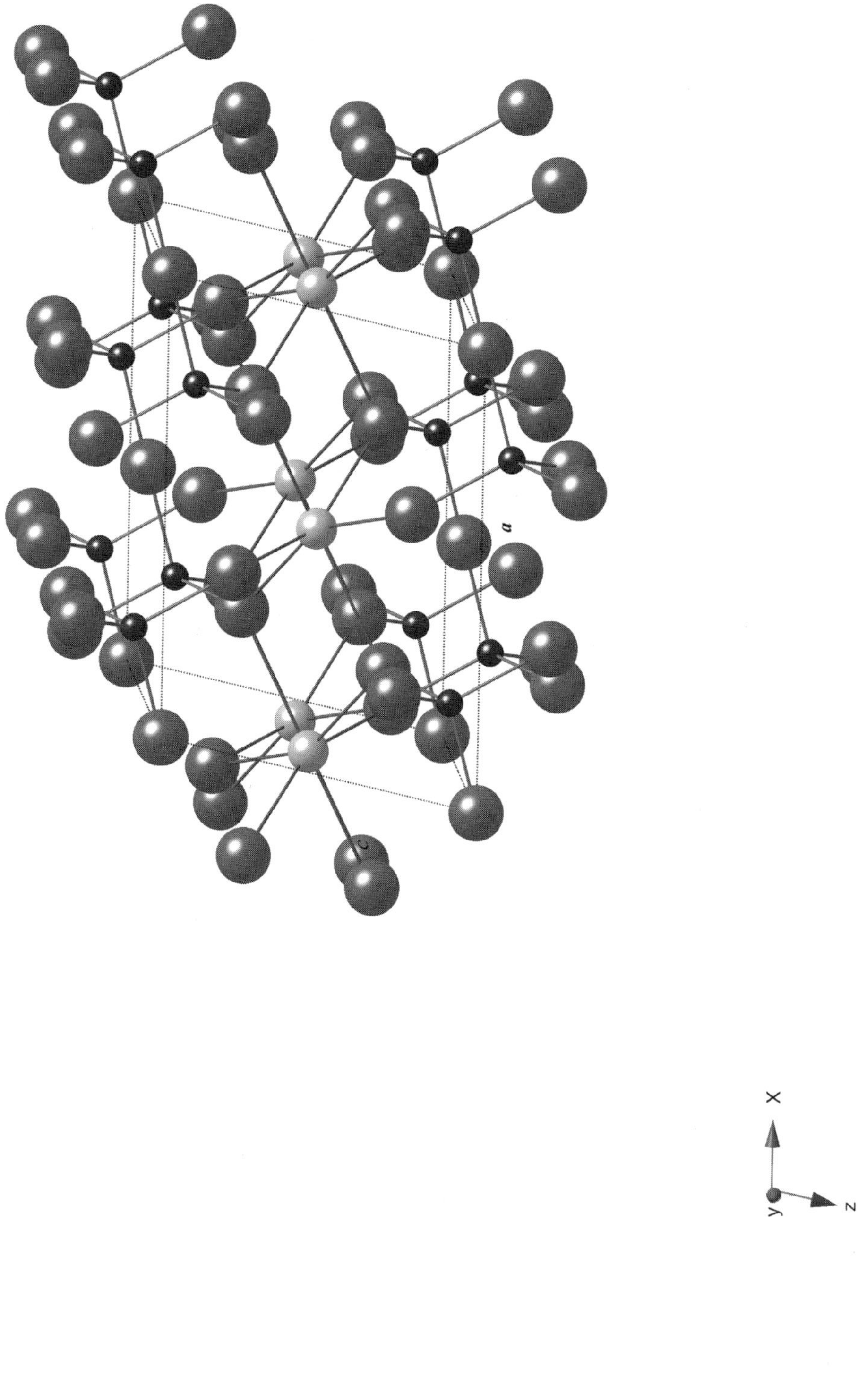

Figure 10.22. The unit cell of thortveitite, $Sc_2Si_2O_7$. Bridging oxygen atoms of $Si_2O_7^{2-}$ are in the corners of the cell. Sc atoms are light circles in octahedral sites.

(*Source: CrystalMaker*, by David Palmer, *CrystalMaker* Software Ltd., Begbroke Science Park, Bldg. 5, Sandy Lane, Yarnton, Oxfordshire, OX51PF, UK.)

between packing layers, but each Si has two oxygen atoms in one adjacent **P** layer and one oxygen in the other **P** layer. The fourth oxygen atom, the bridging oxygen, is in a layer halfway between $\mathbf{P}_A$ and $\mathbf{P}_B$ layers. This is the spacing for an octahedral site and the positions of these oxygen atoms are approximately *C*. We label these as **P** sites although these are not truly packing layers. These sites are one-third occupied. In each **T** layer one-third of sites are occupied by Si. The notation is $\mathbf{2{\cdot}3PT_{1/3}P_{1/3}T_{1/3}PO_{2/3}(m)}$.

10.3.11. The Extraordinary Crystal Structure of Akermanite [$Ca_2MgSi_2O_7$]

Melilite represents a continuous series of minerals from akermanite, $Ca_2MgSi_2O_7$, to gehlenite, $Ca_2Al(AlSi)O_7$, where Al replaces one Ca and one Al replaces one Si to maintain charge balance. The crystal structure is tetragonal, $\mathbf{D}_{2d}^3$, $P\bar{4}2_1m$, $a = 7.84$, and $c = 5.01$ Å with two molecules in the unit cell. For akermanite, the oxygen atoms are in filled distorted close-packed layers in an *AB* sequence. Between adjacent oxygen layers both **T** layers are partially filled by Si and halfway between the oxygen layers Mg atoms occupy tetrahedral sites but with a C_2 axis (Figure 10.23) aligned with the *c* axis (the packing direction). Two oxygen atoms from each layer form the opposite edges of the MgO_4 tetrahedron. These positions are normally occupied by octahedral sites with a C_3 axis aligned with the packing direction. The silicate sheet consists of layers $\mathbf{PT^+T}(C_2)\mathbf{T^-P}$. Halfway between these multiple layer sheets there is a layer of Ca atoms with *CN* 8 in distorted square antiprisms. Terminal oxygen atoms of Si_2O_7 are bonded to two Ca, one Si, and one Mg. The bridging oxygen is bonded to two Ca and two Si. The atoms in the repeating $\mathbf{P}^{O}\mathbf{T}^{Si}\mathbf{T}(C_2)^{Mg}\mathbf{T}^{Si}\mathbf{P}^{O}$Ca layers are in the ratio 7:2:2:2:7:4. This is twice the number of atoms in the formula, corresponding to two molecules in the unit cell. Hence, the

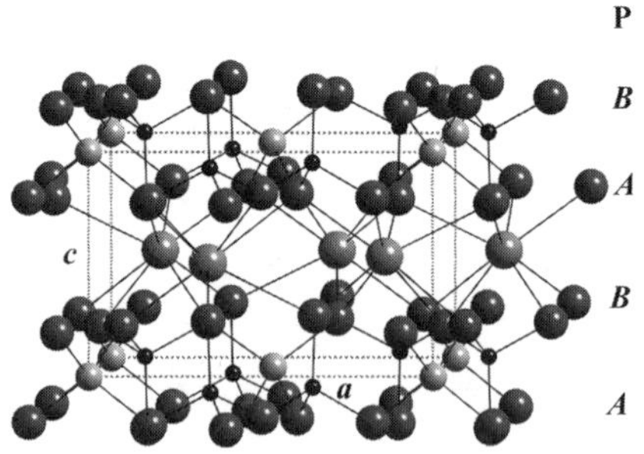

Figure 10.23. The unit cell of akermanite, $Ca_2MgSi_2O_7$. Si (small black circles) occupies $\mathbf{T}^+$ and $\mathbf{T}^-$ sites while Mg (small light circles) occupy **T** sites with a C_2 axis parallel to the *c* axis. Alternating with these layers there are layers occupied by Ca (large circles) with *CN* 8.

(*Source*: *CrystalMaker*, by David Palmer, *CrystalMaker* Software Ltd., Begbroke Science Park, Bldg. 5, Sandy Lane, Yarnton, Oxfordshire, OX51PF, UK.)

stoichiometry requires the fractions in sevenths, giving the notation $\mathbf{2{\cdot}3PT^{Si}_{2/7}T(C_2)^{Mg}_{2/7}T^{Si}_{2/7}PCa_{4/7}(m)}$. This is a very unusual example with three **T** layers between two **P** layers.

Rings

10.3.12. The $1{\cdot}4P^{A(Si6)}_{1/4}\mathbb{P}^{A(Be)(B,C)}_{3/4\ 1/4\,1/4}{}^{(Al)}$ (h) Crystal Structure of Beryl

Beryl, $Be_3Al_2Si_6O_{18}$, is the primary ore for beryllium. The gem emerald is beryl colored by a small amount of Cr^{3+}; aquamarine is also beryl of gem quality. Beryl is hexagonal with two molecules per cell, $\mathbf{D}^2_{6h}$, $P6/mcc$, $a_o = 9.212$, and $c_o = 9.236$Å. There are reflection planes at heights 0 and 50 along c_o. The Si_6O_{18} rings are centered at the corners at 0, 50, and 100. Six Si atoms and six bridging oxygen atoms of each ring are in a reflection plane, and the other two oxygens of the SiO_4 tetrahedra are symmetrically arranged relative to the reflection plane. The centers of the Si_6O_{18} rings and Be^{2+} ions are in *A* positions but in separate planes. AlO_6 octahedra are in the body of the cell at heights 25 and 75 at *B* and *C* positions. A summary of heights and positions is given below:

Height:	0	25	50	75	100
Atom or ring:	Si_6	**Be**	Si_6	**Be**	Si_6
Position:	*A*	*A*	*A*	*A*	*A*
Atom:		Al		Al	
Position:		*BC*		*BC*	

The Be^{2+} and Al^{3+} ions are in planes halfway between the planes of the Si_6O_6 rings in positions providing ***CN*** 4 for Be^{2+} and 6 for Al^{3+}. The BeO_4 tetrahedra involve oxygens from two Si_6O_{18} rings and AlO_6 octahedra share two oxygen atoms with three BeO_4. Oxygen atoms above and below the planes of Si_6O_{18} rings used for BeO_4 and AlO_6, interlock the rings. Treating the Si_6O_{18} rings as units, the notation is $\mathbf{1{\cdot}4P^{A(Si6)}_{1/4}\mathbb{P}^{A(Be)(B,C)}_{3/4\ 1/4\,1/4}{}^{(Al)}}$ **(h)**, that is, Si_6 rings occupy one-quarter of *A* sites, Be occupies three-quarters of *A* sites, and Al occupies one-quarter of *B* sites and one-quarter of *C* sites. Beryl could be considered as a triple layer structure because the layers at heights 25 and 75 have two elements in three positions, *A*, *B*, and *C*. Occupation of three positions in one layer is possible because large Si_6O_{18} rings are treated as units, giving an open structure.

Commonly beryl incorporates Cs, Li, and Na with some Al replacing Si to maintain charge balance. Li and Na occupy the centers of the $Al_2Si_4O_{18}$ rings (for Si_6O_{18} rings) in the same plane. Cs is centered between two rings. Figure 10.24***a*** is a projection along *c* of $NaCsBe_3Al_2(Al_2Si_4O_{18})$ showing the same arrangement of $Al_2Si_4O_{18}$ rings, AlO_6 and BeO_4 as for beryl. The Na and Cs atoms are aligned at the corners of the cell in the centers of rings. Figure 10.24***b*** provides a side view of three $Al_2Si_4O_{18}$ rings with Cs centered between the rings. Na^+ ions are in the center of six bridging oxygen atoms of the rings. Cs has ***CN*** 12 occupying a drum-shaped site, between the staggered rings. It is a hexagonal antiprism. The faces of the antiprisms are shared between Cs sites. The central plane contains $Al_2Si_4O_6$ rings and Na atoms. Oxygen layers adjacent to a plane of rings are eclipsed, but they

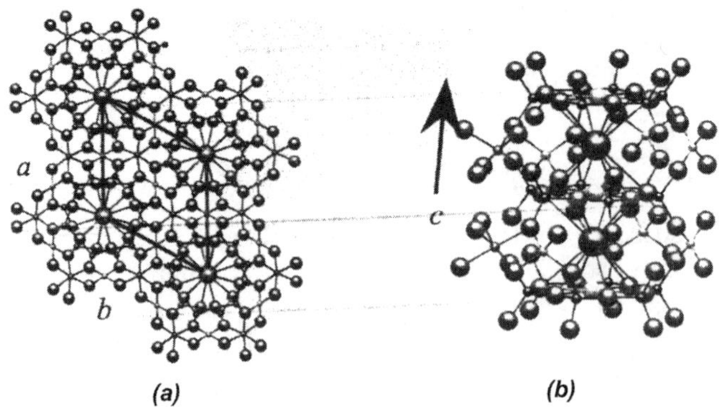

Figure 10.24. (*a*) A projection along *c* of $NaCsBe_3Al_2(Al_2Si_4O_{18})$, $Al_2Si_4O_{18}$ rings replace Si_6O_{18} rings in beryl. Na and Cs are in the rings at the corners. (*b*) A view of Cs between rings and Na in the planes of the rings.

(*Source*: *CrystalMaker*, by David Palmer, *CrystalMaker* Software Ltd., Begbroke Science Park, Bldg. 5, Sandy Lane, Yarnton, Oxfordshire, OX51PF, UK.)

are staggered relative to the next oxygen layer because adjacent rings are staggered along c_0. The Al and Cs atoms are in this "**O**" layer between oxygen-only layers. Be atoms are also centered between the oxygen-only layers and two atoms of each oxygen layer form opposite edges of the BeO_4 tetrahedra. These tetrahedra have a C_2 axis aligned with *c*. The beautiful crystal structure of beryl is easily described and visualized from the description treating Si_6O_{18} rings as units. The layers through the centers of the planes of the rings contain Si and bridging oxygen atoms. The layer between oxygen-only planes is centered, as expected for an octahedral layer. It is occupied by metal ions of AlO_6 octahedra but also CsO_{12} and BeO_4. The positions are *A* for the centers of the rings, BeO_4, Cs, and Na atoms and *B* and *C* for Al.

10.3.13. The Crystal Structures of Cordierite and Indialite [$Mg_2Al_3(AlSi_5O_{18})\cdot H_2O$]

Cordierite, $Mg_2Al_3(AlSi_5O_{18})\cdot H_2O$, has Si_6O_{18}-type rings with one Al substituted for an Si in the ring, similar to those of beryl. It is orthorhombic $\mathbf{D}_{2h}^{20}$, *Cccm*, $a = 17.083$, $b = 9.738$ and $c = 9.335$ Å with four molecules in the unit cell. The Si atoms in the rings are bonded to two oxygen atoms in the ring and to one oxygen in layers above and below. There is no direct bonding between rings, but they are linked by Mg in octahedra and by Al in tetrahedra above and below. The oxygen layers above and below the rings are in the same positions *AA* or *BB*. The Mg of MgO_6 and Al of AlO_4 are between *A* and *B* positions. These AlO_4 tetrahedra have two oxygen atoms from each of *A* and *B* layers, having a C_2 axis aligned with the *c* axis. Figure 10.25***a*** shows a projection of the lower half of the unit cell along the *c* axis. In the upper half of the cell rings are rotated by 30° about axes parallel to the *c* axis. The MgO_6 octahedra and AlO_4 tetrahedra (with a C_2 axis parallel to *c*)

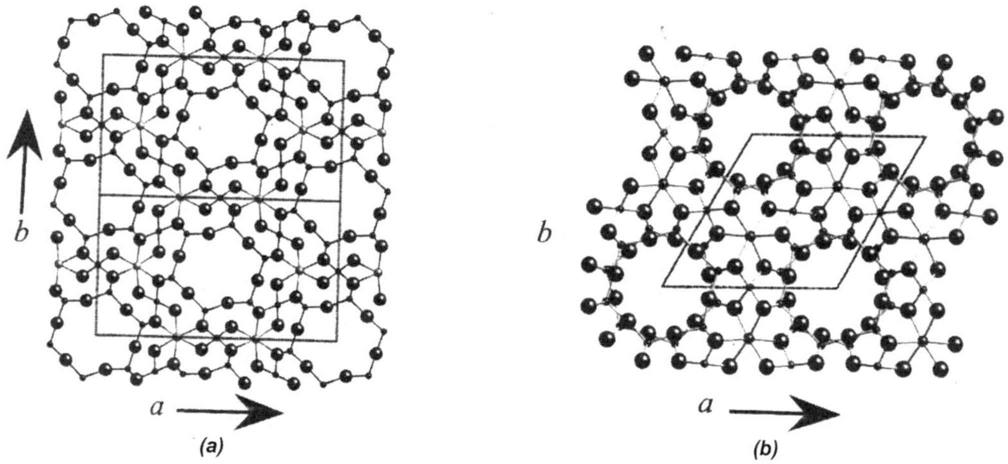

Figure 10.25. (*a*) A projection along *c* of the lower half of the unit cell of low cordierite, $Mg_2Al_3(AlSi_5O_{18})\cdot 2H_2O$ to show the $(AlSiO_4)_6$ rings joined by MgO_6 octahedra and AlO_4 tetrahedra. (*b*) A projection along *c* of the full unit cell of indialite (high cordierite). The staggered $(AlSiO_4)_6$ rings appear as $(AlSiO_4)_{12}$ rings.

(*Source*: *CrystalMaker*, by David Palmer, *CrystalMaker* Software Ltd., Begbroke Science Park, Bldg. 5, Sandy Lane, Yarnton, Oxfordshire, OX51PF, UK.)

are aligned with those in the upper half of the cell. A projection of the whole cell differs only in that the projection of two rings appear as 12 SiO_4 rings.

Above 1450° cordierite converts to the hexagonal indialite (high cordierite) mineral. The hexagonal unit is $\mathbf{D}_{6h}^2$, $P6/mcc$. The structure is very similar to that of beryl. Figure 10.25***b*** shows a projection of the whole cell along the *c* axis showing that the superimposed $(Al,SiO_4)_6$ rings appear as $(Al,SiO_4)_{12}$ rings.

10.3.14. The Crystal Structure of Nepheline [$KNa_3Al_4Si_4O_{16}$]

The formula of nepheline ($KNa_3Al_4Si_4O_{16}$) is often written as $(K,Na)AlSiO_4$, but the common form of the mineral is Na rich. It is described as a stuffed derivative of high tridymite. The structure is simplified by an interesting pattern. These are sheets of $Al_4Si_4O_{16}$ rings (Figure 10.26***a***) with the tetrahedra alternating in pointing upward (Si) or downward (Al) around the ring. The oxygen atoms in the sheets of the rings fill three-quarter of the **P** layer. The apical oxygen atoms fill one-quarter of the adjacent **P** layers giving the sequence $\mathbf{P}_{3/4}^{A}\mathbf{P}_{1/4}^{B}\mathbf{P}_{3/4}^{A}\mathbf{P}_{1/4}^{B}$. The Al and Si atoms fill one-quarter of one **T** layer between **P** layers, the layers alternating as $\mathbf{T}^+$(Si) and $\mathbf{T}^-$(Al) (see Figure 10.26***b***). Each oxygen atom of a $\mathbf{P}_B$ layer is at the apices of two tetrahedra. Some of the pairs of tetrahedra are eclipsed along *c* and others are almost eclipsed. The $\mathbf{P}_A$ layers are slightly puckered because some of the bases are tipped. The K and Na atoms occupy the cavities in the $\mathbf{P}_B$ layers. K atoms are in *A* positions, aligned with centers of the rings and occupying one-sixteenth of *A* sites in the layer. The *CN* of K is 9 as a tricapped trigonal prism. The

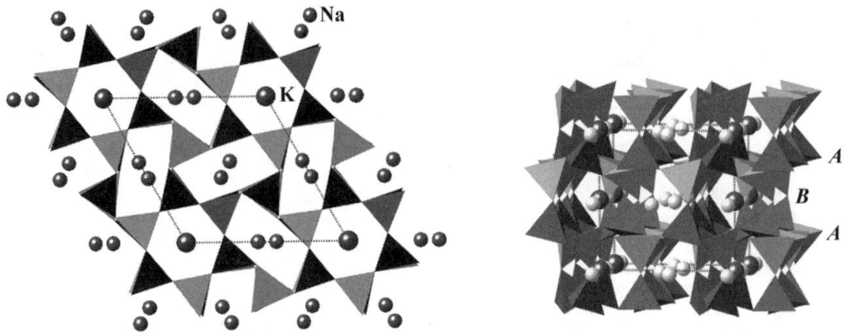

(*a*) (*b*)

Figure 10.26. (*a*) A projection along *c* of nepheline ($KNa_3Al_4Si_4O_{16}$). Tetrahedra of SiO_4 (up) and AlO_4 (down) alternate around the rings at the corners. (*b*) A side view of the cell.

CN of Na is 8, having two triangular groups of adjacent oxygen layers and two close oxygen atoms in the layer of Na. The full sequence is $\mathbf{P}^{A}_{3/4}\mathbf{T}^{+}_{1/4}\mathbf{P}^{B}_{1/4}(\mathrm{K})_{1/16}\mathbf{T}^{-}_{1/4}\mathbf{P}^{A}_{3/4}\mathbf{T}^{+}_{1/4}\mathbf{P}^{B}_{1/4}\mathbf{T}^{-}_{1/4}$ giving the notation $\mathbf{4{\cdot}2P_{3/4}T^{+}_{1/4}P_{1/4}T^{-}_{1/4}}$ The hexagonal cell of nepheline contains two molecules, $\mathbf{C}^{6}_{6}$, $P6_3$, $a_o = 10.05$, and $c_o = 8.38$ Å.

10.3.15. The Complex 3P(h) Crystal Structure of Tourmaline

An idealized chemical formula of tourmaline is $NaLi_3Al_6(OH)_4(BO_3)_3Si_6O_{18}$ or more generally $(Na,Ca)(Li,Mg,Al)_3(Al,Fe,Mn)_6(OH)_4(BO_3)_3Si_6O_{18}$. The composition is even more complex than this because many metal ions can substitute for those shown in wide proportions. The composition can vary within a crystal because of changing concentrations of metal ions during the growth of the crystal. The F^- and O^{2-} ions can substitute for OH^-. Black tourmaline is most common, but tourmaline is seen in every color of the rainbow. Colors can blend into others along the length of a crystal. Watermelon tourmaline has a pink to red core surrounded by green edges. Tourmaline displays pyroelectricity and piezoelectricity, acquiring an electric charge on rubbing, warming or application of pressure. Like quartz, it is used in pressure gauges.

The crystal structure of tourmaline is simpler when viewed as groupings of polyhedra. Tourmalines have in common Si_6O_{18} rings with the apices of the six SiO_4 tetrahedra pointing in the same direction. The six Si atoms are in a plane. Tourmalines also contain three planar BO_3 groups in the cell. The unit cell is trigonal, $C_{3v}{}^{5}$, $R3m$, $a = 15.805$, and $c = 7.084$ Å. The Li, Mg and Al ions are octahedrally coordinated. The coordination of the Na^+ is 9, a polyhedron with a hexagonal face opposite a trigonal face. The base groupings in the cell viewed down the *c* axis is shown in Figure 10.27***a***. The Si_6O_{18} rings are at the corners with LiO_6, AlO_6, and BO_3 and NaO_9 within the cell. At a height about 33, NaO_9 are at the corners with a Si_6O_{18} ring and LiO_6, AlO_6, and BO_3 within the cell (Figure 10.27***b***). At a height about 66 the positions of groups within the cell are reversed.

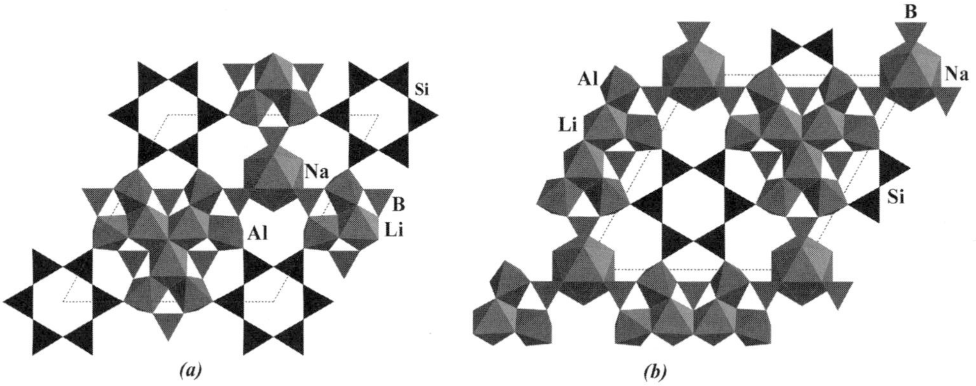

Figure 10.27. (*a*) A projection along *c* of the bottom layer of the unit cell of tourmaline. (*b*) A projection along *c* of the second layer of the unit cell. The tetrahedra for SiO_4, trigonal planes for BO_3, octahedra of Li and Al, and the polyhedra of Na are labeled.

(*Source*: *CrystalMaker*, by David Palmer, *CrystalMaker* Software Ltd., Begbroke Science Park, Bldg. 5, Sandy Lane, Yarnton, Oxfordshire, OX51PF, UK.)

Pyroxenes

10.3.16. The Crystal Structure of Spodumene [$LiAlSi_2O_6$]

Spodumene, $LiAlSi_2O_6$ is a pyroxine containing single chains of SiO_4 tetrahedra extending along the *c* axis. The crystal structure is monoclinic, C_{2h}^6, $C2/c$, $a = 9.468$, $b = 8.412$, $c = 5.224$ Å, and $\beta = 110.05°$, with four molecules in the unit cell. It is an important ore for Li. The packing layers of oxygen atoms are distorted by bonding to Li^+ and Al^{3+} with gaps in octahedral sites and the gaps staggered in adjacent **O** layers (see Figure 10.28). Between the first two **P** layers each of the two **T** layers is 1/3 occupied by Si. Between the next two **P** player 1/3 of octahedral sites are occupied by Al and 1/3 by Li. of octahedral sites are occupied by Al and one-third by Li. The notation is $\mathbf{10PT_{1/3}T_{1/3}PO^{Li}_{1/3}{}^{Al}_{1/3}(m)}$.

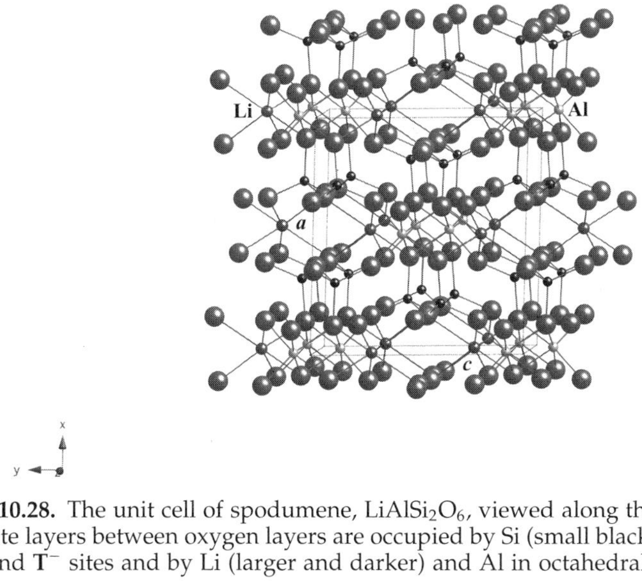

Figure 10.28. The unit cell of spodumene, $LiAlSi_2O_6$, viewed along the *b* axis. Alternate layers between oxygen layers are occupied by Si (small black circles) in T^+ and T^- sites and by Li (larger and darker) and Al in octahedral sites.

(*Source*: *CrystalMaker*, by David Palmer, *CrystalMaker* Software Ltd., Begbroke Science Park, Bldg. 5, Sandy Lane, Yarnton, Oxfordshire, OX51PF, UK.)

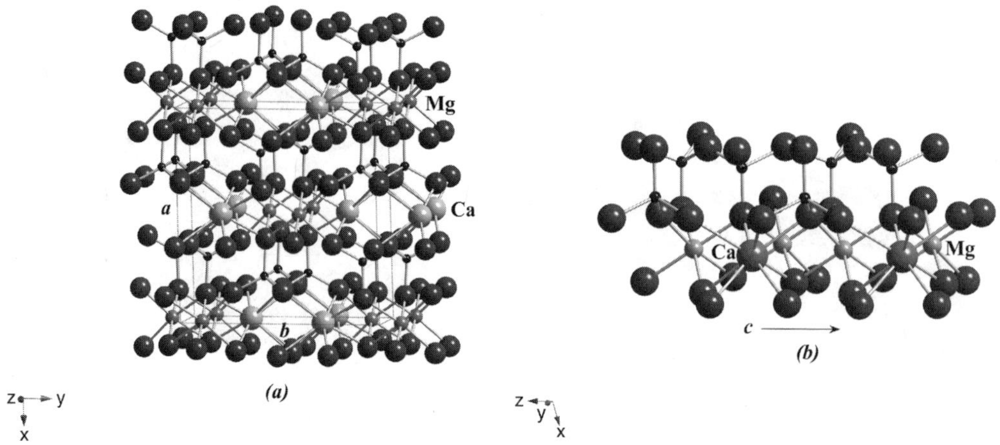

Figure 10.29. (*a*) The extended unit cell of diopside, $CaMgSi_2O_6$. Alternate layers between oxygen layers are occupied by Si (small black circles) in $\mathbf{T}^+$ and $\mathbf{T}^-$ sites and by Mg in octahedral sites and Ca with *CN* 8. (*b*) Linked chains aligned with the *c* axis.

(*Source: CrystalMaker*, by David Palmer, *CrystalMaker* Software Ltd., Begbroke Science Park, Bldg. 5, Sandy Lane, Yarnton, Oxfordshire, OX51PF, UK.)

10.3.17. The Crystal Structure of Diopside [$CaMgSi_2O_6$]

Diopside, $CaMgSi_2O_6$, is a pyroxene similar to spodumene except that Ca, in the **O** layer with Mg, has *CN* 8 (see Figure 10.29***a***). Figure 10.29***b*** shows the linked chains of SiO_4, MgO_6 and CaO_8 along the *c* axis. The basic notation is limited to **P**, **T**, and **O** layers. The Ca^{2+} ions are centered in 1/3 of octahedral sites in the **O** layers with Mg^{2+}. Oxygen layers are distorted by bonding to Ca^{2+} and Mg^{2+} ions which differ in size and *CN*. The octahedral sites are in the *ABC* sequence so the notation is given as $\mathbf{3{\cdot}10PT_{1/3}T_{1/3}PO^{Mg}_{1/3}{}^{Ca}_{1/3}}$(*CN* **8**)(**m**). The crystal structure is monoclinic, $\mathbf{C}^6_{2h}$, $C2/c$, $a = 9.746$, $b = 8.899$, $c = 5.251$ Å, and $\beta = 105.63°$ with four molecules in the cell.

Jadeite, $NaAlSi_2O_6$, is a pyroxene similar to diopside with Na (*CN* 8) replacing Ca in the same layers with Al octahedral sites. Figure 10.30 shows a projection of the oxygen layer above the center of the cell and the Si, Al and Na bonded to the oxygen atoms. We see SiO_4 chains along *c* separated by one AlO_6 chain and two NaO_8 chains.

Amphiboles

10.3.18. Crystal Structure of Tremolite [$NaCa_2Mg_5(Si_7AlO_{22})(OH)_2$] and Paragasite, [$NaCa_2Mg_4Al(Si_6Al_2O_{22})(OH,F)_2$]

The idealized formula of tremolite is $Ca_2Mg_5(Si_8O_{22})(OH)_2$, but usually some Fe replaces Mg, and in the structure reported here there is also an Na atom requiring Al to replace one Si. It is monoclinic, $\mathbf{C}^3_{2h}$, $C2/m$, $a = 9.863$, $b = 18.048$, $c = 5.285$ Å and $\beta = 104.79°$ with two molecules in the unit cell. Tremolite is an amphibole with double chains of SiO_4 joined to form hexagons aligned along the *c* axis (see Figure 10.8). Between packing layers of oxygen atoms double chains

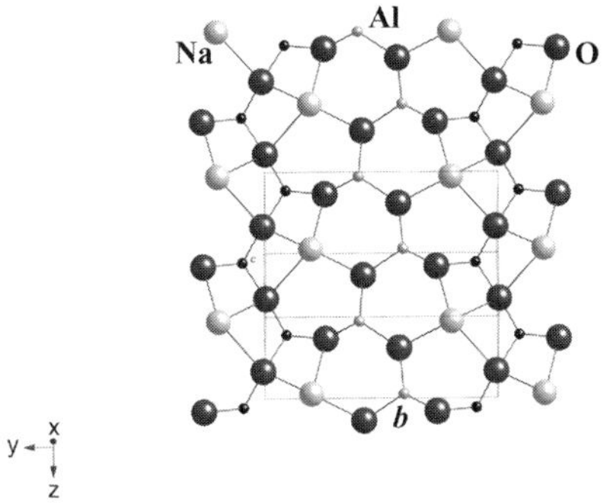

Figure 10.30. A projection along *a* of the jadeite, $NaAlSi_2O_6$, cell of a layer of oxygen atoms bonded to Na, Al, and Si (small black circles) showing the chains along *c* of SiO_4 tetrahedra along the edges, AlO_6 octahedra in the center of the cell and NaO_8 between the SiO_4 and AlO_6 chains.

(*Source*: *CrystalMaker*, by David Palmer, *CrystalMaker* Software Ltd., Begbroke Science Park, Bldg. 5, Sandy Lane, Yarnton, Oxfordshire, OX51PF, UK.)

formed by $\mathbf{T}^+$ Si sites alternate in the *b* direction with those formed by $\mathbf{T}^-$ Si sites (see Figure 10.31***a***). Above MgO_6 octahedra the chains are formed by $\mathbf{T}^-$ Si sites. Between these chains along the *b* axis the chains are $\mathbf{T}^+$ above Na and Ca and $\mathbf{T}^-$ below these ions. The Si_6 rings are centered over and below Na giving ***CN*** 12 for Na. Two pairs of superimposed staggered rings are shown in Figure 10.31***b***. Tremolite is very similar to diopside (Section 10.3.17) for which SiO_4 chains are not linked to form rings of six SiO_4 tetrahedra. Oxygen layers in tremolite are distorted from the usual close-packed arrangement because of interactions with Na, Ca, Mg and Si (or Al), but positions repeat after 20 layers. The notation is $\mathbf{20PO(Mg_{5/12}Ca_{1/6}Na_{1/12})PT_{1/3}T_{1/3}(m)}$. The 24 oxygen atoms are in 2 **P** layers, Si and Al are in one-third of each **T** layer, and Mg, Ca and Na are counted as occupying one-third of the **O** layer even though the *CN* is 8 for Ca and 12 for Na. The metals are centered between

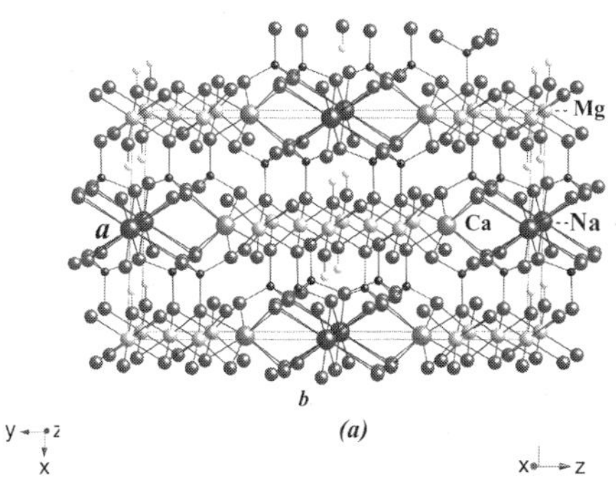

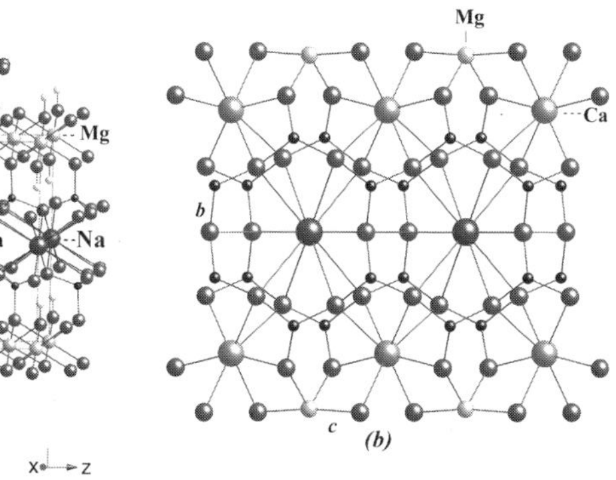

Figure 10.31. (***a***) The unit cell of tremolite, $NaCa_2Mg_5(Si_7AlO_{22})(OH)_2$. (***b***) A projection along *a* of two superimposed chains. Na^+ ions are centered with the rings of the double chains.

(*Source*: *CrystalMaker*, by David Palmer, *CrystalMaker* Software Ltd., Begbroke Science Park, Bldg. 5, Sandy Lane, Yarnton, Oxfordshire, OX51PF, UK.)

oxygen layers corresponding to the octahedral sites for Mg. The oxygen atoms above and below Ca and Na are displaced relative to those for the MgO_6 corresponding to the greater size of Ca and Na. The *CN* of Ca is eight giving a slightly distorted Archimedes antiprism.

Paragasite, $NaCa_2Mg_4Al(Si_6Al_2O_{22})(OH,F)_2$, has the same structure as tremolite with Al replacing one Mg and another Al atom for charge balance. Some OH^- is replaced by F^-. The notation is the same except an Al atom replaces one Mg atom. Tremolite and paragasite are asbestos minerals.

Sheets

10.3.19. The Crystal Structure of Lizardite [$Mg_3Si_2O_5(OH)_4$]

The serpentines, $Mg_3Si_2O_5(OH)_4$, exist as three polymorphs, lizardite, antigorite and chrysotile. They are Type I layer silicates with a Mg octahedral layer sharing the apical oxygen atoms of a tetrahedral layer (**TO**). There is some mismatch between the oxygen layer of the bases of the tetrahedra relative to those of the octahedra. The polymorphs of serpentine differ in the ways of dealing with this mismatch. There is less mismatch for serpentines containing MgO_6 octahedra compared to kaolinite, where AlO_6 contains the smaller Al atom. Of the serpentines, lizardite has least distortion and the layers remain flat with only a rotation of tetrahedra by about 3.5°. For chrysotile the layers are curved, causing them to roll up like a carpet. Chrysotile is the common form of asbestos. Earlier it was thought to be a chain silicate because it occurs as fibers.

Lizardite is hexagonal, $\mathbf{C}_{6v}^3$, $P6_3cm$, $a = 5.318$, and $c = 14.541$ Å, with two molecules in the unit cell. Figure 10.32 shows the **TO** layers with H of OH groups on the nonbonded side of the octahedral layer and for the oxygen atoms not bonded to Si on the other side. There is only hydrogen bonding between **TO** layers. The O—O distance is smaller for the oxygen layers for bases of tetrahedra (2.66 Å) compared to the octahedral layers (3.07 and 3.21 Å). All Si atoms and the lower oxygen layers of the octahedral sheets are at *B* positions. The higher oxygen

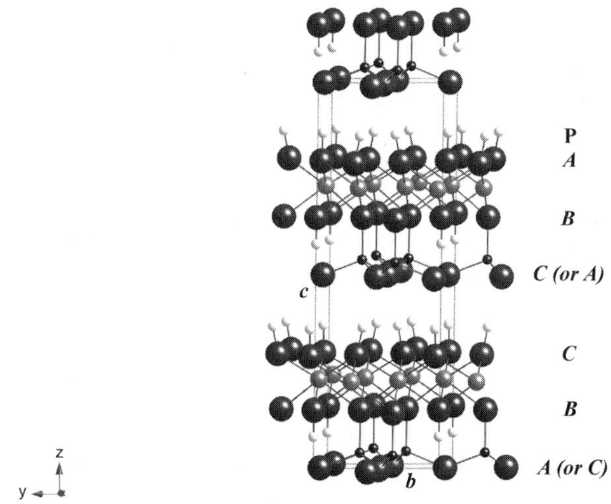

Figure 10.32. The unit cell of lizardite, $Mg_3Si_2O_5(OH)_4$, showing the **TO** layers. Si atoms in **T** sites are small black circles. The larger light circles are Mg atoms in octahedral sites. H atoms are shown as small very light circles.

(*Source*: *CrystalMaker*, by David Palmer, *CrystalMaker* Software Ltd., Begbroke Science Park, Bldg. 5, Sandy Lane, Yarnton, Oxfordshire, OX51PF, UK.)

layers of the octahedral sheets are *C* and *A*, giving the sequence *BC BA* for oxygen layers bonded to Mg. Counting the **P**, **O** and **T** layers there are 10 layers repeating. All layers are filled except for the **T** layer, which is two-thirds filled, giving the notation **10PT$_{2/3}$POP(h)**. The positions of oxygen atoms in the layers forming bases of tetrahedra are displaced from those forming octahedra. They are between *A* and *C* positions, closer to *A* in the base of the cell and closer to *C* in the center of the cell. The oxygen atoms of these layers form hexagons without an oxygen atom at the center and the hexagons are smaller than those of the oxygen layers forming octahedra. These are full layers since there is the same number of oxygen atoms in the unit cell for each layer. Stoichiometry requires full oxygen layers. This is a common feature of TO and TOT structures.

10.3.20. The Crystal Structure of Antigorite [$Mg_3Si_2O_5(OH)_4$]

Antigorite, $Mg_3Si_2O_5(OH)_4$, is one of the three polymorphs of serpentine. It is monoclinic, $\mathbf{C}_s^3$, *Cm*, $a = 5.33$, $b = 9.52$, $c = 14.93$ Å, and $\beta = 101.9°$ with four molecules in the unit cell. The Type I (**TO**) layer silicate structure is smilar to that of lizardite (Section 10.3.19). The octahedra are distorted, with shorter Mg—O distances for oxygen atoms of the oxygen layer not bonded to Si. The tetrahedra are tilted slightly, but unlike chrysotile, the layers do not coil up. In antigorite the tetrahedral sheet inverts orientation periodically to produce a ''corrugated'' stacking. The notation for antigorite is the same as that for lizardite, **10PT$_{2/3}$POP(m)** except the structure is monoclinic. The sequence of packing positions is different but all Mg atoms are at the same positions. The sequence for oxygen layers is *ABA BAB* (Figure 10.33). Actually the positions of oxygen atoms in layers forming bases of tetrahedra are between *A* and *C* lower in the cell and between *B* and *C* higher in the cell.

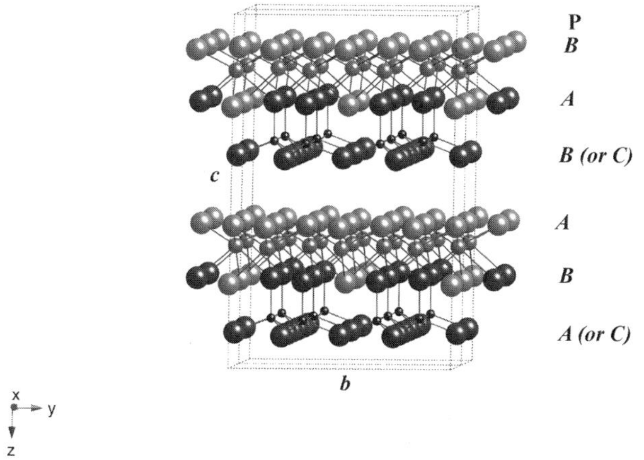

Figure 10.33. The unit cell of antigorite, $Mg_3Si_2O_5(OH)_4$, showing the **TO** layers. Si atoms in **T** sites are black circles and Mg atoms in **O** sites between oxygen layers. H atoms are not shown. Oxygen atoms of OH^- ions are lighter large circles.

(*Source*: *CrystalMaker*, by David Palmer, *CrystalMaker* Software Ltd., Begbroke Science Park, Bldg. 5, Sandy Lane, Yarnton, Oxfordshire, OX51PF, UK.)

10.3.21. The Crystal Structure of Kaolinite [$Al_2Si_2O_5(OH)_4$]

Kaolinite, $Al_2Si_2O_5(OH)_4$, is a clay mineral. It is triclinic, $\mathbf{C}_1$, C1, $a = 5.14$, $b = 8.93$, $c = 7.37$ Å, $\alpha = 91.8°$, $\beta = 104.5°$, and $\gamma = 90.0°$ with two molecules in the cell. The structure is similar to lizardite (Section 10.3.19) and antigorite (Section 10.3.20) with Al replacing Mg. In the **OT** layer the octahedral layer is two-thirds occupied by Al. The **O** sheets consist of hexagonal rings of AlO_6 with two edges shared within a ring and one edge with the joined ring. On one side of the Al layer the **P** layer is filled by OH^- ions. On the other side, one third of the **P** layers are occupied by OH^- ions and two-thirds occupied by the oxide ions at the apices of the tetrahedral layer of SiO_4. Si atoms occupy two-thirds of the **T** layer and the next **P** layer is filled by oxide ions forming the bases of SiO_4 tetrahedra. The **P** layers are in the *ABC CAB BCA* sequence with 15 layers, including **O** and **T** layers, repeating as shown in Figure 10.34. The notation is $\mathbf{3{\cdot}5P^{OH}O_{2/3}P_{2/3}{}^{O}{}_{1/3}{}^{OH}T_{2/3}P^{O}(tri)}$. As for the serpentines there is only hydrogen bonding between layers. The mismatch between tetrahedral and octahedral layers is greater with Mg replaced by the smaller Al atom. This is a factor in limiting crystal size for clays.

10.3.22. The Crystal Structure of $CaAl_2Si_2O_8$ (Omisteinbergite)

The crystal structure of $CaAl_2Si_2O_8$, omisteinbergite, is hexagonal, $\mathbf{D}_{6h}^3$, $P6_3/mcm$, $a_o = 5.10$, and $c_o = 14.72$ Å with two molecules per cell. Figure 10.35 shows that there are double sheets of $(Al,Si)O_4$ tetrahedra joined by the apical oxygen atoms so that all oxygens are shared. These are **TT** layers. Between the doubled **TT** sheets, Ca^{2+} ions are in octahedral sites giving a **TTO** structure. The octahedra, formed

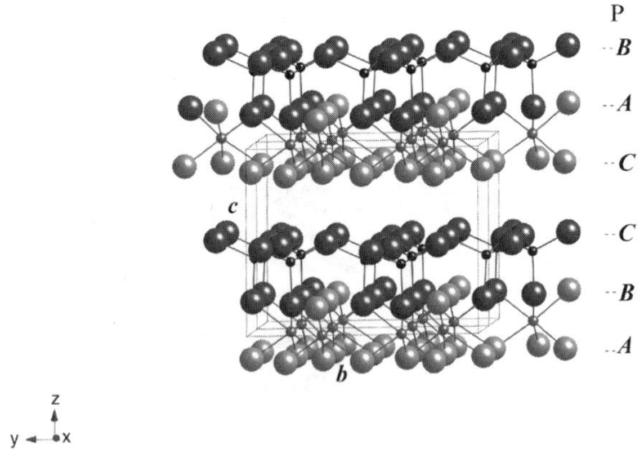

Figure 10.34. The extended unit cell of kaolinite, $Al_2Si_2O_5(OH)_4$, showing the **TO** layers. Si atoms are small black circles. Al atoms are in the bottom and top faces in octahedral sites. Oxygen atoms of OH^- ions are lighter large circles.

(*Source*: *CrystalMaker*, a powerful computer program for the Macintosh, by David Palmer, *CrystalMaker* Software Ltd., Begbroke Science Park, Bldg. 5, Sandy Lane, Yarnton, Oxfordshire, OX51PF, UK.)

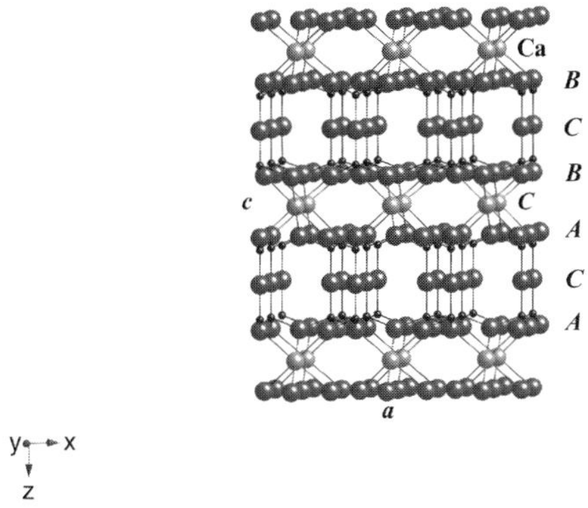

Figure 10.35. Stacking of the **TTO** layers along *c* axis is shown for $CaAl_2Si_2O_8$. (*Source*: *CrystalMaker*, by David Palmer, *CrystalMaker* Software Ltd., Begbroke Science Park, Bldg. 5, Sandy Lane, Yarnton, Oxfordshire, OX51PF, UK.)

by three oxygens from each of the filled $\mathbf{P}_A$ and $\mathbf{P}_B$ layers and each **O** layer, are one-third filled. One Ca^{2+} ion compensates for replacement of 2 Si^{4+} by 2 Al^{3+}. The oxygen atoms are in **P** layers with Al and Si atoms in **T** layers. The oxygens of the filled **P** layers form the bases of the tetrahedra. The bridging oxygens at the apices of the tetrahedra are at the same *C* positions as those of Si and Al. Ca ions are also at *C* positions. These **P** and **T** layers are two-thirds filled. The heights along the *c* axis with positions of oxygen and Ca layers are shown. The (Si, Al)—O—(Si,Al) bridge is linear. This is an unusual pattern for a sheet silicate. The sequence of layers is:

Height:	0	10	14	25	36	40	50	60	64	75	86	90
Atom:	Ca	O	Si,Al	O	Si,Al	O	Ca	O	Si,Al	O	Si,Al	O
Layer:	$\mathbf{O}^C_{1/3}$	$\mathbf{P}^A$	$\mathbf{T}^C_{2/3}$	$\mathbf{P}^C_{2/3}$	$\mathbf{T}^C_{2/3}$	$\mathbf{P}^A$	$\mathbf{O}^C_{1/3}$	$\mathbf{P}^B$	$\mathbf{T}^C_{2/3}$	$\mathbf{P}^C_{2/3}$	$\mathbf{T}^C_{2/3}$	$\mathbf{P}^B$
Spacing:		10	4	11	11	4	10	10	4	11	11	4

The *AC AB CB* sequence of the **P** layers gives the notation $\mathbf{6{\cdot}2PT_{2/3}P_{2/3}T_{2/3}PO_{1/3}[P^{O}T^{Si,Al}O^{Ca}](h)}$. Cymrite, $BaAlSi_3O_8(OH)$ (Section 10.3.23) has a similar double sheet of two **T** layers. The Si and Al and apical oxygen atoms are in one-third of *A* sites or *B* sites giving double layers.

10.3.23. The Crystal Structure of Cymrite [$BaAlSi_3O_8(OH){\cdot}H_2O$]

Cymrite, $BaAlSi_3O_8(OH){\cdot}H_2O$, has double sheets of $(Al,Si)O_4$ tetrahedra joined by the apical oxygen atoms so that all oxygen atoms are shared. Between the doubled sheets there are Ba^{2+} ions at height zero with 3/4 filled oxygen layers at height $\pm$ 20 and OH^- ions and H_2O molecules at $\pm$ 43. The bases of the joined tetrahedra are in *A* positions at $\pm$ 20, Si and Al (3:1 ratio) at $\pm$ 28 and the apex oxygen atoms at height 50 joining the tetrahedral layers (see Figure 10.36). Si and Al

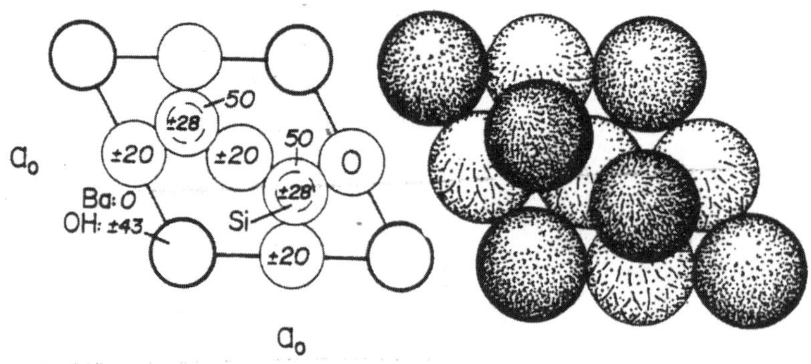

Figure 10.36. A projection along *c* of the unit cell of cymrite, $BaAlSi_3O_8(OH){\bullet}H_2O$, and the same packing view.

(*Source*: R.W.G. Wyckoff, *Crystal Structures*, Vol. 4, 2nd ed., Wiley, New York, 1965, p. 401.)

randomly occupy one-quarter of *B* sites and one-quarter of *C* sites in the same layers at ± 28. The apex oxygen atoms also occupy one-quarter of *B* sites and one-quarter of *C* sites in the double packing layer. This structure is similar to that of $CaAl_2Si_2O_8$ (Section 10.3.20) since they both have two tetrahedral layers joined by apical oxygen atoms. Oxygen layers forming the bases of tetrahedra are filled for $CaAl_2Si_2O_8$ and the apices are all at *C* positions. Cymrite is hexagonal, $\mathbf{D}_{6h}^1$, $P6/mmm$, $a = 5.324$, and $c = 7.662$ Å with one molecule in the unit cell. The notation is adapted as $\mathbf{2{\cdot}4(Ba)}_{1/4}^{A}\mathbf{P}_{3/4}^{A}\mathbf{TT}_{1/4\,1/4}^{B\;\;C}$ $\mathbf{(OH,H_2O)}_{1/4}^{A}\mathbf{PP}_{1/4\,1/4}^{B\;\;C}$ $\mathbf{(OH,H_2O)}_{1/4}^{A}\mathbf{TT}_{1/4\,1/4}^{B\;\;C}\mathbf{P}_{3/4}^{A}\mathbf{(h)}$. This is the first time **TT** is used for a double **T** layer using *B* and *C* positions. In cymrite the bases of six tetrahedra, at *A* positions, form interlocking rings. The Ba, OH^- and water molecules occupy vertical channels at the corners of the cells (*A* positions) and aligned with the centers of the rings of six SiO_4 tetrahedra. giving Ba atoms a hexagonal prism bicapped by OH^- ions. The heights along the *c* direction are:

Height:	0	20	28	43	50	57	72	80	100
Atom:	Ba	O	(Si,Al)	OH, H_2O	O	OH, H_2O	(Si,Al)	O	Ba
Layer:	*A*	P_A	$TT_{B,C}$	*A*	$PP_{B,C}$	*A*	$TT_{B,C}$	P_A	*A*

$Ba(Al_2Si_2O_8)$ has the same crystal structure and same heights in the unit cell as cymrite without OH^- and H_2O. Another Si is replaced by Al to maintain charge balance.

10.3.24. The Crystal Structure of Talc [$Mg_3Si_4O_{10}(OH)_2$]

Talc, $Mg_3Si_4O_{10}(OH)_2$, is a Type II (**TOT**) layer silicate with the octahedral layer filled by Mg. Si occupies two-thirds of **T** layers. Only oxide ions fill the layers forming the bases of tetrahedra. Hydroxide ions occupy one-third of the oxygen layers which form octahedra and two-thirds of the sites are occupied by oxide ions at the apices of tetrahedra. The **TOT** layer is not charged so there is only weak

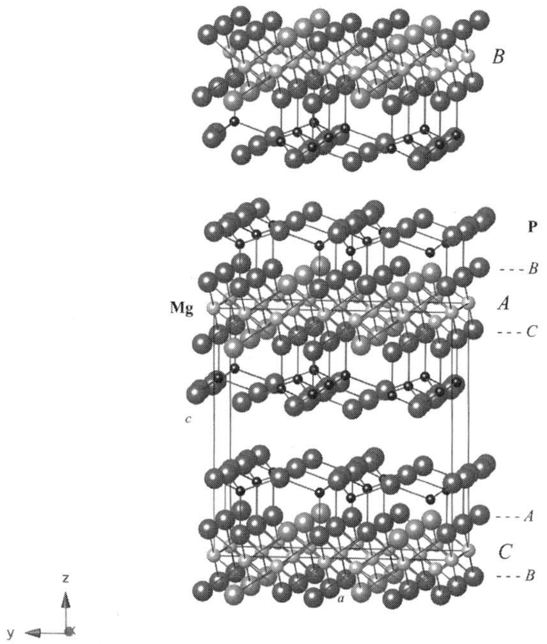

Figure 10.37. The unit cell of talc, $Mg_3Si_4O_{10}(OH)_2$, is shown extended along *c* to show the **TOT** packing. Oxygen atoms of OH^- ions are lighter large circles not bonded to Si.

(*Source*: *CrystalMaker*, by David Palmer, *CrystalMaker* Software Ltd., Begbroke Science Park, Bldg. 5, Sandy Lane, Yarnton, Oxfordshire, OX51PF, UK.)

van der Waals attraction between **TOT** layers. Talc is the softest mineral and it is used as a lubricant, French chalk. Layers of oxygen atoms bonded to Mg have the usual close-packed arrangements, forming regular octahedra. The oxygen atoms forming bases of tetrahedra fall between regular sites. The octahedral sites are in an *ABC* sequence (see Figure 10.37). There are seven layers in the cell requiring 21 layers repeating giving the notation $\mathbf{3{\cdot}7PT_{2/3}POPT_{2/3}P(tri)}$. Talc is triclinic, $\mathbf{C}_i^1$, $P\bar{1}$, $a = 5.29$, $b = 9.173$, $c = 9.46$ Å, $\alpha = 90.46°$, $\beta = 98.68°$, and $\gamma = 90.09°$ with two molecules in the unit cell.

10.3.25. The Crystal Structure of Pyrophyllite [$Al_2Si_4O_{10}(OH)_2$]

Pyrophyllite, $Al_2Si_4O_{10}(OH)_2$, is a Type II (**TOT**) layer silicate similar to talc (Section 10.3.24) with Al replacing Mg (see Figure 10.38). The octahedral layer is only two-thirds occupied by Al to maintain charge balance. The sequence of octahedral sites is *BCA*. There are 14 layers in the cell and 21 layers repeating. The notation similar to that of talc, $\mathbf{3{\cdot}7PT_{2/3}PO_{2/3}PT_{2/3}P(m)}$. As noted for talc, the positions of the sites of oxygen layers forming bases of tetrahedra are between regular positions of octahedral layers. The data presented here are for the monoclinic form, $\mathbf{C}_{2h}^2$, $C2/c$, $a = 5.16$, $b = 8.90$, $c = 18.64$ Å and $\beta = 100.75°$ with four molecules in the unit cell. The O—O distances in the octahedral layers are 2.964 to 2.977 Å, and in the layer of oxygen atoms forming the tetrahedral bases are 2.567 to 2.580 Å. The difference

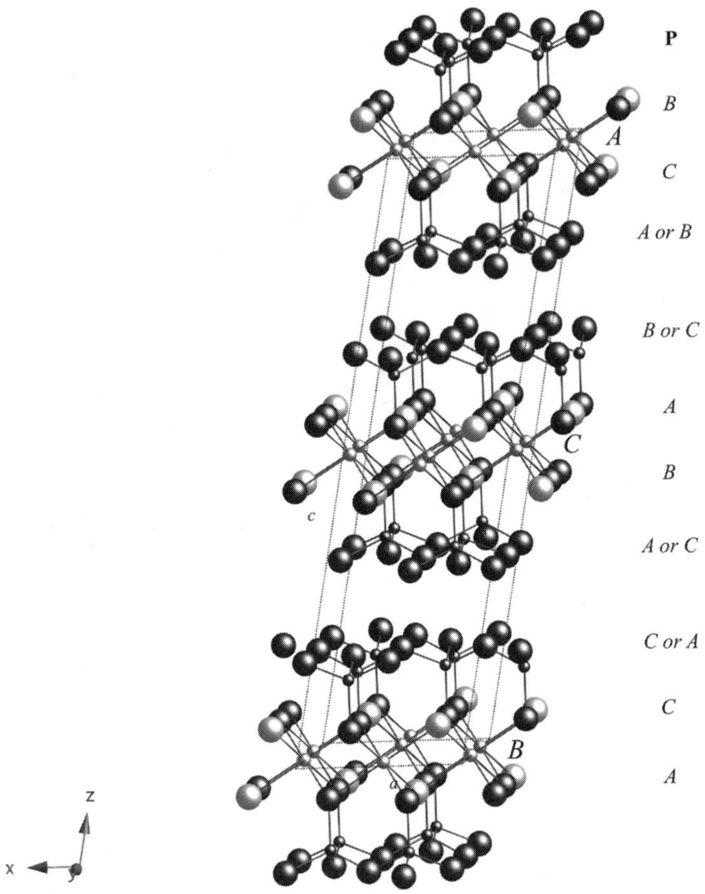

Figure 10.38. The unit cell of pyrophyllite, $Al_2Si_4O_{10}(OH)_2$, is extended along *c* to emphasize the **TOT** packing. Oxygen atoms of OH^- ions are lighter large circles not bonded to Si.

(*Source*: *CrystalMaker*, by David Palmer, *CrystalMaker* Software Ltd., Begbroke Science Park, Bldg. 5, Sandy Lane, Yarnton, Oxfordshire, OX51PF, UK.)

indicates a significant mismatch between oxygen layers forming the faces of octahedra and the bases of tetrahedra.

10.3.26. The Crystal Structure of Montmorillonite [$KMgAlSi_4O_{10}(OH)_2 \cdot nH_2O$]

Montmorillonite, $KMgAlSi_4O_{10}(OH)_2 \cdot nH_2O$, is a clay mineral closely related to pyrophyllite (Section 10.3.25). Mg replaces some Al and K is added between the **TOT** layers. The clay mineral expands when water is absorbed between **TOT** layers. Figure 10.39 shows the unit cell with *ABC* labels for the oxygen layers. It is monoclinic, $\mathbf{C}_{2h}^3$, $C2/m$, $a = 5.2$, $b = 9.2$, $c = 10.15$ Å, $\beta = 99.0°$ with two molecules in the unit cell. Montmorillonite commonly contains Na and Ca. The mismatch of oxygen layers forming octahedral and bases of tetrahedra is shown by differences in O—O distances, 2.818–2.868 Å for oxygen layers bonded to Mg and Al and 2.631–2.655 Å for tetrahedral base

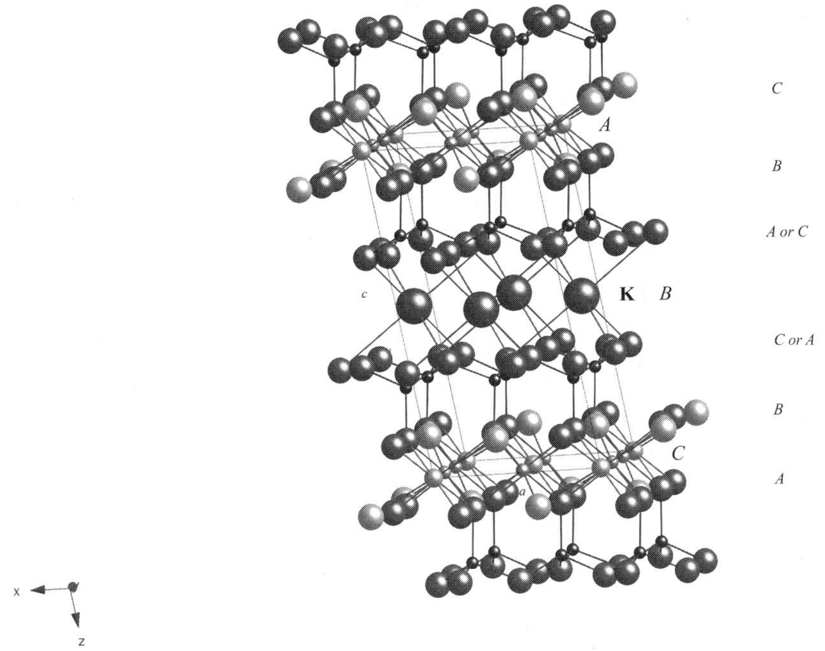

Figure 10.39. The unit cell of montmorillonite, $KMgAlSi_4O_{10}(OH)_2{\cdot}nH_2O$, is extended along *c* to show the **TOT** layers separated by a layer of K. Oxygen atoms of OH^- ions are slightly lighter circles in layers with positions labeled, those not bonded to Si. The largest circles in the layer in the center of the cell are for K.

(*Source*: *CrystalMaker*, by David Palmer, *CrystalMaker* Software Ltd., Begbroke Science Park, Bldg. 5, Sandy Lane, Yarnton, Oxfordshire, OX51PF, UK.)

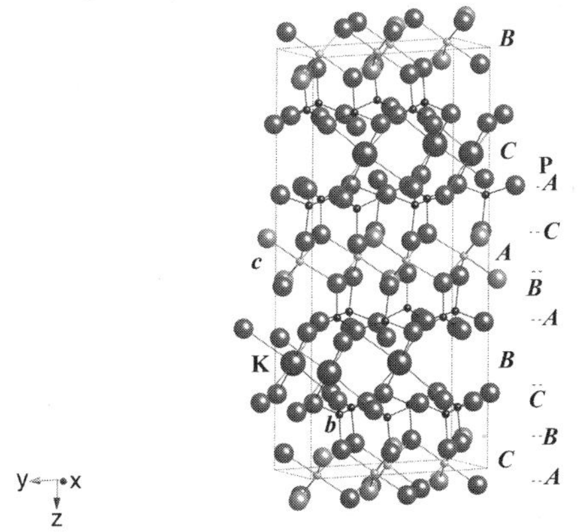

Figure 10.40. The unit cell of muscovite, $KAl_2(AlSi_3O_{10})(OH)_2$, is shown with positions of oxygen layers labeled. Oxygen atoms of OH^- ions are slightly lighter in color than those for O^{2-} and not bonded to Si.

(*Source*: *CrystalMaker*, by David Palmer, *CrystalMaker* Software Ltd., Begbroke Science Park, Bldg. 5, Sandy Lane, Yarnton, Oxfordshire, OX51PF, UK.)

layers. The hydroxide ions are the those bonded to Mg or Al but not bonded to Si. K has *CN* 6, occupying one-third of sites in an **O** layer. The oxide, hydroxide layers bonded to Mg and Al have regular close-packed arrangements. Octahedral layers are in the *CBA* sequence. There are 8 layers in the unit cell and 24 layers repeating. The notation is $\mathbf{3{\cdot}8PT_{2/3}PO^{Mg}_{1/3}{}^{Al}_{1/3}PT_{2/3}PO^{K}_{1/3}(m)}$.

10.3.27. The Crystal Structure of Muscovite [$KAl_2(AlSi_3)O_{10}(OH)_2$]

Muscovite, $KAl_2(AlSi_3)O_{10}(OH)_2$, has the pyrophyllite **TOT** layers (Section 10.3.25) with K^+ ions between them. Charge balance for addition of K is achieved by addition of one Al atom ion replacing Si randomly in each **T** layer. **T** layers are two-thirds occupied and the Al **O** layer is two-thirds occupied. K atoms are halfway between oxygen **P** layers, the normal spacing for an octahedral layer, and K has *CN* 6. K occupies one-third of the **O** sites. This fractional occupancy is determined by stoichiometry, but the large K atoms could not fill all **O** sites in such a structure. Figure 10.40 shows the positions of **P** and **O** layers. The octahedral sites are in the *CBA* sequence, but K and Al atoms occupy alternate **O** layers. The total sequence repeats after 24 layers. The notation is $\mathbf{3{\cdot}8PT_{2/3}PO^{Al}_{2/3}PT_{2/3}PO^{K}_{1/3}(m)}$.

Muscovite is a mica mineral. Its most common polytype is monoclinic, $\mathbf{C}^6_{2h}$, $C2/c$, $a = 5.189$, $b = 8.995$, $c = 20.097$ Å, $\beta = 95.183^\circ$ with four molecules in the unit cell.

10.3.28. The Crystal Structure of Phlogopite [$KMg_3(Si_3Al)O_{10}F_2$]

Phologopite, $KMg_3(Si_3Al)O_{10}F_2$, is a multiple layer Mg rich mica mineral closely related to biotite (Section 10.3.29). F^- ions replace OH^- of biotite. Pholgopite is monoclinic, $\mathbf{C}^3_{2h}$, $C2/m$. $a = 5.31$, $b = 9.21$, $c = 10.13$, $\beta = 100.167^\circ$ with two molecules in the cell. The Mg atoms in the layer at the center of the cell in Figure 10.41***a*** are at *A* positions in octahedral sites between well arranged close-packed oxide-fluoride layers at *B* and *C* positions. Each MgO_4F_2 octahedron has one F in the octahedral face in each O–F layer. Each oxygen atom of each O–F layer is at the apex of a SiO_4 tetrahedron. The oxygen atoms of the next layer form the bases of the tetrahedra and the triangular octahedral faces for K. Each K is in an octahedron formed by three oxygen atoms of each adjacent oxygen layer and each K is aligned with the F atoms in the second nearest packing layer above and below. This requires that these K and F atoms are in the same positions. In the figure the K atoms in the bottom layer of the cell and the O–F layers above and below are at *B* positions. The K layers are in the *BCA* sequence and Mg octahedral sites are in an *ABC* sequence. There are 24 layers (**P, T** and **O**) repeating.

The oxygen layers adjacent to K layers are unusual. In projection along the *c* axis the hexagons of the layers are eclipsed and their oxygen atoms are aligned with the centers of lines of the edges of the hexagons of the adjacent O–F layers. The positions for the oxygen-only layer are not the regular *A*, *B*, or *C* positions corresponding to those of

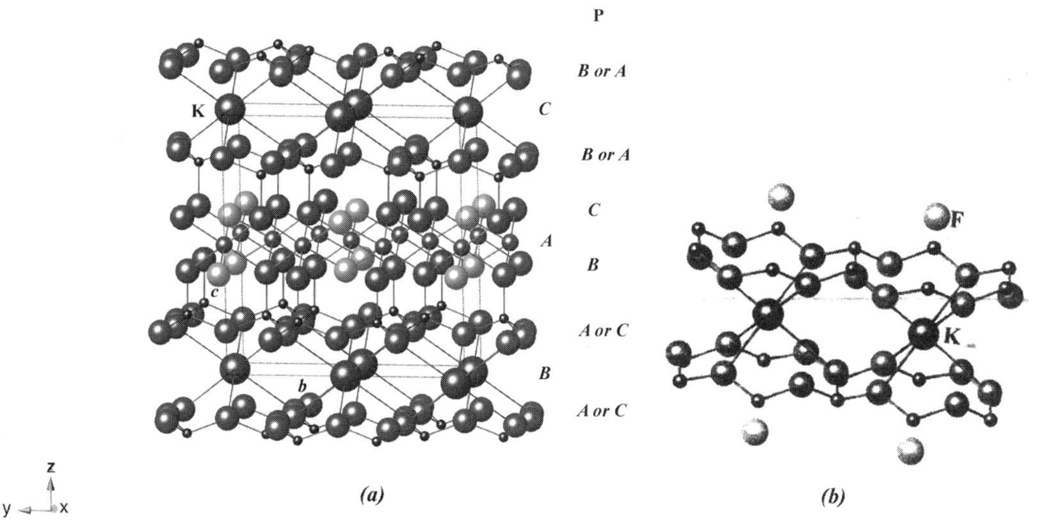

Figure 10.41. (*a*) The unit cell of phlogopite, $KMg_3(AlSi_3)O_{10}F_2$, is shown extended along *c* to show the packing sequence. The positions of potassium and oxygen atoms in octahedral layers are shown. The **TOT** layers are separated by K layers. F atoms are lighter in color than the oxygen atoms. (*b*) The K sites are between eclipsed O_6 hexagons with F atoms centered above and below.

(*Source*: *CrystalMaker*, by David Palmer, *CrystalMaker* Software Ltd., Begbroke Science Park, Bldg. 5, Sandy Lane, Yarnton, Oxfordshire, OX51PF, UK.)

the octahedral layers ($\mathbf{P}_B\mathbf{O}_A\mathbf{P}_C$) at the center of the cell. The oxygen-only layers are unusual because the positions of oxygen atoms form triangular faces of SiO_4 tetrahedra and triangular faces of KO_6 octahedra. Normally two oxygen layers forming octahedral faces are in different positions. Here for KO_6 the oxygen positions are the same for both layers since the hexagons are eclipsed in projection. K atoms are centered relative to the eclipsed hexagons and are bonded to three oxygen atoms alternating in the hexagon of one oxygen layer and the other three alternating oxygen atoms in the other oxygen layer forming a fairly regular octahedron (see Figure 10.41***b***). The K—O bonding distances are 2.986 and 2.998 Å. The K—O distance to the other six oxygen atoms not bonded are ~ 3.277 Å. The K—F distance is 4.008 Å. This is long for significant bonding but alignment of K and F atoms is important in the structure. In the layers of F and O the O—O distances are 3.067 and 3.068 Å and the O—F distances are 3.062 and 3.074 Å. In the oxygen-only layers the O—O distances are 2.699, 2.670, and 2.678 Å, significantly lower. Normally in close-packed structures the O—O distances are the same for layers at the apex and base of SiO_4 tetrahedra. For phlogopite there is an unusual adaption for the octahedron of the large K atom and the mismatch for oxygen layers in octahedral faces and tetrahedral bases. The smaller hexagons of oxygen atoms without an oxygen at the center is a common feature of oxygen layers of the bases of tetrahedra for **TO** and **TOT** structures. Also the positions of oxygen atoms in layers bonded to K are in unusual positions: they are between *A*, *B*, *C* positions. **T** layers are two-thirds filled by Si and Al randomly in a 3:1 ratio. The octahedral sites are filled in the Mg layer and one-third filled in the K layer, effectively using the space available. The octahedral sites have the *BAC* sequence and 24 layers repeat. The notation for phlogopite is $\mathbf{3{\cdot}8PT_{2/3}P^{O}_{2/3}{}^{F}_{1/3}O^{Mg}P^{O}_{2/3}{}^{F}_{1/3}T_{2/3}PO^{K}_{1/3}(m)}$.

10.3.29. The Crystal Structure of Biotite [$K(Mg,Fe)_3(AlSi_3O_{10})(OH)_2$]

Biotite, $K(Mg,Fe)_3(AlSi_3O_{10})(OH)_2$, is the most common of the micas. The structure is similar to that of phlogopite (Section 10.3.28). Some Mg sites of phlogopite are occupied by Fe in biotite and OH^- ions replace F^-. The cell is monoclinic, C^6_{3h}, $C2/c$, $a = 5.357$, $b = 9.245$, $c = 20.234$ Å, and $\beta = 94.978°$. The cell for mica is about as twice as long along the *c* axis compared to phlogopite. There are four molecules in the unit cell. Figure 10.42 shows the structure of biotite. No distinction is shown between Si and Al in tetrahedral sites as they are distributed randomly in the 3:1 ratio. Fe atoms in octahedral layers are darker than Mg. No distinction is shown between O^{2-} and OH^- ions in the layers bonded to Mg and Fe. The OH^- ions are the oxygen

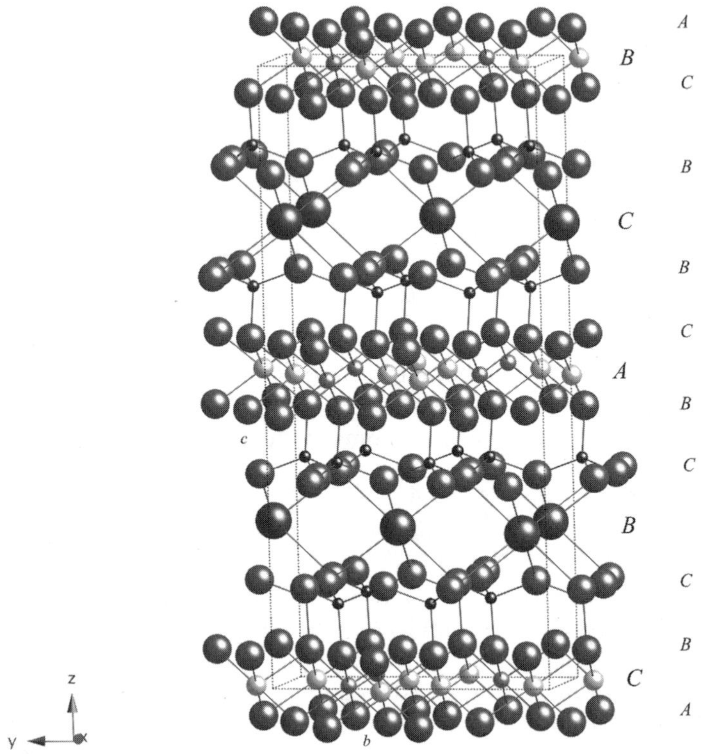

Figure 10.42. The unit cell of biotite, $K(Mg,Fe)_3(AlSi_3O_{10})F_2$, is shown extended along *c* to show the packing sequence. The positions of potassium and oxygen atoms in octahedra layers are shown. Oxygen atoms of OH^- ions are those not bonded to Si.

(*Source*: *CrystalMaker*, by David Palmer, *CrystalMaker* Software Ltd., Begbroke Science Park, Bldg. 5, Sandy Lane, Yarnton, Oxfordshire, OX51PF, UK.)

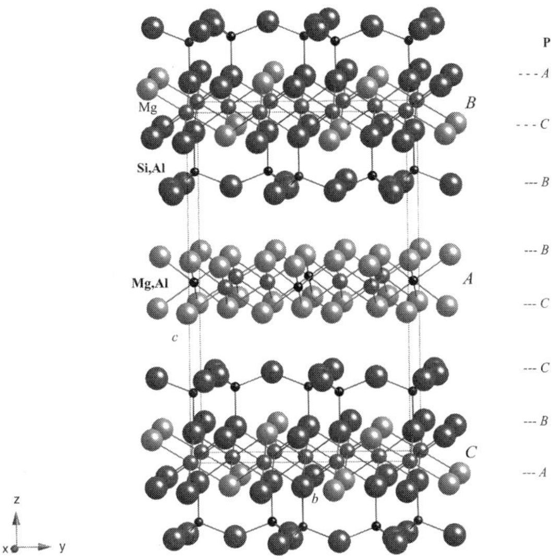

Figure 10.43. The unit cell of chlorite, $Mg_5Al(AlSi_3lO_{10})(OH)_8$, is extended along *c* to show the **TOT O TOT** packing. Packing positions of oxygen layers are shown. Oxygen atoms of OH^- ions are lighter in color and not bonded to Si. (*Source*: *CrystalMaker*, by David Palmer, *CrystalMaker* Software Ltd., Begbroke Science Park, Bldg. 5, Sandy Lane, Yarnton, Oxfordshire, OX51PF, UK.)

atoms not bonded to Si. The O^{2-}/OH^- layers and the intermediate Mg,Fe layers are typical close-packed layers in sequence, *ACB*, *BAC*, and *CBA*. As for phlogopite, the oxygen atoms of oxygen-only layers form the bases of $(Si,Al)O_4$ tetrahedra and form octahedral faces of KO_6. The hexagons of adjacent oxygen-only layers are eclipsed and smaller than the hexagons of the O^{2-}/OH^- layers. K is bonded to three oxygen atoms of the eclipsed hexagons above and below to form the octahedron. As for phlogopite, the eclipsed oxygen hexagons form a hexagonal prism around K. The positions of K atoms are the same as those of nearest O^{2-}/OH^- layers above and below. There are 16 layers in the unit cell and 24 layers repeating since K occupies alternate **O** layers. The notation is similar to that of phlogopite with substitution of OH^- for F^- and Mg and some Fe are in the base, center and top of the cell, $\mathbf{3{\cdot}8\ P^{O}_{2/3}{}^{OH}_{1/3}T_{2/3}PO^{K}_{1/3}PT_{2/3}P^{O}_{2/3}{}^{OH}_{1/3}O^{Mg,Fe}(m)}$.

10.3.30. The Crystal Structure of Chlorite $[Mg_5Al(AlSi_3)O_{10}(OH)_8]$

Chlorite, $Mg_5Al(AlSi_3)O_{10}(OH)_8$, is a type III (**TOT O TOT**) sheet silicate. It is triclinic, $\mathbf{C}_i^1$, $C\bar{1}$, $a = 5.327$, $b = 9.227$, $c = 14.356$ Å, $\alpha = 90.45°$, $\beta = 97.35°$, $\gamma = 89.98°$, and there are two molecules in the cell. Chlorites commonly contain Fe and other metals. The composition of the crystal described requires the 3:1 ratio (Si_3Al) in tetrahedral sites. The **T** layers consist of rings of six tetrahedra. The layers of oxygen atoms bonded to Mg in the center of the **TOT** layers (top and bottom of cell) are atoms at the apices of SiO_4 and AlO_4 tetrahedra and hydroxide ions. Between two **TOT** silicate sheets there is an octahedral layer two-thirds occupied by Mg and 1/3 by Al, a brucite-like layer (see Sec. 5.5.5), see Figure 10.43. There is only hydrogen bonding between the **TOT** and **O** layers. There are 10 layers in the unit cell. Octahedral sites are in the *CAB* sequence, but there are two types of **O** layers so 30 layers repeat. The notation is $\mathbf{3{\cdot}10P^{O}_{2/3}\,{}^{OH}_{1/3}O^{Mg}P^{O}_{2/3}\,{}^{OH}_{1/3}T^{-}_{2/3}P^{O}P^{OH}O^{Mg}_{2/3}\,{}^{Al}_{1/3}P^{OH}P^{O}T^{+}_{2/3}(tri)}$. There is also a monoclinic form of chlorite. Vermiculite has a layered structure similar to that of chlorite except that the intermediate octahedral layer has water coordinated to Mg^{2+}. Heating vermiculite causing the flakes to expand to 20 to 30 times of the original thickness in a direction normal to the cleavage planes (the planes of the silicate layers). Expanded vermiculate is used as a soil conditioner, a filler in light-weight plastics and concretes and as a light packing material. Heating vermiculite slowly does not cause the great increase in volume.

Chapter 11

Structures of Organic Compouds

Many small organic molecules with high symmetry have crystal structures based on close packing schemes. Methane, CH_4, below 89 K, forms cubic crystals with four molecules in the unit cell as expected for a *ccp* or **3P** structure. The **3P** structure is found also for symmetrical compounds such as carbon tetrachloride, 2,2-dimethylpropane, 2,2-dichloropropane, cyclohexane, and the cage-like adamantane, $C_{10}H_{16}$. For large molecules having long chains or extensive branching the structures with efficient packing do not follow the simple patterns of close-packed structures. We cannot expect the same patterns for stacking chairs as for balls. We have seen that large symmetric molecules such as B_{12} (Section 4.4.2, Figure 4.10) and C_{60} (Figure 4.11) give simple close-packed structures. Symmetry is more important than size. The most extensive compilation of organic crystal structures is *Crystal Structures*, Second Edition, Volume 5, *The Structures of Aliphatic Compounds* (1966) and Volume 6 Parts I (1969) and II (1971), *Structures of Benzene Derivatives* by R.W.G. Wyckoff, Wiley, New York.

In this chapter we examine only a few examples for which the **PTOT** system provides good descriptions. The more unsymmetrical the molecules, the more deviation from normal packing with high symmetry is expected in crystals. For inorganic crystal structures we usually think of structures such as NaCl, CaF_2 (fluorite), ZnS, and perovskite ($CaTiO_3$). There are many salts of the same type having each of these structures. We have included many molecular crystal structures of nonmetals such as ice, SiF_4, and B_4Cl_4. Most organic crystals contain molecules for which their shapes are determined by covalent bonding.

11.1. Aliphatic Compounds

11.1.1. The $2{\cdot}2PO_{1/3}$ Crystal Structure of CHI_3 (Iodoform)

The crystal structure of CHI_3 is dominated by the large I atoms. They occupy the **P** layers in an *hcp* arrangement (*A* and *B* positions in Figure 11.1). The CH groups are in $\mathbf{O}_C$ layers, but because the carbon atom is bonded to three I atoms it is closer to the **P** layer containing these

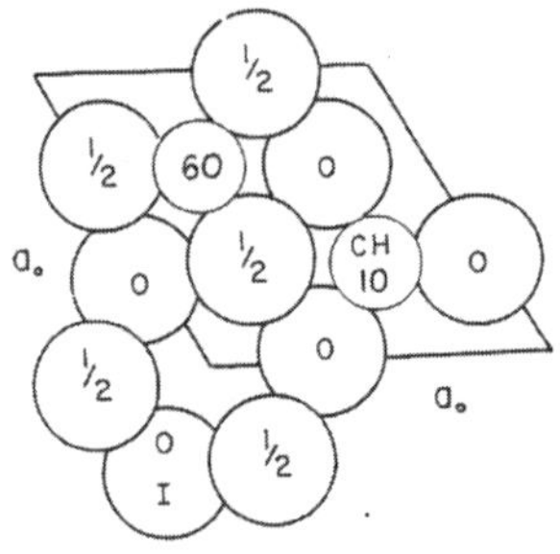

Figure 11.1. The structure of CHI_3 projected along its c_o axis. Numbers give the heights along c_o.

(*Source*: R.W.G. Wyckoff, *Crystal Structures*, Vol. 5, 2nd ed., *The Structures of Aliphatic Compounds*, Wiley, New York, 1965, p. 2.)

I atoms. There is a tetrahedral arrangement of three I and one H for each carbon atom. The **O** layer is normally half the distance between **P** layers. For CHI_3, the height (along c) of the $\mathbf{P}_A$ layer is zero and the height for the layer of carbon atoms is 10. Each carbon atom has an H atom projecting up along c. The next **P** layer ($\mathbf{P}_B$) is at a height of 50 and the carbon layer is at 60. For each carbon atom three I atoms bonded to it in the $\mathbf{P}_A$ layer and three I atoms in the $\mathbf{P}_B$ layer form a badly distorted octahedron. The octahedral arrangement of the six I atoms is not distorted greatly, but each C atom is shifted from the center toward one triangular face formed by the bonded I atoms. One-third of the **O** sites are filled by CH giving the notation $\mathbf{2{\cdot}2PO_{1/3}}$. The unit cell of hexagonal crystals of CHI_3 contains two molecules ($\mathbf{C}_6^6$, $P6_3$, $a_o = 6.818$, $c_o = 7.524$ Å).

11.1.2. The $3{\cdot}3P_{3/4}O_{1/4}(h)$ Crystal Structure of $CHI_3{\cdot}3S_8$

Iodoform (CHI_3, Section 11.1.1) reacts with sulfur to give an addition compound, $CHI_3{\cdot}3S_8$, forming hexagonal crystals with three molecules per unit cell, $\mathbf{C}_{3v}^5$, $R3m$, $a_o' = 24.32$, $c_o' = 4.44$ Å. The large crown-shaped S_8 molecules are close packed with CHI_3 molecules in **O** layers (see Figure 11.2). The C—H bond is directed along the packing direction (c axis). Each **P** layer is three-quarter filled because of the large S_8 molecules, and each **O** layer is one-quarter filled by CHI_3, giving the notation $\mathbf{3{\cdot}3P_{3/4}O_{1/4}(h)}$

11.1.3. The 2P Crystal Structure of Ethane (C_2H_6)

Solid ethane (C_2H_6) at −185°C has a hexagonal unit cell with two molecules (Figure 11.3), $\mathbf{D}_{6h}^4$, $P6_3/mmc$, $a_o = 4.46$, $c_o = 8.19$ Å. The C—C bonds are aligned as shown in the figure. The C—C bond length is 1.54 Å. The notation is **2P** corresponding to an *hcp* arrangement of C_2H_6 molecules. The *A* and *B* packing positions correspond to letters *C* and *B* in the drawings. Diborane, B_2H_6, has the same crystal structure.

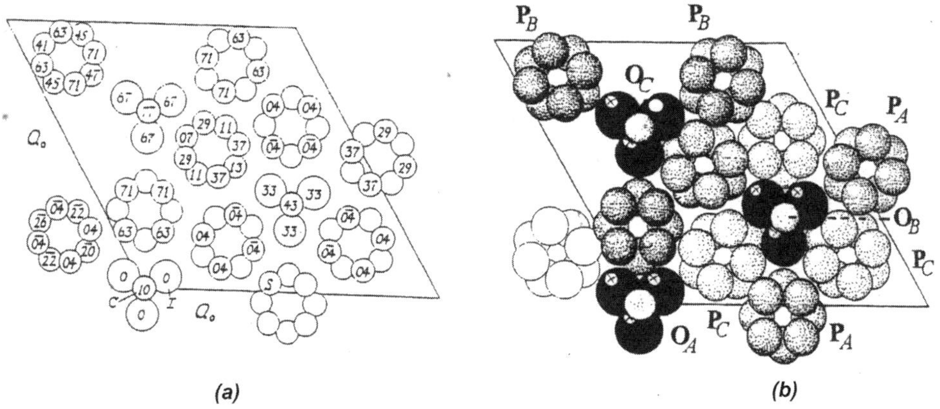

Figure 11.2. The **3·3P$_{3/4}$O$_{1/4}$(h)** structure of $CHI_3 \cdot 3S_8$. (*a*) The structure projected along its c_o axis with heights. (*b*) A packing drawing corresponding to (*a*).

(*Source*: R.W.G. Wyckoff, *Crystal Structures*, Vol. 5, 2nd ed., *The Structures of aliphatic Compounds*, Wiley, New York, 1965, p. 3.)

11.1.4. The 3P(o) Crystal Structure of Cyanogen [$(CN)_2$]

The crystal structure of cyanogen [$(CN)_2$] at −95°C has a close-packed *ABC* arrangement (Figure 11.4). This structure is similar to that of Cl_2 (Section 4.4.1, Figure 4.9). The long linear N ≡ C—C ≡ *N* molecules are staggered as shown in the face-centered cell. The structure is distorted to an orthorhombic cell with four molecules per cell, $\mathbf{D}_{2h}^{15}$, *Pbca*, $a_o = 6.31$, $b_o = 7.08$, $c_o = 6.19$ Å. The notation is **3P(o)**. The N—C bond length is 1.13 Å and that of C—C is 1.37 Å.

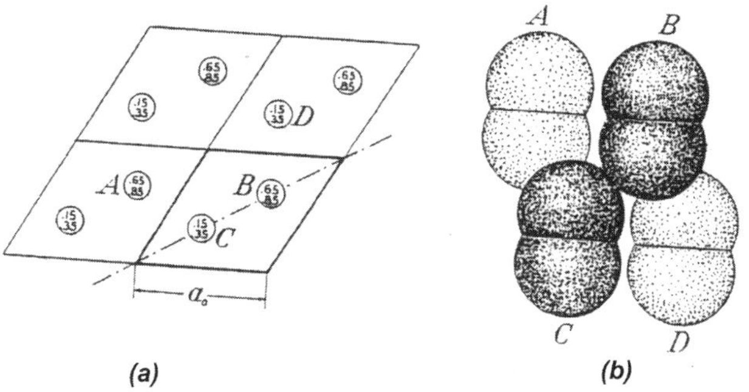

Figure 11.3. The *hcp* structure of ethane. (*a*) A projection on the base for four unit cells with heights along c_o are shown. (*b*) A packing drawing of four C_2H_6 molecules as projected on a vertical plane indicated by a dot-and-dash line in (*a*). Letters identify corresponding molecules in (*a*) and (*b*).

(*Source*: R.W.G. Wyckoff, *Crystal Structures*, Vol. 5, 2nd ed., *The Structures of Aliphatic Compounds*, Wiley, New York, 1965, p. 225.)

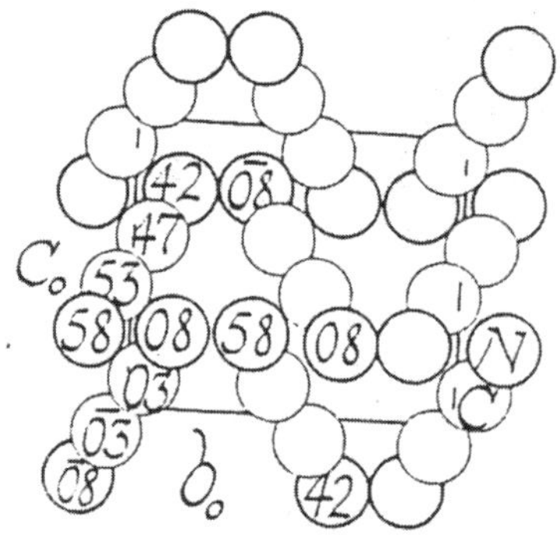

Figure 11.4. The face-centered orthorhombic crystal structure of cyanogen. The structure projected along the $\boldsymbol{a}_o$ axis. Heights along $\boldsymbol{a}_o$ are shown.

(*Source*: R.W.G. Wyckoff, *Crystal Structures*, Vol. 5, 2nd ed., *The Structures of Aliphatic Compounds*, Wiley, New York, 1965, p. 226.)

11.1.5. The 3P Crystal Structure of Acetylene [C_2H_2]

The crystal structure of acetylene, C_2H_2, is cubic with four molecules per cell, $\mathbf{T}_h^6$, $Pa3$, $a_o = 6.14$ Å at −117°C. This corresponds to a *ccp* or face-centered cubic structure or **3P**.

11.1.6. The 3P(o) Crystal Structure of Oxalic Acid [$(COOH)_2$]

There are two crystal modifications of oxalic acid, $(COOH)_2$, orthorhombic and monoclinic. The orthorhombic form, α-$(COOH)_2$, has four molecules per cell, $\mathbf{D}_{2h}^{15}$, $Pbca$, $a_o = 6.546$, $b_o = 7.847$, $c_o = 6.086$ Å. Figure 11.5 shows that the cell is face centered, with the C—C bonds

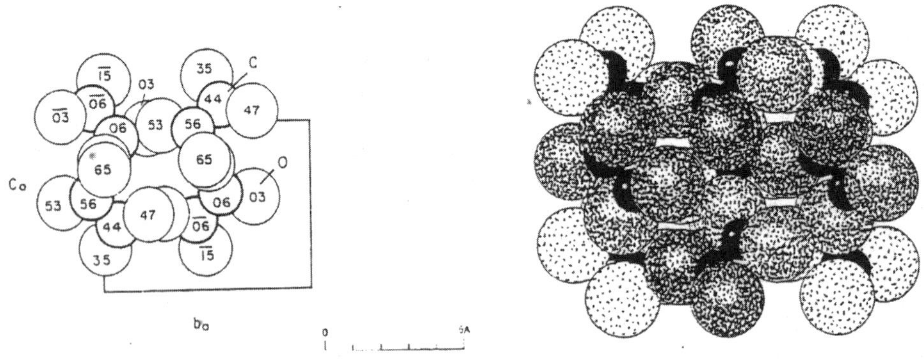

Figure 11.5. The face-centered crystal structure of orthorhombic oxalic acid. On the left is the cell projected along the a_o axis. Only one molecule at a corner and three molecules in the centers of the faces are shown for clarity. On the right, a packing drawing shows all molecules in the cell with the same view.

(*Source*: R.W.G. Wyckoff, *Crystal Structures*, Vol. 5, 2nd ed., *The Structures of Aliphatic Compounds*, Wiley, New York, 1965, p. 397.)

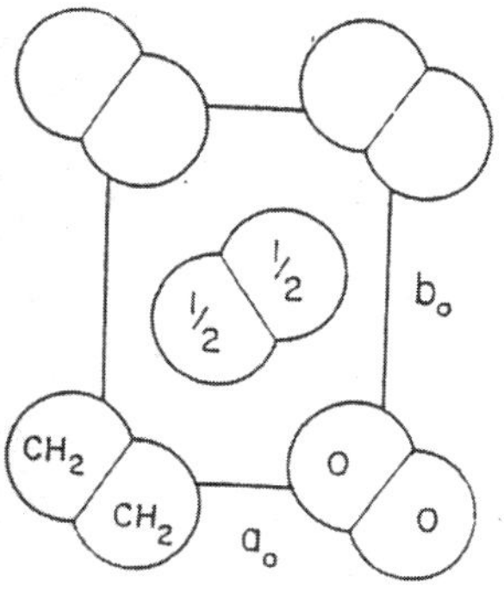

Figure 11.6. The body-centered orthorhombic structure of ethylene projected along the c_o axis. Heights along c_o are shown.

(*Source*: R.W.G. Wyckoff, *Crystal Structures*, Vol. 5, 2nd ed., *The Structures of Aliphatic Compounds*, Wiley, New York, 1965, p. 258.)

centered at 0 (and 100) at the corners and in the top and bottom faces. The other four molecules are centered (C—C bonds centered at 50) in the other four faces. The notation is **3P(o)**.

11.1.7. The Body-Centered [3·2PTOT(o)] Crystal Structure of Ethylene

The planar ethylene (C_2H_4) molecules form a different type of crystal structure than that for ethane. As shown in Figure 11.6, the structure is body-centered **[3·2PTOT(o)]** with two molecules in the orthorhombic unit cell, $\mathbf{D}_{2h}^{12}$, *Pnnm*, $a_o = 4.87$, $b_o = 6.46$, $c_o = 4.14$ Å. The cell is elongated along b_o because the C═C bonds are skewed in two orientations in this direction. The orientation of the C═C bond in the molecule in the center of the cell differs from those at the corners. The C═C bond length is 1.33 Å.

11.1.8. The Body-Centered Crystal Structure of $(CN)_2C{=}C(CN)_2$

Tetracyanoethylene has a body-centered structure **[3·2PTOT(m)]** similar to that of ethylene as seen in Figure 11.7. Like ethylene, $(CN)_2C{=}C(CN)_2$ is planar. The planes of the molecules are tipped relative to the *b* axis. The molecule in the center is tipped in the opposite direction relative to those at the corners. The monoclinic cell contains two molecules, $\mathbf{C}_{2h}^{5}$, $P2_1/b$, $a_o = 7.51$, $b_o = 6.21$, $c_o = 7.00$ Å, and $\beta = 97.17°$.

11.1.9. The Body-Centered [3·2PTOT(t)] Crystal Structure of Urea

Urea, $(NH_2)_2CO$, has a tetragonal structure with two molecules per cell, $\mathbf{D}_{2d}^{3}$, $P\bar{4}2_1m$, $a_o = 5.582$, and $c_o = 4.686$ Å, at room temperature. Figure 11.8 shows the body-centered structure. The C atoms at ±50

Figure 11.7. The body-centered monoclinic structure of $C_2(CN)_4$ projected along the $\boldsymbol{b}_o$ axis. Heights along $\boldsymbol{b}_o$ are shown.

(*Source*: R.W.G. Wyckoff, *Crystal Structures*, Vol. 5, 2nd ed., *The Structures of Aliphatic Compounds*, Wiley, New York, 1965, p. 259.)

form the corners of the cell with one at 0 in the center of the cell. The notation is **3·2PTOT(t)**.

11.1.10. The Body-Centered Crystal Structure of $(NH_2C(O)N)_2$

The structure of azidocarbonamide, $NH_2\overset{\overset{O}{\|}}{C}N{=}N\overset{\overset{O}{\|}}{C}NH_2$, is shown in Figure 11.9. The crystal is monoclinic with two molecules per cell, $\mathbf{C}_{2h}^5$, $P2_1/b$, $a_o = 3.57$, $b_o = 9.06$, $c_o = 7.00$ Å, $\beta = 94°50'$. Each half of the molecule is essentially planar with a small rotation about the C—N= bond. The molecules are thought to be connected by hydrogen bonding (N—H—N and N—H—O) to form sheets parallel to the 101 plane. Sheets are linked by van der Waals forces. The structure is body centered or **3·2PTOT(m)**.

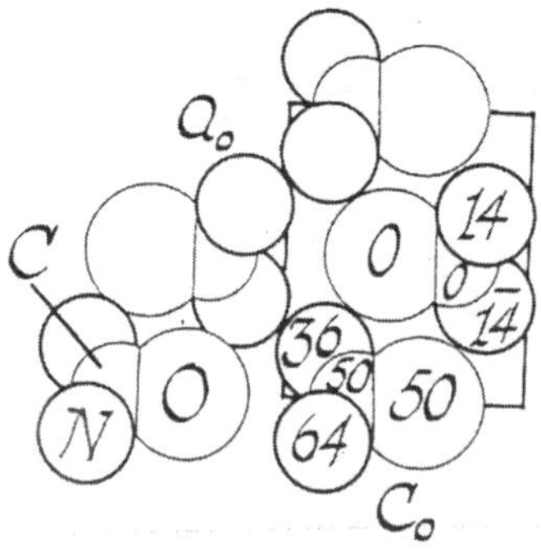

Figure 11.8. A projection along an a_o axis of the tetragonal cell of urea. The carbon atoms at ± 50 and 0 form a body-centered tetragonal cell.

(*Source*: R.W.G. Wyckoff, *Crystal Structures*, Vol. 5, 2nd ed., *The Structures of Aliphatic Compounds*, Wiley, New York, 1965, p. 163.)

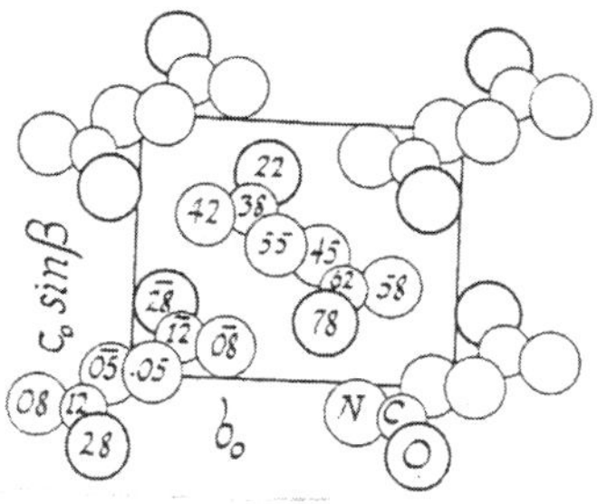

Figure 11.9. The cell of the body-centered monoclinic structure of $(NH_2C(O)N)_2$ projected along its $\boldsymbol{a}_o$ axis. Heights along $\boldsymbol{a}_o$ are shown.

(*Source*: R.W.G. Wyckoff, *Crystal Structures*, Vol. 5, 2nd ed., *The Structures of Aliphatic Compounds*, Wiley, New York, 1965, p. 30.)

11.1.11. The Crystal Structures (Body-Centered and *ccp* Types) of Tetrakis(hydroxymethyl)methane, $C(CH_2OH)_4$

Tetrakis(hydroxymethyl)methane, [$C(CH_2OH)_4$, also known as pentaerythritol], is one of the simpler substitution products of methane for which structures have been determined. The tetragonal cell contains two molecules, $\mathbf{S}_4^2$, $I\bar{4}$, $a_o = 6.083$, $c_o = 6.726$ Å. The structure shown in Figure 11.10 is body centered, **3·2PTOT(t)**. At higher temperatures, between 179.5 and 260.5°C, this body-centered structure is transformed to a face-centered cubic structure. The cubic cell ($a_o = 8.963$ Å at 230°C) contains four molecules. It has been proposed that at the higher temperature the molecules achieve a statistical orientation that makes them effectively spherical crystallographically. The notation is **3P**.

For metals and some other solids transformation from *ccp* to *bcc*, or reverse, under different conditions, is common. It is less common among molecular solids, but this example indicates that these transformations are generally applicable.

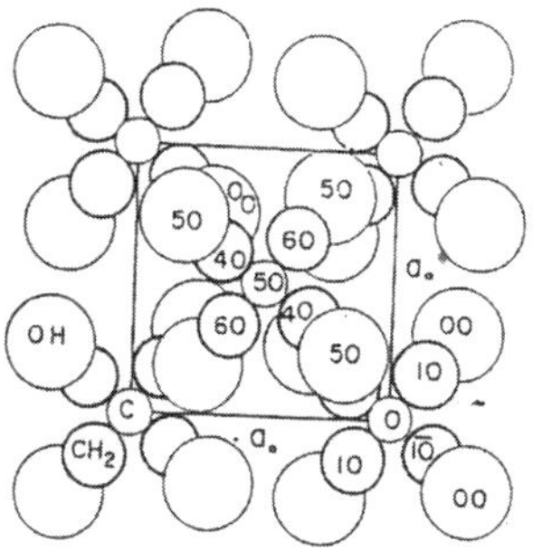

Figure 11.10. The body-centered tetragonal structure of $C(CH_2OH)_4$ projected along the $\boldsymbol{c}_o$ axis. Heights along c_o are shown.

(*Source*: R.W.G. Wyckoff, *Crystal Structures*, Vol. 5, 2nd ed., *The Structures of Aliphatic Compounds*, Wiley, New York, 1965, p. 8.)

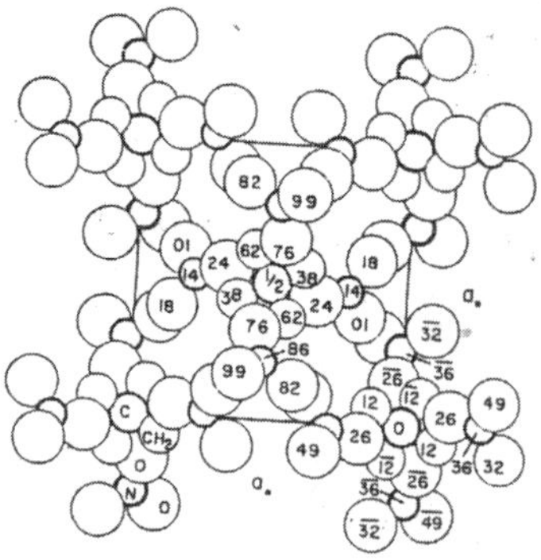

Figure 11.11. The body-centered tetragonal structure of $C(CH_2ONO_2)_4$ projected along the c_o axis. Heights along c_o are shown.

(*Source*: R.W.G. Wyckoff, *Crystal Structures*, Vol. 5, 2nd ed., *The Structures of Aliphatic Compounds*, Wiley, New York, 1965, p. 10.)

11.1.12. The Body-Centered Crystal Structure of $C(CH_2ONO_2)_4$

The structure of $C(CH_2ONO_2)_4$, like $C(CH_2OH)_4$, the alcohol from which it is formed, has a bimolecular tetragonal unit cell, $\mathbf{D}_{2d}{}^4$, $P\bar{4}2_1c$, $a_o = 9.33$, $c = 6.66$ Å. The body-centered structure, **3·2PTOT(t),** is shown in Figure 11.11. The compact molecule with high symmetry packs efficiently in the body-centered structure.

11.1.13. The 3·2PTOT(t) Body-Centered Crystal Structure of $C(SCH_3)_4$

There are three modifications of $C(SCH_3)_4$ for which structures have been determined. Modification I, stable below 23.2°C, is tetragonal with two molecules in the cell, $\mathbf{D}_{2d}{}^4$, $P\bar{4}2_1c$, $a_o = 8.536$, and $c = 6.949$ Å. The structure is shown in Figure 11.12. It is body centered, **3·2PTOT(t)**. Modification II, stable between 23.2 and 45.5°C, is also tetragonal with two molecules in the cell. The cell is shorter in the a_o direction and longer in the c_o direction. The structure is also body centered. Modification III, stable from 45.5°C to the melting point at 66°C, is cubic (*bcc*), with two molecules in the cell ($a_o = 8.15$ Å).

11.1.14. The $3{\cdot}4P_{1/16}O_{1/16}(t)$ Crystal Structure of $(C_3H_7)_4NBr$

Tetra(*n*-propyl)ammonium bromide, $(C_3H_7)_4NBr$, as shown in Figure 11.13, has a filled body-centered tetragonal structure of Br^- ions. A body-centered cell is the **3·3PTOT** structure with all sites occupied. Adding the cations requires partial filling of layers. With the Br^- ions filling one-sixteenth of **P** sites places, $N(C_3H_7)_4^+$ ions in one-sixteenth of **O** sites, with **T** layers vacant, giving the notation $\mathbf{3{\cdot}4P_{1/16}O_{1/16}(t)}$. The low occupancy of sites is unusual. The structure is not very open. The partial occupancies locate the positions of all sites and the spacings. Each ion is at the center of a tetrahedron formed by four counterions. The propyl groups are arranged to be equally distant

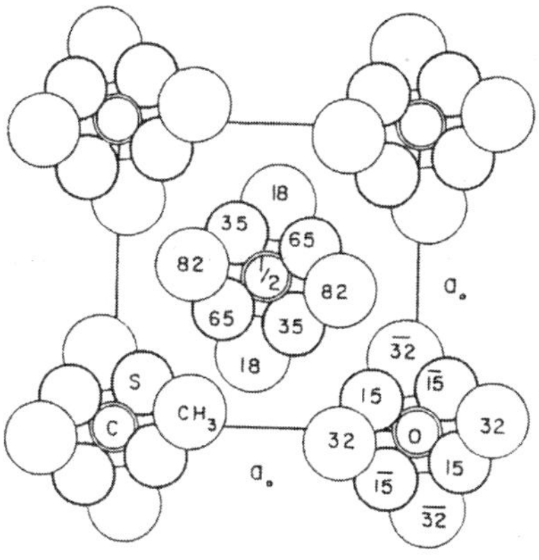

Figure 11.12. The body-centered tetragonal structure of $C(SCH_3)_4$ projected along the c_o axis. Heights along c_o are shown.

(*Source*: R.W.G. Wyckoff, *Crystal Structures*, Vol. 5, 2nd ed., *The Structures of Aliphatic Compounds*, Wiley, New York, 1965, p. 13.)

from Br atoms. We can see in the figure that the Br atoms are in a body-centered arrangement. The tetragonal cell contains two molecules, $\mathbf{S}_4^2$, $I\bar{4}$, $a_o = 8.24$, and $c_o = 10.92$ Å.

11.2. Aromatic Compounds

11.2.1. The 3P(o) Crystal Structure of Benzene

Benzene freezes at 5.5°C. The crystals are orthorhombic with four molecules per cell, $\mathbf{D}_{2h}^{15}$, *Pbca*, $a_o = 7.460$, $b_o = 9.666$ and $c_o = 7.034$ Å.

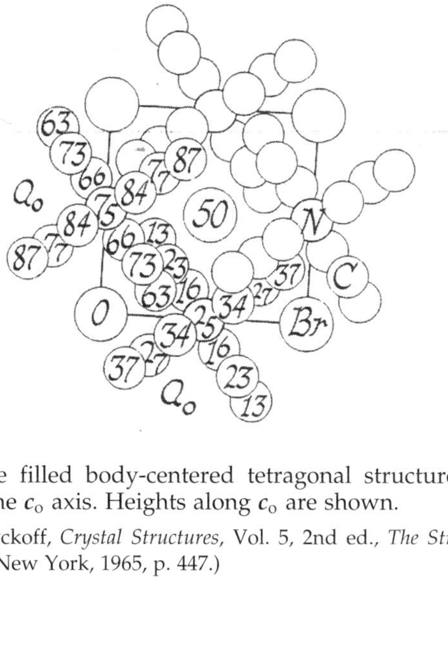

Figure 11.13. The filled body-centered tetragonal structure of $(C_3H_7)_4NBr$ projected along the c_o axis. Heights along c_o are shown.

(*Source*: R.W.G. Wyckoff, *Crystal Structures*, Vol. 5, 2nd ed., *The Structures of Aliphatic Compounds*, Wiley, New York, 1965, p. 447.)

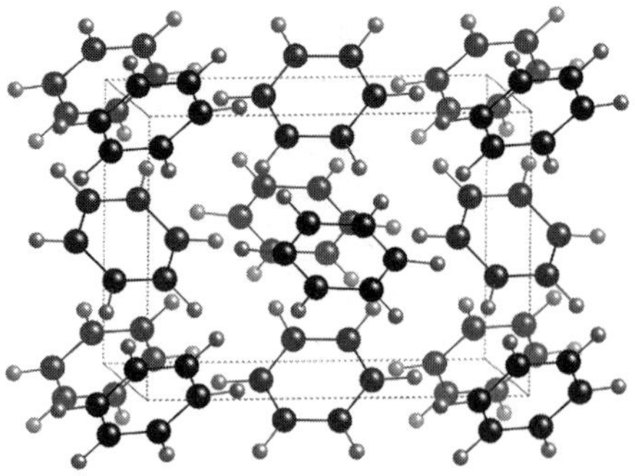

Figure 11.14. The face-centered unit cell of benzene is shown.

(*Source*: *CrystalMaker*, by David Palmer, *CrystalMaker* Software Ltd., Begbroke Science Park, Bldg. 5, Sandy Lane, Yarnton, Oxfordshire, OX51PF, UK.)

at −3°C. Benzene rings are tipped relative to the cell faces (Figure 11.14). The planar C_6H_6 molecules are a *ccp*-type (*ABC*, face centered) arrangement designated as **3P(o)**.

11.2.2. The $3P_{1/2}$(m) Crystal Structure of Naphthalene

Naphthalene crystals are monoclinic with two molecules per cell, $\mathbf{C}_{2h}^5$, $P2_1/b$, $a_o = 8.235$, $b_o = 6.003$, $c_o = 8.658$ Å, and $\beta = 122.92°$. The planes of the $C_{10}H_8$ molecules are nearly planar in the *bc* plane (Figure 11.15). The average positions of eight molecules are at the corners of the cell with molecules centered in two opposite aces (*ab*). These are

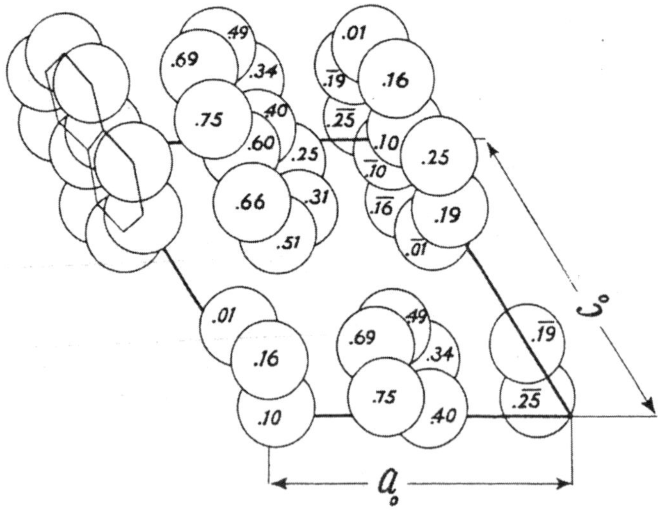

Figure 11.15. A projection along the b_o axis of the monoclinic cell of naphthalene. The naphthalene molecules are centered at the corners and centers of two faces.

(*Source*: R.W.G. Wyckoff, *Crystal Structures*, Vol. 6, Part 1, *The Structures of Aromatic Compounds*, Wiley, New York, 1966, p. 383.)

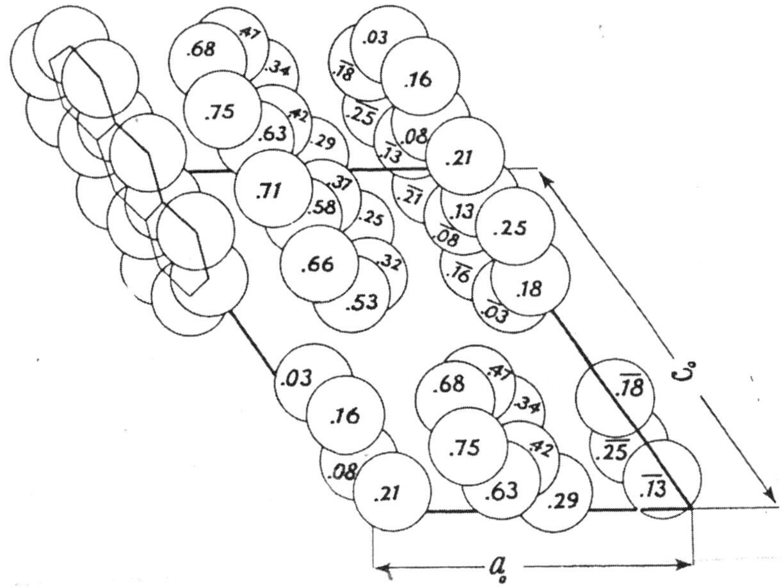

Figure 11.16. A projection along the b_o axis of the monoclinic cell of anthracene. Like naphthalene, the molecules are centered at the corners and centers of two faces.

(*Source*: R.W.G. Wyckoff, *Crystal Structures*, Vol. 6, Part 1, *The Structures of Aromatic Compounds*, Wiley, New York, 1966, p. 455.)

ABC positions, but the layers are partially (one-half) filled. The notation is $\mathbf{3P_{1/2}(m)}$.

11.2.3. The $3P_{1/2}(m)$ Crystal Structure of Anthracene

Anthracene is also monoclinic, with two molecules per cell with the same space as naphthalene, C_{2h}^5, $P2_1/b$, $a_o = 8.562$, $b_o = 6.038$, $c_o = 11.184$ Å, and $\beta = 124°42'$ at 290 K. Figure 11.16 shows that the structure corresponds to that of naphthalene. The planar $C_{14}H_{10}$ molecules are tipped in the same way as naphthalene. The notation is the same $\mathbf{3P_{1/2}(m)}$.

11.2.4. The $3P_{1/2}(m)$ Crystal Structure of *p*-Diphenylbenzene

The crystal structure of *p*-diphenylbenzene (Figure 11.17) is very similar to those of naphthalene and anthracene. It is monoclinic, with two molecules per cell, C_{2h}^5, $P2_1b$, $a_o = 8.08$, $b_o = 5.60$, $c_o = 13.59$ Å, and $\beta = 91°55'$. The long planar $C_{18}H_{14}$ molecules are tipped in a similar way to naphthalene and anthracene. The notation is $\mathbf{3P_{1/2}(m)}$.

11.2.5. The $3P_{1/2\ 1/2}(m)$ Crystal Structure of the Addition Compound Benzene–Chlorine

Benzene forms an addition compound with chlorine, $C_6H_6{\cdot}Cl_2$. The crystal structure is monoclinic, with two molecules per cell,

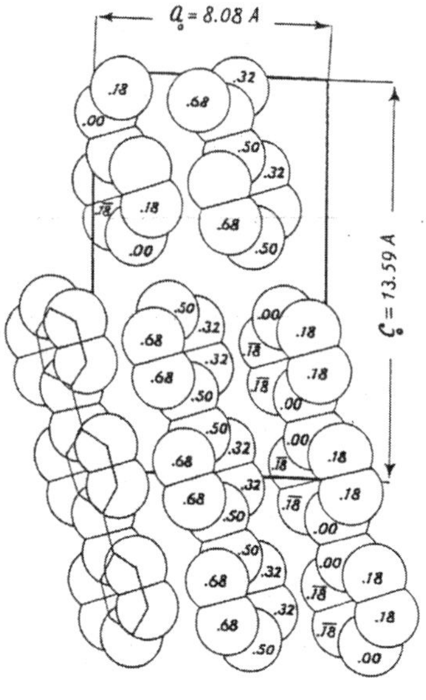

Figure 11.17. A projection along the b_o axis of the monoclinic cell of *p*-diphenylbenzene. Like naphthalene and anthracene, the molecules are centered at the corners and centers of two faces.

(*Source*: R.W.G. Wyckoff, *Crystal Structures*, Vol. 6, Part 1, *The Structures of Aromatic Compounds*, Wiley, New York, 1966, p. 196.)

C_{2h}^3, B2/*m*, $a_o = 7.41$, $b_o = 8.65$, $c_o = 5.65$ Å, and $\beta = 99°30'$at $-90°C$. The structure is very interesting. Figure 11.18 shows it as b_o vs. $a_o \sin\beta$ to show the positions clearly. The unit cell is face centered, with Cl_2 molecules at the corners and in the top and bottom faces (average positions 0 and 100). C_6H_6 molecules are in the centers of the other four faces (at 50). This face-centered cell corresponds to an *ABC*

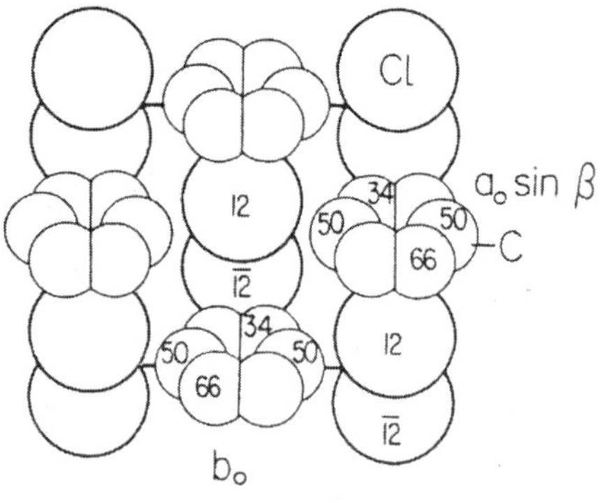

Figure 11.18. A projection along the c_o axis of the monoclinic cell of the addition compound $C_6H_6{\cdot}Cl_2$. The cell is face centered, with C_6H_6 molecules in the side faces.

(*Source*: R.W.G. Wyckoff, *Crystal Structures*, Vol. 6, Part 1, *The Structures of Aromatic Compounds*, Wiley, New York, 1966, p. 3.)

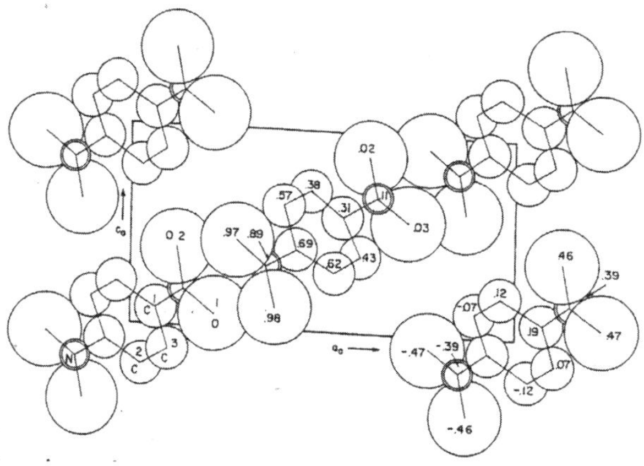

Figure 11.19. A projection along the b_o axis of the body-centered monoclinic cell of *p*-dinitrobenzene.

(*Source*: R.W.G. Wyckoff, *Crystal Structures*, Vol. 6, Part 1, *The Structures of Aromatic Compounds*, Wiley, New York, 1966, p. 115.)

sequence with Cl_2 and C_6H_6 molecules sharing each **P** layer giving the notation $\mathbf{3P_{1/2\ 1/2}(m)}$. The Cl—Cl bonds are normal to the planes of benzene molecules. The Cl_2 bond distance is 1.99 Å, and the distance between a Cl atom and the center of the C_6H_6 ring is 3.28 Å. The Br_2 compound, $C_6H_6{\cdot}Br_2$, has the same structure.

11.2.6. The 3·2PTOT(m) Crystal Structure of *p*-Dinitrobenzene

The crystals of *p*-dinitrobenzene ($O_2NC_6H_4NO_2$) are monoclinic, with two molecules per cell, $\mathbf{C}_{2h}^5$, $P2_1b$, $a_o = 11.05$, $b_o = 5.42$, $c_o = 5.65$ Å, and $\beta = 92°18'$. Figure 11.19 shows that the benzene rings are centered at the corners and the center of the cell giving a body-centered cell, **3·2PTOT(m)**. The long direction of the molecule in the center is inclined in the opposite direction compared to those at the corners.

Chapter 12

Predicting Structures and Assigning Notations

12.1. General Considerations for Solids

The crystal state is the most dense state except in extraordinary situations such as for ice. In ice, the boomerang-shaped H_2O molecules are not stacked in the most compact arrangement because of hydrogen bonding. Each oxygen atom has two bonded H atoms and two hydrogen-bonded H atoms of neighboring water molecules. For hard spheres, the most compact arrangements are close packed: cubic close packed (*ccp*) and hexagonal close packed (*hcp*). For both cases, each atom has a hexagonal planar arrangement of six neighbors and three triangular arrangements of neighbors above and below the plane of the hexagon. These two triangles are eclipsed for *hcp* and staggered for *ccp* (Figure 3.2). The *ccp* and *hcp* structures are common for metals. The coordination number (*CN*) for *ccp* and *hcp* is 12, but metal atoms have too few electrons for forming ordinary bonding to 12 neighbors atoms. Valence electrons of metals are delocalized, providing high electrical conduction. The bonding in metals is treated using the band theory. The energy levels are essentially molecular orbitals delocalized over many atoms. The body-centered cubic (*bcc*) structure is encountered for many metals, especially at higher temperature or pressure. The *bcc* packing of hard spheres is less efficient in packing compared to *ccp* and *hcp*. However, metal atoms are compressible; they are not hard spheres, and *bcc* modifications, in some cases, have comparable or even higher densities than *ccp* or *hcp* modifications (Section 4.3.3).

Intermetallic compounds can be "valence" compounds, with structures corresponding to those of NaCl, CaF_2, etc., or compounds of various compositions with all atoms in close-packed layers. Because of the deficiencies of electrons and delocalized bonding, metals are not limited by the valence rules for ionic and covalent compounds. Some intermetallic compounds have structures found only for metals (Chapter 9).

Close-packed structures are found for noble gases and nearly spherical or highly symmetrical molecules, for example, H_2, CH_4, OsO_4, $Cr(C_6H_6)_2$, and C_{60} (Section 4.5).

For ionic crystals there is strong attraction between cations and anions, and strong repulsion between ions having the same charge. These interactions determine structures because ions must be shielded from those with the same charge. The relative sizes of the ions are important in determining the *CN*. Removal of electron(s) decreases the size of a cation relative to the atom and addition of electron(s) increases the size of an anion relative to the atom. Commonly, for an MX compound the anion is larger than the cation and the anions are close packed in crystals with cations in octahedral or tetrahedral sites.

The relative sizes of ions are not the only factors in determining structures. The covalence or polarization can be important also, because electron density is transferred from the anions to the cation. Fajans' rules (Douglas *et al*, pp. 200–202, see Appendix A) give guidance in judging the extent of covalence of bonds. Fajans did not believe in covalent bonding so we referred to nonpolar or covalent character. His description of CH_4 would have four H^- ions being highly polarized by C^{4+}. (1) Covalent character increases with decreasing cation size or increasing cation charge—the higher charge density of the cation, the greater the polarization of anions. (2) Covalent character increases with an increase in anion size or anion charge—polarization increases with size and charge of an anion because the electrons are held more loosely. (3) Covalent character is greater for cations with nonnoble gas [18- or (18 + 2)-electron] configurations than for noble-gas type (eight-electron) cations of the same size. Cations with ns^2np^6 configurations cause less distortion of anions, and undergo less distortion themselves than those with 18-electron configurations. We usually consider polarization of anions; however, large cations and particularly 18-electron cations are also polarizable. Cs_2O has the anti-$CdCl_2$ layer structure (Section 5.5.1), Cs^+ is a very large eight-electron cation. The *CN* of Cs^+ is 3. The Cs–O–Cs sandwich layers are stacked with no anions between the sandwiches. There is only van der Waals interaction between Cs layers of adjacent sandwiches.

Polarization of anions reduces the effective charge of cations. This favors lower *CN* of cations because of the charge transfer per neighboring anion. Polarization of S^{2-} is important in limiting the *CN* of Zn in ZnS to 4. AlF_3 is an ionic compound with *CN* 6 in crystals, Al^{3+} ions occupy one-third of **O** sites in each layer. With the larger Cl^- ion, $AlCl_3$ exists as molecular Al_2Cl_6 in the vapor state. In crystals, $AlCl_3$ has an ionic structure with *CN* 6. Al_2Br_6 has a molecular crystal structure (Figure 6.23). HgF_2 has the typical ionic fluorite structure (CaF_2, Sections 6.3.1 and 12.3.2), $HgCl_2$ forms discrete $HgCl_2$ molecules, and $HgBr_2$ and HgI_2 have layer sandwich structures.

The shapes of covalent molecules are determined by the number of valence electrons and orbitals available, giving H_2, BF_3, CH_4, NH_3, PH_3, H_2O, HF, ClF, PF_5, SF_6, and IF_7. Carbon has four valence electrons and four orbitals, so tetrahedral sp^3 bonding dominates the chemistry of carbon. This matches the three-dimensional bonding in zinc blende (**3•2PT**), and it is the structure of diamond with carbon in **P** and **T** layers.

Lone pairs (electron pairs occupying nonbonding orbitals) can be important in determining molecular shape. Lone pairs act as bonded atoms, actually as large bonded atoms (Douglas *et al*, pp. 80–81, see Appendix A), thus :NH_3 is pyramidal, $\ddot{\underset{..}{C}}lF_3$ is T-shaped (two lone pairs in equatorial positions of a trigonal pyramidal arrangement of five electron pairs), :BrF_5 is square pyramidal (the lone pair is in one position of the octahedral arrangement of six electron pairs), :$\ddot{\underset{..}{X}}eF_2$ is linear (the three lone pairs are in equatorial positions of the trigonal pyramidal arrangement of five electron pairs), and $\ddot{\underset{..}{X}}eF_4$ is planar (the two lone pairs are in axial positions of an octahedral arrangement of six electron pairs). Because of the lone pair on Pb, PbO contains square pyramidal :PbO_4 units sharing oxygen atoms. Without the lone pair the :PbO_4 unit would be expected to be planar. The pyramidal arrangement is important in the crystal structure (Figure 6.17).

12.2. General Features of Close-Packed Structures

All close-packed layers have the same hexagonal pattern, a hexagon of atoms with another atom at the center (Figure 3.1***a***). There are only three packing positions, *A*, *B*, and *C*. The two common sequences for layers are *AB AB* $\cdots$ for *hcp* and *ABC ABC* $\cdots$ for *ccp*. The unit cell for *ccp* is face-centered cubic (*fcc*). Halfway between any two close-packed packing layers (**P**) there is a layer of octahedral sites (**O** layers). Also, there are two layers of tetrahedral sites (**T** layers) between adjacent **P** layers, one at one-quarter and one at three-quarters of the distance between **P** layers. Thus, the **T** layers are halfway between **P** and **O** layers. Figure 3.4 shows the relative positions for **O** and **T** layers between two **P** layers. The octahedra and tetrahedra are aligned with their C_3 axes parallel to the packing direction ($\perp$ to layers). These spacings are not approximations, being determined by geometry. For many ionic compounds one set of ions is in **P** layers and the counter ions are in **O** and/or **T** layers. If these counterions are larger than the **O** and **T** sites, the structure must open, increasing the separation between **P** layers, but the relative spacings of the **P**, **O**, and **T** layers are the same. The sequence is **PTOT**. For *hcp* structures it is $\mathbf{P}_A$**TOT** $\mathbf{P}_B$**TOT**, and for *ccp* structures it is $\mathbf{P}_A$**TOT** $\mathbf{P}_B$**TOT** $\mathbf{P}_C$**TOT**. For *hcp* and *ccp* metals **T** and **O** layers are empty, giving $\mathbf{P}_A\mathbf{P}_B$ and $\mathbf{P}_A\mathbf{P}_B\mathbf{P}_C$, respectively.

Thousands of compounds result from filling or partial filling of **P, O,** and/or **T** layers. All filled layers have the same hexagonal arrangement so there is the same number of sites in each **P, O,** or **T** layer. In the repeating **PTOT** unit there are equal numbers of **P** and **O** sites, but twice as many **T** sites because there are two **T** layers. For a **T** site, three atoms of the closer **P** layer (assume $\mathbf{P}_A$) form the base of the tetrahedron and an atom of the more distant **P** layer (assume $\mathbf{P}_B$) is at the apex of the tetrahedron (Figure 3.2). The packing positions of a **T** site is the same as that of the apex atom; for the case cited it is $\mathbf{T}_B$. For an **O** site, three atoms of each adjacent layer form a triangular face of the octahedron. The packing positions of the **O** sites are different from those of

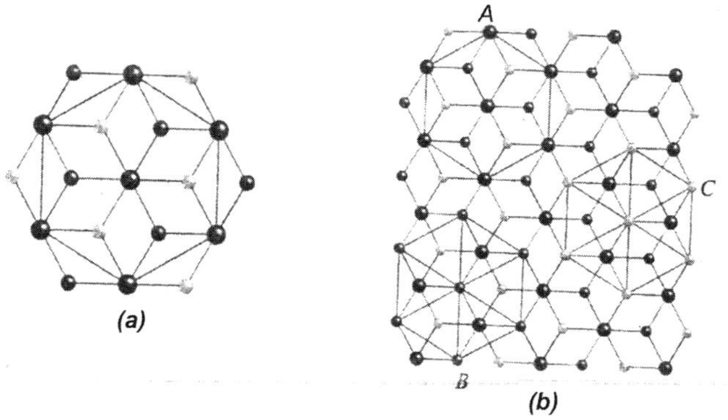

Figure 12.1. (*a*) Hexagonal pattern for a packing layer ($\mathbf{P}_A$, large atoms) with positions of $\mathbf{T}_B$ (small dark atoms), and $\mathbf{O}_C$ (small light atoms) sites. (*b*) The network extended to show hexagonal patterns for all sites.

(*Source*: *CrystalMaker*, by David Palmer, *CrystalMaker* Software Ltd., Begbroke Science Park, Bldg. 5, Sandy Lane, Yarnton, Oxfordshire, OX51PF, UK.)

adjacent **P** layers (Figure 3.3), for example, $\mathbf{P}_A\mathbf{O}_C\mathbf{P}_B$ and $\mathbf{P}_B\mathbf{O}_A\mathbf{P}_C$. Figure 3.5 shows the positions of $\mathbf{P}_A$, $\mathbf{T}_B$, and $\mathbf{O}_C$ layers.

Figure 12.1***a*** shows a projection along the packing direction of three layers *A*, *B*, and *C*. *A* positions are large dark balls, *B* positions are small dark balls, and *C* positions are light balls. These *A*, *B*, and *C* positions could represent three **P** layers or one $\mathbf{P}_A$ layer with $\mathbf{T}_B$ and $\mathbf{O}_C$ layers. Figure 12.1***b*** shows the network. The hexagonal arrangement is shown around each position. The projections in Figure 12.1 are those for *ccp* or *hcp* structures. For a *ccp* structure the packing layers are perpendicular to a body-diagonal of the cubic cell.

Structures with an *ABC* sequence of packing layers are not necessarily cubic or orthorhombic, and those with an *AB* sequence are not necessarily hexagonal. For a tetrahedral site, the line joining the **P** and **T** sites at the same positions is aligned with a C_3 axis of the tetrahedron and perpendicular to the packing layers. There are a few examples in this book of structures with tetrahedra orientated with a C_2 axis aligned with the packing direction (Sections 10.1.3, 10.3.3, 10.3.6, and 10.3.11). Because there are two anions, forming a tetrahedral edge, of each **P** layer, the center of the tetrahedron is halfway between **P** layers. These unusual cases require an extension of the **PTOT** system, which normally has only **T**(C_3) sites.

12.3. Predicting Structures

12.3.1. Crystal Structures of MX Compounds

The common structures for ionic MX compounds are NaCl (**3•2PO**, Section 5.1.1), CsCl (**3•2PTOT**, Section 7.2.1), and ZnS (**3•2PT**, Section 6.1.1, and **2•2PT**, Section 6.2.1). The radius ratio (r_+/r_-) is important in determining the structure. The NaCl structure is found for over 200 compounds, including over 50 intermetallic compounds. The Cl^- ions are *ccp* and Na^+ ions in octahedral sites are well shielded from Na^+ ion neighbors. The larger Cs^+ ions would not be well shielded by six Cl^- ions so the ***CN*** increases to 8 for CsCl, in which the Cs^+ ions are surrounded by a cubic arrangement of Cl^- ions. For ZnS, the repulsion would be too great among six S^{2-} ions around Zn^{2+} and the ***CN*** for Zn is 4 in the **T** sites. There are two structures for ZnS: zinc blende (sphalerite, **3·2PT**) with the S atoms in *ccp* positions, and wurtzite (**2·2PT**) with S atoms in *hcp* positions. ZnO has the wurtzite structure, suggesting that more ionic character favors the *hcp* structure. Burdett showed a plot of $r+$ vs.r for 98 known MX compounds. (J. K. Burdett, *Chemical Bonding in Solids*, Oxford University Press, Oxford, 1995, p. 195.) Thirty-eight of these compounds fall in the wrong area. The radius ratio rule is imperfect.

Relative sizes of ions and the degree of covalence are important in determining structures of MX compounds. Mooser and Pearson plotted the average principal quantum number of the ions vs. the difference in electronegativites of the elements to give reasonable separation of CsCl, NaCl, ZnS (*hcp*), and ZnS (*ccp*) structures. (E. Mooser and W.B. Pearson, *Acta Crystallogy* 1959; *12*: 1015.) This plot is more successful than the radius ratio plots.

Some coordination compounds such as $[Co(NH_3)_6][TlCl_6]$ (Section 5.1.5) have the NaCl structure, with the octahedral cations and anions occupying the positions of Na^+ and Cl^-. CaC_2 (Section 5.1.2) is an MX compound with Ca^{2+} and $C_2{}^{2-}$ (acetylide) ions. The structure is the NaCl structure with the cell elongated (tetragonal) because the $C–C^{2-}$ ions are aligned along the *c* axis. Linear ions are commonly aligned for most efficient packing. Pyrite, FeS_2 (Section 5.1.2) has the NaCl structure without elongation. The $S_2{}^{2-}$ ions are tipped and staggered to avoid distortion. Calcite, $CaCO_3$, and $NaNO_3$ (Section 5.1.6) have structures related to the NaCl structure with distortion to accommodate the trigonal planar $CO_3{}^{2-}$ and $NO_3{}^-$ ions. The planes of the $CO_3{}^{2-}$ or $NO_3{}^-$ ions are parallel and the structures are related to an NaCl cell depressed along the threefold axis perpendicular to the packing layers.

Some $MM'O_2$ compounds have structures based on the NaCl structure. $NaFeO_2$ (Section 5.1.7) has such a structure with Na^+ and Fe^{3+} in alternate **O** layers. The oxide ions are in **P** layers with expected different spacings for O—Fe—O and O—Na—O layers. Many oxide, sulfide, and selenide compounds have this structure (Table 5.3).

NiAs (**2·2PO**, Section 5.2.1) is an MX compound with a structure based on an *hcp* arrangement of As atoms. The Ni atoms are in **O** sites, but for an *hcp* structure all **O** sites are at packing *C* positions. Thus, the Ni atoms are aligned along the *c* axis without shielding. NiAs is a one-dimensional metal with good electrical conductance along one direction. This structure is unusual, and it occurs only for compounds with significant covalent character. The alignment of **O** sites does not occur for *ccp* structures because all **P**, **O**, and **T** sites are staggered at *A*, *B*, and *C* positions. γ′-MoC and TiP (**4·2PO** Section 5.2.5) have carbon or phosphorus in **P** layers in a sequence $|\mathbf{P}_A\mathbf{O}_C\mathbf{P}_B\mathbf{O}_C\mathbf{P}_A\mathbf{O}_B\mathbf{P}_C\mathbf{O}_B|$. The *ABAC* sequence for **P** layers is called double hexagonal. Here, Mo and Ti atoms are aligned in pairs in *C–C* and *B–B* layers.

Structures of some MX compounds differing from the typical ionic MX structures are not surprising when the chemistry of the elements is considered. CuS (Section 8.2.7) is not $Cu^{2+}S^{2-}$, because copper in CuS exists as Cu^+ and Cu^{2+}, and S exists as S^{2-} and S_2^{2-}. TlSe (Section 7.2.14) is not expected, as it suggests Tl^+Se^- (Se^{2-} is the usual anion) or $Tl^{2+}Se^{2-}$ (Tl^+ and Tl^{3+} are common oxidation states); the crystal contains Tl^+, Tl^{3+}, and S^{2-}.

12.3.2. Crystal Structures of MX_2 and M_2X Compounds

The important ionic crystal structure of MX_2 compounds is fluorite, CaF_2. Here, the cations are in a *ccp* arrangement of **P** layers with F^- ions filling both **T** layers (Section 6.3.1, **3·3PTT**, Figure 6.12). Ca^{2+} ions are at the center of a cube formed by F^- ions. Table 6.4 lists compounds with the fluorite structure. The Li_2O structure is known as the antifluorite structure. The roles of the ions are reversed with oxide ions in **P** layers and Li^+ ions filling both **T** layers (Figure 6.11). Table 6.3 lists compounds with the Li_2O structure.

It is not clear from the formula that MgAgAs might have the fluorite structure, but the As atoms are in a *ccp* arrangement of **P** layers with Ag in one **T** layer and Mg in the other **T** layer (Section 6.3.2, Figure 6.13). K_2PtCl_6 has the Li_2O (antifluorite) structure with $PtCl_6^{2-}$ ions in a *ccp* arrangement of **P** layers with K^+ ions in both **T** layers (Section 6.3.3, Figure 6.14). $[N(CH_3)_4]_2SnCl_6$ also has the antifluorite structure (Section 6.3.4).

The other important crystal structure for MX_2 compounds is rutile (TiO_2). Rutile [Section 5.4.1, **2·2PO$_{1/2}$(t)**, Figure 5.30] has an *hcp* sequence of **P** layers filled by oxide ions, with Ti^{4+} ions in one-half of each **O** layer. Because all **O** sites are at *C* positions for an *hcp* structure, partial filling of **O** layers avoids having Ti^{4+} ions in adjacent **O** sites. Compounds with the rutile structure are listed in Table 5.8. The structure is favored for transition metal difluorides and dioxides.

There are *no* examples of **PTT** structures based on an *hcp* arrangement of **P** layers except for a few **TPT** layered structures (MoS_2 and ReB_2, Sections 6.3.16 and 6.3.17). For an *hcp* structure the **T** layers just above and below a **P** layer are at the same positions (*A* or *B*) and very close (Figure 6.10). The compounds with structures of this type have *partial* filling of **T** layers to avoid occupancies of close **T** sites. Thus, for MX_2 compounds the common structure found is **3·3PTT** (*ccp*), for a *ccp* structure **P**, **T**, and **O** sites are at *A*, *B*, and *C* positions. For structures using only **P** and **O** layers the stoichiometry requires only one-half of the **O** sites, as for rutile. Structures of MX_2 compounds with one-half of **O** sites occupied based on a *ccp* arrangement of **P** layers are not common. Anatase, another allotropic form of TiO_2, has a *ccp* pattern for oxide ions with one-half of **O** sites occupied, but as **3·4PO$_{2/3}$PO$_{1/3}$** (Section 5.3.1).

The structures of SiO_2 are very important, but they are rare for other MX_2 compounds. There are several SiO_2 minerals, and all common ones are based on SiO_4 tetrahedra sharing apices. The metal silicates are the most important minerals in the earth's crust. Chapter 10 is devoted to silica and silicates.

12.3.3. MX_2 Layer Structures

For MX_2 compounds with covalent character, layer structures are encountered. $CdCl_2$ has a *ccp* arrangement of Cl^- ions with Cd^{2+} ions filling *alternate* **O** layers (Section 5.5.1, **3•3POP(h)**, Figure 5.55). The **POP** sandwiches are stacked without cations between them. Because of the polarization of the Cl^- ions by the Cd^{2+} ions, there is only van der Waals attraction between these adjacent Cl^- layers. CdI_2 has a similar structure (Section 5.5.4, **(2•3/2)POP**, Figure 5.58) with I^- ions in an *hcp* arrangement. Brucite [$Mg(OH)_2$, Section 5.5.5, Figure 5.59] has the CdI_2 structure. Other compounds with this structure are included in Section 5.5.

Molybdenite (MoS_2, **2•3TPT**, Section 6.3.16, Figure 6.28) has a hexagonal cell stacked similar to CdI_2 with significant differences. For each S—Mo—S sandwich these layers of S atoms are at the *same* positions (*AA* and *BB*) and the Mo atoms are in an *ABAB* sequence. The Mo atoms are at centers of trigonal prisms formed by six S atoms. We consider the layers of Mo atoms as **P** layers in an *hcp* sequence. The S atoms are in **T** layers and for an *hcp* structure the sequence is $\mathbf{T}_B\mathbf{P}_A\mathbf{T}_B\ \mathbf{T}_A\mathbf{P}_B\mathbf{T}_A$. This is an unusual structure and requires interactions of the close S atoms; presumably some bonding is involved.

A plot by Mooser and Pearson of average quantum number vs. difference in electronegativities of the M and X atoms was reasonably successful in separating CaF_2, MX_2 layer structures, TiO_2 and SiO_2 (E, Moser and W.B. Pearson, *Acta Crystallogy*. 1959; 12:1015.)

12.3.4. Crystal Structures of M_2X_3 Compounds

The most common structure of M_2X_3 compounds is that of corundum, α-Al_2O_3. Corundum [Section 5.4.7, $\mathbf{2{\cdot}6PO_{2/3}(h)}$, Figure 5.36] has oxide ions filling **P** layers in an *hcp* pattern. Al^{3+} ions occupy two-thirds of **O** layers. The **O** sites occupied are staggered because all **O** sites are at *C* positions. This requires 12 (**2•6**) layers in the repeating unit. The occupancy of two-thirds of **O** sites requires that each Al^{3+} ion has an Al^{3+} ion close neighbor on one side. This causes Al^{3+} ions to be displaced from the normal positions of **O** layers halfway between **P** layers.

Hematite (α-Fe_2O_3, Section 5.4.8) has a structure similar to that of corundum. Ilmenite [$FeTiO_3$, $\mathbf{2{\cdot}6PO^{Ti}_{2/3}PO^{Fe}_{2/3}(tri)}$, Section 5.4.9] has the corundum structure with Ti and Fe in alternate **O** layers. The structure of $LiSbO_3$ [Section 5.4.10, $\mathbf{2{\cdot}2PO_{1/3\ 1/3}(o)}$, Figure 5.39] is also similar to that of corundum with Li and Sb sharing the **O** layers. The SbO_6 octahedra form chains with edges shared. LiO_6 octahedra alternate on sides of the chains of SbO_6 octahedra. The perovskite structure (next section) is more common for $MM'O_3$ compounds.

The structure of La_2O_3 [Section 5.5.10, **3•5/3POPPO(h)**, Figure 5.64) is known as the A-type M_2O_3 or A-type rare-earth structure. It is rather complex. The unit cell is hexagonal, but oxide ions are in an *A*, *B*, *C* sequence. La^{3+} ions fill two **O** layers with the next **O** layer vacant. This is a layer structure because the sequence is **POP POP**, but the first and sixth **P** layers are identical; thus, five layers |**POP PO**| repeat. La has *CN* 7; the seventh oxide ion is from the next **POP** sandwich.

Compounds with the La_2O_3 structure are listed in Table 5.12. Tl_2O_3 has the A-type M_2O_3 structure (Section 6.3.7, $\mathbf{3{\cdot}3PT_{3/4}T_{3/4}}$). It is a fluorite-like structure with vacancies in oxygen layers.

It is not a common structure, but In_2Se_3 (Section 6.2.4) has the $\mathbf{2{\cdot}2PT_{2/3}(h)}$ structure with Se atoms in an *hcp* pattern and In atoms occupying two-thirds of one **T** layer.

12.3.5. Perovskite and Spinel Structures

Perovskite ($CaTiO_3$, $\mathbf{3{\cdot}2P_{1/4\,3/4}O_{1/4}}$, Section 5.3.4, Figure 5.21) is the most important structure for $MM'O_3$ compounds. The structure of perovskite has oxide *and* Ca^{2+} ions in *ccp* **P** layers in a 3:1 ratio. The Ti^{4+} ions occupy one-quarter of the **O** layers; those sites occupied are surrounded by six O^{2-} ions. In the unit cell Ti is at the center of a cube formed by Ca^{2+} ions at the corners and with O^{2-} ions in the faces, forming a TiO_6 octahedron. Compounds with the perovskite structure are important because of their ferroelectric, ferromagnetic, and superconducting properties. Some of the hundreds of examples are listed in Table 5.6. It is interesting that without Ca, the structure is the same as that of ReO_3 (Section 5.3.7). Spinel ($MgAl_2O_4$, $\mathbf{3{\cdot}6PO_{3/4}PT_{1/4}O_{1/4}T_{1/4}}$, Section 7.3.3, Figure 7.26) is the structure of hundreds of MM'_2X_4 compounds (Table 7.4). For spinel, the oxide ions are in **P** layers in a *ccp* pattern, with Mg^{2+} ions in **T** sites and Al^{3+} ions in **O** sites. $MgAl_2O_4$ is a **normal spinel** and $MgFe_2O_4$ is an **inverse spinel**. $MgFe_2O_4$ can be written as $Fe(MgFe)O_4$ to show that there are two roles for Fe, half in **T** sites and half in **O** sites. Larger cations prefer **O** sites, and preferences of some transition metal cations can be predicted from ligand field effects (Douglas *et al*, p. 402, see Appendix A). The structure is complex because between two **P** layers there is a three-quarters filled **O** layer and between the next two **P** layers **O** and both **T** layers are one-quarter filled. This occupancy agrees with the atom ratios of $MgAl_2O_4$, 2**P**, 1**O**, and 0.5**T**.

12.4. Assignment of Notation for Structures Involving Close Packing

For an *hcp* structure the packing layers are perpendicular to the *c* axis of the cell. Figure 12.2 shows a projection along *c* of a hexagonal cell. The *A* and *B* positions might represent $\mathbf{P}_A$, $\mathbf{P}_B$, and $\mathbf{O}_C$ sites, or they might represent $\mathbf{P}_A$, $\mathbf{T}_B$, and $\mathbf{O}_C$ sites. For an *hcp* structure of a metal (**2P**, Figure 4.2) the **O** and **T** sites are vacant. NiAs has the **2·2PO** structure (Figures 5.11). A **PO** structure indicates a 1:1 compound such as NiAs (**2·2PO**) and NaCl (**3·2PO**). The first number of the index is usually **2** (for *AB* sequence) or **3** (for *ABC* sequence) and the product gives the number of layers repeating. There are many *hcp* compounds with one **T** layer filled or partially filled. The stoichiometry of the compound suggests the likely fraction of **T** sites occupied. For GeS_2 the notation is $\mathbf{2{\cdot}2PT_{1/2}}$ (Section 6.2.3) and for In_2Se_3 it is $\mathbf{2{\cdot}6PT_{2/3}}$ (Section 6.2.4). However, the expectation of fractional filling must be verified. Figure 6.8 shows the positions of the atoms in the

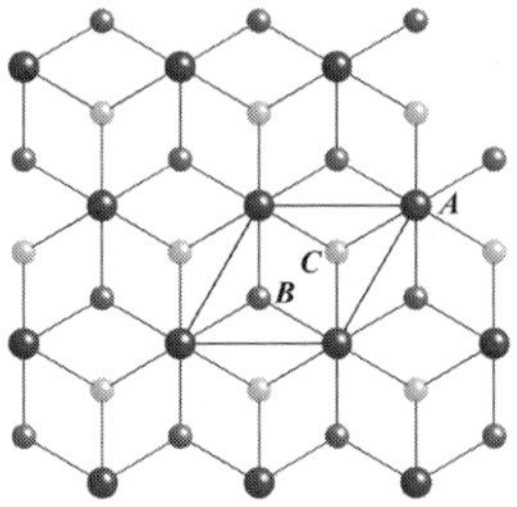

Figure 12.2. A projection along *c* showing the *A*, *B*, and *C* positions for a hexagonal cell.

(*Source*: *CrystalMaker*, by David Palmer, *CrystalMaker* Software Ltd., Begbroke Science Park, Bldg. 5, Sandy Lane, Yarnton, Oxfordshire, OX51PF, UK.)

unit cell with the heights of layers for In_2Se_3. Figure 12.3 shows labels for the positions in the cell. The cell leans to the left and that in Figure 12.2 leans to the right; the orientation is arbitrary. The large dark circles at corners and inside the cell locate vacant *C* sites. The cell could be chosen with *A* or *B* sites at the corner. In the projection of the cell of In_2Se_3 all *A* and *B* positions are occupied. All sites in the Se layers are filled, but the In layers are partially filled. At height 13 there are In atoms in each edge labeled *B*, but the *B* site inside the cell is vacant; thus $4 \times 1/2 = 2$ In sites are occupied, representing a **T** layer two-thirds filled. $LiIO_3$ (Section 5.4.20, Figure 5.50) is an example of an *hcp* structure with *A*, *B*, and *C* sites occupied in a similar cell. $\mathbf{P}_A$ and $\mathbf{P}_B$ layers are filled by oxygen and $\mathbf{O}_C$ layers are one-third filled by Li and one-third filled by I, $\mathbf{2{\cdot}2PO_{1/3\ 1/3}}$.

The computer program *CrystalMaker* makes it easy to examine individual layers or any combination of layers. Colors of atoms in layers can be changed to identify them in projection. The cell can be reoriented to find and examine the packing layers, looking for the expected spacing of **P** and **O** and/or **T** layers. The cell can be rotated to examine chosen layers to verify that they have close-packed arrangements and determine the occupancy for partially filled layers. For most *hcp* structures a projection of several layers is simple, verifying the repetition of the simple cell. Cubanite, $CuFe_2S_3$, is an example of a compound with an *hcp* structure with partial filling of both **T** layers ($\mathbf{2{\cdot}3PT_{1/6\ 2/6}\ T_{1/6\ 2/6}}$, Figure 6.25, Section 6.3.13).

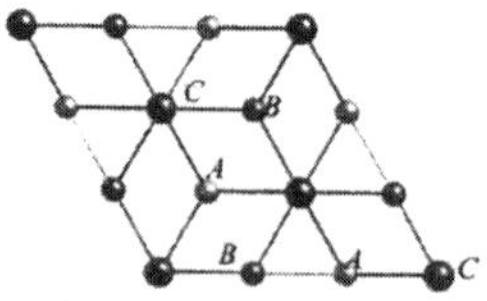

Figure 12.3. A projection along *c* showing the *A*, *B*, and *C* positions for a larger hexagonal cell.

(*Source*: *CrystalMaker*, by David Palmer, *CrystalMaker* Software Ltd., Begbroke Science Park, Bldg. 5, Sandy Lane, Yarnton, Oxfordshire, OX51PF, UK.)

For structures based on *ccp* there are no problems involving interactions of atoms in **P**, **T**, or **O** sites because each type of site is staggered in *A*, *B*, and *C* positions. Assigning notation is more complex in some cases because all sites use *A*, *B*, and *C* positions so in projection perpendicular to the packing layers all sites are aligned and eclipsed. The packing direction is along the body diagonal of the cubic cell so it can be difficult to evaluate packing layers from the usual views of models. For these cases the *CrystalMaker* is extremely helpful in examining individual and any combination of layers. Atoms can be added to extend layers to see filling patterns and selected groupings can be isolated to examine polyhedra. Orientation can be changed easily. Many of the complex structures such as the silicates could not have been included in this book without the help of *CrystalMaker*. Many structures were added after *CrystalMaker* was available. Continuous rotation of a structure reveals internal features that are obscured in even several fixed views.

Figure 3.12 shows the simple structures obtained from several combinations of **T** and **O** sites in a *ccp* arrangement of **P** layers. For cubic structures, often one can identify the **T** and/or **O** sites with the general pattern shown in Figure 3.8. ReO_3 (Figure 5.23***a***) has alternating Re and O along edges of the cell. The Re atoms are in octahedral sites but there are no atoms in faces or at the center of the cell. It is a **PO** structure with vacant sites. There are one Re atom and three O atoms in the cell, the expected 3:1 ratio. For the NaCl cell (expanded in Figure 5.1) there are four Na atoms and four Cl atoms in the cell, so for ReO_3 the Re atoms fill one-quarter of **O** sites and oxygen atoms occupy three-quarters of the **P** sites. The notation is $\mathbf{3{\cdot}2P_{3/4}O_{1/4}}$. It also corresponds to the perovskite ($\mathbf{3{\cdot}2P_{1/4\ 3/4}O_{1/4}}$) structure without Ca.

Using the **PTOT** cell of Figure 3.12 as the reference, we can assign notations for many cubic structures. This **PTOT** cell represents BiF_3 with Bi in all **P** sites and F in all **O** and **T** sites. It is the general case, with all **P**, **T**, and **O** layers filled. For NaCl, the **T** sites are vacant. For CaF_2, the **O** sites are vacant and for zinc blende (ZnS) **O** sites and all sites of one **T** layer are vacant. The unit cell of CsCl corresponds to one of the octants in the **PTOT** figure with Cs at the center and Cl at the corners. The **PTOT** roles of the ions are not obvious from the CsCl unit cell, but viewed as eight cells or octants in the **PTOT** figure we can recognize the Cl^- positions as **P** and **O** sites with Cs^+ in all **T** sites (filling both **T** layers). If all atoms are the same, this is the body-centered cubic structure, so with all sites occupied it is a **3·2PTOT** structure. The *bcc* structure is common for metals. It is not considered as a close-packed structure, but we see it corresponds to the general case for close packing in terms of the **PTOT** system. For an ideal close-packed structure **O** sites are smaller than **P** sites and **T** sites are even smaller. Large atoms occupying **T** and **O** sites require the structure to expand or open, but the relative spacing of the layers is unchanged.

In many cases involving partial filling of layers the overall pattern is not clear from the unit cell. Distortions can result from vacancies. Examination of spacings between layers is helpful in identifying **P**, **T**, and **O** layers. In the cases of $CdCl_2$, CdI_2, $CuBr_2$, and $CrCl_3$

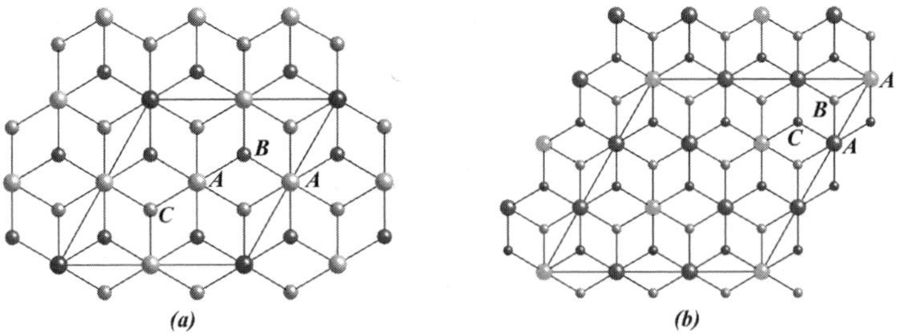

Figure 12.4. The projection along *c* of a hexagonal cell with *A*, *B*, and *C* positions shown. (***a***) The large dark balls represent a one-quarter filled layer and large light balls represent a three-quarter filled layer. (***b***) The large light balls represent a one-third filled layer and the large dark balls represent a two-third filled layer.

(*Source*: *CrystalMaker*, by David Palmer, *CrystalMaker* Software Ltd., Begbroke Science Park, Bldg. 5, Sandy Lane, Yarnton, Oxfordshire, OX51PF, UK.)

(Section 5.5), spacings of layers make it clear that alternate **O** layers are vacant. The partially filled layers of spinel ($\mathbf{3{\cdot}6PO_{2/4}PT_{1/4}O_{1/4}T_{1/4}}$) are easily identified by spacings of the layers (Figure 7.26***a***, Section 7.3.3).

Figure 6.4***a*** for In_2CdSe_4 (Section 6.1.5) looks like a *ccp* cell (with some tetragonal distortion) with **T** sites filled and two **P** sites vacant in top and bottom faces. The **P** and **T** designations can be reversed, giving **P** layers filled and **T** sites three-quarters filled ($4 \times 1/2 = 2$ in faces and $8 \times 1/8 = 1$ in corners); on this basis the **T** sites are one-quarter filled by Cd and one-half filled by In, giving $\mathbf{3{\cdot}4PT_{2/4\ 1/4}(t)}$. Figure 12.4 shows the patterns for (***a***) $\mathbf{P_{1/4}}$ (large dark atoms) or $\mathbf{P_{3/4}}$ (large light atoms), and (***b***) $\mathbf{P_{1/3}}$ (large light atoms) or $\mathbf{P_{2/3}}$ (large dark atoms) in reference to an *hcp* cell. The same patterns apply to **P** layers in a *ccp* structure.

K_2PtCl_6 illustrates the versatility of the **PTOT** system. It can be handled as a 2:1 salt (Li_2O structure) treating $PtCl_6^{2-}$ as a unit (**3•3PTT**, Section 6.3.3, Figure 6.14). If we consider it as a **PO** structure, the **P** layers are three-quarters filled by Cl and one-quarter by K. The Pt atoms are in alternate layers, occupying one-quarter of the **O** sites, only those sites surrounded by Cl atoms. This gives the notation $\mathbf{3{\cdot}3P_{1/4\ 3/4}P_{1/4\ 3/4}O_{1/4}}$.

CuS is an interesting compound (Section 8.2.7). The structure is more complicated than expected from the simple formula. There are two roles for Cu, Cu^{I}, and Cu^{II}, and two roles for S, S^{2-}, and S_2^{2-}. A projection of the hexagonal cell looks like Figure 12.2. The S_2^{2-} ions are in *A* positions and Cu and S^{2-} are in *B* and *C* positions. Atoms at the same positions are eclipsed, but changes of the view by a few degrees reveal the atoms covered. In Figure 8.12 we see that Cu^{II} atoms are in **T** sites and Cu^{I} and S^{2-} are in a double layer consisting of Cu_3S_3 hexagons. CuS_3 groups are trigonal planar and the S^{2-} ions in SCu_5

have trigonal bipyramidal arrangements. The structure can be described as $\mathbf{8P}^{S_2}\ \mathbf{T}^{Cu}\ \mathbb{P}^{S,Cu}\ \mathbf{T}^{Cu}$, which requires the planar double layer to have filled *B* and *C* positions. We can avoid overfilling the double layer by considering each single layer to be one-quarter filled and the *B* and *C* positions one-quarter filled for the double layer, $\mathbf{8P}_{1/4}{}^{S_2}\ \mathbf{T}_{1/4}{}^{Cu}\mathbb{P}_{1/4\,1/4}{}^{S,Cu}\ \mathbf{T}_{1/4}{}^{Cu}$. We have treated structures with small atoms such as AlB_2 (Section 8.2.1) and β-Be_3N_2 (Section 8.2.17) as having double layers with two positions filled. We treated the structure of $CaIn_2$ (Section 9.1.24) as having two filled positions in a puckered double layer. The structure of $InNi_2$ was treated as having two positions filled by In and Ni in a double layer. This is a formalism, and all of these cases can be considered as having positions filled for all layers including double layers.

The patterns for positions in a hexagonal cell can vary, depending on the occupancies of layers. Four patterns are shown in Figures 12.2, 12.3, and 12.4***a*** and ***b***. Usually corner positions are labeled *A*. These positions were labeled *C* in Figure 12.3 to agree with Figure 6.8, where corners are not occupied.

12.5. Summary

Many books discuss structures in terms of octahedral and tetrahedral sites and build models by stacking these polyhedra. Adams, Pearson, West, and Wells (see books cited in Appendix A) often discussed structures in terms of close packing. Pearson gave heights and packing positions of layers for many structures. The **PTOT** scheme was published (S.-M. Ho and B.E. Douglas, *J. Chem. Educ.* 1969; *46:* 208) and was used in second and third editions of the book by Douglas, McDaniel, and Alexander. Adams cited the **PTOT** paper and used Figure 3.4. Some books make the distinction between tetrahedra pointed upward ($\mathbf{T}^+$) along the packing direction and those pointed downward ($\mathbf{T}^-$) without showing any figures showing the difference. The usual views of cells such as fluorite (CaF_2, **3•3PTT**) do not show that there are two different **T** layers.

Books using the close-packing description missed the important overall general **PTOT** pattern. The **P**, **T**, and **O** layers are obvious in *hcp* structures when we look for them, because the layers are packed along the *c* axis. The layers are not obvious in *ccp* structures because the common models and drawings show one axis horizontal and one vertical. Layers are packed along the body diagonal of the cell.

The general **PTOT** scheme predicts similar stabilities of *hcp* and *ccp* structures using **P** layers only and those using **P** layers and one set of the two **T** layers. For many metals the *ccp* (**3P**) and *hcp* (**2P**) structures have nearly the same stability and have both structures under different conditions. For ZnS, both zinc blende (**3•2PT**, *ccp*) and wurtzite (**2•2PT**, *hcp*) occur as minerals. Because of similar stabilities of *ccp* (**3P**) and *hcp* (**2P**) structures of metals and for **3PT** and **2PT** structures of ZnS, we might expect similarities in other combinations of layers for *ccp* and *hcp* structures, but this is not the case. The expected compounds using both **T** layers are common for the *ccp* structure of fluorite (CaF_2, **3·3PTT**) and antifluorite (Li_2O, **3·3PTT**) but not for corresponding *hcp* structures, because two adjacent **T** layers are so close (Figure 6.10) they cannot be fully occupied unless the atoms involved are bonded (MoS_2). The **PTOT** scheme also explains that **3·2PO** (*ccp*, NaCl) is a common structure, but the **2·2PO** (*hcp*, NiAs) structure is found only for compounds for which close, poorly shielded cations can be tolerated. Structures such as $\mathbf{2{\cdot}2PO_{1/2}}$ (*hcp* for **P** layers) are common because partial filling of **O** layers avoids problems of close cation alignment.

From the overall **PTOT** pattern it is clear that BiF_3 (or $BiLi_3$) represents the general case with Bi filling all **P** layers and F^- filling **O** and both **T** layers, $\mathbf{3{\cdot}2P^{Bi}T^{F}O^{F}T^{F}}$. It is also a description of CsCl, $\mathbf{3{\cdot}2P^{Cl}T^{Cs}O^{Cl}T^{Cs}}$, and *ccp* metals with the M atoms filling all **P**, **T**, and **O** layers. This interpretation of the *bcc* structure has not been recognized generally because it is well known that for hard spheres *bcc* is less dense than *ccp* or *hcp*. Atoms are not hard spheres. Structures using all of the filled **P**, **T**, and **O** layers are limited to those based on *ccp* structures because of problems for *hcp* structures with aligned **O** sites and very close adjacent **T** layers.

The general **PTOT** scheme applies to many compounds and all combinations of occupancies of the four layers. The expected distortions of the cubic NaCl (**3·2PO**) structure occur for nonspherical ions such as C_2^{2-}(CaC_2) and CO_3^{2-}($CaCO_3$). For hard spheres the relative radii of **P**, **O**, and **T** sites are 1:0.414:0.225. Structures based on the **PTOT** scheme occur for a range of sizes of ions in all sites. Expansion of a structure caused by ions "too large" for **O** and **T** sites does not change the relative spacings of the layers. In this book we have extended the basic descriptions of the **PTOT** scheme for rare cases of double layers and for those where the orientation of tetrahedra has C_2 axes (rather than the usual C_3 axes) aligned with the packing direction. There are many different structures encountered for metal silicates, but the general **PTOT** scheme helps us to see the patterns. The obvious layering for some silicates are expected from stacking octahedral and tetrahedral layers. The investigation and recognition of patterns for layers of metal silicates require close examination of individual and combinations of layers possible only using *CrystalMaker*. It seems that the **PTOT** scheme applies for the most efficient packing in crystals except for particular chemical bonding requirements or molecular shapes.

Figure 12.5 shows a projection of three close-packed positions, *A*, *B*, and *C*. The octahedra are shown for $\mathbf{P}_A\mathbf{O}_C\mathbf{P}_B$, but the pattern for the projection pattern is the same for $\mathbf{P}_A\mathbf{P}_C\mathbf{P}_B$ or $\mathbf{P}_A\mathbf{P}_C\mathbf{T}_B$ (the **T** sites between $\mathbf{P}_A$ and $\mathbf{P}_B$ are $\mathbf{T}_B$ and $\mathbf{T}_A$.) Three cells are shown with *A* positions (dark balls) at corners and edges. The simplest cell is shown in two equivalent orientations. The other cell has *A* positions at corners with *B* and *C* positions in the edges. This plot is very helpful in assigning positions of atoms in a cell. Unless there is distortion, it should be possible to account for the position of each atom in each layer. For structures in the system, *CrystalMaker* makes it possible to

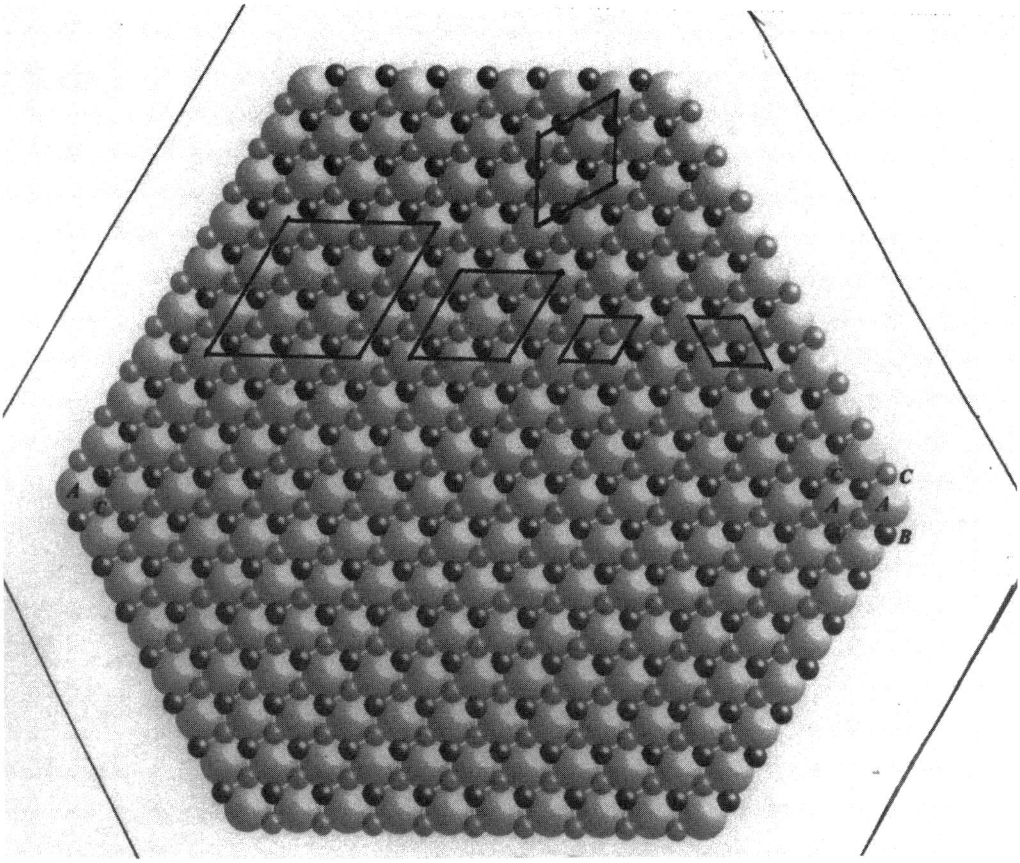

Figure 12.5. A projection of three closed-packed layer at the *A*, *B* and *C* positions.
(*Source*: *CrystalMaker*, by David Palmer, *CrystalMaker* Software Ltd., Begbroke Science Park, Bldg. 5, Sandy Lane, Yarnton, Oxfordshire, OX51PF, UK.)

examine each layer or any combination of layers. Colors (or darkness for grayscales) can be changed for making distinctions. Atoms at the same positions are eclipsed in the projection of two layers. This can be verified by shifting the stucture by a few degrees to reveal the hidden atoms.

The CD included with the book helps in visualization of sturctures and relationships among similar structures. Color and large size make the figures much more appealing and effective. The approach presented aids in interpretation of sturctures and determincation of notations, especially in case of partially filled layers. Part IV allows the user to manipulate structures, add atoms or remove atoms to examine the environment of atoms using *CrystalMaker*. Autorotation reveals features not clear, even from models.

Appendix A

Further Reading

D.M. Adams, *Inorganic Solids*, Wiley, New York, 1974. Very good older book with excellent figures. It emphasizes close packing.

L.V. Azaroff, *Introduction to Solids*, McGraw-Hill, New York, 1960.

L. Bragg, *The Crystalline State*, G. Bell and Sons, London, 1965.

L. Bragg and G.F. Claringbull, *Crystal Structures of Minerals*, G. Bell, London, 1965.

P.J. Brown and J.B. Forsyth. *The Crystal Structure of Solids*, E. Arnold, London, 1973.

M.J. Buerger, *Elementary Crystallography*, Wiley, New York, 1956.

J.K. Burdett, *Chemical Bonding in Solids*, Oxford University Press, Oxford, 1992.

Cambridge Structural Data Base (CSD). Cambridge Crystallographic Data Centre, University Chemical Laboratory, Cambridge, England.

C.R.A. Catlow, Ed., *Computer Modelling in Inorganic Crystallography*, Academic Press, San Diego, 1997.

A.K. Cheetham and P. Day, *Solid-State Chemistry, Techniques*, Clarendon, Oxford, 1987.

P.A. Cox, *Transition Metal Oxides*, Oxford University Press, Oxford, 1992.

CrystalMaker, A powerful computer program for the Macintosh and Windows by David Palmer, *CrystalMaker* Software Ltd., Yarnton, Oxfordshire, UK. This program was used for many figures and it aided greatly in interpreting many structures for this book and accompanying CD.

B.D. Cullity, *Elements of X-ray Diffraction*, Addison-Wesley, Reading, MA, 1956.

J. Donohue, *The Structure of The Elements*, Wiley, New York, 1974. The most comprehensive coverage of the structures of elements.

B.E. Douglas, D.H. McDaniel, and J.J. Alexander, *Concepts and Models of Inorganic Chemistry*, 3rd ed., Wiley, New York, 1994. The **PTOT** system is discussed and applied briefly.

F.S. Galasso, *Structure, Properties and Preparation of Perovskite-Type Compounds*, Pergamon, Oxford, 1969.

F.S. Galasso, *Structure and Properties of Inorganic Solids*, Pergamon, Oxford, 1970. Excellent figures to help to visualize structures.

C. Hammond, *Introduction to Crystallography*, Oxford University Press, Oxford, 1990.

N.B. Hannay, *Solid-State Chemistry*, Prentice-Hall, Englewood Cliffs, NJ, 1967.

R.M. Hazen and L.W. Finger, *Comparative Crystal Chemistry*, Wiley, New York, 1984.

W. Hume-Rothery, R.E. Smallman and C.W. Haworth, *The Structure of Metals and Alloys*, Institute of Metals and the Institution of Metallurgists, London, 1969.

B.G. Hyde, and S. Anderson, *Inorganic Crystal Structures,* Wiley, New York, 1989.

Inorganic Crystal Structural Data Base (ICSD). Fachinformationszentrum Karlsruhe, Germany.

International Tables for X-Ray Crystallography, Vol. 1, *Symmetry Groups,* N.F.M. Henry and K. Lonsdale, Eds, *International Union of Crystallography,* Kynoch Press, Birmingham, 1952. The complete source for space groups and crystallographic information.

W.D. Kingery, *Introduction to Ceramics,* Wiley, New York, 1967.

H. Krebs, *Fundametals of Inorganic Crystal Chemistry,* McGraw-Hill, London, 1968.

M.F.C. Ladd, *Structure and Banding in Solid State Chemistry,* Wiley, New York, 1979.

F. Liebau, *Structural Chemistry of Silicates,* Springer-Verlag, Berlin, 1985.

Y. Matsushita, *Chalcogenide Crystal Structure Data* library, Version 5.5B, 2004, Institute for Solid State Physics, University of Tokyo. A library of about 10,000 structures including many other than chalcogenides.

H.D. Megaw, *Crystal Structures, A Working Approach,* Saunders, Philadelphia, 1973.

Metals Crystallographic Data File (CRYSTMET). National Research Council of Canada, Ottawa.

U. Müller, *Inorganic Structural Chemistry,* Wiley, New York, 1993.

I. Naray-Szabo, *Inorganic Crystal Chemistry,* Akademiai Kiado, Budapest, 1969.

R.E. Newnham, *Structure–Property Relations,* Springer-Verlag, New York, 1975.

W.B. Pearson, *The Crystal Chemistry and Physics of Metals and Alloys,* Wiley, New York, 1972. An excellent book for intermetallic compounds, excellent figures, gives occupancies and spacings for close-packed layers for many structures.

D. Pettifor, *Bonding and Structure of Molecules and Solids,* Oxford University Press, Oxford, 1995.

F.C. Phillips, *An Introduction to Crystallography,* 4th ed., Wiley, New York, 1971.

A. Putnis, A., *Introduction to Mineral Sciences,* Cambridge University Press, Cambridge, 1992. Good background on experimental methods and excellent coverage of metal silicates.

G.V. Raynor, *The Structure of Metals and Alloys,* Institute of Metals, London, 1954.

R. Roy, Ed., *The Major Ternary Structural Families,* Springer-Verlag, New York, 1974.

D.F. Shriver and P.W. Atkins, *Inorganic Chemistry,* 3rd. ed., Freeman, New York, 1999.

L. Smart and E. Moore, *Solid State Chemistry,* Chapman and Hall, London, 1992.

A.R. Verma and P. Krishna, *Polymorphism and Polytypism in Crystals,* Wiley, New York, 1966.

M.T. Weller, *Inorganic Materials Chemistry,* Oxford University Press, Oxford, 1994.

A.F. Wells, *Structural Inorganic Chemistry,* 5th ed., Oxford University Press, Oxford, 1984. The most complete one-volume coverage of inorganic structures.

A.R. West, *Basic Solid State Chemistry,* 2nd ed., Wiley, New York, 1999.

A. Wold and R. Dwight, *Solid State Chemistry,* Chapman and Hall, London 1993.

R.W.G. Wyckoff,. *Crystal Structures*, Vols. 1–6, 2nd ed., Wiley, New York, 1963–1968. The most comprehensive coverage of crystal structures with fine figures, space groups, unit cell constants and atom coordinates. Vols. 1–4, inorganic compounds and Vols. 5 and 6, organic compounds.

Appendix B

Polyhedra in Close-Packed Structures

Lattice Types

The **Seven Systems of Crystals** are shown in Figure 2.2. The relationship between the trigonal and rhombohedral systems is shown in Figure B.1***a***. The possibilities of body-centered and base-centered cells give the **14 Bravais Lattices**, also shown in Figure 2.2. A face-centered cubic (*fcc*) cell can be represented as a 60° rhombohedron, as shown in Figure B.1***b***. The *fcc* cell is used because it shows the high symmetry of the cube.

Polyhedra

Figure B.2 shows polyhedra commonly encountered. The five Platonic (or regular) solids are shown at the top. Beside the octahedron and cube, the octahedron is shown inside a cube, oriented so the symmetry elements in common coincide. These solids are conjugates: one formed by connecting the face centers of the other. The tetrahedron is its own conjugate, because connecting the face centers gives another tetrahedron. The icosahedron and pentagonal dodecahedron are conjugates. The square antiprism and trigonal

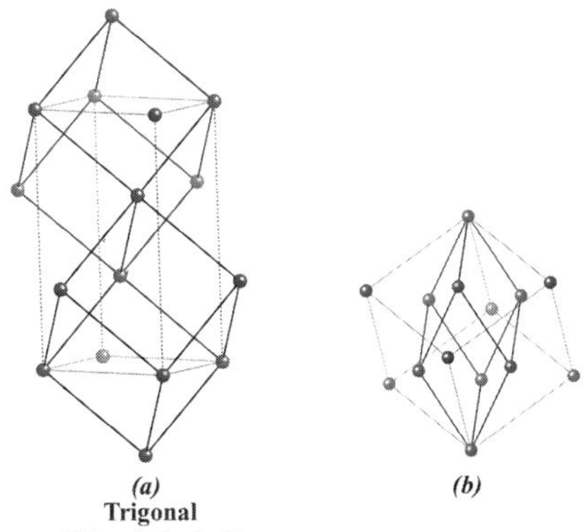

Figure B.1 (*a*) The relationship of a hexagonal cell to trigonal (rhombohedral) cells. (*b*) the 60° rhombohedral cell related to a face-centered cubic cell. (*CrystalMaker*)

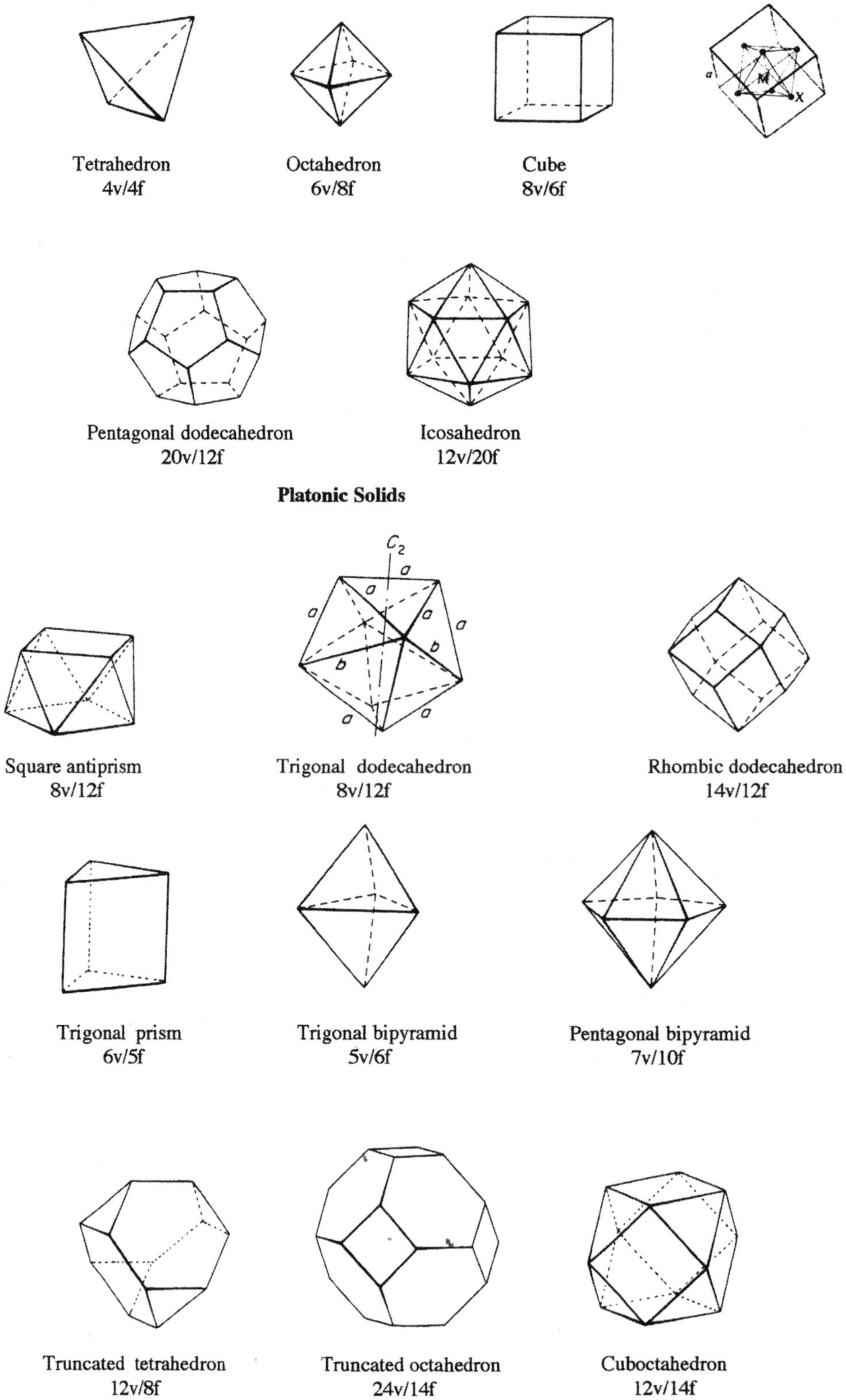

Figure B.2. Platonic and other solids. The numbers of vertices (v) and faces (f) are shown.

dodecahedron are common for coordination compounds with *CN* 8. The cuboctahedron is encountered in cubic close-packed structures. A cuboctahedron is formed by eight ReO_6 octahedra in ReO_3 (Figure 5.23*b*). The truncated tetrahedron is encountered in Laves phases (MgM_2, Figure 9.43). The bucky ball (not shown in Figure B.2) is the structure of C_{60} (Figure 4.11). Figures B.3 and B.4 provide cutouts for some polyhedra. Enlarged copies work well.

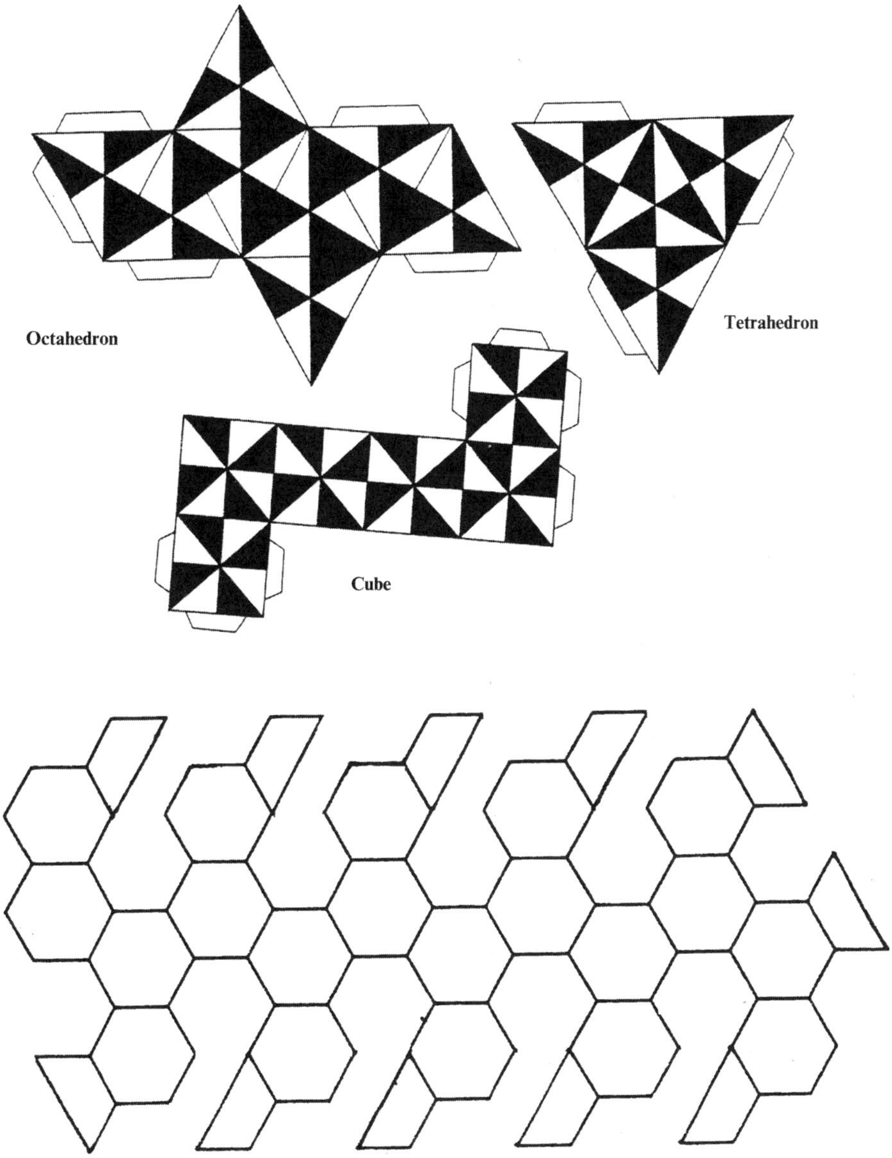

Figure B.3. Cutouts for an octahedron, tetrahedron, cube, and buckyball.

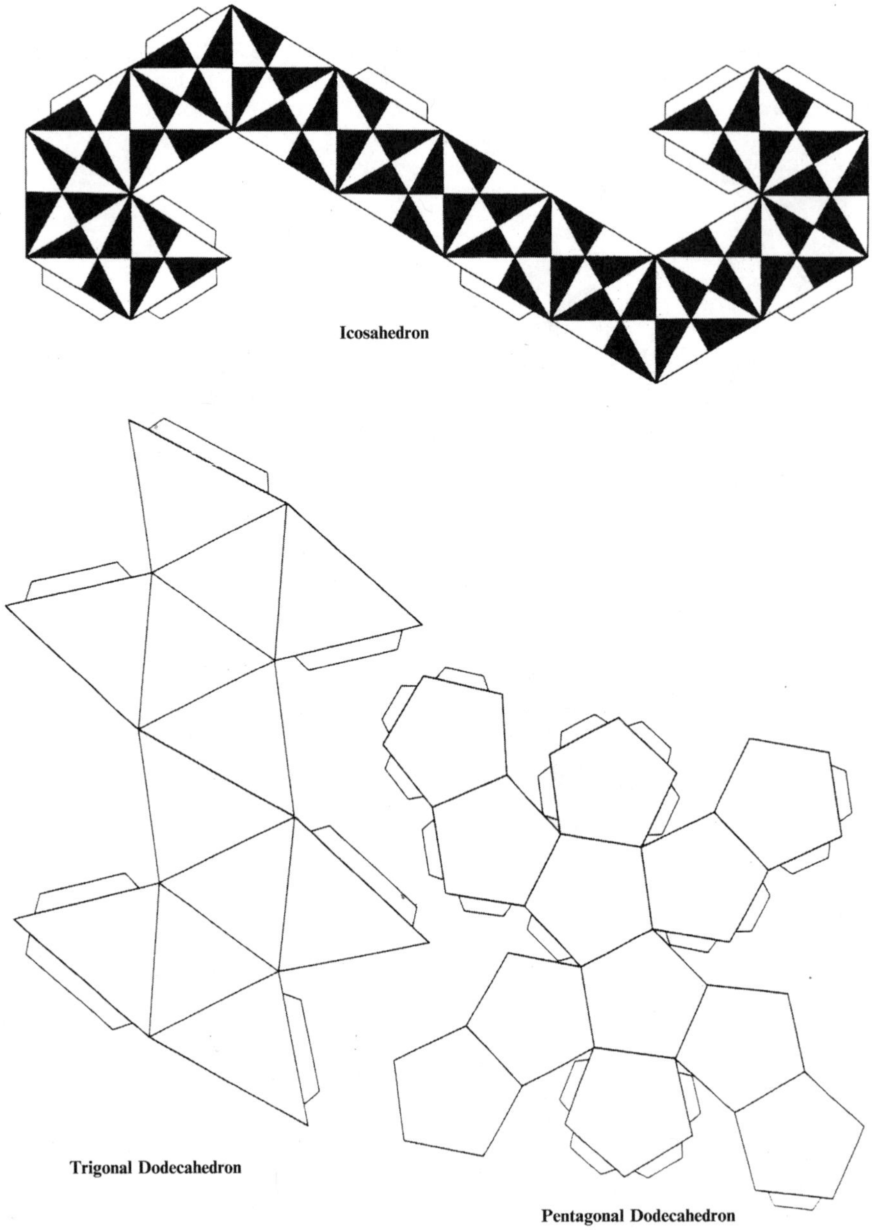

Figure B.4. Cutouts for an icosahedron, trigonal dodecahedron, and pentagonal dodecahedron.

Polyhedra in Cubic Close-Packed (*ccp*) and Hexagonal Close-Packed (*hcp*) Structures

For a cubic close-packed (*ccp*) structure, each atom is surrounded by 12 atoms forming a cuboctahedron (Figure B.5). The six octahedral sites (**O**) form an octahedron around the central atom and the eight tetrahedral sites (**T**) form a cube (Figure 4.6). For a hexagonal close-packed (*hcp*) structure, the polyhedron is shown in Figure B.5. The square and trigonal faces above and below the central plane of the hexagonal plane are aligned.

Models from Stacking Polyhedra

Good models of many crystal structures can be built by stacking tetrahedra or octahedra. Such models can be helpful in visualizing the structure. Figure B.6 shows the wurtzite (ZnS, **2•2PT**) structure with tetrahedra stacked along the *c* axis of the hexagonal cell. One of the two sets of **T** layers between two **P** layers is filled ($\mathbf{T}^+$ or tetrahedra pointing upward as shown here). The S atoms (dark balls) are in an *AB* sequence. The zinc blende (or sphalerite, ZnS, **3•2PT**, structure) is shown with tetrahedra stacked along the body diagonal of the cubic cell. The S atoms are in an *ABC* sequence, a *ccp* arrangement. The *A*, *B*, or *C* positions of Zn are the same as those of the S atom at the upward apex of each tetrahedron. Another view of the cell shows the positions of the tetrahedra in each cubic cell. The Zn atoms form a tetrahedron within the cubic cell. Fluorite (CaF_2, **3•3PTT**) has Ca in **P** layers with both sets of **T** layers filled by F. In Figure B.6 the fluorite cell is shown with Ca as spheres forming the *fcc* cell and F in tetrahedra pointed upward and downward.

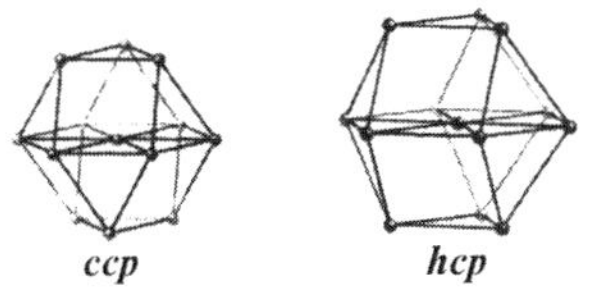

Figure B.5. The polyhedra of close neighbors of an atom in *ccp* and *hcp* structures.

(*Source*: *CrystalMaker*, by David Palmer, *CrystalMaker* Software Ltd., Begbroke Science Park, Bldg. 5, Sandy Lane, Yarnton, Oxfordshire, OX51PF, UK.)

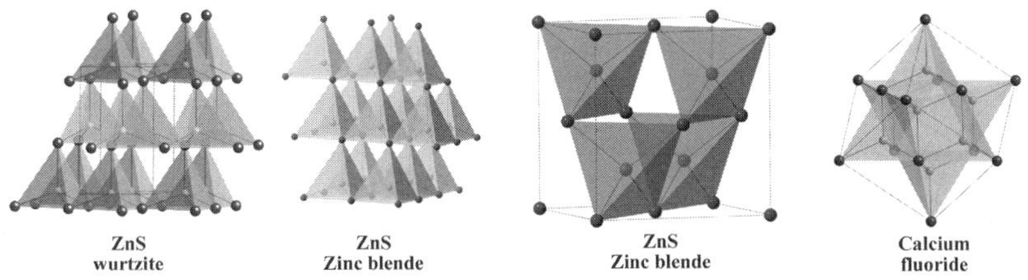

Figure B.6. Structures built from stacking tetrahedra: ZnS, wurtzite, **2•2PT**; ZnS, zinc blends, **3•2PT**; and CaF_2, fluorite **3•dPTT**.

(*Source*: *CrystalMaker*, by David Palmer, *CrystalMaker* Software Ltd., Begbroke Science Park, Bldg. 5, Sandy Lane, Yarnton, Oxfordshire, OX51PF, UK.)

The structure of NaCl (**3•2PO**) is shown in Figure B.7 as $NaCl_6$ octahedra stacked along the body diagonal of the cubic cell. The Cl^- ions are in an *ABC* sequence and the Na^+ ions in **O** sites are in an *ABC* sequence. The NiAs (**2•2PO**) structure is shown also. The As atoms are in an *AB* sequence and Ni atoms in **O** sites are all at *C* positions, aligned along the *c* axis of the cell. The $CdCl_2$ structure [**3•3POP(h)**] shows layers of octahedra with gaps between them. The **P** layers are filled by Cl atoms in an *ABC* sequence, with alternate **O** layers vacant. There are no atoms between the layers of octahedra; there is only van der Waals attraction between these Cl layers. The CdI_2 structure [**2•3/2POP(h)**] (not shown) is similar except the octahedral layers are identical. The I atoms are in an *AB* sequence with Cd atoms at *C* positions.

BiF_3 (or $BiLi_3$, **3•4PTOT**) has a cubic structure with Bi atoms in **P** layers (*ABC* positions) and F (or Li) filling **O** and both **T** layers. The structure can be built by stacking octahedra (Figure B.8). The tetrahedra fill spaces between the octahedra, sharing faces with the octahedra. The structure can also be built by stacking tetrahedra, with all sites for tetrahedra pointing upward and downward are filled. Octahedra fill spaces between tetrahedra. Either choice is satisfactory as the stacked octahedra create the tetrahedra and stacked tetrahedra create the octahedra. In the center and top of the figure of the stacked tetrahedra, there are cavities for octahedra. The third figure shows an octahedron surrounded by tetrahedra with the **P**, **T**, and **O** layers identified. In CaF_2, both filled **T** layers form octahedra between tetrahedra, but the octahedral sites are not occupied.

The CsCl Structure Described in a Close-Packed System

The CsCl structure corresponds to a body-centered cubic (*bcc*) cell, but the atom at the center is different from those at the corners. It is not truly close packed because there is more empty space for hard spheres for *bcc* compared to *ccp* and *hcp* structures under similar conditions. Metals with a *bcc* structure are expected to be less dense than those with *ccp* or *hcp* structures. In Section 4.3.3, we see that in many cases metals with *bcc* structures have comparable, or even greater, density than those with *ccp* or *hcp* structures under similar conditions. Metal atoms are not hard spheres. Figure B.9***a*** shows eight CsCl cells. This large cell can be considered as an *fcc* cell with the dark spheres in **P** sites (corners and centers of faces) and light spheres in **O** sites (center of the cube and centers of centers of edges). Here, the Cl^- atoms fill **P** and **O** sites

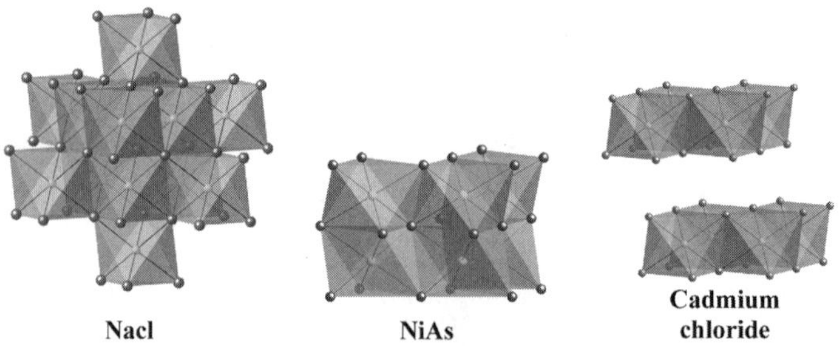

Figure B.7. Structures built from stacking octahedra: NaCl **3•2PO**, NiAs **2•2PO**, and $CdCl_2$ **3•3POP(h)**.

(*Source*: *CrystalMaker*, by David Palmer, *CrystalMaker* Software Ltd., Begbroke Science Park, Bldg. 5, Sandy Lane, Yarnton, Oxfordshire, OX51PF, UK.)

with Cs^+ ions in all **T** sites. Each **T** site is in a cube formed by four **P** sites and four **O** sites; two tetrahedra of four Cl^- ions in **P** sites are shown. The notation is **3·2PTOT**. Figure B.9*b* shows a view of octahedra and tetrahedra with C_3 axes aligned with the body diagonal of the cell. The **P**, **T**, and **O** layers are labeled. The spacings represent accurately those of the CsCl cell. No approximations are required.

Spacial Relationships in Close-Packed Structures

An octahedral site is at the center of an octahedron. In a close-packed structure the octahedra have the C_3 axes aligned perpendicular to the close-packed layers. This is the packing direction. Opposite staggered triangular faces of an octahedron are in each of the packing layers (**P**). The octahedral site is exactly halfway between these faces and the packing layers. In Figure B.10*a* a tetrahedron is shown within a cube and the edge of the cube is a. In Figure B.10*b* the tetrahedron is viewed with M, X_1, and X_2 in a plane and $X_1—X_2$ is an edge of the tetrahedron. The $X_1—X_2$ edge is a face diagonal of the cube, $X_1—X_2 = \sqrt{2}a$. M—X are bond lengths, one-half of the body diagonal of the cube (Figure B.9*a*), $M—X_1 = M—X_2 = a\sqrt{3}/2$. The bond angle $\langle XMX$ is $109.48°$, so $\langle MX_2X_1 = (180° - 109.48°)/2 = 35.26°$, $\langle X_1X_2Y = 90° - 35.26° = 54.74°$, the angle, $\langle YX_2M = 54.74° - 35.26° = 19.48°$. The distance $Y—X_1$ is the distance between layers.

$$\sin 19.48° = 0.333 = (M—Y)/(M—X_2) = (M—Y)/a\sqrt{3}/2 \qquad \sin 54.74° = 0.816 = (Y—X_1)/(X_1—X_2) = (Y—X_1)/\sqrt{2}a$$

$$M—Y = 0.333 \times a\sqrt{3}/2 = 0.288a. \qquad Y—X_1 = 1.155a$$

$$(M—Y)/(Y—X_1) = 0.288a/1.155a = 0.25$$

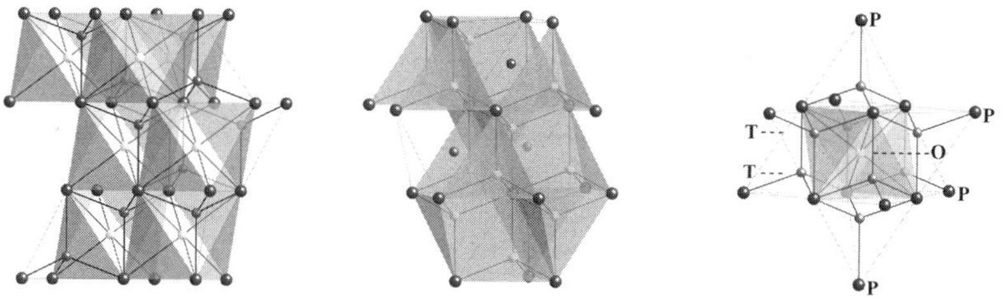

Figure B.8. BiF_3 models built from stacking octahedra and tetrahedra and a figure showing the relationships of octahedra, tetrahedra and **P**, **T**, and **O** layers.

(*Source:CrystalMaker*, by David Palmer, *CrystalMaker* Software Ltd., Begbroke Science Park, Bldg. 5, Sandy Lane, Yarnton, Oxfordshire, OX51PF, UK.)

(a) CsCl *(b)* CsCl

Figure B.9. (*a*) Eight CsCl cells, Cs^+ ions are in the centers of cells (**T** sites). Cl^- ions, at the corners, are in **P** sites (dark spheres) and **O** sites (light spheres). (**b**) A side view showing the spacings of the layers.

(*Source*: *CrystalMaker*, by David Palmer, *CrystalMaker* Software Ltd., Begbroke Science Park, Bldg. 5, Sandy Lane, Yarnton, Oxfordshire, OX51PF, UK.)

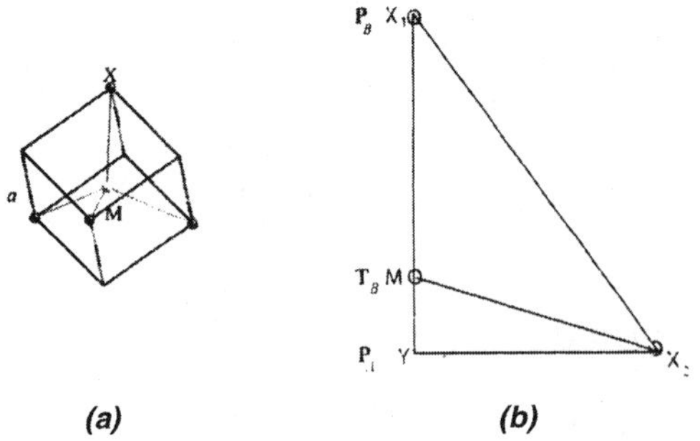

Figure B.10. (*a*) A MX_4 tetrahedron within a cube. (*b*) A view of the tetrahedron with M and two X atoms in a plane.

(*Source*: *CrystalMaker*, by David Palmer, *CrystalMaker* Software Ltd., Begbroke Science Park, Bldg. 5, Sandy Lane, Yarnton, Oxfordshire, OX51PF, UK.)

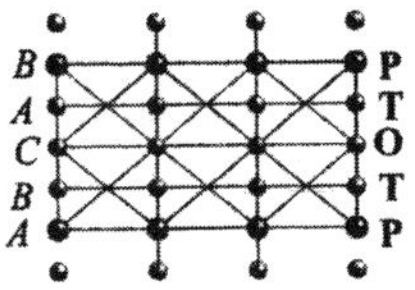

Figure B.11. A side view showing the spacings of **P, T**, and **O** layers in a close-packed structures.

(*Source*: *CrystalMaker*, by David Palmer, *CrystalMaker* Software Ltd., Begbroke Science Park, Bldg. 5, Sandy Lane, Yarnton, Oxfordshire, OX51PF, UK.)

Thus, the height of the center of the tetrahedron is one-quarter of the distance between packing layers.

The spacings of **O** and **T** layers are seen easily from geometry. Figure B.11 shows a view of the $BiLi_3$ structure showing the **P**, **T**, and **O** layers with atoms aligned. The diagonals from **P** sites cross at **O** sites so the **O** sites are halfway between **P** layers. The diagonals connecting sites in **P** and **O** layers cross at the level of **T** sites so the **T** sites must be halfway between **P** and **O** layers, or one-fourth and three-fourths of the distance between **P** layers.

Subject Index

A

A, B, C close-packing positions 22
Acetylene 282
Adamantane 279
Adams, D. M. 19, 303
Akermanite, $Ca_2MgSi_2O_7$ 239, 258F
Albite, $NaAlSi_3O_8$ 243
Alumina, Al_2O_3, see Corundum
Aluminum boride, AlB_2 176, 177F, 303
  structures 177
Aluminum bromide 135, 136F
Aluminum chloride 78, 79F
Aluminum-chromium, $AlCr_2$ 217F
Aluminum fluoride 101
Aluminum hydroxide oxide 91, 92F
Aluminum-manganese-copper, $AlMnCu_2$ 215, 216F
Aluminum zinc sulfide, Al_2ZnS_4 123
Amethyst, quartz 237
Ammonia 294
Ammonium chloride 157F
Ammonium chlorite 159F
Ammonium cyanide 158F
Ammonium hexafluorosilicate 111F, 184F
  structures 112
Ammonium hydrogen fluoride 157, 158F
Amphibole, double chain silicate 240
Anatase, TiO_2 78, 79F
Andalusite, Al_2SiO_5 239, 247, 249, 252F
Anorthite, $CaAl_2Si_2O_8$ 243
Anthracene, $C_{14}H_{10}$ 289F
Anti-fluorite, M_2X 124, 125F
  structures 126
Antigorite, serpetine 242, 267F
Antimony 55F
Antimony pentachloride 60F
Antimonytrimethyl dibromide 61F
Antimony plumbate 106, 107F
Antimony pentachloride 60F
Apophyllite, sheet silicate 241
Aquamarine, beryl 239
Aragonite, $CaCO_3$ 74, 75F
  carbonate positions 75F
Arsenic 54, 55F
Asbestos, silicate chains or rolled sheets 240
Assigning notation 25, 299
Atacamite, $Cu_2Cl(OH)_3$ 86, 87F
A-type, M_2O_3 or rare-earth 114
Avicennite, Tl_2O_3 130F
Azakoff, L. V. 19
Azidocarbonamide 284, 285F

B

Band theory 198
Bardeen, J. 81
Barium ferrate, $BaFe_{12}O_{19}$ 191, 192F
Barium manganate 98, 99F
Barium iron molybdenate, Ba_2FeMoO_6 84
Barium nickelate 97F
Barium perchlorate 180, 181F
Barium strontium tantalate, $Ba_3SrTa_2O_9$ 84, 85F
Barium tantalate, $Ba_5Ta_4O_{15}$ 115F, 116
Barium titanate 97, 98F
Barlow, W. 2, 15
Barlow's structures 3F
bcc 43, 292
  structure 35F

Bednorz, G. 81
Benitoite, $BaTiSi_3O_9$ 239
Benzene 287, 288F
Benzene-chlorine, $C_6H_6{\bullet}Cl_2$ 289, 290F
Dibenzenechromium 56, 292
Bergman, T. 6
Beryl, $Be_3Al_2Si_6O_{18}$ 239, 259, 260F
Beryllium aluminate, $BeAl_2O_4$ 165, 167F
Beryllium nitride 190, 191F
Biotite, layered silicate 241, 276F
Bismuth-indium, BiIn 200, 201F
Bismuth pentafluoride 87, 88F
Bismuth trifluoride 148F, 156, 216F, 315F
Bismuth triiodide 192, 193F
Bismuth-lithium $BiLi_3$ 215, 216F
Bismuth-magnesium (2:3) 225, 226F
Body-centered cells 3F, 6
Body-centered cubic, see *bcc*
Pentaborane 149, 150F
Boron 49F
Boron trichloride 60
Diborontetrachloride 57, 58
Tetraborontetrachloride 149, 150F
Boron trifluoride 9F
Bragg, L. 19, 51
Brass 197, 199, 213
Bravais, A. 1
Bravais lattices 1, 6, 8, 9F
Brewer, L. 36
Brookite, TiO_2 78
Brucite, $Mg(OH)_2$ 111F
Buckminsterfullerene 50F
Buckyball, C_{60} 50F
Buerger, M. J. 19
Burdett, J. K. 296

C
Cadmium chloride 108F, 300, 315F
Cadmium iodide 110F
Calaverite, $AuTe_2$ 112F
Calcite, $CaCO_3$ 2F, 68F, 295
  carbonate positions 75F
Calcium boride, CaB_6 155F
  structures 156
Calcium carbide 56, 65F
  structures 197
Calcium-copper, $CaCu_5$ 210, 211F
Calcium carbonate, see Calcite, Aragonite
Calcium-indium, $CaIn_2$ 213F, 303
Calcium silicide 128, 129F
Calcium titanate, see Perovskite
Calcium tungstate, $CaWO_4$ 132, 133F
Calcium uranyl dioxide 138F
Carbon 51F, 52F
  C_{60} 50F
Carbon dioxide 56, 57F
Carbon monoxide 56
Carbon tetrachloride 279
Carborundum, SiC 143, 145F, 146F
Cassiterite, SnO_2 89
ccp 34, 294
  different sites 148F
  filling layers 32F
  neighbors 44F, 313F
  P layers 29F
  structures 3F, 29F, 32F, 34F
  T sites 162F
ccp to *bcc* conversion 42F, 45
CD included with book 305
Celsian, $BaAl_2Si_2O_8$ 243
Center of symmetry 7
Cesium chloride 3F, 153, 154F, 301, 315F
  structures 155, 196
Cesium suboxide 99, 100F
Chabazite, zeolite 243, 244F
Chains, silicates 240F
Chalcopyrite, $CuFeS_2$ 119F
Chevrel phases 156
Chlorine 48F
Chlorite, layered silicate 248, 277F
Tetrachloroplatinate(II) 9F
Dichloropropane (2,2) 279
Chromium(III) chloride 109F
Chromium sulfides 103F, 104F
Chromium-sulfur phases 103
Chrysoberyl, $BeAl_2O_4$ 165, 167F
Chrysotile, $Mg_3Si_2O_5(OH)_4$ 240, 242
Chu, P. 81
Cinnabar, HgS 172, 173F
Citrine, quartz 237
Clay minerals, layered silicates 242, 268
Close-packed layers 21, 188
  3 layers 22F, 305F
  networks 295F, 305F
  positions 22F, 24F, 25F
CN, coordination number 26
Cobalt(II) iodide hydrate 105, 106F
Cobalt-tin, CoSn 211, 212F
Conversion, *ccp* to *bcc* 39, 42F
Cooper, L. 81
Cooperite, PtS 130F
Coordination compounds

[Co(NH$_3$)$_6$]TlCl$_6$ 67
Cu[C$_6$H$_4$O(CHO)]$_2$ 153F
[Fe(CH$_3$NC)$_6$]Cl$_2$•3H$_2$O 175F
K$_2$NiF$_4$ 88, 89F
K$_2$PtCl$_4$ 164F
K$_2$PtCl$_6$ 127, 128F, 302
[N(CH$_3$)$_4$]$_2$SnCl$_6$ 128
(NH$_4$)$_2$SiF$_6$ 111F, 112, 184F
Ni[(CH$_3$)$_2$C$_2$N$_2$O$_2$H]$_2$ 61, 62F
[Ni(H$_2$O)$_6$]SnCl$_6$ 160F
TlAlF$_4$ 156F
Coordination number, ***CN*** 26
Copper-aluminum, CuAl$_2$ 218, 220F
Copper(II) bromide 112, 113F
Copper cesium chloride 87, 88F
Copper chloride hydroxide 86, 87F
Copper family 38
Copper gallinate 170F
Copper germanate 169, 170F, 171
Copper hydroxide 91, 92F
Copper mercury iodide, Cu$_2$HgI$_4$ 120F
Copper(I) nitride 84, 86F
Copper(I) oxide 135F
Copper sulfide 182F, 302
Copper-titanium, CuTi 221F
Copper tungstate 74F
Copper-zinc
CuZn 213
CuZn$_3$ 199
system 197F
Cordierite, Mg$_2$Al$_4$Si$_5$O$_{18}$•H$_2$O 260, 261F
Corundum, Al$_2$O$_3$ 93, 94F, 298
Covellite, CuS 182F
Cristobalite, SiO$_2$ 233, 235, 236F
Crystal classes 13, 17
Crystal systems 8, 9F, 17, 18
CrystalMaker® 300
C-type M$_2$O$_3$ structures 130, 132F
Cubanite, CuFe$_2$S$_3$ 137F, 300
Cubic, simple 8, 9F
Cubic close-packed, see *ccp*
Cubic system 6
Cuboctahedron 44, 310
Cuprite, Cu$_2$O 135F
Cuproscheelite, CuWO$_4$ 73, 74F
Cutouts, polyhedra 311F, 312F
Tetracyanoethylene 283, 284F
Cyanogen, (CN)$_2$ 281, 282F
Cyclohexane 279
Cymrite, BaAlSi$_3$O$_8$(OH)•H$_2$O 269, 270F

D
Densities of metals 40, 45
Diamond 51F
cubic 53
hexagonal 53
Diaspore, AlO(OH) 91, 92F
Diatomic molecules 47
Dihedral groups 13
Dimethylglyoxime, Ni complex 62F
Dinitrobenzene 291F
Diphenylbenzene 290F, 291
Diopside, CaMgSi$_2$O$_6$ 240, 264F
Donohue, J. 36, 41
Double hexagonal 22, 26
Double layers 172, 176
Douglas, B. E. 293, 303

E
Electron compounds 198
Elements 34, 37, 40
Emerald, beryl 239
Enargite, Cu$_3$AsS$_4$ 123, 124F
Energy states, 3 electrons/2 atoms 198F
Enstatite, pyroxene, MgSiO$_3$ 240
Ethane 280, 281F
Ethylene 283F

F
Face-centered cells 6
Face-centered cubic, *fcc* 8
Fajans' rules 293
Fayalite, (Fe, Mg)$_2$SiO$_4$ 244
fcc 45F
fcc to 60° rhombohedral cell 309F
Federov, E. S. 15
Feldspars 243
monoclinic 243
triclinic 243
Ferrocene 9F
Ferromagnetic compounds 84
Fluorite 4, 124, 126F, 169F, 313F
structures 127
Forsterite, Mg$_2$SiO$_4$ 238, 244, 245F
Framework silicates 242

G
Gadolinium formate 59F
Gallium 39
Gallium-platinum (3:5) 208, 209F
Gallium selenide 143, 144F
Gallium sulfide 39
GaS 142F
Ga$_2$S$_3$ 117

Garnets, $M_3^{II}M_2^{III}(SiO_4)_3$ 239
Gehlenite, $Ca_2Al(AlSi)O_7$ 239, 258
Gems 51, 93, 239, 246, 248, 260
Germanium 36, 42
Germanium sulfide 122F
Glide plane 10
Gold-beryllium, $AuBe_5$ 229
Gold-copper, $AuCu_3$ 203F
Gold telluride 112F
Graphite 51, 52F
  to diamond 53F

H
Halite, see sodium chloride
Haüy, R. J. 2
hcp 3F, 23F, 35F, 294
  all layers 28F, 125F
  Partial layers 302F
  neighbors 313F
Hematite, Fe_2O_3 94, 95F, 298
Hermann-Mauguin
  notation 16
  symbols 18
Heulandite, zeolite 243, 244F
Hexagonal cell, *A, B, C*
    positions 300F
  layers 300F
  partial filling 302F
Hexagonal close-packed, see *hcp*
Hexagonal system 6
Hexagonal to trigonal cells 7F, 309F
Horizontal plane 7
Hume-Rothery phases 198
Hydrazine 61F
Hydrogen 47
Tetrakis(hydroxomethyl)
    methane 285F

I
Ice I_c, H_2O 117, 119F
Ice I_h 121, 122F
Ilmenite, $FeTiO_3$ 94, 96F, 298
Improper axes 11
Improper rotation 8, 10F
Index of notation 26
Indialite, $Mg_2Al_4Si_5O_{18}\bullet H_2O$
    260, 261F
Indium cadmium selenide,
    In_2CdSe_4 119, 120F, 302
Indium-nickel, $InNi_2$ 223F
Indium selenide 123F
Insulators 195, 198
Intermetallic compounds
    195, 292
Intermetallic compounds, CsCl
    structures 196
International symbol 17
Interstitial sites 22
Interstitial solid solutions 197
Iodine heptafluoride 57, 58F
Iodoform 279, 280F
Iridium-vanadium, IrV 214F
Iron 36
Iron carbide, FeC 78, 80F
Iron-isocyanomethyl complex 174,
    175F
Iron nitride, Fe_4N 109, 110F
Iron oxide, Fe_2O_3 94, 95F
Iron titanate 94, 96F

J
Jadeite, $Na(Al, Fe)Si_2O_6$ 264, 265F

K
Kaolinite, clay mineral 242, 268F
Kepler, J. 1
Kyanite, Al_2SiO_5 247, 249, 251F

L
L (layer) 26
  spacing layers 24F, 316F
$\mathbf{L}_{1/2}$ layers 32F
$\mathbf{L}_{1/4}$ and $\mathbf{L}_{3/4}$ layers 31F
$\mathbf{L}_{1/3}$ and $\mathbf{L}_{2/3}$ layers 30F
Lanthanum boride, LaB_6 155F
Lanthanum oxide 114F, 298
  structures 115
Lattice symbols 4
Laumonite, zeolite 243, 244F
Laves phases 226–232
  frameworks 227F
  $MgCu_2$ 229F, 230F
  $MgNi_2$ 232F
  $MgZn_2$ 231F
  structures 228, 231
  truncated tetrahedron 227F
Laves' principles 21
Layer structures 108–113, 140, 298
Lead-lithium (2:7) 224, 225F
Lead-lithium (3:8) 218F
Lead-magnesium, $PbMg_2$ 222F
Lead oxide 130, 131F, 294
Lithium aluminum silicate,
    $LiAlSiO_4$ 254F
Lithium antimonate, $LiSbO_3$ 95, 96F
Lithium ferrate, $LiFeO_2$ 71F
Lithium iodate 104F
Lithium oxide 124, 125F

Lithium peroxide 184, 185F
Lithium-tin, LiSn 208, 210F
Lizardite, $Mg_3Si_2O_5(OH)_4$ 266F
Lone pairs 294

M
Magnesium-cadmium, $MgCd_3$ 199F
Magnesium-copper, $MgCu_2$ 227F, 228, 229F, 230F
Magnesium hydroxide 111F
Magnesium-nickel, $MgNi_2$ 227F, 231, 232F
Magnesium silver arsenide, MgAgS 126, 127F
Magnesium-zinc, $MgZn_2$ 227F, 230, 231F
Manganese 39
Manganese-gold (2:5) 208, 209F
Marcasite, FeS_2 90F
Martensite, FeC 78, 80F
Melilite, akermanite to gehlenite 258
Mercury 39
Mercury peroxide 66, 67F
Mercury-sodium, HgNa 201F
Mercury sulfide, cinnabar 172, 173F
Metals 35, 291
  densities 40, 45
Methane 9F, 292
Trimethylamine borontrifluoride 59F
Tetramethylammonium chloride 174F
Tetramethylammonium dichloroiodate(1–) 159, 160F
Trimethylantimony dibromide 60, 61F
Dimethylpropane (2,2) 278
Miller index 11F, 12F
Mirror planes 7
Moissanite SiC 143
Molecular crystal structures 48, 56, 149
Molecular sieves 243
Molecules, large discrete 48
Molybdenite, MoS_2 139, 140F, 193, 298
Molybdenum-aluminum (3:8) 200F
Molybdenum-aluminum (1:12) 214F
Molybdenum carbide 77F
Molybdenum-nickel, $MoNi_4$ 205, 206F
Molybdenum-platinum, $MoPt_2$ 207F
Molybdenum sulfide, see Molybdenite
Monoclinic system 6
Monticellite, $CaMgSiO_4$ 244
Montmorillonite, layered silicate 272, 273F
Mooser, E. and Pearson, W. B. 296, 298
Mordenite, zeolite 243, 244F
Morganite, beryl 239
Müller, A. 81
Multiple layers 172, 176, 192
Muscovite, layered silicate 273F, 274
MX compounds 64, 155, 295
MX_2 compounds 127, 297, 298
M_2X compounds 126, 297
$M_2M'X_4$ compounds 165, 166
M_2X_3 compounds 298

N
Naphthalene 288F
Nepheline, $(KNa_3)Al_4Si_4O_{16}$ 261, 262F
Niccolite (nickeline), NiAs 72F, 295, 314F
Nickel-aluminum (2:3) 225, 226F
Nickel arsenide 72F, 295, 315F
  structures 73
Nickel dimethylglyoxime 61, 62F
Trinickel diselenide 194F
Niobium-nickel, $NbNi_8$ 212F
Niobium oxide 85, 86F
Niobium-rhodium, NbRh 201, 202F
p-dinitrobenzene 291F
Nitrogen 47F
Nitrogen family 53
Dinitrogenpentaoxide 180, 181F
Nitrous oxide 56
Noble gases 34
Nonmetals 47
Notation, **PTOT** 25, 299

O
O sites 22, 23F, 25F, 44F
O and **T** layers, spacing 24F, 316F
Octahedra, stacking 315F
Octahedral site 22, 23F, 25F, 44F
Olivine, $(Mg, Fe)_2SiO_4$ 244, 245F
Omisteinbergite, layered silicate 268, 269F
Onnes, K. 81
Optical activity 236
Orthoclase, $KAlSi_3O_8$ 243
Orthorhombic system 6
Orthosilicates 239, 244
Osmium tetraoxide 58, 292

Oxalic acid 282F
Oxygen, O_2 48
Oxygen family 55

P
P layers 3
  ccp 29F
P, **T** and **O** layers 24F
  pattern 295F, 305F
  spacings 24F, 316F
Tetrapalladium selenide 152, 154F
Paragasite, amphibole 264, 266
Partially filled layers 30F, 31F, 32F, 302F
Pearls 74
Pearson, W. B. 19, 303
Pearson's symbols 19
Peridot, olivine 244
Perovskite, $CaTiO_3$ 79, 80F, 299
  structures 82, 84
Phenakite, Be_2SiO_4 239, 254, 255F
p-diphenylbenzene 289, 290F
Phillips, F. C. 19
Phlogopite, a mica 242, 254, 274, 275F
Phosphorus 53, 54F
Piezoelectricity 237
Plagioclase feldspars 243
Platinum-mercury, $PtHg_4$ 217F
Platinum sulfide 130F
Platonic solids 310F
Plutonium 36, 38, 39
PO crystal structures 27, 63
PO *hcp* crystal structure 72
Point groups 7, 13, 18
Polarization 292
Polonium 56
Polyhedra 311F, 312F
  ccp 44F, 314F, 315F
  cutouts 312F, 313F
  hcp 314F
  stacking 314F, 315F
Potassium azide 160, 161F
Potassium hexachloroplatinate(IV) 127, 128F, 302
  structures 129
Potassium tetrachloroplatinate(II) 164F
Potassium tetrafluoronickelate(II) 88, 89F
Potassium fullerene, K_3C_{60} 161, 162F
Potassium nonahydridorhenate 189F
Predicting structures 295
Proper rotation 7, 8, 10F
Tetra(*n*-propyl)ammonium bromide 286, 287F
Proustite, Ag_3AsS_3 187F
PT structures 27
PTOT framework 24
PTOT notation 25
PTOT structures 26, 35F, 149, 153, 163
PTOT system 24F, 25F, 316F
PTT structures 28, 124
Pyrargyrite, Ag_3SbS_3 187
Pyrite, FeS_2 65, 66F
  structures 66, 196
Pyrochlore, $Ca_2Ta_2O_7$ 133, 134F
  structures 134
Pyrophyllite, $AlSi_2O_5(OH)$ 242, 271, 272F
Pyrosilicates 256
Pyroxene, single chain silicate 240

Q
Quartz, SiO_2 233, 235, 236F, 237
  α and β 237F
  optical activity 236

R
Radius ratio 199, 295, 296
Regular solids 310F
Rhenium diboride 141F
Rhenium triboride 163F
Rhenium dioxide 91F
Rhodium trioxide 84, 85F, 300
Rhombohedral-hexagonal cells 7F
Rock salt, see sodium chloride
Rose quartz 237
Rotation, proper and improper 8, 10F
Rotational groups 8
Rotation-inversion 10
Rotation-reflection 9
Ruby, Al_2O_3 93
Rutile, TiO_2 89, 90F, 297
  structures 89

S
Bis(salicylaldehydato)copper(II) 152, 153F
Sandwich compounds 108F, 109F, 110F, 111F, 140F, 298
Sanmartinite, $ZnWO_4$ 74F
Sapphires, Al_2O_3 93
Scheelite, $CaWO_4$ 132, 133F
  structures 133
Schoenflies, A. M. 15

Schoenflies notation 9
Schrieffer, R. 81
Screw axis 10
Selenium 55
Semiconductors 195
Serpentines, sheet silicates 242, 266
Sheet silicates 240
Silica, SiO_2 233
Silicates
  chains 240F
  cyclic 239F
  frameworks of tetrahedra 234F
  layered 242F
  sheets 240, 241F
Silicon 36, 42
Silicon carbide 143, 145F, 146F
Silicon tetrafluoride 149F
Silicon nitride 188F
Sillimanite, Al_2SiO_5 239, 247, 250F
Silver arsenic trisulfide, Ag_3AsS_3 187F
Silver bismuth diselenide 70F, 71
Silver cesium iodide, Ag_2CsI_3 107F
Silver copper sulfide, AgCuS 105F
Silver iodide 167
  T′ sites 167F
Silver mercury iodide, Ag_2HgI_4 120F
Sizes of ions 292
Snow flakes 2F
Sodalite, $Na_4(Si_3Al_3)O_{12}Cl$ 243F
Sodium arsenide 178, 179F
Sodium chloride 3F, 27F, 63, 65F, 162F, 315F
  structures 64
Sodium ferrate 69F, 296
  structures 70
Sodium nickelate 71, 72F
Sodium peroxide 185, 186F
Sodium-platinum, $NaPt_2$ 221, 222F
Sodium-tantalum, NaTl 215F
Sodium titanyl silicate 256F
Solid solutions 197
Space group 3, 14, 18
Space lattice 14
Special positions 12
Sphalerite, ZnS, see zinc blende
Spinel, $MgAl_2O_4$ 17, 165F, 166F, 299
  inverse 165, 299
  normal 165, 299
  structures 166
Spodumene, $LiAlSi_2O_6$ 240, 263F
Square planar 131F
Square pyramid 131F
Stannite, Cu_2FeSnS_4 119F
Stereograms
  rotations 13, 14F
  rotations and planes 5F, 16F
Stereographic projection 12
Stishovite, SiO_2 237, 238F
Stromeyerite, AgCuS 105F
Structure symbols 19
Substitutional solid solutions 197
Sulfur 50, 51F
Sulfur dioxide 57F
Sulfur nitride, S_4N_4 150, 151F
Superconductors 81
Symmetry operations 7, 8
Symmetry planes 7
Systems of crystals 6, 8, 9F

T

T layers 22
  spacing 22, 24F, 315
T sites 22, 23F, 25F, 44F, 131F
$\mathbf{T}^+$ and $\mathbf{T}^-$ sites 23
Talc, clay mineral 241, 270, 271F
Tantalum nitride, TaN 190F
Tellurium 55, 56F
Tephroite, Mn_2SiO_4 244
Tetragonal system 6, 8
Tetrahedron 131F
  in cube 316F
  spacing 316F
  stacking 234F, 314F
Thallium tetrafluoroaluminate 156F
Thallium(I) oxide 113F
Thallium(III) oxide 130, 132F
Thallium selenide, TlSe 162F
Thortveitite, $(Sc, Y)_2Si_2O_7$ 239, 256, 257F
Tin 42, 43F
  densities 43
Tin(IV) iodide 136F
Tin-nickel, $SnNi_3$ 202F
Tin-niobium, $SnNb_3$ 223, 224F
Titanium-aluminum, $TiAl_3$ 203, 204F
Titanium-copper, $TiCu_3$ 203F
Titanium-copper, 2:3, 3:4, 218, 219F
Titanium-nickel, $TiNi_3$ 199
Titanium dioxide. see Rutile and Anatase
Titanium phosphide, TiP 77, 78F
Topaz, $Al_2SiO_4(F, OH)_2$ 239, 246F
TOT layers 242F
TOT O TOT layers 242F
Tourmaline, sheet silicate 262, 263F
Tremolite, layered silicate 240, 265F

Triclinic system 6, 8
Tridymite, SiO_2 233, 234, 235F
Trigonal system 6, 8
Triple layers (II**P**) 192
Tungsten-aluminum, WAl_5 210
Tungsten (β) structures 224
Tungsten carbide 173F
Types of sheet silicates 242F
  OT 242
  TOT 242
  TOT O TOT 242

U
Ultramarines, basketlike silicates 243
Unit cell 6
Uranium dodecaboron 67F
Uranium hexachloride 100F
Uranium hexafluoride 101F
Urea 283, 284F

V
Vanadium trifluoride 101, 102F
Vanadium-zinc (4:5) 216F
van der Waals interaction 108, 298
Vertical planes 7

W
Weiss index 11
Wells, A. F. 19, 303
West, A. R. 303
Willemite, Zn_2SiO_4 254
Wollastonite, $CaSiO_3$ 239
Wu, M.-K. 81
Wurtzite, ZnS 27, 120, 121F, 296, 313F
  stacking tetrahedra 314F
  structures 121
Wyckoff, R. W. G. 19

X
Xenon fluorides 152, 153F

Y
Ytterbium borate 177, 178F
Yttrium hydroxide 177, 180F

Z
Zeolites, framework silicates 243, 244F
Zinc antimonate, $ZnSb_2O_6$ 92, 93F
Zinc blende 27, 117F, 126F, 313F
  stacking tetrahedra 313F
  structures 118
Zinc iodide 139F
Zinc sulfide, see wurtzite and zinc blende
Zinc tungstate 74F
Zintl phases 199
Zircon, $ZrSiO_4$ 239, 247, 248F
Zirconium-aluminum, $ZrAl_3$ 204, 205F
Zirconium-gallium, $ZrGa_2$ 207, 208F
Zirconium-gold, $ZrAu_4$ 205, 206F
Zirconium trihalides 101, 102F

Minerals and Gems*

Akermanite, $Ca_2MgSi_2O_7$ 258
Albite, $NaAlSi_3O_8$ 243
Alumina, Al_2O_3 94
Amethyst, quartz 237
Amphibole, double chain 240
Anatase, TiO_2 79
Andalusite, $Al_2O_3{\bullet}SiO_2$ 252
Anorthite, $CaAl_2Si_2O_8$ 243
Antigorite, $Mg_3Si_2O_5(OH)_4$ 267
Antimony 55
Apophyllite, $KCa_4[F(Si_4O_{10})_2]{\bullet}8H_2O$ 241
Aquamarine, beryl 239
Aragonite, $CaCO_3$ 75
Asbestos 240
Atacamite, $Cu_2Cl(OH)_3$ 87
Avicennite, Tl_2O_3 130

Benitoite, $BaTiSi_3O_9$ 239
Beryl, $Be_3Al_2Si_6O_{18}$ 260
Biotite, layered silicate 276
Bismuth 55
Brookite, TiO_2 78
Brucite, $Mg(OH)_2$ 111

Calaverite, $AuTe_2$ 112
Calcite, $CaCO_3$ 68
Cassiterite, SnO_2 89
Celsian, $BaAl_2Si_2O_8$ 243
Chabazite, zeolite 244
Chalcopyrite, $CuFeS_2$ 119
Chlorite, layered silicate 277
Chrysoberyl, Al_2BeO_4 167
Chrysotile, $Mg_3Si_2O_5(OH)_4$ 240
Cinnabar HgS 172
Citrine, quartz 237
Cooperite, PtS 130
Copper 38
Cordierite, $Mg_2Al_4Ai_5O_{18}$ 261
Corundum, α-Al_2O_3 94
Covellite, CuS 182
β-Cristobalite, SiO_2 236
Cubanite, $CuFe_2S_3$ 137
Cuprite, Cu_2O 135
Cuproscheelite, $CuWO_4$ 73
Cymrite, $Ba(Si, Al)_4O_8{\bullet}H_2O$ 270

Diamond 51
Diaspore, AlO(OH) 92
Diopside, $CaMgSi_2O_6$ 264

Emerald, beryl 239
Enargite, Cu_3AsS_4 123
Enstatite, pyroxene, $MgSiO_3$ 240

Fayalite, Fe_2SiO_4 244
Feldspars, 3-dimensional silicates 243
Fluorite, CaF_2 6, 169, 314
Forsterite, Mg_2SiO_4 245

Garnets, $M_3^{II}M_2^{III}(SiO_4)_3$ 239
Gehlenite, $Ca_2Al(AlSi)O_7$ 239
Gold 38
Graphite 52

Halite, NaCl 27, 65, 315
Hematite, α-Fe_2O_3 95
Heulandite, zeolite 244

*E.H. Nickel and M.C. Nichols, *Mineral Reference Manual*, van Nostrand Reinhold, New York, 1991

Ice, I_c H_2O 117, 119F
Ice, I_h 122
Ilmenite, $FeTiO_3$ 96
Indialite, $Mg_2Al_3(AlSi_5O_{18})\bullet H_2O$ 261
Iron 36

Jadeite, $Na(Al, Fe)Si_2O_6$ 265

Kaolinite, $Al_2Si_2O_5(OH)_4$ 268
Kyanite, $Al_2O_3\bullet SiO_2$ 251

Laumonite, zeolite 244
Lizardite, $Mg_3Si_2O_5(OH)_4$ 266

Marcasite, FeS_2 90
Martensite, FeC 80
Melilite, series of pyrosilicates 258
Moissanite, SiC 143
Molybdenite, MoS_2 140
Monticellite, $CaMgSiO_4$ 244
Montmorillonite, layered silicate 273
Mordenite, zeolite 244
Morganite, beryl 239
Muscovite 273

Nepheline, $KNa_3Al_4Si_4O_{16}$ 262
Niccolite, NiAs 72, 314

Olivine, $(Mg, Fe)_2SiO_4$ 245
Omisteinbergite, $CaAl_2Si_2O_8$ 268
Orthoclase, $KAlSi_3O_8$ 243

Paragasite, amphibole 266
Pearls 74
Peridot, olivine 244
Perovskite, $CaTiO_3$ 80
Phenakite, Be_2SiO_4 255
Phlogopite, $KMg_3(Si_3Al)O_{10}F_2$ 275
Plagioclase feldspars 243
Proustite, Ag_3AsS_3 187
Pyrargyrite, Ag_3SbS_3 187
Pyrite, FeS_2 66
Pyrochlore, $Ca_2Ta_2O_7$ 134
Pyrophyllite, $Al_2Si_4O_{10}(OH)_2$ 272
Pyroxine, chain silicate 240

β-Quartz, SiO_2 236, 237

Rose quartz 237
Ruby, Al_2O_3 93
Rutile, TiO_2 90

Sanmartinite, $ZnWO_4$ 74
Sapphire, Al_2O_3 93
Scheelite, $CaWO_4$ 133
Serpentines, sheet silicates 242
Silica, SiO_2 233
Sillimanite, $Al_2O_3\bullet SiO_2$ 250
Silver 38
Sodalite, $Na_4(Si_3Al_3)O_{12}Cl$ 243
Spinel, $MgAl_2O_4$ 165, 166
Spodumene, $LiAlSi_2O_6$ 263
Stannite, Cu_2FeSnS_4 119
Stishovite, SiO_2 238
Stromeyerite, AgCuS 105
Sulfur 50, 51

Talc, $Mg_3Si_4O_{10}(OH)_2$ 271
Tephroite, Mn_2SiO_4 244
Thortveitite, $Sc_2Si_2O_7$ 257
Topaz, $[Al(F, OH)]_2SiO_4$ 246
Tourmaline, sheet silicate 263
Tremolite, layered silicate 265
β-Tridymite, SiO_2 235

Ultramarines 243

Willemite, Zn_2SiO_4 254
Wollastonite, $Ca_3Si_3O_9$ 239
Wurtzite, ZnS 121, 314

Zeolites, framework silicates 244
Zinc Blende, ZnS 117, 126, 314
Zircon, $ZrSiO_4$ 248

Formula Index

A
AM_2 195
AcOF 127
Ac_2O_3 115
AgB_2 177
$AgBe_2$ 228
$AgBiS_2$ 71
$AgBiSe_2$ 70, 71, 171
$AgBiTe_2$ 71
AgBr 64
AgCd 155, 196, 198
AgCe 155, 196
AgCl 64
$AgCrO_2$ 70, 171
AgCuS, stromeyerite 105
AgF 64
$AgFeO_2$ 70, 171
AgI 118, 121, 167
$AgIO_4$ 133
$AgIn_2$ 219
AgLa 155, 196
AgLi 196
AgMg 155, 196
AgNd 196
$AgPt_3$ 204
$AgReO_4$ 133
$AgSbS_2$ 71
$AgSbSe_2$ 71
$AgSbTe_2$ 71
$AgTh_2$ 219
AgY 196
AgYb 196
AgZn 155, 196, 198
AgZnAs 128
$AgZnF_3$ 82
$AgZn_3$ 198
Ag_2CsI_3 107
Ag_2Er 197
Ag_2HgI_4 120
Ag_2Ho 197
Ag_2MoO_4 166
Ag_2O 135
Ag_2Yb 197
Ag_3Al 198
Ag_3AsS_3, proustite 187
Ag_3Pt 204
Ag_3SbS_3, pyrargyrite 187
Ag_5Al_3 198
Ag_5Zn_8 198
AlAs 118
AlB_2 176, 177, 303
$AlBiO_3$ 82
$AlCl_3$ 78, 79
$AlCl_3{\bullet}6H_2O$ 79
AlCo 196
$AlCo_3$ 204
$AlCr_2$ 197, 217
$[Al(F, OH)]_2SiO_4$, topaz 246
AlF_3 101
AlFe 196
$AlFeNiO_4$ 166
$AlMnCu_2$ 215, 216
$AlMo_3$ 224
AlN 121
$AlNb_3$ 224
AlNd 155, 196
AlNi 155, 196
$AlNi_3$ 204
AlO(OH), diaspore. 91, 92
AlP 118
AlPd 196
AlSb 118
$AlSbO_4$ 89
AlSc 196
$AlTh_2$ 219
$AlZr_3$ 204

Al_2Au 127
$Al_2Be_3[Si_6O_{18}]$, beryl. 260
Al_2Br_6 136
Al_2CdO_4 166
Al_2CoO_4 166
Al_2CrS_4 166
Al_2CuO_4 166
Al_2O_3, corundum 94, 297
$Al_2O_3 \bullet SiO_2$, andalusite 249, 252
$Al_2O_3 \bullet SiO_2$, kyanite 249,251
$Al_2O_3 \bullet SiO_2$, sillimanite 247, 250
Al_2Pt 127
Al_2S_3 115
Al_2SiO_5 247, 250–252
$Al_2Si_2O_5(OH)_4$, kaolinite 242, 268
$Al_2Si_4O_{10}(OH)_2$, pyrophyllite 271, 272
Al_2ZnS_4 123
Al_3Er 204
Al_3Ho 204
Al_3Np 204
Al_3U 204
Al_3Yb 204
Am 35
AmO 64
AmO_2 127
Am_2O_3 115
As 55
AsLiZn 127
AsV_3 224
As_2Pd 196
As_2Pt 196
As_3GeLi_5 127
As_3Li_5Si 127
As_3Li_5Ti 127
AuB_2 177
$AuBe_2$ 229
AuCd 155, 196
$AuCu_3$ 203, 204
$AuEr_2$ 197
$AuGa_2$ 127
$AuHo_2$ 197
$AuIn_2$ 127
AuMg 155, 196
AuMn 196
$AuNa_2$ 219
$AuNb_3$ 224
$AuPb_2$ 219
$AuSb_2$ 66, 196
AuSn 73
$AuTa_3$ 224
$AuTe_2$ 112
$AuTh_2$ 219
$AuTi_3$ 224
AuV_3 224
AuYb 196
AuZn 155, 198
$AuZr_3$ 224
Au_2Yb 197
Au_3Al 198
Au_3Pt 204

B

B 49
BCl_3 60
BF_3 9
BN 118
BP 118
B_2Cl_4 57, 58
B_4Cl_4 149, 150
B_5H_9 150
B_{12} 49
$B_{12}H_{12}^{2-}$ 8
Ba 37
$Ba[Al_2Si_2O_8]$, celsian 243
$BaAlSi_3O_8(OH) \bullet H_2O$, cymrite 269, 270
BaCd 196
$BaCeO_3$ 82
$Ba(ClO_4)_2 \bullet 3H_2O$ 180, 181
BaF_2 127
$BaFeO_3$ 82
$BaFe_{12}O_{19}$ 191, 192
$BaGa_2$ 177
BaHg 196
$BaIn_2O_4$ 166
$BaMg_2$ 231
$BaMnO_3$ 98, 99
$BaMoO_3$ 82, 132
$BaMoO_4$ 133
BaNH 64
$BaNiO_3$ 97
BaO 64
BaO_2 65
$BaPbO_3$ 82
$BaPd_2$ 228
$BaPrO_3$ 82
$BaPt_2$ 228
$BaPuO_3$ 82
$BaRh_2$ 228
BaS 64
BaSe 64
$BaSnO_3$ 82
BaTe 64
$BaThO_3$ 82
$BaTiO_3$ 82, 97–99
$BaTiS_3$ 97
$BaTi[Si_3O_9]$, benitoite 239

$BaUO_3$ 82
$BaWO_4$ 133
$BaZrO_3$ 82
$BaZrS_3$ 82
Ba_2CoOsO_6 99
Ba_2CrOsO_6 99
Ba_2CrTaO_6 99
Ba_2ErIrO_6 99
Ba_2FeMoO_6 84
Ba_2FeOsO_6 99
Ba_2FeReO_6 84
Ba_2FeSbO_6 99
Ba_2NiOsO_6 99
Ba_2RhUO_6 99
$Ba_3CoTi_2O_9$ 99
$Ba_3FeTi_2O_9$ 99
$Ba_3IrTi_2O_9$ 99
$Ba_3MoCr_2O_9$ 99
$Ba_3OsTi_2O_9$ 99
$Ba_3SrTa_2O_9$ 84, 85
$Ba_3UCr_2O_9$ 99
$Ba_3WCr_2O_9$ 99
$Ba_5Ta_4O_{15}$ 115, 116
Be 7, 38
$BeAl_2O_4$, chrysoberyl 165, 167, 244
BeCo 155, 196
BeCu 155, 196
BeNi 196
BeO 121
BePd 155, 196
BeS 118
BeSe 118
$BeTa_2$ 219
BeTe 118
Be_2B 127
Be_2C 126, 127
Be_2SiO_4, phenakite 254, 255
$Be_3Al_2Si_6O_{18}$, beryl 239, 259, 260
Be_3N_2 190, 191
Bi 55
$BiAsO_4$ 133
$BiAu_2$ 228
BiF_3 148, 156, 216, 301, 311, 315
BiF_5 87, 88
BiI_3 192, 193
BiIn 200, 201
$BiLi_3$ 215, 216, 314
$BiNb_3$ 224
BiSe 64
BiTe 64
Bi_2Mg_3 225, 226
Bi_2Pd 196
Bi_2Pt 196
BrF_5 294

C
C 51
C, diamond 51, 53
C, graphite 51, 52
C, graphite to diamond 53
$C(CH_2OH)_4$ 285
$C(CH_2ONO_2)_4$ 286
CCl_4 279
CH_4 9, 292
CHI_3 279, 280
$CHI_3\bullet 3S_8$. 280, 281
$(CH_3)_3N\bullet BF_3$ 59
$(CN)_2$ 281, 282
$(CN)_2C = C(CN)_2$ 284
CO 56
CO_2 56, 57
$(COOH)_2$, oxalic acid 282
$C(SCH_3)_4$ 286, 287
C_2H_2 282
C_2H_4 283
C_2H_6 280, 281
$(C_3H_7)_4NB$ 286, 287
$C_6H_4(NO_2)_2$, *p*-dinitrobenzene 291
C_6H_6 287, 288
$C_6H_6\bullet Cl_2$ 289, 290
$C_{10}H_8$, napthalene 288
$C_{10}H_{16}$, adamantane 279
$C_{14}H_{10}$, anthracene 289
$C_{18}H_{14}$, *p*-diphenylbenzene 289, 290
C_{60} 50, 292
Ca 38
$CaAg_2$ 231
$CaAl(AlSi)O_7$, gehlenite 239, 258
$CaAl_2$ 228
$CaAl_2Si_2O_8$, anorthite 243
$CaAl_2Si_2O_8$, omisteinbergite 268, 269
$CaAl_2Si_4O_{12}\bullet 4H_2O$, laumonite 243, 244
$CaAl_2Si_4O_{12}\bullet 6H_2O$, chabazite 243, 244
CaB_6 155, 156
$CaCN_2$ 171
$CaCO_3$, aragonite 74, 75
$CaCO_3$, calcite 2, 68, 296
CaC_2 65, 304
CaCd 199
$CaCd_2$ 231
$CaCeO_3$ 82
$CaCu_5$ 210, 211
CaF_2, fluorite 126, 169, 297, 311, 313
$CaGa_2$ 177
$CaHg_2$ 177
$CaIn_2$ 213, 302

$CaIn_2O_4$ 166
$CaLi_2$ 231
$CaMgSiO_4$, monticellite 244
$CaMgSi_2O_6$, diopside 263, 264
$CaMg_2$ 231
$CaMnO_3$ 82
$CaMoO_3$ 82
$CaMoO_4$ 133
CaNH 64
$CaNa_2Al_2Si_7O_{18}\bullet 6H_2O$,
heulandite 243, 244
$CaNi_5$ 211
CaO 64
CaO_2 65
$CaPb_3$ 204
$CaPd_2$ 228
$CaPt_2$ 228
CaS 64
CaSe 64
$CaSi_2$ 128, 129
$CaSn_3$ 204
$CaSnO_3$ 82.
CaTe 64
$CaThO_3$ 82
CaTi 196
$CaTiO_3$, perovskite 79, 80, 82, 84, 299
CaTl 155
$CaTl_3$ 204
$Ca(UO_2)O_2$ 138
$CaVO_3$ 82
$CaWO_4$, scheelite 132, 133
$CaVO_3$ 82
$CaZn_2$ 199
$CaZrO_3$ 82
$Ca_2Al(Si, Al)O_7$, gehlenite 239, 258
Ca_2CrMoO_6 84
Ca_2CrReO_6 84
Ca_2CrWO_6 84
Ca_2FeMoO_6 84
Ca_2FeReO_6 84
$Ca_2MgSi_2O_7$, akermanite 258
$Ca_2Mg_5[(OH)_2|(Si_4O_{11})_2]$,
tremolite. 240, 265
$Ca_2Sb_2O_7$ 134
$Ca_2Ta_2O_7$, pyrochlore 133, 134
$Ca_3[Si_3O_9]$, wollastonite 239
$Ca_4K[F|(Si_4O_{10})_2]\bullet 8H_2O$,
apophyllite 241
Cd 38
$Cd(CN)_2$ 135
CdCe 155, 196
$CdCeO_3$ 82
$CdCl_2$ 108, 298, 300, 311, 315
$CdCr_2O_4$ 166
CdCuSb 128
CdEu 196
CdF_2 127
$CdFe_2O_4$ 166
$CdGa_2O_4$ 166
CdI_2 110
$CdIn_2O_4$ 166
CdLa 155, 196
$CdMg_3$ 199
$CdMn_2O_4$ 166
$CdMoO_4$ 133
CdO 64
CdO_2 66
CdPd 155
CdPr 155, 196
$CdPt_3$ 204
$CdRh_2O_4$ 166
CdS 118, 121
CdSe 121
$CdSnO_3$ 82
CdTe 118
$CdThO_3$ 82
$CdTiO_3$ 82
CdV_2O_4 166
$Cd_2Nb_2O_7$ 133
$Cd_2Sb_2O_7$ 133
$Cd_2Ta_2O_7$ 133
Cd_3Nb 204
Ce 35
$CeAl_2$ 228
$CeAlO_3$ 82
CeAs 64
CeB_6 156
CeBi 64
$CeCo_2$ 228
$CeCo_5$ 211
$CeCrO_3$ 82
$CeFe_2$ 228
$CeFeO_3$ 82
$CeGa_2$ 177
$CeGaO_3$ 82
$CeGeO_4$ 133
CeH_2 127
CeHg 196
$CeIn_3$ 204
$CeIr_2$ 228
CeMg 196
$CeMg_2$ 228
CeN 64
$CeNi_2$ 228
CeO_2 127
CeOF 127
$CeOs_2$ 228
CeP 64

$CePb_3$ 204
$CePt_3$ 204, 228
$CeRh_2$ 228
$CeRu_2$ 228
CeS 64
CeSb 64
CeSe 64
$CeSn_3$ 204
CeTe 64
$CeVO_3$ 82
CeZn 196
Ce_2O_3 115
Ce_3In 204
Cl_2 48
ClF_3 293
CmO_2 127
Cm_2O_3 132
Co 38
CoAl 198
CoAsS 66
CoF_2 89
CoFe 196
CoFeGe 224
CoGeMn 224
$[Co(H_2O)_6][SiF_6]$ 161
$CoI_2{\bullet}6H_2O$ 105, 106
CoMnSb 128
$[Co(NH_3)_4(H_2O)_2][Co(CN)_6]$ 161
$[Co(NH_3)_4(H_2O)_2][TlCl_6]$ 68
$[Co(NH_3)_5H_2O][Co(CN)_6]$ 161
$[Co(NH_3)_5H_2O][Fe(CN)_6]$ 161
$[Co(NH_3)_6][BiCl_6]$ 68
$[Co(NH_3)_6][Co(CN)_6]$ 161
$[Co(NH_3)_6][PbCl_6]$ 68
$[Co(NH_3)_6][TlBr_6]$ 68
$[Co(NH_3)_6][TlCl_6]$ .67
CoNiSb 224
CoNiSn 224
CoO 64
$CoPt_3$ 204
CoS 73
CoS_2 66
CoSb 73
CoSc 196
$CoSc_2$ 219
CoSe 73
$CoSe_2$ 66
$CoSi_2$ 127
CoSn 211, 212
$CoSn_2$ 219
CoTe 73
CoV_3 224
$CoZn_3$ 198
Co_2B 219
Co_2CuO_4 166
Co_2CuS_4 166
Co_2GeO_4 166
Co_2MgO_4 166
Co_2MnO_4 166
Co_2NiO_4 166
Co_2SnO_4 166
Co_2TiO_4 166
Co_2ZnO_4 166
Co_3O_4 166
Co_3S_4 166
Co_3Se_4 166
CrB_2 177
$CrBe_2$ 231
$CrBiO_3$ 82
$Cr(C_6H_6)_2$ 56, 292
$CrCl_3$ 109
$CrIr_3$ 204
$CrMg_2$ 231
CrN 64
$CrNbO_4$ 89
CrO_2 89
$CrRhO_3$ 95
CrS 73, 103
CrSb 73
$CrSbO_4$ 89
CrSe 73
CrSr 196
$CrTaO_4$ 89
CrTe 73
Cr_2B 219
Cr_2CdO_4 166
Cr_2CdS_4 166
Cr_2CdSe_4 166
Cr_2CoO_4 166
Cr_2CoS_4 166
Cr_2CuS_4 166
Cr_2CuSe_4 166
Cr_2CuTe_4 166
Cr_2FeO_4 166
Cr_2FeS_4 166
Cr_2HgS_4 166
Cr_2MnO_4 166
Cr_2MnS_4 166
Cr_2NiO_4 166
Cr_2S_3 103
Cr_2ZnO_4 166
Cr_2ZnS_4 166
Cr_2ZnSe_4 166
Cr_3Pt 204
Cr_3S_4 103
Cr_5S_6 103
Cr_7S_8 103
$CsBi_2$ 228

CsCN 155
$CsCaF_3$ 82
$CsCdBr_3$ 82
$CsCdCl_3$ 82
CsCl 3, 154, 155, 196, 215, 295, 314, 316
$CsCrO_3F$ 133
CsF 64
CsH 64
$CsHgBr_3$ 82
$CsHgCl_3$ 82
CsI 155
$CsIO_3$ 82
$CsMgF_3$ 82
$CsNH_2$ 155, 174
CsO_2 65
$CsPbBr_3$ 82
$CsPbCl_3$ 82
$CsPbF_3$ 82
CsSH 155
CsSeH 155
$CsZnF_3$ 82
Cs_2CoF_6 129
Cs_2CrCl_6 129
Cs_2CrF_6 129
Cs_2GeCl_6 129
Cs_2GeF_6 129
Cs_2HfF_6 112
Cs_2MnF_6 129
Cs_2MoBr_6 129
Cs_2MoCl_6 129
Cs_2NiF_6 129
Cs_2O 109
Cs_2PbCl_6 129
Cs_2PdBr_6 129
Cs_2PdCl_6 129
Cs_2PdF_6 129
Cs_2PoBr_6 129
Cs_2PoI_6 129
Cs_2PtBr_6 129
Cs_2PtCl_6 129
Cs_2PtF_6 112
Cs_2PuF_6 112
Cs_2ReF_6 112
Cs_2RuF_6 112
Cs_2SeCl_6 129
Cs_2SiF_6 129
Cs_2SnBr_6 129
Cs_2SnCl_6 129
Cs_2SnI_6 129
Cs_2TeBr_6 129
Cs_2TeCl_6 129
Cs_2TeI_6 129
Cs_2ThF_6 112
Cs_2TiCl_6 129
Cs_2TiF_6 64, 129
Cs_2UF_6 112
Cs_2ZrF_6 112
Cs_3O 99, 100
Cu family 38
$CuAl_2$ 218–220
$CuAlO_2$ 70, 171
$CuBe_2$ 228
CuBr 118, 121
$CuBr_2$ 112, 113
$Cu[C_6H_4O(CHO)]_2$ 153
CuCl 118, 121
$CuCl_2$ 112
$CuCoO_2$ 70, 171
$CuCrO_2$ 70, 171
CuEu 196
CuF 118
$CuFeO_2$ 70, 171
$CuFeS_2$, chalcopyrite 118, 119
$CuFe_2S_3$, cubanite 137, 300
$CuGaO_2$ 170, 171
$CuGeO_3$ 169–171
CuH 121
CuI 118, 121
CuMgBi 128
CuMgSb 128
$Cu(OH)_2$ 91, 92
CuPd 196
$CuRhO_2$ 70, 171
CuS, covellite 182, 297, 302
CuSn 73
$CuTh_2$ 219
CuTi 221
$CuWO_4$ 73, 74
CuY 196
CuZn 155, 196, 198, 213
Cu-Zn system 197
$CuZn_3$ 198, 199
$Cu_2Cl(OH)_3$, atacamite 86, 87
Cu_2CsCl_3 87, 88
Cu_2FeSnS_4, stannite 118, 119
Cu_2HgI_4 120
Cu_2O, cuprite 135
Cu_2SnS_4 166
Cu_3Al 198
Cu_3AsS_4 123, 124
Cu_3N 84, 86
Cu_3Pd 204
Cu_3Pt 204
Cu_3Si 198
Cu_3Sn 198
Cu_5Sn 198

Cu_5Zn_8 198
Cu_9Al_4 198
$Cu_{31}Sn_8$ 198

D
$DyAlO_3$ 82
DyAs 64
DyB_6 156
$DyCo_2$ 228
$DyFe_2$ 228
$DyFeO_3$ 82
DyH_2 127
$DyIr_2$ 228
$DyMn_2$ 228
$DyMnO_3$ 82
DyN 64
$DyNi_2$ 228
$DyPt_2$ 228
$DyRh_2$ 228
DySb 64
DyTe 64
$Dy_2Ru_2O_7$ 134
$Dy_2Sn_2O_7$ 134
$Dy_2Tc_2O_7$ 134
$Dy_2Ti_2O_7$ 134

E
ErAs 64
ErB_6 156
$ErCo_2$ 228
$ErFe_2$ 228
ErH_2 127
$ErIn_3$ 204
$ErIr_2$ 228
$ErMn_2$ 228
ErN 64
$ErNi_2$ 228
$ErPt_3$ 204
$ErRh_2$ 228
ErSb 64
ErTe 64
ErTl 196
$ErTl_3$ 204
Er_2O_3 132
$Er_2Ru_2O_7$ 134
$Er_2Sn_2O_7$ 134
$Er_2Tc_2O_7$ 134
$Er_2Ti_2O_7$ 134
$EuAlO_3$ 82
EuB_6 156
$EuCrO_3$ 82
EuF_2 127
$EuFeO_3$ 82
$EuGa_2$ 177
$EuHg_2$ 177
$EuIr_2$ 228
EuN 64
EuO 64
EuS 64
EuSe 64
EuTe 64
$EuTiO_3$ 82
EuZn 196
Eu_2O_3 132
$Eu_2Ru_2O_7$ 134
$Eu_2Sn_2O_7$ 134

F
F_2 48
Fe 36
FeAl 198
$FeAlMgO_4$ 166
$FeBe_2$ 228, 231
$FeBiO_3$ 82
FeC, martensite 78, 80
$[Fe(CH_3NC)_6]Cl_2 \bullet 3H_2O$ 175
$Fe(C_5H_5)_2$ 9
$FeCrMnO_4$ 166
FeF_2 89
$FeGe_2$ 219
FeGeMn 224
FeGeNi 224
$(Fe, Mg)_2SiO_4$, fayalite 244
$FeNbO_4$ 89
$FeNi_3$ 204
FeO 64
$FePd_3$ 204
$FeRhO_3$ 95
FeS 73
FeS_2, marcasite 90
FeS_2, pyrite 65, 66, 296
FeSb 73
$FeSbO_4$ 89
FeSe 73
FeSn 73
$FeSn_2$ 219
$FeTaO_4$ 89
FeTe 73
$FeTiO_3$, ilmenite 94, 96
$FeVO_3$ 95
Fe_2B 219
Fe_2CdO_4 166
Fe_2CoO_4 166
Fe_2CuO_4 166
Fe_2GeO_4 166
$Fe_2(Mg, Mn, Fe)O_4$ 166
Fe_2MgO_4 166
Fe_2MnO_4 166

Fe_2MoO_4 166
Fe_2NiO_4 166
Fe_2O_3 94, 95, 298
Fe_2PbO_4 166
Fe_2TiO_4 166
Fe_2ZnO_4 166
Fe_3O_4 166
Fe_3S_4 166
Fe_3Sb_2 224
Fe_4N 109, 110

G
Ga 39
GaAs 118
$GaCr_3$ 224
$GaMo_3$ 224
GaN 121
$GaNb_3$ 224
GaNi 196
$GaNi_3$ 204
GaP 118
GaRh 196
GaS 142
GaSb 118
$GaSbO_4$ 89
GaSe 143
GaV_3 224
Ga_2CdO_4 166
Ga_2CoO_4 166
Ga_2CrS_4 166
Ga_2CuO_4 166
Ga_2MgO_4 166
Ga_2MnO_4 166
Ga_2NiO_4 166
Ga_2Pt 127
Ga_2ZnO_4 166
Ga_2S_3 117
Ga_3Pt_5 208, 209
Ga_3U 204
$GdAlO_3$ 82
GdAs 64
GdB_6 156
$GdCo_2$ 228
$GdCrO_3$ 82
$Gd(HCOO)_3$ 59
$GdFe_2$ 228
$GdFeO_3$ 82
GdH_2 127
$GdIr_2$ 228
$GdMnO_3$ 82
$GdMn_2$ 228
GdN 64
$GdNi_2$ 228
$GdPt_2$ 228
$GdRh_2$ 228
GdSb 64
GdSe 64
$Gd_2Ru_2O_7$ 134
$Gd_2Sn_2O_7$ 134
$Gd_2Ti_2O_7$ 134
Ge 36, 42
$GeCr_3$ 224
$GeLi_5N_3$ 127
$GeLi_5P_3$ 127
$GeMg_2$ 127
GeMnNi 224
$GeMo_3$ 224
$GeNb_3$ 224
$GeNi_3$ 204
GeO_2 89
GeS_2 122, 299
GeV_3 224
Ge_2Mo 197
Ge_3Pu 204
Ge_3U 204

H
H_2 47
H_2O, I_c ice 117, 119
H_2O, I_h ice 122
HfB_2 177
HfC 64
$HfCo_2$ 228
$HfCr_2$ 228
$HfFe_2$ 228, 231
$HfGeO_4$ 133
$HfMo_2$ 228
$HfOs_2$ 231
$HfRe_2$ 231
HfV_2 228
HfW_2 228
Hg 39
$HgBr_2$ 293
$HgCl_2$ 293
HgF_2 127, 293
HgI_2 293
HgLi 196
HgMg 196
HgMn 196
HgNa 201
HgNd 196
$HgNiF_3$ 82
HgO_2 67
HgPr 196
HgS 173
HgSe 118
HgSr 196
HgTe 118

$HgTi_3$ 204, 224
$HgZr_3$ 224
Hg_2Mg 197
Hg_3Zr 204
HoAs 64
HoB_6 156
HoBi 64
$HoCo_2$ 228
$HoFe_2$ 228
HoH_2 127
HoIn 196
$HoIn_3$ 204
$HoIr_2$ 228
$HoMn_2$ 228
HoN 64
$HoNi_2$ 228
HoOF 127
HoP 64
$HoPt_3$ 204
$HoRh_2$ 228
HoS 64
HoSb 64
HoSe 64
HoTe 64
HoTl 196
$HoTl_3$ 204
Ho_2O_2Se 115
$Ho_2Ru_2O_7$ 134
$Ho_2Sn_2O_7$ 134

I
IF_7 58, 59
InAs 118
$InCu_2$ 224
InLa 196
$InLa_3$ 204
InN 121
$InNb_3$ 224
$InNd_3$ 204
$InNi_2$ 223
InP 118
InPd 196
InPr 196
$InPr_3$ 204
$InPu_3$ 204
InSb 118
InTm 196
InYb 196
In_2CaS_4 166
In_2CdO_4 166
In_2CdS_4 166
In_2CdSe_4 119, 120, 302
In_2CoS_4 166
In_2CrS_4 166
In_2FeS_4 166
In_2HgS_4 166
In_2MgO_4 166
In_2MgS_4 166
In_2NiS 166
In_2Pt 127
In_2S_3 166
In_2Se_3 123
In_3Pr 204
In_3Pu 204
In_3Sc 204
In_3Tb 204
In_3Tm 204
In_3U 204
In_3Yb 204
$IrCr_3$ 224
IrLu 196
$IrMn_3$ 204
$IrMo_3$ 224
$IrNb_3$ 224
IrO_2 89
IrSb 73
$IrSn_2$ 127
IrTe 73
$IrTe_2$ 66
$IrTi_3$ 224
IrV 214
IrV_3 224
Ir_2P 126, 127

K
$K[AlSi_3O_8]$, orthoclase 243
$KAl_2(AlSi_3)O_{10}(OH)_2$, muscovite 242, 273
KAs_4O_6I 183
KBe_2 228
KBi_2 228
$KBiS_2$ 71
$KBiSe_2$ 71
KCN 64
$KCaF_3$ 82
$KCdF_3$ 82
KCl 64
$KCoF_3$ 82
$KCrF_3$ 82
$KCrO_3F$ 133
$KCrS_2$ 171
$KCuF_3$ 82
KF 64
$KFeF_3$ 82
$KFeF_4$ 156
KH 64
KI 64
KIO_3 82

$KMgAlSi_4O_{10}(OH)_2 \bullet nH_2O$, montmorillonite 271, 273
$KMgF_2$ 82
$K(Mg, Fe)_3[(OH)_2|AlSi_3O_{10}]$, biotite 241, 276
$KMg_3(Si_3Al)O_{10}F_2$, phlogopite 242, 274, 275
$KMnF_3$ 82
KN_3 160, 161
$K(NF_2)$ 153
KNO_3 76
$(K, Na)AlSiO_4$, nepheline 261, 262
KO_2 65
KNa_2 231
$KNbO_3$ 82
$KNiF_3$ 82
KO_2 66
KOH 64
KPb_2 231
$KReO_4$ 133
$KRuO_4$ 133
KSH 64
$KSeH$ 64
$KTaO_3$ 82
$KTlO_2$ 171
KVF_4 156
$KZnF_3$ 82
$K_2Cd(CN)_4$ 166
K_2CrF_6 129
K_2GeF_6 112
$K_2Hg(CN)_4$ 166
K_2MnF_6 112, 129
K_2MoCl_6 129
K_2NiF_4 88, 89
K_2NiF_6 129
K_2O 126
K_2OsBr_6 129
K_2OsCl_6 129
$K_2Pb_2Ge_2O_7$ 168
K_2PdBr_6 129
K_2PdCl_6 129
K_2PtBr_6 129
K_2PtCl_4 164
K_2PtCl_6 128, 297, 302
K_2PtF_6 112
K_2ReBr_6 129
K_2ReCl_6 129
K_2ReF_6 112
K_2ReH_9 189
K_2RuCl_6 129
K_2RuF_6 112
K_2S 126
K_2Se 126
K_2SeBr_6 129
K_2SiF_6 129
K_2SnBr_6 129
K_2SnCl_6 129
K_2TeCl_6 129
K_2TiCl_6 129
K_2TiF_6 112
$K_2Zn(CN)_4$ 166
K_3C_{60} 161, 162

L
La 35
$LaAlO_3$ 82
LaAs 64
LaB_6 155
$LaBO_3$ 76
LaBi 64
$LaCoO_3$ 82
$LaCrO_3$ 82
$LaFeO_3$ 82
$LaGa_2$ 177
$LaGaO_3$ 82
$LaHg_2$ 177
$LaIr_2$ 228
LaMg 196
$LaMg_2$ 228
$LaMnO_3$ 82
LaN 64
$LaNi_2$ 228
$LaNiO_3$ 82
LaOF 127
$LaOs_2$ 228
LaP 64
$LaPb_3$ 204
$LaPd_3$ 204
$LaPt_2$ 228
$LaPt_3$ 204
$LaRh_2$ 228
$LaRhO_3$ 82
$LaRu_2$ 228
LaS 64
LaSb 64
LaSe 64
$LaSn_3$ 204
LaTe 64
$LaTiO_3$ 82
$LaVO_3$ 82
LaZn 196
$LaZn_5$ 211
$La_2Hf_2O_7$ 134
La_2O_2Se 115
La_2O_3 114, 115, 298
$La_2Sn_2O_7$ 134
Li 37, 38
LiAg 155

$LiAlF_4$ 156
$LiAlMnO_4$ 166
$LiAlO_2$ 70, 171
$LiAlSiO_4$ 253, 254
$LiAlSi_2O_6$ spodumene 263
$LiAlTiO_4$ 166
$LiBaF_3$ 82
$LiBiS_2$ 71
LiBr 64
LiCl 64
$LiCoMnO_4$ 166
$LiCoO_2$ 70, 171
$LiCoSbO_4$ 166
$LiCoVO_4$ 166
$LiCrGeO_4$ 166
$LiCrMnO_4$ 166
$LiCrO_2$ 70, 171
$LiCrTiO_4$ 166
LiD 64
$LiDyF_4$ 133
$LiErF_4$ 133
$LiEuF_4$ 133
LiF 64
$LiFeO_2$ 71
$LiFeTiO_4$ 166
$LiGaTiO_4$ 166
$LiGdF_4$ 133
$LiGeRhO_4$ 166
LiH 64
$LiHF_2$ 71, 171
LiHg 155
$LiHoF_4$ 133
LiI 64
$LiIO_3$ 104
$LiLuF_4$ 133
LiMgBi 128
LiMgN 128
LiMgP 127
LiMgSb 128
$LiMnTiO_4$ 166
$LiNO_2$ 171
$LiNbO_3$ 95
$LiNiO_2$ 70, 171
$LiNiVO_4$ 166
$LiOs_2$ 231
LiPb 196
$LiRhGeO_4$ 166
$LiRhMnO_4$ 166
$LiRhO_2$ 70, 171
$LiSbO_3$ 95, 96
LiSn 208, 210
$LiTbF_4$ 133
LiTi 196
$LiTiO_2$ 71
$LiTiRhO_4$ 166
LiTl 155
$LiTlO_2$ 71
$LiTmF_4$ 133
$LiUO_3$ 82
$LiVO_2$ 171
$LiYF_4$ 133
$LiYbF_4$ 133
LiZn 199
LiZnN 128
LiZnP 128
$LiZnSbO_4$ 166
Li_2NH 126
Li_2NiF_4 166
Li_2O 124–126
Li_2O_2 184, 185
Li_2S 126
Li_2Se 126
Li_2Te 126
Li_5P_3Si 127
Li_5P_3Ti 127
Li_5SiN_3 127
Li_5TiN_3 127
$LiVO_2$ 70
Li_xWO_3 82
LuB_2 177
LuB_6 156
$LuCo_2$ 228
$LuFe_2$ 228
LuH_2 127
$LuMn_5$ 231
LuN 64
$LuNi_2$ 228
LuRh 196
$LuRh_2$ 228
$LuRu_2$ 231
$Lu_2Ru_2O_7$ 134
$Lu_2Sn_2O_7$ 134

M
MAs 195
MBi 195
$MM'O_3$ 82, 84
$MM'X_3$ 81
MMnO 75
MSb 195
MSn 195
MX 64, 73, 155, 295
MX_2 127, 297, 298
$M_2M'X_4$ 165, 166
M_2X 126, 296
M_2X_3 297
$M_3^{II}M_2^{III}[SiO_4]_3$, garnet 239
$M_xMo_6X_8$ 157

Mg 37, 38
MgAgAs 126–128
$MgAl_2O_4$, spinel 17, 165, 166, 299
MgB_2 177
$MgCd_3$ 199
MgCe 155
$MgCeO_3$ 82
$MgCu_2$ 227–232
MgF_2 89
$MgFe_2O_4$ 166
$MgGeO_3$ 95
$[Mg(H_2O)_6][SiF_6]$ 161
$[Mg(H_2O)_6][SnF_6]$ 161
$[Mg(H_2O)_6][TiF_6]$ 161
MgHg 155
MgLa 155
MgLiAs 128
MgNiZn 228
$MgNi_2$ 227–232
MgO 64
MgO_2 66
$Mg(OH)_2$, brucite 111
MgPr 155, 196
$MgPu_2$ 127
MgS 64
MgSc 196
MgSe 64
$Mg[SiO_3]$, enstatite 240
MgSr 155, 196
MgTe 121
MgTi 196
MgTl 155
$MgZn_2$ 227, 230–232
$Mg_2Al_3(AlSi_5O_{18})\bullet H_2O$, cordierite 260, 261
$Mg_2Al_3(AlSi_5O_{18})\bullet H_2O$, indialite 260, 261
$(Mg, Fe)_2[SiO_4]$, olivine 244, 245
Mg_2GeO_4 166
Mg_2Pb 127, 227
Mg_2Si 127
$Mg_2[SiO_4]$, forsterite 244, 245
Mg_2Sn 127
Mg_2SnO_4 166
Mg_2TiO_4 166
Mg_2VO_4 166
Mg_3Bi_2 115
$Mg_3[(OH)_2|Si_4O_{10}]$, talc 242, 270, 271
$Mg_3[(OH)_4|Si_2O_5]$, chrysotile 240, 242
Mg_3Sb_2 115
$Mg_3Si_2O_5(OH)_4$, serpentines
  antigorite 267
  chrysotile 240, 242
  lizardite 266
$Mg_5Al(Si_3Al)O_{10}(OH)_8$, chlorite 242, 277
Mn 39
MnAs 73
MnB_2 177
$MnBe_2$ 231
MnBi 73, 195
MnF_2 89
$[Mn(H_2O)_6][SiF_6]$ 161
MnO 64
MnO_2 89
MnPd 196
$MnPt_3$ 204
MnRh 196
MnS 64, 118, 121
MnS_2 66
MnSb 73
MnSe 64, 118,121
$MnSe_2$ 66
$MnSn_2$ 219
MnTe 73, 121
$MnTe_2$ 66
Mn_2Au_5 208, 209
Mn_2B 219
Mn_2CuO_4 166
Mn_2LiO_4 166
Mn_2MgO_4 166
Mn_2NiO_4 166
Mn_2O_3 132
Mn_2SiO_4, tephroite 244
Mn_2SnO_4 166
Mn_2TiO_4 166
Mn_3O_4 166
Mn_3Pt 204
Mn_3Rh 204
$MoAl_{12}$ 214
MoB_2 177
$MoBe_2$ 231
MoC 77
$MoFe_2$ 231
$MoNi_4$ 205, 206
MoO_2 89
$MoPt_2$ 207
MoS_2, molybdenite 139, 140, 193, 297
$MoSi_2$ 197
Mo_2B 219
Mo_3Al_8 200
$Mo_6Cl_8^{4+}$ 157
$[Mo_6Cl_8(OH)_4(H_2O)_2]\bullet 12H_2O$ 176
$Mo_6S_8^{4-}$ 157

N
N_2 47
$N(CH_3)_4Br$ 174
$N(CH_3)_4Cl$ 174
$N(CH_3)_4I$ 174
$N(CH_3)_4ICl_2$ 159, 160
$N(CH_3)_4MnO_4$ 174
$[N(CH_3)_4]_2SnCl_6$ 128
ND_4Br 155
ND_4Cl 155
$(NH_2C(O)N)_2$ 284, 285
$(NH_2)_2CO$, urea 283, 284
NH_3 293
NH_4Br. 64, 155, 174
NH_4CN 158
NH_4Cl 64, 157
NH_4ClO_2 159
NH_4CoF_3 82
NH_4F 121
NH_4HF_2 157, 158
NH_4I. 64, 155, 174
NH_4IO_4 133
NH_4MnF_3 82
NH_4NiO_3 82
NH_4ReO_4 133
NH_4SH 174
$(NH_4)_2GeF_6$ 112, 129
$(NH_4)_2IrCl_6$ 129
$(NH_4)_2PbCl_6$ 129
$(NH_4)_2PdBr_6$ 129
$(NH_4)_2PdCl_6$ 129
$(NH_4)_2PoBr_6$ 129
$(NH_4)_2PoCl_6$ 129
$(NH_4)_2PtBr_6$ 129
$(NH_4)_2SeCl_6$ 129
$(NH_4)_2SiF_6$ 111, 184
$(NH_4)_2SnBr_6$ 129
$(NH_4)_2SnCl_6$ 129
$(NH_4)_2TeBr_6$ 129
$(NH_4)_2TeCl_6$ 129
$(NH_4)_2TiBr_6$ 129
$(NH_4)_2TiCl_6$ 129
$(NH_4)_2TiF_6$ 112
N_2H_4, hydrazine 61
N_2O 56
N_2O_5 180, 181
Na 37, 38
$NaAlF_4$ 156
$NaAlO_3$ 82
$NaAlSi_2O_6$, jadeite 264
$Na[AlSi_3O_8]$, albite 243
$NaAlSi_4O_{12}\bullet 3H_2O$, mordenite 243, 244
$NaAu_2$ 228
$NaBiO_3$ 95
$NaBiS_2$ 71
$NaBiSe_2$ 71
NaBr 64
NaCN 64
$NaCa_2Mg_4Al(Si_6Al_2O_{22})(OH, F)_2$, paragasite 265, 266
$NaCa_2Mg_5(Si_7AlO_{22})(OH)_2$, tremolite 265
NaCl 3, 27, 63, 65, 162, 315
$NaCrO_2$ 70, 171
$NaCrS_2$ 70, 171
$NaCrSe_2$ 70, 171
NaF 64
$NaFeO_2$ 69, 70, 296
NaH 64
$NaHF_2$ 70
$NaH(PO_3NH_2)$ 76
NaI 64
$NaIO_4$ 133
$NaInO_2$ 70, 171
$NaInS_2$ 70, 171
$NaInSe_2$ 70, 171
$NaLi_3Al_6(OH)_4(BO_3)_3Si_6O_{18}$, tourmaline 262, 263
$NaMgF_3$ 82
$NaMnF_3$ 82
$NaNO_3$ 296
$NaNbO_3$ 82
$NaNiO_2$ 70, 72, 171
NaO_2 66
$NaPb_3$ 204
$NaPt_2$ 221, 222
$NaReO_4$ 133
$(NSF)_4$ 152
NaSH 64
NaSeH 64
$NaSbO_3$ 95
$NaTaO_3$ 82
$NaTcO_4$ 133
$NaTiO_2$ 70, 171
NaTl 215
$NaTlO_2$ 70, 171
$NaVO_2$ 70, 171
$NaWO_3$ 82
NaZnAs 128
$NaZnF_3$ 82
Na_2MoF_6 129
Na_2O 126
Na_2O_2 185, 186
Na_2PuF_6 112
Na_2S 126
Na_2Se 126
Na_2Te 126

$Na_2(TiO)SiO_4$ 256
Na_2WO_4 166
Na_3As 178, 179
$Na_4(Al_3Si_3)O_{12}Cl$ sodalite 243
$Na_{31}Pb_8$ 198
NbB_2 177
NbC 64
$NbFe_2$ 231
NbH_2 127
$NbMn_2$ 231
NbN 121
$NbNi_8$ 212
NbO 64, 85, 86
NbO_2 89
NbRh 201, 202
NbS 73, 173
Nb_3Si 204
$NdAlO_3$ 82
NdAs 64
NdB_6 156
NdBi 64
$NdCo_2$ 228
$NdCoO_3$ 82
$NdCr_2$ 228
$NdCrO_3$ 82
$NdFeO_3$ 82
$NdGaO_3$ 82
NdH_2 127
$NdIr_2$ 228
$NdMnO_3$ 82
NdN 64
$NdNi_2$ 228
NdOF 127
NdP 64
$NdPt_2$ 228
$NdPt_3$ 204
$NdRh_2$ 228
$NdRu_2$ 228
NdS 64
NdSb 64
NdSe 64
$NdSn_3$ 204
NdTe 64
$NdVO_3$ 82
$Nd_2Hf_2O_7$ 133
Nd_2O_2Se 115
Nd_2O_3 115
$Nd_2Ru_2O_7$ 133
$Nd_2Sn_2O_7$ 133
$Nd_2Zr_2O_7$ 133
Ni 37
NiAl 198
NiAs 72, 296, 315
NiAsS 66
NiBi 195
$Ni[(CH_3)_2C_2N_2O_2H]_2$ 61, 62
NiF_2 89
$[Ni(H_2O)_6][SiF_6]$ 161
$[Ni(H_2O)_6][SnCl_6]$ 160, 161
NiMgBi 128
NiO 64
NiS 73
NiS_2 66
NiSb 73
NiSbS 66
NiSe 73
$NiSe_2$ 66
$NiSi_2$ 127
NiSn 73
NiTe 73
NiV_3 224
Ni_2Al_3 225, 226
Ni_2B 219
Ni_2FeS_4 166
Ni_2GeO_4 166
Ni_2SiO_4 166
Ni_3S_4 166
Ni_3Si 204
Ni_3Se_2 194
Ni_3Sn_2 224
$NpAl_2$ 228
NpN 64
NpO 64
NpO_2 127

O

O_2 48
OsB_2 177
$OsCr_3$ 224
$OsMo_3$ 224
$OsNb_3$ 224
OsO_2 89
OsO_4 58, 291
OsS_2 66
$OsSe_2$ 66
$OsTe_2$ 66
OsTi 196

P

P 54
PCl_4ICl_2 155
PH_4Br 174
PH_4I 174
PRh_2 127
P_4 53
PaO 64
PaO_2 127
Pb 36

$PbAu_2$ 228
$PbCeO_3$ 82
PbF_2 127
$PbMg_2$ 222
$PbMo_6S_8$ 157
$PbMoO_4$ 133
$PbNb_3$ 224
PbO 130, 131, 294
PbO_2 89
$PbPd_3$ 204
$PbPu_3$ 204
PbS 64
PbSe 64
$PbSnO_3$ 82
PbTe 64
$PbThO_3$ 82
$PbTiO_3$ 82
PbV_3 224
$PbWO_4$ 133
$PbZrO_3$ 82
Pb_2Li_7 224, 225
Pb_2O 135
Pb_2Pd_3 224
Pb_3Li_8 218
Pb_3Pr 204
$PdAs_2$ 66, 196
$PdBi_2$ 66, 196
PdF_2 89
PdH 64
$PdPb_2$ 219
PdSb 73
$PdSb_2$ 66
PdSn 73
PdTe 73
$PdTh_2$ 219
PdV_3 224
Pd_3Sn 204
Pd_3Sn_2 224
Pd_3U 204
Pd_3Y 204
Pd_4Se 152, 154
Po 56
PoO_2 127
Pr 35
$PrAlO_3$ 82
PrAs 64
PrB_6 156
PrBi 64
$PrCo_2$ 228
$PrCoO_3$ 82
$PrCrO_3$ 82
$PrFeO_3$ 82
$PrGa_2$ 177
$PrGaO_3$ 82
PrH_2 127
$PrIr_2$ 228
$PrMg_2$ 228
$PrMnO_3$ 82
PrN 64
$PrNi_2$ 228
PrOF 127
$PrOs_2$ 228
PrP 64
$PrPt_2$ 228
$PrPt_3$ 204
$PrRh_2$ 228
$PrRu_2$ 228
PrS 64
PrSb 64
PrSe 64
$PrSn_3$ 204
PrTe 64
$PrVO_3$ 82
PrZn 196
Pr_2O_3 115, 132
Pr_2O_2Se 115
$Pr_2Ru_2O_7$ 134
$Pr_2Sn_2O_7$ 134
Pt metals 38
$PtAs_2$ 66
PtB 73
PtBi 195
$PtBi_2$ 66, 196
$PtCl_4^{2-}$ 9
$PtCr_3$ 224
$PtHg_4$ 217
$PtMg_2$ 228
$PtNb_3$ 224
PtP_2 66
PtS, cooperite 130
PtSb 73
$PtSb_2$ 66, 196
PtSn 73
$PtSn_2$ 127
$PtTi_3$ 224
PtV_3 224
Pt_3Sc 204
Pt_3Sm 204
Pt_3Sn 204
Pt_3Tb 204
Pt_3Ti 204
Pt_3Tm 204
Pt_3Y 204
Pt_3Yb 204
Pt_3Zn 204
Pt_5Zn_{21} 198
Pu 38, 39
$PuAl_2$ 228

$PuAlO_3$ 82
$PuAs$ 64
PuB 64
PuB_2 177
PuB_6 156
PuC 64
$PuFe_2$ 228
$PuMn_2$ 228
$PuMnO_3$ 82
PuN 64
$PuNi_2$ 228
PuO 64
PuO_2 127
$PuOF$ 127
PuP 64
$PuRu_2$ 228
PuS 64
$PuSn_3$ 204
$PuTe$ 64
$PuZn_2$ 228
Pu_2O_3 115, 132

R
Ra 38
RaF_2 127
$RbBr$ 64
$RbCaF_3$ 82
$RbCN$ 64
$RbCl$ 64, 155
$RbCoF_3$ 82
$RbCrS_2$ 70, 171
$RbCrSe_2$ 70, 171
RbF 64
$RbFeF_4$ 156
RbH 64
RbI 64
$RbIO_3$ 82
$RbIO_4$ 133
$RbMgF_3$ 82
$RbMnF_3$ 82
$RbNH_2$ 64
RbO_2 65
$RbReO_4$ 133
$RbSH$ 64
$RbSeH$ 64
$RbTlO_2$ 70, 171
$RbZnF_3$ 82
Rb_2CrF_6 129
Rb_2GeF_6 112
Rb_2HfF_6 112
Rb_2MnF_6 129
Rb_2MoBr_6 129
Rb_2MoCl_6 129
Rb_2NiF_6 129
Rb_2O 126
Rb_2PbCl_6 129
Rb_2PdBr_6 129
Rb_2PdCl_6 129
Rb_2PtBr_6 129
Rb_2PtCl_6 129
Rb_2PtF_6 112
Rb_2ReF_6 112
Rb_2RuF_6 129
Rb_2S 126
Rb_2SbCl_6 129
Rb_2SeCl_6 129
Rb_2SiF_6 129
Rb_2SnBr_6 129
Rb_2SnCl_6 129
Rb_2SnI_6 129
Rb_2TeCl_6 129
Rb_2TiCl_6 129
Rb_2TiF_6 112, 129
Rb_2ZrCl_6 129
Rb_2ZrF_6 112
ReB_2 141
ReB_3 163
$ReBe_2$ 231
ReO_2 91
ReO_3 84, 85, 300
$ReSi_2$ 197
$RhBi$ 73, 195
$RhCr_3$ 224
$RhNb_3$ 224
$RhNbO_4$ 89
RhP_2 126
$RhPb_2$ 219
RhS_2 66
$RhSbO_4$ 89
$RhSe_2$ 66
$RhSn$ 73
$RhTaO_4$ 89
$RhTe$ 73
$RhTe_2$ 66
RhV_3 224
$RhVO_4$ 89
RhY 196
Rh_2CdO_4 166
Rh_2CoO_4 166
Rh_2CuO_4 166
Rh_2MgO_4 166
Rh_2MnO_4 166
Rh_2NiO_4 166
Rh_2ZnO_4 166
Rh_3Sc 204
Rh_5Zn_{21} 198
RuB_2 177
$RuCr_3$ 224

RuO_2 89
RuS_2 66
$RuSe_2$ 66
RuSi 196
$RuSn_2$ 66, 196
$RuTe_2$ 66
RuTi 196
Ru_3U 204

S
SF_6 8
SO_2 57
S_4N_4 151
S_6 50, 51
S_8 50
Sb 55
$Sb(CH_3)_3Br_2$ 60, 61
$Sb(CH_3)_3Cl_2$ 60
$Sb(CH_3)_3I_2$ 60
$SbCl_5$ 60
SbF_5 60
$SbNb_3$ 224
SbTi 196
$SbTi_3$ 224
SbV_3 224
Sb_2PbO_6 106, 107
ScAs 64
ScB_2 177
ScH_2 127
$ScMn_2$ 231
ScN 64
$ScOs_2$ 231
ScRh 196
ScSb 64
Sc_2O_3 132
$Sc_2Si_2O_7$, thortveitite 239, 256, 257
Se 55
Si 36, 42
SiB_6 156
SiC 143, 145, 146
SiF_4 149
SiO_2, cristobalite 235, 236
SiO_2, quartz 235,236
SiO_2, stishovite 237, 238
SiO_2, tridymite 234, 235
$SiTa_2$ 219
$SiZr_2$ 219
Si_2W 197
Si_3N_4 188
Si_3U 204
Sm 38
$SmAlO_3$ 82
SmAs 64
SmB_6 156
SmBi 64
$SmCoO_3$ 82
$SmCrO_3$ 82
$SmFeO_3$ 82
SmH_2 127
$SmIr_2$ 228
SmN 64
SmO 64
SmOF 127
SmP 64
$SmRu_2$ 228
SmS 64
SmSb 64
SmSe 64
SmTe 64
$SmVO_3$ 82
Sm_2O_2Se 115
Sm_2O_3 132
$Sm_2Ru_2O_7$ 134
$Sm_2Sn_2O_7$ 134
$Sm_2Tc_2O_7$ 134
Sn 42, 43
SnAs 64
$SnCl_2$ 43
$SnCl_4$ 43
SnI_4 136
$SnMn_2$ 224
$SnMo_3$ 224
$SnNb_3$ 223, 224
$SnNi_3$ 202
SnO_2, cassiterite 89
SnSb 64
SnSe 64
SnTe 64
$SnTi_2$ 224
SnV_3 224
$Sn_2Ti_2O_7$ 134
Sn_3U 204
Sr 38
SrB_6 156
$SrCO_3$ 76
$SrCeO_3$ 82
$SrCl_2$ 127
$SrCoO_3$ 82
SrF_2 127
$SrFeO_3$ 82
$SrGa_2$ 177
$SrHfO_3$ 82
$SrHg_2$ 177
$SrIr_2$ 228
$SrMg_2$ 231
$SrMoO_3$ 82
$SrMoO_4$ 133
SrNH 64

SrO 64
SrO_2 65
$SrPbO_3$ 82
$SrPd_2$ 228
$SrPt_2$ 228
$SrRh_2$ 228
SrS 64
$SrSe$ 64
$SrSnO_3$ 82
$SrTe$ 64
$SrThO_3$ 82
$SrTi$ 196
$SrTiO_3$ 82
$SrTiS_3$ 97
$SrTl$ 155
$SrWO_4$ 133
$SrZrO_3$ 82
Sr_2CrMoO_6 84
Sr_2CrReO_6 84
Sr_2CrWO_6 84
Sr_2FeMoO_6 84
Sr_2FeReO_6 84

T
TaB_2 177
TaC 64
$TaCo_2$ 228, 231
$TaCoCr$ 231
$TaCoTi$ 231
$TaCoV$ 231
$TaCr_2$ 228, 231
$TaCuV$ 228
$TaFe_2$ 231
$TaFeNi$ 228
$TaMn_2$ 231
TaN 190
TaO 64
TaO_2 89
$TaSnO_3$ 82
Ta_2B 219
$TbAs$ 64
TbB_6 156
$TbBi$ 64
$TbFe_2$ 228
TbH_2 127
$TbIr_2$ 228
$TbMn_2$ 228
TbN 64
$TbNi_2$ 228
TbO_2 127
TbP 64
TbS 64
$TbSb$ 64
$TbSe$ 64
$TbTe$ 64
$TbTl_3$ 204
Tb_2O_3 132
$Tb_2Ru_2O_7$ 133
$Tb_2Sn_2O_7$ 133
Te 55, 56
TeO_2 89
$TeTh$ 196
$ThAl_2$ 177
ThB_6 156
ThC 64
$ThCo_5$ 211
$ThCu_2$ 177
$ThGeO_4$ 133
$ThIr_2$ 228
$ThMg_2$ 228
$ThMn_2$ 231
$ThNi_2$ 177
ThO_2 127
$ThOs_2$ 228
$ThRu_2$ 228
ThS 64
$ThSb$ 64
$ThSe$ 64
$ThTe$ 155
$ThZn_5$ 211
Th_2N_3 115
$TiAl_3$ 203, 204
TiB_2 177
$TiBe_2$ 228
TiC 64
$TiCo_2$ 228
$TiCr_2$ 228, 231
$TiCu_3$ 203
$TiFe_2$ 231
$TiMn_2$ 231
TiN 64
$TiNi_3$ 199
TiO 64
TiO_2, anatase 78, 79
TiO_2, rutile 89, 90, 297
TiP 77, 78
TiS 73
$TiSb_2$ 219
$TiSe$ 73
$TiTe$ 73
TiV_2 177
$TiZn_3$ 204
Ti_2CuS_4 166
Ti_2Cu_3 218, 219
Ti_2MgO_4 166
Ti_2MnO_4 166
Ti_3Cu_4 218, 219
$TlAlF_4$ 156

$TlBi$ 155
$TlBiS_2$ 71
$TlBiTe_2$ 70, 171
$TlBr$ 155
$TlCl$ 155
$TlCN$ 155
$TlCoF_3$ 82
$TlCrF_4$ 156
$TlGaF_4$ 156
TlI 155
$TlIO_3$ 82
$TlReO_4$ 133
$TlSb$ 155
$TlSbS_2$ 71
$TlSbTe_2$ 70, 171
$TlSe$ 162
$TlTm$ 196
Tl_2MoCl_6 129
Tl_2O 113
Tl_2O_3 130, 132
Tl_2PtCl_6 129
Tl_2SiF_6 129
Tl_2SnCl_6 129
Tl_2TeCl_6 129
Tl_2TiF_6 112, 129
$TmAs$ 64
TmB_6 156
$TmCo_2$ 228
$TmFe_2$ 228
TmH_2 127
$TmIr_2$ 228
TmN 64
$TmNi_2$ 228
$TmRh$ 196
$TmSb$ 64
$TmTe$ 64
Tm_2O_3 132
$Tm_2Ru_2O_7$ 134
$Tm_2Sn_2O_7$ 134

U
UB_{12} 67
UAl_2 228
UAs 64
UB_2 177
UBi 64
UC 64
UCl_6 100
UCo_2 228
UF_6 101
UFe_2 228
$UGeO_4$ 133
UMn_2 228
UN 64
UN_2 127
UNi_2 231
$UNiFe$ 228
UO 64
UO_2 127
UOs_2 228
UP 64
US 64
USb 64
USe 64
USi_2 177
UTe 64

V
V 38
VB_2 177
VBe_2 231
VC 64
VF_3 101, 102
VHg_2 177
VIr_2 228
VN 64
VNi_2 231
VO 64
VP 73
VS 73
VSb_2 219
VSe 73
VTe 73
V_2CdO_4 166
V_2CuS_4 166
V_2FeO_4 166
V_2LiO_4 166
V_2MgO_4 16
V_2MnO_4 166
V_2ZnO_4 166
V_4Zn_5 216

W
WAl_5 210
WBe_2 231
WC 173
WFe_2 231
WO_2 89
WS_2 141
W_2B 219

X
XeF_2 153, 294
XeF_4 152, 294

Y
YAl_2 228
$YAlO_3$ 82

YAs 64
YBO_3 76, 78
YCo_2 228
$YCrO_3$ 82
YFe_2 228
$YFeO_3$ 82
YH_2 127
YIr_2 228
YMn_2 228
YN 64
$YNbO_4$ 133
YNi_2 228
YOF 127
$Y(OH)_3$ 177, 180
YPt_2 228
YRh_2 228
YTe 64
$Y_2Ru_2O_7$ 133
$Y_2Sn_2O_7$ 133
$Y_2Ti_2O_7$ 133
$Y_2Zr_2O_7$ 133
YbAs 64
YbB_6 156
$YbBO_3$ 177, 178
$YbCo_2$ 228
$YbIr_2$ 228
YbN 64
$YbNi_2$ 228
YbO 64
YbSb 64
YbSe 64
YbTe 64
Yb_2O_2Se 115
Yb_2O_3 132
$Yb_2Ru_2O_7$ 134
$Yb_2Sn_2O_7$ 134
$Yb_2Ti_2O_7$ 134
$Yb_2Zr_2O_7$ 134

Z
Zn 38
$Zn(CN)_2$ 135
ZnCe 155
$[Zn(H_2O)_6][SiF_6]$ 161
$[Zn(H_2O)_6][SnF_6]$ 161
$[Zn(H_2O)_6][TiF_6]$ 161
$[Zn(H_2O)_6][ZrF_6]$ 161
ZnI_2 139
ZnLa 155
ZnO 121
ZnO_2 66
ZnPr 155
ZnS, see Wurtzite and Zinc blende
ZnS, wurtzite 27, 120, 121, 296, 314
ZnS, zinc blende 27, 117, 126, 314
$ZnSb_2O_6$ 93
ZnSe 118, 121
ZnTe 118
$ZnTh_2$ 219
$ZnWO_4$ 74
Zn_2SiO_4, willemite 254
Zn_2SnO_4 166
Zn_2TiO_4 166
$Zn_7Sb_2O_{12}$ 166
$ZrAl_3$ 204, 205
$ZrAu_4$ 205, 206
ZrB 64
ZrB_2 177
$ZrBe_2$ 177
ZrC 64
$ZrCl_3$ 102
$ZrCo_2$ 228
$ZrCr_2$ 231
$ZrFe_2$ 228
$ZrGa_2$ 207, 208
$ZrGeO_4$ 133
ZrI_3 101
$ZrIr_2$ 228, 231
$ZrMn_2$ 231
$ZrMo_2$ 228
ZrN 64
ZrO 64
$ZrOs_2$ 231
ZrP 64
$ZrRe_2$ 231
$ZrRu_2$ 231
ZrS 64
$ZrSiO_4$, zircon 247, 248
ZrTe 73
ZrV_2 228, 231
ZrW_2 228
$ZrZn_2$ 228